Metal Metabolism in
Aquatic Environments

2. Ecotoxicology in Theory and Practice
V.E. Forbes and T.L. Forbes
1994, reprinted 1994, xiii+248pp, 33 illus.
Hardback: 0-412-43530-6

3. Interconnections Between Human and Ecosystem Health
R.T. DiGiulio and E. Monosson (eds)
1996, xiv+276pp, 14 line, 2 half tone illus.
Hardback: 0-412-62400-1

4. ECOtoxicology: Ecological Dimensions
D.J. Baird, P.E.I. Douben, P. Greig-Smith and L. Maltby (eds)
1996, xi+90pp, 18 line illus.
Hardback: 0-412-75470-3
Paperback: 0-412-75490-8

5. Ecological Risk Assessment of Contaminants in Soils
N.M. van Straalen and H. Løkke (eds)
1997, xvii+334pp, 77 line illus.
Hardback: 0-412-75900-4

6. Structure-Activity Relationships in Environmental Sciences
M. Nendza
1998, xiv+270pp, 65 line illus.
Hardback: 0-412-56430-0 (1998)

7. Metal Metabolism in Aquatic Environments
W.J. Langston and M. Bebianno (eds)
1998, xxi+448pp, 25 line illus.
Hardback: 0-412-80370-4

Forthcoming

Molluscs as Aquatic Biomonitors
J.G. Wilson and R.F. McMahon
Hardback: 0-412-71080-3

Metal Metabolism in Aquatic Environments

Edited by

William J. Langston
Centre for Coastal and Marine Sciences
Plymouth Marine Laboratory
Plymouth
UK

and

Maria João Bebianno
Department of Science and Technology
of Aquatic Resources
University of the Algarve
Portugal

**Published by Chapman & Hall, an imprint of Thomson Science,
2–6 Boundary Row, London SE1 8HN, UK**

Thomson Science, 2-6 Boundary Row, London SE1 8HN, UK

Thomson Science, 115 Fifth Avenue, New York, NY 10003, USA

Thomson Science, Suite 750, 400 Market Street, Philadelphia, PA 19106, USA

Thomson Science, Pappelallee 3, 69469 Weinheim, Germany

First edition 1998

© 1998 Chapman & Hall Ltd

Thomson Science is a division of International Thomson Publishing

Typeset in 10/12pt Times by Saxon Graphics Ltd, Derby

Printed in Great Britain by TJ International, Padstow, Cornwall

ISBN 0 412 80370 4

A catalogue record for this book is available from the British Library

∞ Printed on acid-free text paper, manufactured in accordance with
ANSI/NISO Z39.48-1992 (Permanence of Paper).

Contents

8 Metal handling strategies in molluscs   219
William J. Langston, Maria João Bebianno and Gary R. Burt

9 Phylogeny of trace metal accumulation in crustaceans   285
Philip S. Rainbow

10 Mechanisms of heavy metal accumulation and toxicity in fish   321
Per-Erik Olsson, Peter Kling and Christer Hogstrand

11 Influence of ecological factors on accumulation of metal mixtures   351
Claude Amiard-Triquet and Jean-Claude Amiard

Contributors

Jean-Claude Amiard
Service d'Ecotoxicologie
Faculté de Pharmacie
Université de Nantes
1 rue Gaston Veil
44035 Nantes, Cedex 01
France

Claude Amiard-Triquet
Service d'Ecotoxicologie
Faculté de Pharmacie
Université de Nantes
1 rue Gaston Veil
44035 Nantes, Cedex 01
France

Maria João Bebianno
UCTRA
University of the Algarve
Faro
Portugal

Murray T. Brown
Marine Biology and Ecotoxicology
Group
Plymouth Environmental Research
Centre
University of Plymouth
Drake Circus
Plymouth, Devon PL4 8AA
UK

Gary R. Burt
Centre for Coastal and Marine
Sciences
Plymouth Marine Laboratory
Citadel Hill
Plymouth, Devon PL1 2PB
UK

Hing Man Chan
Centre for Indigenous Peoples'
Nutrition and Environment (CINE)
and School of Dietetics and Human
Nutrition
Macdonald Campus of McGill
University
21,111 Lakeshore Road
Ste Anne de Bellevue
Quebec H9X 3VP
Canada

Michael H. Depledge
Marine Biology and Ecotoxicology
Group
Plymouth Environmental Research
Centre
University of Plymouth
Drake Circus
Plymouth, Devon PL4 8AA
UK

John S. Edmonds
Western Australian Marine Research
Laboratories
PO Box 20
North Beach
Western Australia 6020
Australia

Kevin A. Francesconi
Institute of Biology
Odense University
DK-5230 Odense M
Denmark

Christer Hogstrand
T.H. Morgan School of Biological
Sciences
101 Morgan Building
University of Kentucky
Lexington, KY 40506-0225
USA

Togwell A. Jackson
Aquatic Ecosystem Restoration
Branch
National Water Research Institute
PO Box 5050
Burlington
Ontario L7R 4A6
Canada

David Barrie Johnson
School of Biological Sciences
University of Wales
Bangor, Gwynedd LL57 2UW
UK

Peter Kling
Department of Cellular and
Developmental Biology
Umeå University
S-901 87 Umeå
Sweden

William J. Langston
Centre for Coastal and Marine
Sciences
Plymouth Marine Laboratory
Citadel Hill
Plymouth, Devon PL1 2PB
UK

James A. Nott
Plymouth Marine Laboratory
Citadel Hill
Plymouth, Devon PL1 2PB
UK

Per-Erik Olsson
Department of Cellular and
Developmental Biology
Umeå University
S-901 87 Umeå
Sweden

Philip S. Rainbow
The Natural History Museum
Cromwell Road
London
SW7 5BD
UK

Gerhardt F. Riedel
The Academy of Natural Sciences
Estuarine Research Center
10545 Mackall Road
St Leonard, MD 20685
USA

James G. Sanders
The Academy of Natural Sciences
Estuarine Research Center
10545 Mackall Road
St Leonard, MD 20685
USA

Ken Simkiss
School of Animal and Microbial
Sciences
University of Reading
Reading, Berkshire RG6 6AJ
UK

Bjorn Sundby
INRS-Océanologie
Université du Québec
Rimouski
Québec G5L 3A1
Canada

Carlos Vale
Instituto Portugues de Investigaçao
Maritima
Avenida Brasilia
1400 Lisbon
Portugal

Erratum

When reading the Series Foreword and Preface please note the following.

Pages xvii and xviii should follow page xix.
Page xx should follow page xvi.

Metal Metabolism in Aquatic Environments
Edited by W.J. Langston and M.J. Bebianno
ISBN 0 412 80370 4

Preface

A DEFINITION

Since elements cannot be resolved by chemical means into simpler substances, they cannot, by strict biological definition, undergo metabolic degradation. We have adopted the term 'metal metabolism' throughout this volume to encompass the sum total of the constructive and destructive changes in form that can take place in a cell, tissue or organ system (or in the environment) through the actions of that organism.

CONTENTS AND PURPOSE

This book aims to provide an insight into how diverse forms of life interact with and are influenced by metals and how they vary in their ability to bioaccumulate, utilize, store, transform, detoxify and redistribute metals in the aquatic environment. Emphasis has been placed on those elements and species that are currently of major concern and that have been well characterized. It cannot, of course, be comprehensive: the range of taxa reviewed extends from microbes to humans, but the selected examples represent only a modest fraction of the total biosphere. Nevertheless, the contents of this book illustrate the variability that has been achieved through evolutionary adaptation to the presence of both essential and non-essential elements. It is also evident that some metal uptake and handling systems have been conserved remarkably consistently, from the lowest to the highest branches of the evolutionary tree. Chapter 1, for example, explains some of the generalized and apparently paradoxical concepts of how metals move from what is, in effect, a dilute solution in the surrounding environment to the more concentrated medium within the cell.

Of course the importance of metal–organism interactions extends beyond direct ecotoxicological considerations. The activities of aquatic animals and plants can strongly influence the fate of most elements; and for marine systems in particular, this has repercussions on a large scale. Biological cycling of metals, involving sediments and overlying water, is in fact a central theme linking a number of the earlier contributions (Chapters 2 to 6). Microorganisms (bacteria and phytoplankton) are recognized as major players in these dynamic exchanges (Chapters 3 and 4): they, along with a number of other groups, are also intimately involved in toxicologically important transformations of the elements mercury and arsenic, each of which merits individual review here (Chapters 5 and 6).

Several authors address the fundamental issue of **chemical** and **physical** form (speciation) and the consequences for bioavailability and assimilation of metals, in different taxonomic groups: topics included range from entry and loss of metals at the cellular level to a consideration of uptake pathways from water, sediment and food, and subsequent transfer along food chains, including humans. However, the major emphasis throughout is placed on **biological** characteristics that mediate accumulation, homeostasis and toxicity (overviewed in Chapter 7). A further objective of this volume is to illustrate how basic research into the underlying mechanisms of regulation, storage and detoxification – involving organic (e.g. metallothionein) and inorganic (e.g. intracellular metal-binding granules) pathways – may have rewarding applications in biomonitoring. Thus, the relatively recent (on evolutionary timescales) anthropogenic metal contributions to aquatic environments often interact with metal-metabolizing systems in a predictable way, compromising their ability to control intracellular ions that are in excess of requirements. The potential for using these indices (and similar diagnostic features) to identify metal-stressed individuals and populations is a frequently revisited theme, particularly with reference to the more commonly used bioindicators such as molluscs, crustaceans and fish (Chapters 8, 9 and 10). It is no accident that the latter groups include species that are among the best-studied, since they are often of direct commercial and nutritional significance and thus form the focus of attention for ecotoxicological concerns expressed by the public and environmental managers. A consideration of the degree to which metal–metal interactions modify bioavailability, body burdens and toxicity has been included (Chapter 11). This aspect is frequently overlooked in environmental assessment exercises, yet it is clearly an important factor in natural systems which are subjected to contamination by several metals.

The final section of this book deals with the transfer of metals between members of aquatic food chains and illustrates that the form of chemical complexation in food (as in water and sediment) is of paramount importance in determining metal bioavailability to consumers (Chapter 12). In conclusion, Chapter 13 reviews the summation and consequences of the varied metal metabolism strategies observed in the aquatic environment, for the ultimate consumer – ourselves.

INTENDED READERS

This synthesis of current understanding of the biology of metals in marine and freshwater organisms, and its application in the expanding field of metal ecotoxicology, is intended as a source of information for graduate and final-year environmental managers, protection officers, technical staff and consultants. The ever-increasing use of new chemicals places further demands on government agencies and industries who are required by law to evaluate potential toxicity and likely environmental impacts. The environmental manager's problem is that he needs rapid answers to current questions concerning a very broad range of chemical effects and also information about how to control discharges, so that legislative targets for *in situ* chemical levels can be met. It is not surprising, therefore, that he may well feel frustrated by more research-based ecotoxicological scientists who constantly question the relevance and validity of current test procedures and the data they yield. On the other hand, research-based ecotoxicologists are often at a loss to understand why huge amounts of money and time are expended on conventional toxicity testing and monitoring programmes, which may satisfy legislative requirements, but apparently do little to protect ecosystems from long-term, insidious decline.

It is probably true to say that until recently ecotoxicology has been driven by the managerial and legislative requirements mentioned above. However, growing dissatisfaction with laboratory-based tests for the prediction of ecosystem effects has enlisted support for studying more fundamental aspects of ecotoxicology and the development of conceptual and theoretical frameworks.

Clearly, the best way ahead for ecotoxicological scientists is to make use of the strengths of our field. Few sciences have at their disposal such a well-integrated input of effort for people trained in ecology, biology, toxicology, chemistry, engineering, statistics, etc. Nor have many subjects such overwhelming support from the general public regarding our major goal: environmental protection. Equally important, the practical requirements of ecotoxicological managers are not inconsistent with the aims of more academically-orientated ecotoxicologists. For example, how better to validate and improve current test procedures than by conducting parallel basic research programmes *in situ* to see if controls on chemical discharges really do protect biotic communities?

More broadly, where are the major ecotoxicological challenges likely to occur in the future? The World Commission on Environment and Development estimates that the world population will increase from *c.* 5 billion at present to 8.2 billion by 2025. 90% of this growth will occur in developing countries in subtropical and tropical Africa, Latin America and Asia. The introduction of chemical wastes into the environment in these regions is likely to escalate dramatically, if not due to increased industrial output, then due to the use of pesticides and fertilizers in agriculture and the disposal of damaged, unwanted or obsolete consumer goods supplied from industrialized countries. It may be many years before resources become available to implement effective waste-recycling programmes in countries with poorly developed infra-structures, constantly threatened by natural disasters and poverty. Furthermore, the fate, pathways and effects of chemicals in subtropical and tropical environments have barely begun to be addressed.

Whether knowledge gained in temperate ecotoxicological studies is directly applicable in such regions remain to be seen.

The Chapman & Hall Ecotoxicology Series brings together expert opinion on the widest range of subjects within the field of ecotoxicology. The authors of the books have not only presented clear, authoritative accounts of their subject areas, but have also provided the reader with some insight into the relevance of their work in a broader perspective. The books are not intended to be comprehensive reviews, but rather accounts which contain the essential aspects of each topic for readers wanting a reliable introduction to a subject or an update in a specific field. Both conceptual and practical aspects are considered. The Series will be constantly added to and books revised to provide a truly contemporary view of ecotoxicology. I hope that the Series will prove valuable to students, academics, environmental managers, consultants, technicians, and others involved in ecotoxicological science throughout the world.

Michael Depledge
University of Plymouth, UK

Ecotoxicology is a relatively new scientific discipline. Indeed, it might be argued that it is only during the last 5–10 years that it has come to merit being regarded as a true science, rather than a collection of procedures for protecting the environment through management and monitoring of pollutant discharges into the environment. The term 'ecotoxicology' was first coined in the late sixties by Prof. Truhaut, a toxicologist who had the vision to recognize the importance of investigating the fate and effects of chemicals in ecosystems. At that time, ecotoxicology was considered a sub-discipline of medical toxicology. Subsequently, several attempts have been made to portray ecotoxicology in a more realistic light. Notably, both Moriarty (1988) and F. Ramade (1987) emphasized in their books the broad basis of ecotoxicology, encompassing chemical and radiation effects on all components of ecosystems. In doing so, they and others have shifted concern from direct chemical toxicity to humans, to the far more subtle effects that pollutant chemicals exert on natural biota. Such effects potentially threaten the existence of life on earth.

Although I have identified the sixties as the era when ecotoxicology was first conceived as a coherent subject area, it is important to acknowledge that studies that would now be regarded as ecotoxicological are much older. Wherever people's ingenuity has led them to change the face of nature significantly, it has not escaped them that a number of biological consequences, often unfavourable, ensue. Early waste disposal and mining practices must have alerted the practitioners to effects that accumulated wastes have on local natural communities; for example, by rendering water supplies undrinkable or contaminating agricultural land with toxic mine tailings. As activities intensified with the progressive development of human civilizations, effects became even more marked, leading one early environmentalist, G. P. Marsh, to write in 1864: 'The ravages committed by Man subvert the relations and destroy the balance that nature had established'.

But what are the influences that have shaped the ecotoxicological studies of today? Stimulated by the explosion in popular environmentalism in the sixties, there followed in the seventies and eighties a tremendous increase in the creation of legislation directed at protecting the environment. Furthermore, political restructuring, especially in Europe, has led to the widespread implementation of this legislation. This currently involves enormous numbers of undergraduate students in environmental sciences. In addition, academic and research scientists may find benefits for their own work from the breadth of viewpoints represented in this comparative volume: many issues that need to be resolved in the future are indicated, and an insight into the advantages and limitations of existing lines of research may be gleaned from the experiences of the present contributors.

Examples of diagnostic features and predictive approaches contained in this book should also be of interest to those involved with monitoring, prediction and regulation of metal impact. An increased understanding of the fundamental mechanisms that dictate contaminant metal behaviour will, at the very least, reduce our uncertainty concerning biological effects and, it is hoped, will assist those charged with the management of our environment in the selection and application of improved monitoring and assessment schemes.

W.J.L. and M.J.B.
January 1998

1   *Mechanisms of metal uptake*

KEN SIMKISS

1.1 INTRODUCTION

Roughly 70 years ago Henderson (1927) wrote a book entitled *The Fitness of the Environment* in which he identified the physical and chemical properties of the environment that facilitated the evolution of life. One of the aspects that he focused on was the unique properties of water that enabled life to survive in this medium. These included its specific heat, latent heats of melting and evaporation, thermal conductivity, surface tension, expansion before freezing, and solvent and ionization properties. Fifty years afterwards, Tanford (1978) described the hydrophobic effect and its influence on the organization of living matter. The essence of his argument was that this is a unique organizing force based on repulsion by the solvent. As such it is responsible for the assembly of most of the cellular and subcellular membrane compartments that form the basis of cellular organization. Clearly water has not been a passive influence in the evolution of life. Despite these great insights and the fact that cells consist of 75–85% water, its role in biology remains controversial and poorly understood. Discussions range from modelling the state of water around the polyelectrolytes in the cell to its role in numerous reactions. As an example of such controversies we can consider osmosis.

Most animal biologists describe osmosis as the movement of water from a dilute solution into a more concentrated one – referring, of course, to the solute, not the solvent. It would be more sophisticated to point out that, in a solution containing high concentrations of solute, the water itself is less concentrated. As a result the osmotic effect could be considered as water moving from a region where it is more concentrated (i.e. in the 'dilute' solution) to that where it is more diluted (i.e. in the 'concentrated' solution). In fact, adding salt to a solution has relatively little effect on the concentration of water *per se*. The osmotic effect has been interpreted, therefore, as being caused by a reduction in water ion activity induced by hydration shells around the solute (Figure 1.1).

Metal Metabolism in Aquatic Environments. Edited by William J. Langston and Maria João Bebianno. Published in 1998 by Chapman & Hall, London. ISBN 0 412 80370 4

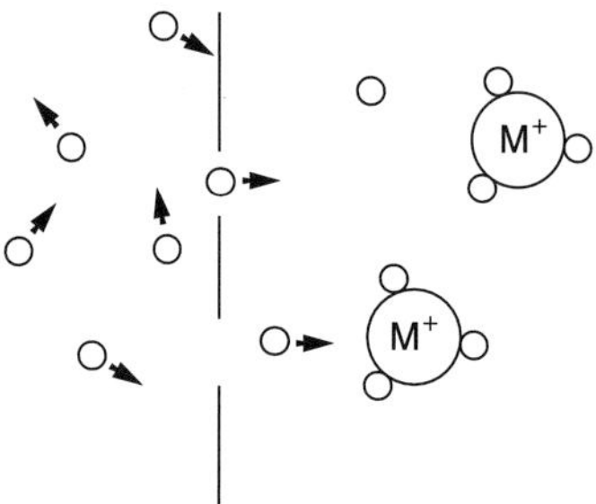

Figure 1.1 The osmotic effect depicted as a reduction in free-water activity. Metal ions (M^+) are retained on the right-hand side of the semi-permeable membrane because of their size but are able to attract a hydration shell because of their charge.

These models do not fit the facts particularly well. The osmotic force is a colligative property, i.e. it depends solely on the number of particles dissolved in the water rather than their type, whereas the hydration of molecules is well known to be influenced by factors such as ionic charge. All metal ions are charged; so, too, is water. The attraction of electrons from the oxygen atom of a water molecule by the positive charge of a metal ion can cause a proton to detach from the water. In this situation the metal is acting as a Lewis acid, facilitating the production of a proton (Figure 1.2). Many of the components of living systems such as proteins and membranes are composed of zwitterions, i.e. molecules with both positive and negative charges. By binding to these groups, protons can greatly influence their configuration. Thus the ionization of water and its ability to dissociate into both protons (H^+) and hydroxyl ions (OH^-) is a crucial property with enormous implications for metal ion uptake. But it does little to explain osmosis.

In an attempt to provide an alternative explanation for osmosis, Zeuthen (1995) has emphasized the crucial role of the pores in the semi-permeable membrane. In a bulk solution a solute molecule is capable of moving randomly with an equal chance of it going 'left or right'. If such a molecule approaches a hole in a membrane, through which it is incapable of progressing, it must return in the direction from which it has come (Figure 1.3). This phenomenon will occur most frequently in the solution containing the most solute molecules; consequently there will be a net transfer of momentum

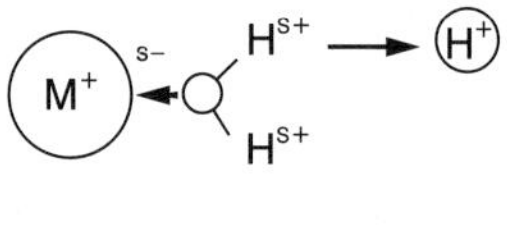

Figure 1.2 A metal ion acting as a Lewis acid by attracting electrons from a water molecule, thereby facilitating the release of a proton.

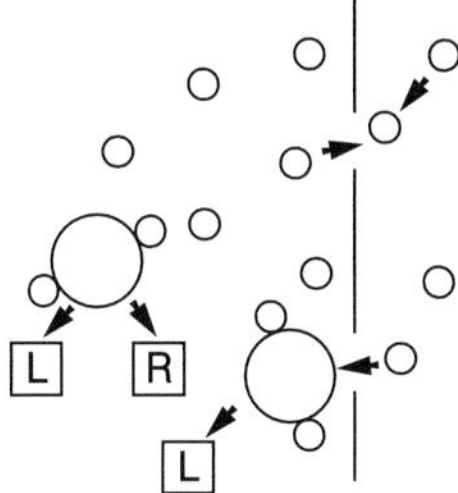

Figure 1.3 The osmotic effect depicted as an asymmetry in the movement of water molecules across a semi-permeable membrane due to the restriction in the movement of solute molecules near the surface. L, left-hand movement; R, right-hand movement possible. (After Zeuthen, 1995.)

towards the more concentrated side. Zeuthen (1995) concluded that although there is still no realistic physical description of osmosis the explanation for this force must reside in this asymmetrical transfer of momentum, drawing water across the membrane.

This chapter will consider the uptake of metals by living organisms in an aquatic environment. In the vast majority of cases these metals will be in an ionic form dissolved in the water. Their dissolution from the solid state will not have been an entirely physical event and the metal ions will be variously hydrated and may react with the water to release protons or hydroxyl ions. These metal ions may enter the living system by passing through pores in cell membranes but, as we have already seen, this may be an incompletely understood phenomenon and we must expect the pores also to be interacting with water. Thus the metal, the solvent and the cell membrane form a complex set of interactions.

The composition of natural waters is the result of acid–base effects, gas solution processes, coordination reactions, oxidation–reduction effects, adsorption–desorption processes and precipitation–dissolution influences (Stumm and Morgan, 1970). The majority of metal ions have entered the system from crystalline salts dissolving in water. Ionic salts have high lattice energies and separating the ions in a crystal is energetically very difficult (784 kJ mol^{-1} for NaCl) but may be overcome by the energetically favourable hydration of the ions (788 kJ mol^{-1} for Na^+ and Cl^-). The dissolution therefore proceeds because of the small 'energetically favourable' difference between these two large forces (4 kJ mol^{-1}). This emphasizes the strength of the ion–water interaction (Burgess, 1988) but the actual size of this effect depends on the surface charge density. Thus, increasing the radius of a monovalent cation reduces the size of the hydration layer; whilst increasing the charge increases it (Figure 1.4). The final effect, therefore, is also influenced by the shielding from inner shells of electrons. The effects of hydration and

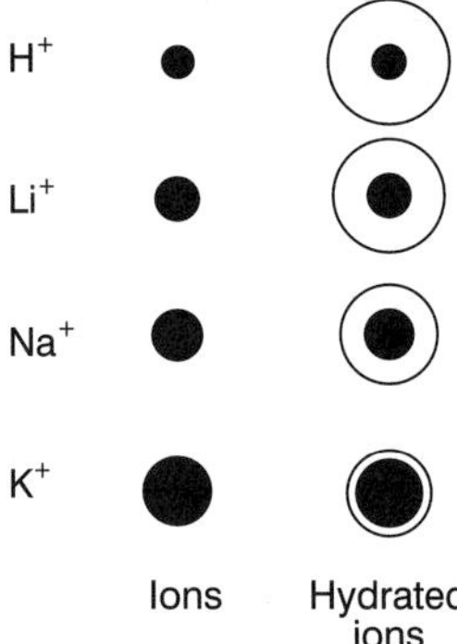

Figure 1.4 Diagrammatic representation of the size of monovalent cations (left) and their hydrated equivalents (right). Note that the size of the ions increases from H$^+$ to K$^+$ but that the size sequence of the hydrated ions is reversed.

the additional complications due to interaction with other ions in solution (i.e. the speciation of the metals) is discussed in several chapters in this volume. These factors dominate the understanding of the uptake of metals by cell membranes.

1.2 THE STRUCTURE OF THE CELL MEMBRANE

1.2.1 FLUID MOSAIC MODEL

The generally accepted structure of the cell membrane is based on the fluid mosaic model of Singer and Nicholson (1972). This accounts for the facts that the 'typical' membrane (e.g. red blood cell) consists of roughly equal weights of lipids and proteins, that the lipids are largely phospholipids in sufficient amounts to form a fluid bilayer over the cell surface and that the proteins are embedded in this to form a mosaic structure (Figure 1.5). The lipid composition is variable between different tissues and different species. A number of membranes have been analysed in great detail and there are clearly both important and subtle differences (Kotyk *et al.*, 1988). The organization of the lipid in the membrane is dictated by four important influences:

1. The phospholipids, being amphipathic, contain both hydrophilic (polar) and hydrophobic (non-polar) groups and the bilayer is therefore arranged with the polar groups facing outwards on the two exposed surfaces of the membrane.
2. This results in relatively free movement of the lipids in the plane of the membrane (i.e. it is fluid) but with virtually no flipping of lipid molecules from one surface to the other unless catalysed by specific (flip-flop) enzymes (Zachowski and Deveaux, 1990).

3. Since some of the membrane lipids are synthesized in the cytoplasm, this also results in a different composition on the two surfaces of the membrane – i.e. it is asymmetrical.
4. Other lipids may be present that influence the fluidity of the structure. This is partly due to unsaturated bonds, which may be incorporated as specific adaptations to influence membrane fluidity in response, for example, to cold (Hazel, 1988), or it may be due to the inclusion of dietary products such as cholesterol, which may also stabilize the membrane and modify its permeability (Skrtic and Eanes, 1992).

A variety of protein molecules float within this fluid lipid bilayer. Although lipids and proteins may be present in roughly equal weights there are approximately 50 lipid molecules for each of the proteins. Simple extraction procedures have established that some of these proteins are relatively easy to remove, suggesting that they are peripheral or embedded in the surface of the lipid bilayer. Such proteins may be quite mobile with diffusion coefficients in the membrane of around 10^{-10} cm^2 s^{-1}, i.e. roughly 1% of the value of lipids (Carruther and Melchior, 1988), although others – for example, on the inner surface of the membrane – may attach to the cytoskeleton and thus be very rigid.

In addition to these peripheral proteins there is another group that is very difficult to separate from the membrane. These integral proteins are thought to penetrate frequently across the total thickness of the membrane, providing for the possibility of water-filled channels across the structure. Singer (1990)

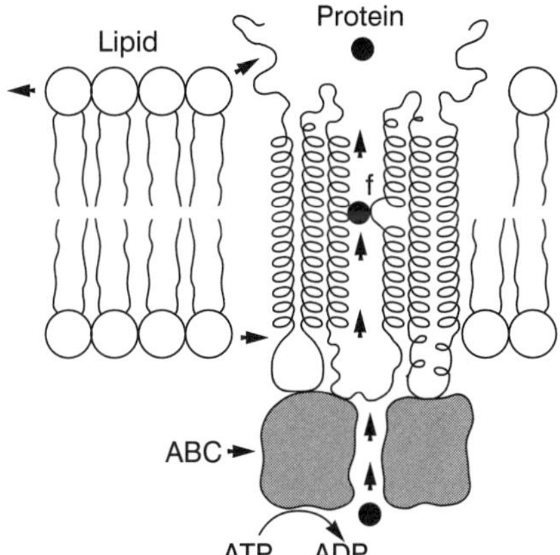

Figure 1.5 Interpretation of the fluid-mosaic model of a cell membrane consisting of a bilayer of lipid molecules in which a protein molecule is embedded. The lipid molecules are fluid in that they can move in the surface. The protein forms a mosaic structure which in this example is a channel pump with an ABC (ATP-binding cassette) component. A metal ion (filled circle) is depicted crossing the membrane by this route, which is impeded by a selectivity filter (f) that constricts the channel.

has considered the possible orientation of such proteins and there are detailed calculations on the properties of amino acids in relation to their insertion into hydrophobic or hydrophilic environments (Kotyk *et al.*, 1988). As a result it is thought that the outer transmembrane sequences of such proteins may consist of stretches of 15 to 25 amino acids uninterrupted by any ionic sequences, whereas the inner water-filled channels are lined by hydrophilic amino acids. Long polypeptide chains may therefore traverse the membrane many times to form cylindrical channels, the outer layers of which are hydrophobic or embedded in the lipid bilayer, while the inner layers are hydrophilic and provide aqueous routes between the extracellular and intracellular fluids (Figure 1.5). Much of the phenomenon of metal ion uptake will depend on the properties of these routes.

1.2.2 CHANNEL SELECTIVITY

When working on the properties of glass electrodes, Eisenman (1962) observed that of the 120 possible series in which Li^+, Na^+, K^+, Rb^+ and Cs^+ could be arranged, only 11 seemed to occur. One of these ($Li^+ < Na^+ < K^+ < Rb^+ < Cs^+$) corresponded to the size of the ionic radii while another ($Li^+ > Na^+ > K^+ > Rb^+ > Cs^+$) followed the size of the hydrated ionic radii (Figure 1.4). Working on this basis he concluded that there must be two forces controlling the movement of these ions through the glass. One would be the attraction of the ion to water whilst the other would reflect the binding of the ion to sites in the glass. The energy of interaction was found to be inversely proportional to the ionic radii of the ion (r_c) and the binding site (r_{site}) i.e. $1/(r_c + r_{site})$. This is effectively a measure of the reactivity of the ion to water when in competition with binding sites in the glass channels. On the basis of these interactions, Eisenman (1962) was able to explain the 11 commonly occurring series for the monovalent ions. This explanation assumed greater significance with the observation that these 11 series were also the most common sequences for the rates of movement of these ions through biological membranes. The same explanations were therefore advanced – namely, that the selectivity in the movement of ions through biological membrane, was based upon an interaction between the hydration/dehydration effect and the binding/release of the ion from ligands lining the channel pore. In order to appreciate the way that membrane channels work, it is necessary to understand the kinetics of these reactions.

If an ion such as Na^+ was to diffuse through a pore 0.6 nm wide and 1 nm long, it would emerge in about 0.4 ns. If in doing so it went through one dehydration/rehydration cycle, an extra 1 ns would be required. The channels across cell membranes are about 10 nm long and so equivalent ion binding reactions would take 5–10 ns; this reaction would then dominate the flux rate of ions that bind water tightly. Correspondingly, the movement of Mg^{2+} or Mn^{2+} ions would take hundreds or thousands of times longer than diffusion in order to pass through a discrimination channel. Thus selectivity carries a heavy cost in terms of the rates at which ions can move through channels. This cost can be reduced if most of the channel is wide and the region of discrimination is reduced to a small narrow selectivity filter (Figure 1.5).

It will be apparent that channels with a large ion flux through them will need to have a lower selectivity than those with smaller fluxes. As an example it can be noted that a single Na^+ channel can pass 10^7 ions s^{-1} (Hille, 1992) but such a channel may not be very selective. In the case of the sodium channels in a muscle end-plate five monovalent, nine divalent and 41 organic cations were all shown experimentally to be at least 10% as permeable as sodium (Adams *et al.*, 1980; Dwyer *et al.*, 1980). Such design criteria are not the only factors that influence selectivity. For example, Na^+ ions may pass rapidly through Ca^{2+} channels as long as there are no divalent cations present, but if trace amounts (1 μmol dm^{-3}) of calcium are present the flow of sodium ions ceases immediately. The explanation is that the calcium ion binds to the channel wall, blocking sodium ions from entering it and repelling other cations from approaching this positively charged region. For this reason the calcium channel may function in an enormously more selective way than one might expect from its shape, because of these operational influences.

Traditionally transporters, carrier proteins, channels and ion pumps have all been regarded as systems that are capable of facilitating the movement of ions across cell membranes. Kinetically it has often proved impossible to distinguish between the mechanisms implied by these descriptions (Hille, 1992) but ion pumps typically depend on the presence of adenosine triphosphate (ATP) to facilitate the uphill transport of ions against electrochemical gradients. Over 100 channels of this type are known, transporting amino acids, sugars, proteins and inorganic ions with varying degrees of specificity. All of them share a common domain with four components, two of which are located on the cytosol surface and hydrolyse ATP during the solute transport cycle. Recently it has become apparent that this ATP-binding cassette (ABC) may itself be part of a channel or, even more surprisingly, it may regulate other channels by mechanisms that are not clearly understood (Higgins, 1995) (Figure 1.5). The implication is that the activity of one type of ion-transporting channel may influence the activity of a quite separate channel. The physiological significance of this is not clear but it could have important implications in metal ion toxicity.

1.3 TRAVERSING THE MEMBRANE, TRAVERSING THE CELL

It is a fundamental property of all living systems that the composition of the intracellular (cytosol) fluids is different from those of the extracellular solution. The difference may be due to the synthesis or retention of cell proteins, leading to a Donnan equilibrium of ions. Alternatively it may have been the evolution of a phosphate-based metabolism that stopped these highly ionized metabolites from diffusing across the cell membrane and led to cellular com-

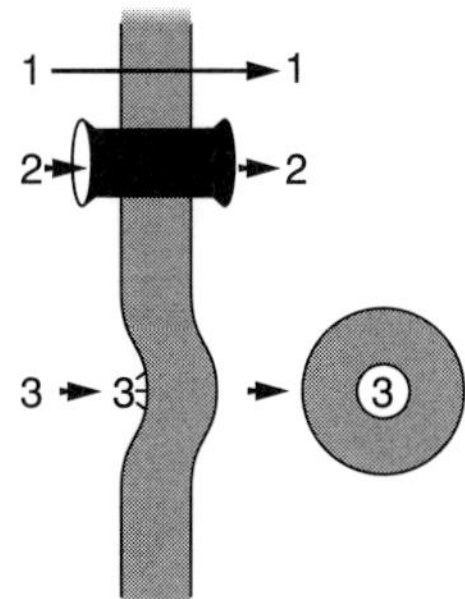

Figure 1.6 Three routes of entry into the cytosol. Hydrophobic lipid-soluble materials may pass directly through the cell membrane (route 1). Hydrophilic materials with the correct size and configuration may pass through aqueous channels (route 2). Materials that attach to the cell membrane and cause it to form vesicles may be carried into the cytosol in this form (route 3). Their release into the cytosol will depend on the subsequent fate of this vesicle. Membrane shown as a shaded layer.

partmentalization (Davis, 1958). In all cases, it is the properties of the cell membrane that eventually regulate water and metabolite flow into and out of the cell. This regulation is based upon controlling three routes of entry across the cell membrane, and ecotoxicology is largely concerned with the effect of anthropogenic influences on these systems (Simkiss, 1996a). A highly stylized view of these three routes is shown in Figure 1.6.

1.3.1 ROUTE I: HYDROPHOBICITY

The cell membrane is a negatively charged bilayer of lipids covered on each side with a cloud of counter-ions that differ between the extracellular and intracellular surfaces.

Molecules that are not charged are able to move close to this bilayer. If they contain few or no exposed polar groups they will then be able to leave the aqueous environment and enter the lipid environment of the membrane. If they are very lipophilic they may be retained there, but for molecules with an appreciable hydrophilic attraction they will emerge into the cytosol. It is this property that toxicologists measure by determining the oil/water partitioning of drugs and a parabolic relationship is found for the biological response if a large range of lipid solubilities is studied (Figure 1.7).

In an aqueous environment most metal ions are hydrated and charged and are therefore unlikely to cross a cell membrane by this route. There are two types of situation where this is not true. In many natural waters metal ions can form complex species such as MCl^+, MCl_2^0, MCl_3^-. It will be apparent that some of these complexes, such as MCl_2^0, are uncharged and can therefore approach the membrane surface. Such uncharged complexes may be heavily hydrated and therefore will not penetrate further. An exception to this is $HgCl_2^0$ which,

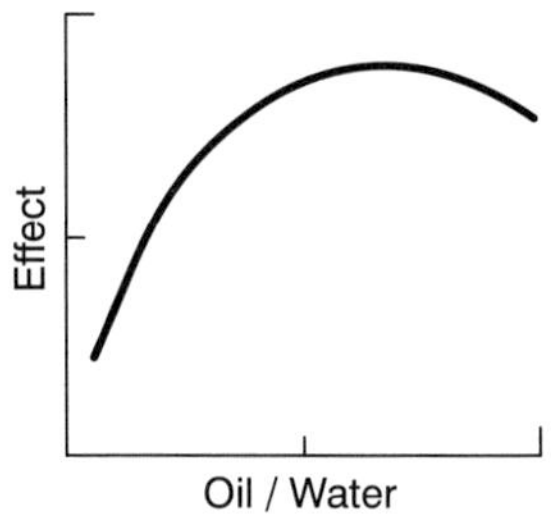

Figure 1.7 For many pharmacologically active molecules the rate at which they enter the cytosol depends upon their lipid solubility (oil/water coefficient) but if they are extremely lipid soluble they may be retained in the membrane, giving rise to a parabolic relationship.

because mercury has a coordination number of only 2, is not hydrated. In an interesting study of this complex and its penetration of an artificial cell membrane, Gutknecht (1981) found a diffusion coefficient of 10^{-4} m s^{-1}, which is over a million times the permeability of Na$^+$, K$^+$ or Cl$^-$ ions and close to the rate of diffusion of many ions in water. The implication is that such a metal complex would diffuse equally well through water or through a cell membrane.

A second set of situations that enable metal ions to enter cells across the lipid bilayer occurs when metal ions become incorporated into lipophilic molecules. The most important of these are ionophores, which may occur naturally (as in certain antibiotics) or which may be synthesized artificially. These ionophores can be of very diverse structures and may have very specific properties, but they effectively surround the metal ion with a lipophilic shell and facilitate its entry across the cell membrane.

1.3.2 ROUTE II: CHANNEL PENETRATION

There is undoubtedly a great variety of membrane-traversing proteins that facilitate the movement of ions. The terms pores, channels, transporters, and pumps emphasize various properties of particular examples but they are difficult to define in an exclusive way. A specific example is therefore probably the best way of discussing these systems.

Because of its enormous importance in cell signalling, the calcium channel is currently the best choice. Most models of the cell envisage a relatively constant intracellular calcium ion concentration $[Ca_i]$ of about 10^{-7} mol l^{-1} maintained at that level by:

- a slow conducting calcium channel (3×10^6 Ca^{2+} s^{-1});
- an outwardly directed pump with a high affinity ($K_m < 1$ μmol l^{-1}) but low capacity for calcium;
- a Na$^+$/Ca^{2+} exchanger with a low affinity ($K_m > 2$ μmol l^{-1}) but high capacity (Figure 1.8).

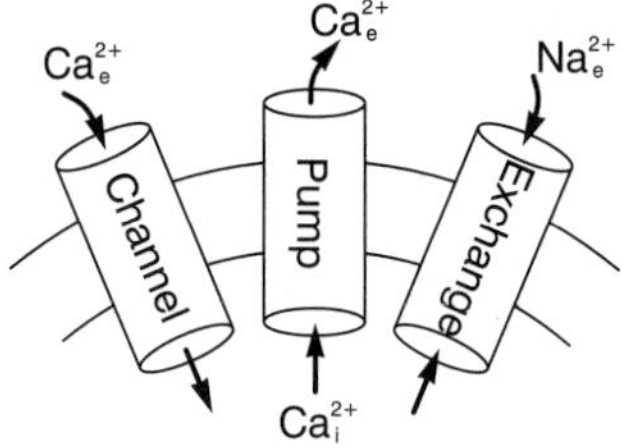

Figure 1.8 The intracellular calcium ion activity (Ca$_i^{2+}$) is thought to be regulated by an inwardly directed channel, an outwardly directed ABC pump and a Na/Ca exchange system. Clearly if the influx channels were on an apical surface and the efflux pumps and ion exchanges were on a basal surface, such a cell would transport calcium ions. (From Simkiss, 1996b.)

Calcium-sensitive proteins such as calmodulin (Scharff and Foder, 1993) monitor [Ca$_i^{2+}$] and control the activity of the Ca^{2+} ATPase pump by allosteric changes in shape (Tsien and Tsien, 1990). This normally balances the influx from the calcium-rich extracellular fluids ([Ca$_e^{2+}$] = 10^{-3} mol l^{-1}), but if [Ca$_i^{2+}$] rises the Na$^+$/Ca^{2+} exchanger becomes operative. Thus a simple calcium channel, an ATPase-activated pump and a channel that exchanges ions in opposite directions (Na$^+$ in / Ca^{2+} out) are all capable of interacting to maintain intracellular homeostasis. There are additional calcium-accumulating and calcium-releasing systems in the mitochondria and the endoplasmic reticulum that contain calcium pumps and calcium channels. The whole calcium system is carefully regulated both for homeostasis and to function as an intracellular signalling system (Berridge, 1993).

It will be apparent from the arrangement of the calcium channels in the cell membrane (Figure 1.8) that if these components were asymmetrically distributed around the cell they would take on a vectorial component. A predominance of influx channels in the apical membrane and an accumulation of efflux calcium pumps in the basal membrane would produce a calcium-transporting cell. There is considerable evidence that this occurs in a number of ion-transporting cells in the gills of the crab, *Carcinus maenas* (Flik *et al.*, 1994) and the trout (Hogstrand *et al.*, 1994).

It is implicit in these theories of ion-transporting cells that a gradient must exist across the cytoplasm and that diffusion occurs from one 'ion-accumulating' surface to the other 'ion-extruding' region. Calculations on the size of these gradients have raised questions as to whether cytotoxic ions, such as calcium, could be tolerated at these concentrations (Bronner, 1991). As a consequence it has been suggested that it might be necessary either to bind the transported ions to an intracellular protein during this process (Feher *et al.*, 1992) or to concentrate them within some intracellular vesicle (Atkins and Tuan, 1993; Simkiss, 1996b). In the latter case it may again be necessary to postulate additional membrane pumps in these cytoplasmic organelles. At the end of this process the ions would be extruded from the cell.

1.3.3 ROUTE III: ENDOCYTOSIS

The phenomenon of endocytosis is widespread and occurs in several forms. At its simplest it probably represents one of the basic systems of nutrition where a food particle becomes surrounded by the cell membrane and carried into the cytoplasm. In other situations the system is extremely specific and involves a particular ion or carrier molecule binding to a protein receptor molecule in the membrane surface. In both cases the membrane complex becomes associated with 'coated-pit' regions on the cell surface and subsequently attached to a collection of clathrin molecules. These form a lattice structure around the membrane vesicle that breaks away from the cell surface and fuses with a variety of vesicles such as lysosomes, in the cytoplasm. These organelles can transport protons into the lumen of the vesicle, acidifying it and facilitating the release of receptors and clathrin molecules back to the cell surface. Lysosomes also contain a variety of enzymes that break down organic material and release it to the cytosol. This frequently results in the transfer of materials from the extracellular fluids into the cell interior, though it is not clear at what stage they actually pass across the cell membrane and enter the cytosol.

The importance of this system is most clearly seen in filter-feeding organisms where phytoplankton, or bacteria on similar sized particles, are the main components of the diet. These particles are typically assimilated by cells of the alimentary tract, endocytosed and subjected to intracellular digestion. The phenomenon may be important in transporting pollutants into such sediment-living organisms because these particles are the main food source. Unfortunately such sediments are also major reservoirs of pollutants (Bryan and Langston, 1992) and the endocytosis process can carry such surface-bound metals directly into the cytoplasmic compartment (Davies and Simkiss, 1996).

1.4 SYSTEMS OF POLLUTANT METAL UPTAKE

1.4.1 HYDROPHOBIC ROUTES

One of the worst examples of metal toxicity arose in 1953 when an outbreak of neurological disorders in the Japanese fishing village of Minamata was traced to the discharge of mercury from a plastics factory. The causative agent was identified 15 years later when a crash in the population of fish-eating birds in Sweden was traced to mercurial fungicides in a woodpulp factory. The toxic agent was methylmercury, formed in both cases by bacteria as a by-product of their normal metabolic pathways (Jenson and Jerneløv, 1969). Two methylating agents – S-adenosylmethionine (SAM) and methylcobalamin – are capable of converting arsenic and selenium (SAM) or mercury, tin, lead and antimony into organic forms (Thayer, 1993).

Other organometallic compounds, such as tributyltin, are also powerful biocides – often with other unexpected biological effects as in the induction of imposex in the dogwhelk, *Nucella lapillus* (Bryan *et al.*, 1987). As a general rule the bioavailability of these organometallic materials depends on their hydrophobicity, and the progressive introduction of organic groups increases both their oil/water partitioning coefficient and their toxicity (Laughlin *et al.*, 1985). These organometallics appear, therefore, to be particular forms of the metal in which the introduction of an organic component produces a molecule that is able to penetrate the lipid bilayer of the cell membrane.

Similar effects can be produced by ionophores. These are both synthetic and naturally occurring compounds, such as some antibiotics. They either penetrate the cell membrane to form an aqueous channel, through which metal ions can pass, or bind metal ions into a lipid-soluble basket-type structure and thus transport it across the membrane (Easwaran, 1991). Their ability to kill cells is due to the enhanced influx of metal ions that can overwhelm the regulatory ability of the cell.

1.4.2 HYDROPHILIC ROUTES

As has already been discussed, most hydrated ions enter the cell through an interaction between the hydration shell and the ligands lining a hydrophilic channel in the lipid bilayer. Even a cursory reading of the literature indicates the enormous range of such channels that have been defined by operational experiments (e.g. the recently described copper channel; Mercer *et al.*, 1993) and genetic analysis shows that many of these systems are closely related (Vulpe *et al.*, 1993). This is not surprising, since a change in one amino acid is sufficient to convert a sodium channel into a calcium channel (Heinemann *et al.*, 1992). The questions that are pertinent to this book, however, are whether particular trace metals travel through specific channels or whether trace metal requirements are likely to be met by non-specific leakage through the variety of ion channels that occur in all cells. One of the best examples of recent studies in this area comes from the work of Hogstrand *et al.* (1994) on the gills of rainbow trout. Calcium is thought to enter the fish gills along an electrochemical gradient, entering cells through channels in the apical epithelium and exiting through a basolateral high-affinity Ca ATPase pump, i.e. similar to the model shown in Figure 1.8. Zinc ions apparently inhibit this process, leading to hypocalcaemia in the fish. A kinetic analysis of the results of such an experiment is shown in Figure 1.9, where changes in K_m indicate that there are competitive interactions while changes in J_{max} imply non-competitive effects. These results show that zinc enters the gill by competing with calcium ions for access to this calcium channel. Subsequent work led to the suggestion that in the recovery phase there was a change in the apical permeability to calcium, and the K_m of the channel was modified so as to reduce zinc binding without greatly affecting calcium influx. It is interesting to note that some of these modifications are similar to the regulatory features attributed to some of the ATP binding cassette systems (section 1.2.2).

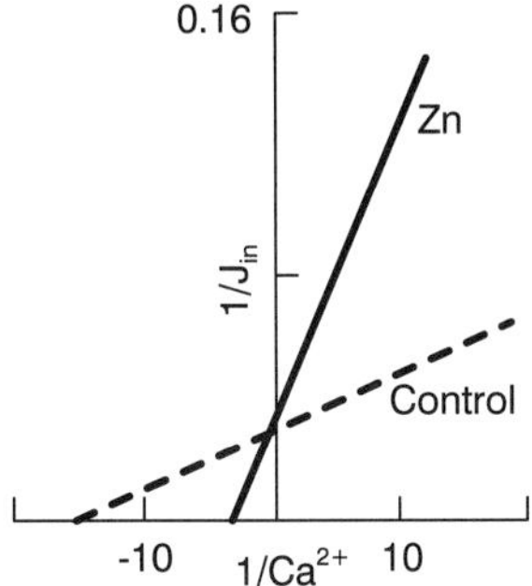

Figure 1.9 Kinetic evidence showing that zinc enters the gills of rainbow trout by competing with calcium to enter the calcium channel. (After Hogstrand *et al.*, 1994.)

Similar results have been obtained on the electrogenic $2Na^+/H^+$ antiporter system in the gastrointestinal epithelium of the starfish *Pycnopodia hebianthoides*, where Ca^{2+} ions act as a competitive inhibitor. A number of other ions (Mn^{2+}, Cu^{2+}, Fe^{2+} and Mg^{2+}) also inhibit this process, though the mechanism of action is less clear (Zhuang *et al.*, 1995). There are, of course, two possible cellular lesions that could be involved in these channel interactions. In the first, a potentially toxic metal could block the channel of an essential ion (such as the Ca^{2+} pump), leading to deleterious effects identified as hypocalcaemia. In the second interpretation, a potentially toxic metal could be transported across the membrane via the channel of an essential ion (such as the Ca^{2+} pump), leading to toxicological effects once it poisons the cytoplasmic proteins.

In terms of normal environmental interactions it is possible to identify the three most common factors that affect metal uptake through channels. The first of these is metal speciation, particularly where strong covalent bonds are involved. It is generally agreed that channels are largely impermeable to metal complexes so that ionic species dominate uptake. The second is calcium availability, which may affect intercellular linkages and thus the potential for paracellular movement. In addition, however, calcium provides the most common divalent cationic channel and thus the most likely route for entry of a number of metal pollutants. Finally, there is environmental pH or proton activity. Because of their small size and great mobility, protons can penetrate many protein structures and, in the case of ion channels, can compete with other cations for attachment to various ligand sites in the channel (Borgmann, 1983). Changes in water pH may therefore have a great effect upon the cellular uptake of a variety of metal ions, either by increasing the number of free metal ions that are available in solution or by modifying the configuration of channel proteins and thus their ability to transport these ions.

1.4.3 PARTICULATE

Metals that exist either as elemental particles (e.g. Fe) or that precipitate as insoluble salts (e.g. Fe (OH)$_3$) may enter cells by the direct endocytotic route. The clearest demonstration of this was with the mussel *Mytilus edulis* that was exposed to a pulse label of ^{59}Fe (George *et al.*, 1976). The metal became localized in specific regions of the gills and iron-containing endocytotic vesicles could be demonstrated in the cytoplasm just beneath the microvilli. The metals were transported across these epithelial cells and exocytosed into the blood, where they were collected and transported around the body in amoebocytes. It was suggested that this is a special adaptation enabling marine molluscs to obtain an essential element from an environment in which it has an exceedingly low solubility (George *et al.*, 1977).

There are many other more general examples where endocytosis influences metal uptake. The most obvious case is with filter-feeding organisms which use endocytosis as part of the normal intracellular digestive process (Chapter 8). Such organisms survive on small food particles or on the organic adsorbed layers that occur on the surfaces of many inorganic deposits (e.g. bacterial exopolymer films; Harvey and Luoma, 1985). These particles may also be major sources of metals that enter the organism along this route. Animals that burrow in sediments or filter suspended particles out of the water may be exposed to much larger concentrations of metals because of adsorption on to these surfaces. Where the organism is simply exposed to pore water that has equilibrated with the sedimentary particles, uptake will be via the processes already discussed; but if the particles are endocytosed it is very difficult to determine their bioavailability. A number of extraction techniques have been used, including ammonium acetate, acetic acid, concentrated nitric acid, and various chelating agents or 'acid-soluble sulphides', but none of these chemical techniques appears to reflect the biological process (Luoma, 1989). Part of the difficulty arises from the fact that organisms may select particles for digestion on such subtle criteria as presence or absence of protein films. In order to overcome this problem Davies and Simkiss (1996) fed *M. edulis* on metal-loaded artificial sediments where the surface binding effects could be precisely characterized. The particles were of a similar size to normal food items and they were endocytosed into the digestive cells of the hepatopancreas. Under these conditions a simple relationship between metal uptake and the dissociation constant of the particle surface could be demonstrated.

Although particle assimilation and endocytosis are two of the less well understood processes of metal uptake, it appears that previous studies have been poorly focused on the problem. Naturally occurring particles in the aquatic environment have a remarkably simple surface chemistry consisting largely of carboxyl and phenolic groups (Hunter and Liss, 1982). This is because they become coated with biogenic films which therefore dictate the chemistry of the subsequent adsorption of metal ions. Once assimilated into digestive vacuoles these metals are soon released and then appear to be treated in at least an analogous way to metals that enter cells via lipophilic or channel-mediated pathways.

1.5 CONCLUSIONS

As Henderson (1927) pointed out, it is the unique properties of water that have been exploited in the evolution of life. One of the aspects that he omitted to emphasize was the development of hydrophobic partitions that facilitated the development of cells and the accumulation of particular elements. Within this system the 'natural selection of the chemical elements' (Williams, 1981) occurred and led to the diversity of processes recognized as biochemistry.

The ability of the cell membrane to contain this variety of materials and reactions depends to a large extent on the hydrophilic nature of the cell contents and the hydrophobicity of the surrounding membranes. In the case of metal ions there are very few examples of direct entry across this lipid barrier and it is to a large extent the aqueous channels that dictate which ions will enter and leave the cytoplasm. The specificity of these routes is essential for life and to a large extent depends on competition between the hydration of a metal ion and its dehydration during the passage down this channel. The success of this discriminatory process depends on the sophistication of the design of the channel and the operational conditions under which it exists. The recent discovery of a proton-coupled channel that is capable of transporting Fe^{2+}, Zn^{2+}, Mn^{2+}, Co^{2+}, Cd^{2+}, Cu^{2+}, Ni^{2+} and Pb^{2+} (Gunshin *et al*, 1997) does, however, raise the possibility of general metal ion transferring systems, and emphasizes the cells need to also contain efflux systems and detoxification mechanisms. When anthropogenic activity increases the bioavailability of some metal ions, it changes the competitive environment under which they exist, or modifies the configuration of the channel route, and poses a fundamental challenge to the nature of these selectivity processes.

REFERENCES

Adams, D.J., Dwyer, T.M. and Hille, B. (1980) The permeability of end plate channels to monovalent and divalent metal cations. *J. Gen. Physiol.* **75**, 493–570.

Atkins, R.E. and Tuan, R.S. (1993) Transepithelial calcium transport in the chick chorioallantoic membrane II compartmentalization of calcium during uptake. *J. Cell Sci.* **105**, 381–385.

Berridge, M.J. (1993) Inositol triphosphate and calcium signalling. *Nature* **361**, 315–325.

Borgmann, V. (1983) Metal speciation and toxicity of free metal ions to aquatic biota, in *Advanced Environmental Science and Technology* (ed. J.O. Nriago) **13**, 47–72.

Bronner, F. (1991) Calcium transport across epithelia. (1991) *Int. Review Cytol.* **131**, 169–212.

Bryan, G.W. and Langston, W.J. (1992) Bioavailability accumulation and effects of heavy metals in sediments with special reference to United Kingdom estuaries: a review. *Environ. Pollution* **76**, 89–131.

Bryan, G.W., Gibbs, P.E., Burt, G.R. and Hummerstone, L.G. (1987) The effects of tributyl tin (TBT) accumulation on adult dogwhelks *Nucella lapillus*, long term field and laboratory experiments. *J. Marine Biol. Assoc.* **67**, 525–544.

Burgess, J. (1988) *Ions in solution: Basic Principles of Chemical Interactions*, Ellis Horwood, Chichester, 191 pp.

Carruther, A. and Melchior, D.L. (1988) Effects of lipid environment on membrane transport. The human erythrocyte sugar transport protein/lipid bilayer system. *Ann. Rev. Biochem.* **50**, 257–271.

Davies, N.A. and Simkiss, K. (1996) The uptake of zinc from artificial sediments by *Mytilus edulis. J. Mar. Biol. Assoc.* **76**, 1073–1079.

Davis, B.D. (1958) The importance of being ionized. *Arch. Biochem. Biophys.* **78**, 497–509.

Dwyer, T.M., Adams, D.J. and Hille, B. (1980) The permeability of the end plate channel to organic cations in frog muscle. *J. General Physiol.* **75**, 469–492.

Easwaran, K.R.K. (1991) Ionophores: structure and interaction in relation to transmembrane ion-transport, in *Molecular Conformation and Biological Interactions*, (eds P. Balaram and S. Ramaseshan), Indian Academy of Science, Bangalore, pp. 671–685.

Eisenman, G. (1962) Cation selective glass electrodes and their mode of operation. *Biophys. J.* **2** (Suppl. 2), 259–323.

Feher, J.J., Fullmer, C.S. and Wasserman, R.H. (1992) Role of facilitated diffusion of calcium by calbindin in intestinal calcium absorption. *Amer. J. Physiol.* **262**, C517–526.

Flik, G., Verbost, P.M. and Atsma, W. (1994) Calcium transport in gill plasma membranes of the crab *Carcinus maenas*: evidence for carriers driven by ATP and a Na^+ gradient. *J. Exptl Biol.* **195**, 109–122.

George, S.G., Pirie, B.J.S. and Coombs, T.L. (1976) The kinetics of accumulation and excretion of ferric hydroxide in *Mytilus edulis* (L) and its distribution in the tissue. *J. Exptl Mar. Biol. Ecol.* **23**, 71–84.

George, S.G., Pirie, B.J.S. and Coombs, T.L. (1977) Absorption, accumulation and excretion of iron–protein complexes by *Mytilus edulis. Proc. Intern. Conf. Heavy Metal Environ. Can.*, Natural Resources Council of Canada, Publication no. 2, Toronto, 887–900.

Gunshin, H., Mackenzie, B., Berger, U.V., Gunshin, Y., Romero, M.F., Boron, W.F., Nussberger, S., Gollan, J.L. and Hediger, M.A. (1977) Cloning and characterization of a mammalian proton-coupled metal-ion transporter. *Nature*, **388**, 482–488.

Gutknecht, J. (1981) Inorganic mercury (Hg^{2+}) transport through lipid bilayer membrane. *J. Memb. Biol.* **61**, 61–66.

Harvey, R.W. and Luoma, S.N. (1985) Effect of adherent bacteria and bacterial extracellular polymers upon assimilation by *Macoma balthica* of sediment-bound Cd, Zn and Ag. *Mar. Ecol. Prog. Series* **22**, 281–289.

Hazel, J.R.(1988) Homeoviscous adaptation in animal cell membranes, in *Physiological Regulation of Membrane Fluidity* (eds R.C. Aloia, C.C. Curtain and L.M. Gordon), Alan R. Liss, New York, pp. 149–188.

Heinemann, S.H., Terlau, H., Stuhmer, W. *et al.* (1992) Calcium channel characteristics conferred on the sodium channel by single mutations. *Nature* **356**, 441–443.

Henderson, L.T. (1927) *The Fitness of the Environment*, MacMillan, New York, 317 pp.

Higgins, C.F. (1995) The ABC of channel regulation. *Cell* **82**, 693–696.

Hille, B. (1992) *Ionic Channels of Excitable Membranes*, 2nd edn, Sinauer Associates, Sunderland, Massachusetts, 607 pp.

Hogstrand, C., Wilson, R.W., Polgard, D. and Wood, C.M. (1994) Effects of zinc on the kinetics of branchial calcium uptake in freshwater rainbow trout during adaptation to waterborne zinc. *J. Exptl. Biol.* **186**, 55–73.

Hunter, K.A. and Liss, P.S. (1982) Organic matter and the surface charge of suspended particles in estuarine waters. *Limnol. Oceanogr.* **27**, 322–335.

Jensen, S. and Jerneløv, A. (1969) Biological methylation of mercury in aquatic organisms. *Nature* **223**, 753–754.

Kotyk, A., Janacek, K. and Koryta, J.H. (1988) *Biophysical Chemistry of Membrane Functions*, John Wiley, Chichester.

Laughlin, R.B., French, W. and Guard, H.E. (1985) On the correlation between acute and sub-lethal stress in mud crab larvae *Rhithropanopeus larrisii*. The example of triorganotin compounds, in *Marine Biology of the Polar Regions and Effects of Stress on Marine Organisms*, Proceedings of the 18th European Marine Biology Symposium, (eds J.S. Gray and M.E. Christiansen), John Wiley, Chichester, pp. 513–526.

Luoma, S.N. (1989) Can we determine the biological availability of sediment-bound trace elements? *Hydrobioligia* **176/177**, 379–396.

Mercer, J.F.B., Livingston, J., Hall, B. *et al.* (1993) Isolation of a partial candidate gene for Menkes disease by positional cloning. *Nature* **3**, 20–25.

Scharff, O. and Foder, B. (1993) Regulation of cytosolic calcium in red blood cells. *Physiol. Rev.* **73**, 547–582.

Simkiss, K. (1996a) Ecotoxicants at the cell-membrane barrier, in *Ecotoxicology, a Hierarchical Approach*, (eds M.C. Newman and C.H. Jagoe), Lewis, Boca Raton, pp. 59–83.

Simkiss, K. (1996b) Calcium transport across calcium-regulated cells. *Physiol. Zool.* **69**, 343–350.

Singer, S.J. (1990) The structure and insertion of integral proteins in membranes. *Ann. Rev. Cell Biol.* **6**, 247–296.

Singer, S.J. and Nicholson, G.L. (1972) The fluid mosaic model of the structure of cell membranes. *Science* **175**, 720–731.

Skrtic, D. and Eanes, E.D. (1992) Effect of different phospholipid-cholesterol membrane compositions on liposome-mediated formation of calcium phosphates. *Calcif. Tissue Int.* **50**, 253–260.

Stumm, W. and Morgan, J.J. (1970) *Aquatic Chemistry. An Introduction Emphasizing Chemical Equilibria in Natural Waters*, Wiley-Interscience, New York, 583 pp.

Tanford, C. (1978) The hydrophobic effect and the organization of living matter. *Science* **200**, 1012–1018.

Thayer, J.S. (1993) Global bioalkylation of the heavy elements, in *Metal Ions in Biological Systems*, (eds H. Sigel and A. Segel), M. Dekker, New York, pp. 1–36.

Tsien, R.W. and Tsien, R.Y. (1990) Calcium channels, stores and oscillations. *Ann. Rev. Cell. Biol.* **6**, 715–760.

Vulpe, C, Levinson, B., Whitney, S. *et al.* (1993) Isolation of a candidate gene for Menkes disease and evidence that it encodes a copper-transporting ATPase. *Nature Genet.* **3**, 7–13.

Williams, R.J.P. (1981) Natural selection of the chemical elements. *Proc. Royal Soc.* **213B**, 361–397.

Zachowski, A. and Devaux, P.F. (1990) Transmembrane movement of lipids. *Experientia* **46**, 644–656.

Zeuthen, T. (1995) Molecular mechanisms for passive and active transport of water. *Int. Review Cytol.* **160**, 99–161.

Zhuang, Z., Duerr, J.M. and Ahearn, G.A. (1995) Ca^{2+} and Zn^{2+} are transported by the electrogenic $2Na^+/H^+$ antiporter in echinoderm gastrointestinal epithelium. *J. Exptl Biol.* **198**, 1207–1217.

2 *The interactions between living organisms and metals in intertidal and subtidal sediments*

CARLOS VALE AND BJORN SUNDBY

2.1 INTRODUCTION

Although it may not be readily apparent to the casual observer, intertidal and subtidal sediments are highly structured chemical environments. In the absence of physical disturbances, the chemical structure is simple and consists of layers of gradually changing composition parallel to the sediment–water interface. Benthic animals and rooted plants disrupt this simple structure by burrowing into the sediment or by growing roots. This affects the rates of transport of gases, solutes and particulate matter within the sediment and between the sediment and the overlying water.

This chapter introduces the subject of how plants and animals can affect the distribution and cycling of particulate and dissolved metals in fine-grained marine sediments. It compares the way in which rooted plants, dominant in the upper littoral zone, and benthic invertebrates, dominant in the sublittoral, interact with sediments. For a general treatment of sediment–water interactions in the marine environment, refer to Berner (1980) and for soil–plant interactions refer to Nye and Tinker (1977). Pearson and Rosenberg (1978), Tenore and Coull (1980) and McCall and Tevesz (1982) discuss the ecology of benthic animals, and Rhoads (1974) discusses the physical effects of animals on sediments. Discussions of the effects of benthic animals on sediment chemistry can be found in Aller (1977, 1978, 1982, 1988). Tinker and Barraclough (1988) discuss the effects of soil–root interactions on the soil system.

Metal Metabolism in Aquatic Environments. Edited by William J. Langston and Maria João Bebianno. Published in 1998 by Chapman & Hall, London. ISBN 0 412 80370 4

2.2 CHEMICAL ZONATION IN SEDIMENTS

The concept of chemical zonation refers to the difference in chemical composition and redox potential between the interior of the sediment and the interface between the sediment and the overlying water or air. The interface can be represented by the sediment surface, the root–sediment interface, or the wall of a water-filled burrow created by a benthic animal. In each case a chemical zonation is created through the flux of solutes and dissolved gases across the interface and their consumption within the sediment. The sediment composition changes rapidly on a scale of millimetres in organic-rich coastal sediments or more slowly over tens of centimetres in deep-sea sediments.

To appreciate the concept of chemical zonation, it may be helpful to carry out a hypothetical experiment. If one were to remove all large organisms from a sample of anoxic sediment, mix it thoroughly, and cover it with oxygenated seawater, the composition of the sediment would immediately start to change. The change would begin at the sediment–water interface where dissolved oxygen, nitrate and sulphate migrate from the overlying water into the pore water of the sediment. The oxygen is consumed in the oxidation of organic matter and reduced pore-water constituents such as ammonia and sulphide. Nitrate and sulphate produced in the reactions with oxygen are not consumed in the presence of oxygen but migrate deeper into the sediment, where they serve as electron acceptors in the anaerobic oxidation of organic matter by microorganisms. Oxygen also reacts with reduced forms of iron and manganese which are precipitated as insoluble oxides (Chapter 3). As a result of these reactions, the composition of both the pore water and the solid sediment phases changes and, after some time, a steady state develops in which the flux of oxygen and other electron acceptors from the overlying water is balanced by their consumption within the sediment. The electron acceptors that are transported into the sediment (oxygen, nitrate, metal oxides, sulphate) are consumed in a sequence that is determined by the free-energy yield of the reaction between organic matter and the particular electron acceptor (e.g. Froelich *et al.*, 1979).

A close look at the steady-state sediment composition would reveal the existence of a number of zones, each of which corresponds to the consumption of one particular electron acceptor. The first zone, just below the sediment–water interface, corresponds to reactions involving oxygen reduction; the next zone corresponds to nitrate reduction and manganese oxide reduction, followed by iron oxide reduction; below this is a zone of sulphate reduction, followed by a zone of CO_2 reduction. The boundaries between these zones are not located at a fixed distance from the interface but vary spatially and temporally because of variations in the temperature and chemical composition of the bottom water and variations in the input of organic carbon.

The chemical zonation affects the solubility and therefore the distributions and fluxes of a number of metal species. The most important solubility changes are due to the different solubilities of reduced and oxidized species, such as in the case of manganese and iron, and to the insolubility of many metal sulphides. When the concentration of a metal changes away from an interface, the metal ions migrate along the resulting concentration gradient. For example, the solubility of manganese and iron increase away from an interface because the reduced forms of these metals are more soluble than their oxidized forms. Therefore, dissolved manganese and iron migrate back towards the interface, and if there is oxygen present they will be re-precipitated as iron and manganese oxides. The overall result of this is that the sediment will be enriched in manganese and iron oxides near an interface and depleted away from it.

The principle of a chemical zonation based on the thermodynamically controlled use of terminal electron acceptors is well established in the literature (Froelich *et al.*, 1979) and is readily observed in many sediments. However, the relative importance of each of the principal reactions will depend on the particular sedimentary environment, and the overall result can also be influenced by the many secondary reactions that take place. For example, dissolved sulphide can reduce the oxides of manganese and iron as well as react with oxygen (Aller and Rude, 1988); and reduced iron can be oxidized by manganese oxides as well as by oxygen, in addition to reacting with sulphide. In sediments that contain high concentrations of organic matter or are subject to high ambient temperatures, the reaction rates may be so high that the complete reaction sequence may be compressed into a very thin layer near the interface and the overall process of organic matter mineralization may be completely dominated by anaerobic processes, particularly sulphate reduction. In the extreme case the interface may be inhabited by sulphur-oxidizing bacteria that use the oxygen from the atmosphere to oxidize directly the sulphide produced by sulphate reduction.

2.3 INTERACTIONS OF PLANTS AND ANIMALS WITH SEDIMENTS

Plants and animals have many similar effects on sediments (Table 2.1). The most important of these, insofar as it affects the chemistry of the sediment, is to improve the exchange of material between the interior of the sediment and the overlying water or air. The roots of marsh plants are efficient conduits of atmospheric oxygen to the sediment because of their well developed aerenchyma system (Anderson, 1974), which allows oxygen to diffuse from the leaves to the roots (Teal and Kanwisher, 1966). The diffusion takes place in the gas phase, which is 10^3 times faster than in the aqueous phase. The oxygen not consumed by root respiration is available for diffusion into the surrounding sediment (Armstrong, 1970). Many benthic animals construct burrows that are connected to the overlying water. The exchange of burrow

Table 2.1 A comparison of the effects of plants and animals on sediments

Action	Plants	Animals
Connecting the sediment interior with ambient water or air	Grow roots	Construct tubes
Transport of O_2 to the interface	Diffusion within the plant to the root	Irrigation of the burrow with overlying water
Transport of O_2 to the sediment interior	Diffusion of O_2 from the roots into the sediment	Diffusion of O_2 from the wall of the burrow into the sediment
Mixing sediment particles across concentration gradients	Plants do not mix the sediment	Animals transport particles by several mechanisms, including burrow excavation and feeding
Mixing pore water across concentration gradients	The pore water is mixed by dispersion as it moves through the porous sediment towards the root surface	Animals mix the pore water by periodical replacement of burrow water with overlying water
Sedimentation accumulation	Plants slow down currents and trap fine suspended matter at the sediment surface, thereby raising the height of the bed	Animals do not contribute to net sediment accumulation
Stabilization of the sediment	Plants add structural stability to a sediment by permeating it with roots	Animals may stabilize sediments by excreting mucus
Detoxification of sediments and succession of organisms over time	Pioneering species invade anaerobic sulphidic sediments and oxidize it, concentrating metal oxides and transmetals into oxidized microenvironments surrounding roots; they add stability and raise the level of the bed, preparing it for the invasion of a succession of species of greater diversity	Pioneering species invade anaerobic sediments, construct high densities of tubes, accelerate the flushing of the pore water of toxic soluble species, deepen the redox boundary, and prepare the sediment for a succession of species of lower tolerance and higher diversity

water with bottom water allows for the transport of oxygen to the interior of the burrows, whence it can diffuse into the anoxic sediment. Aller (1978) points out that the diffusion geometries of burrows become similar to those of a root system in the case where large numbers of burrows permeate the sediment and that in some respects burrows play a functional role similar to that of roots: sediments are tapped for dissolved nutrients and advected into the overlying water, where photosynthesis takes place.

The presence of oxygen at the root–sediment and burrow–sediment interfaces creates local oxidizing conditions in otherwise reducing sediments. This affects in particular the chemistry of iron and manganese, whose soluble reduced forms diffuse towards these interfaces, where they are precipitated as insoluble iron and manganese oxides. Aller and Yingst (1978) have shown that iron, manganese and zinc accumulate at the burrow interface as a result of intense decomposition processes in the burrow lining and mobilization of the metals from the adjacent sediment. Likewise, metal oxides have been observed to coat the surface of roots of a number of aquatic plant species (Bartlett, 1961; Taylor *et al.*, 1984; Otte *et al.*, 1989; Caçador *et al.*, 1996a,b); this has been suggested to be a mechanism for increasing waterlogging tolerance (Green and Etherington, 1977) and for reducing the uptake of excessive amounts of potentially toxic reduced species of iron, manganese and sulphur (Bartlett, 1961; Armstrong, 1967; Green and Etherington, 1977). The metal oxide plaque on the roots of many salt marsh species has the form of thin coatings or stains (Otte *et al.*, 1989). During winter, when the plant is inactive, these oxide plaques may be destroyed and release reduced metal ions to the sediment. In regions where salt marsh plants remain active most of the year, much thicker iron oxide coatings around the roots have been observed (Vale *et al.*, 1990; Caçador, 1994). These coatings, which have the shape of hollow cylinders, 1–2 mm in diameter, are called rhizoconcretions. They have also been observed in areas where salt marshes existed in the past. There are no indications that these concretions dissolve during winter.

Rooted plants increase the average redox potential of the sediment by conducting oxygen to the interior of sediment (Armstrong, 1978). In the same way, burrows lead to the deepening of the average depth of the oxic–anoxic boundary (Aller, 1980). The effect of oxygen on the sediment is not limited to the zone surrounding the individual root or burrow, but can include the composition of the bulk sediment. An illustration of this is the effect of vegetation on the pH, the redox potential, and the sulphur system of a salt marsh sediment. Note that the sediments between the roots are more acidic and more oxidizing than non-vegetated sediments (Figure 2.1). Note also that sediments with high root density contain more sulphate and less acid volatile sulphide (AVS) than non-vegetated sediments (Figure 2.2). In some salt marshes pyrite, which may be formed during the winter when the plants are inactive, may be destroyed by oxidation during the growing season (Luther and Church, 1992).

Plants and animals can affect the accumulation rate of trace metals as well as their distribution within the sediment. In the intertidal zone, the presence of plants slows down the currents and favours the settling of fine-grained particulate matter. This increases the accretion rate relative to non-vegetated areas.

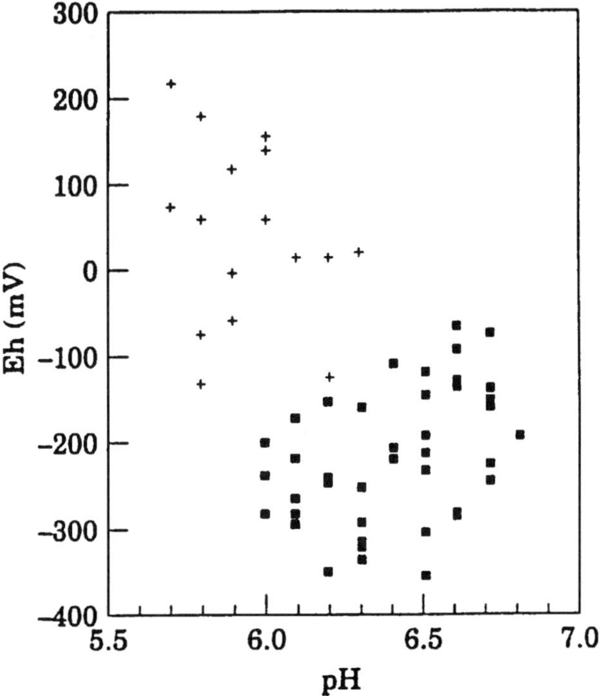

Figure 2.1 Redox potential (Eh) vs. pH in sediments from a Tagus estuary salt marsh (Caçador *et al.*, 1996a): (+), sediment between the roots in the 5–15 cm depth interval at a vegetated site; (■) sediments from a non-vegetated site and from below the root zone at the vegetated site.

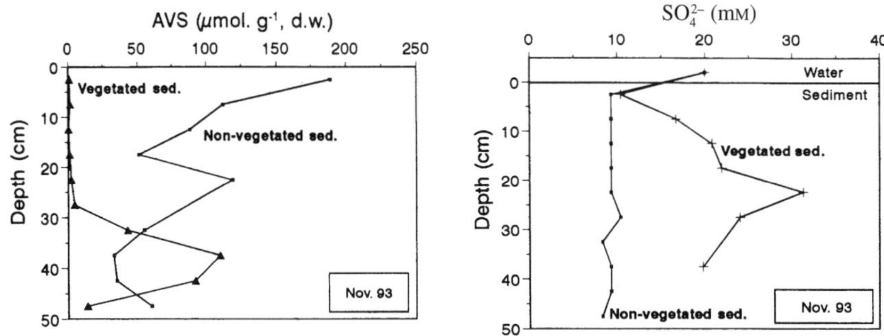

Figure 2.2 The vertical distribution of acid volatile sulphur (AVS) and sulphate in vegetated (▲) and non-vegetated (■) sediments from a Tagus estuary salt marsh (Madureira *et al.*, 1997).

It also increases the trace metal accumulation rate because the higher specific surface area of fine-grained material often leads to higher contents of adsorbed trace metals. Likewise, much of the accreting matter consists of plant debris, also enriched in trace metals (Orson *et al.*, 1992). In the Tagus estuary salt marshes the sediment in vegetated areas is enriched in Zn, Pb and Cu (Caçador *et al.*, 1996a). The vertical distribution of these metals in the vegetated area also shows a subsurface maximum, the origin of which is not known.

In highly sulphidic sediments trace metals are effectively immobilized and their toxicity neutralized because of the low solubility of many metal sulphides (Di Toro *et al.*, 1990; see also Chapter 8). The oxidation of the sediment can change this (Gobeil *et al.*, 1987) and make it possible to have an increased flux of trace metals towards the root surface or the burrow wall. It has been shown that iron plaque on the roots of a salt marsh plant can act as a barrier against the uptake of Zn (Otte *et al.*, 1989), possibly because of the adsorption on to metal oxides. The metal oxide-rich rhizoconcretions in these sediments are rich in trace metals compared with the surrounding sediment (Vale *et al.*, 1990), supporting the hypothesis that iron plaque can indeed be a barrier to trace metal uptake by plants.

2.4 SEDIMENT WATER INTERACTION UNIQUE TO BENTHIC ANIMALS: BIOTURBATION

Unlike plants, benthic animals actually manipulate particles. In addition to simply pushing them aside as they move through the sediment, animals transport particulate matter when they excavate burrows and when they feed. The process is known as bioturbation. Evidence for bioturbation can be seen readily in the mounds of worm castings that occur in abundance on tidal flats. More subtle evidence for bioturbation can be obtained from the vertical distribution of a naturally occurring radioisotope such as Pb-210: in many sediments the actvity of this isotope, which has a half-life of 23 years, is constant in the upper 5–6 cm because of the mixing that results from bioturbation. The expected exponential decay curve is only observed below this mixed, biologically reworked layer.

The way that benthic animals move particles depends to a large extent on their feeding mode. Deep deposit feeders feed below the sediment surface and deposit their faeces on the surface; surface deposit feeders deposit their faeces either on the surface or within excavated burrows; and so on. In environments of great biological diversity all of these modes may be present, and the net result of the diverse particle transport modes may be indistinguishable from random mixing (Aller, 1982). In this case, the process of particle transport is analogous to diffusion. Using the formalism of Fick's first law of diffusion, the diffusion coefficient is replaced by a biological mixing coefficient D_b (Berner, 1980). Particle reactive tracers such as Pb-210, Cs-137, Th-234, and Be-7 have been used successfully to estimate the magnitude of the biological mixing coefficent

(Goldberg and Koide, 1962; Aller and Cochran, 1976; Schink and Guinasso, 1977; Silverberg *et al.*, 1986).

In other environments one or a few organisms may dominate the biological community and the particle transport mode. A well studied example is that of the polychaete *Scoloplos* sp. (Rice, 1986). The feeding mode of this organism is analogous to that of a conveyor belt: it feeds at a depth of 4–6 cm near the redox potential discontinuity and egests particles on the sediment surface. Populations of such species may transport enormous quantities of particles over a vertical distance of several centimetres. This mode of particle transport cannot be described by the random mixing model since it involves bioadvection and subduction: a given sediment layer moves downward as material from the feeding depth is deposited on top of it; the subducted sediment is returned to the sediment surface when the layer enters the feeding zone. Using a tracer, Rice (1986) calculated an average surface subduction rate of 7 cm per annum – almost an order of magnitude greater than the net sediment accumulation rate.

Bioturbation has several important consequences for trace metals in sediments. When particles are mixed across a trace metal concentration gradient, the result is a net transport of those trace metals in the direction of the concentration gradient from high to low concentrations. When the gradient spans two or more chemical zones, the metals are subjected to an abrupt change in chemical environment. For example, mixing manganese oxide-rich surface sediment with a manganese-poor subsurface sediment brings manganese oxide into a reducing environment, where it is unstable. Manganese dissolves and diffuses back to the surface, where it is reoxidized and precipitated. Mixing cadmium sulphide subsurface sediment into the oxidizing surface layer leads to the reoxidation of the sulphide and the release of dissolved cadmium, which can then diffuse into the overlying water as well as back into the sulphidic sediment. The net result of mixing is therefore to cycle metals back and forth across the redox boundary in the sediment, with the possible escape of metals to the overlying water (Sundby and Silverberg, 1985; Gobeil *et al.*, 1987).

A second consequence of bioturbation on trace metals is somewhat analogous to the effect of plants on sediment and trace metal accumulation. The distribution of benthic organisms such as *Scoloplos* sp. is patchy and it appears that the patch density is maintained over long periods of time. For a given concentration of organic matter at the sediment surface, the amount of organic matter subducted into the sediment is proportional to the patch density. Dense patches are in a sense self-sustaining (Rice, 1986). Other reactive sediment components, such as trace metals, that are associated with organic matter and with fine-grained particles will likewise be subducted preferentially into dense patches of conveyor-belt organisms. The result is higher concentrations of trace metals (Aller and Cochran, 1976; Rice and Whitlow, 1985).

2.5 CONCLUSIONS

Plants and animals modify the chemical composition and the physical structure of sediment deposits. They do so in order to meet their needs for nutrition and structural support in the case of plants, and food and shelter in the case of animals. Benthic organisms have evolved strategies that allow them to function in the presence of sediment constituents that would otherwise be toxic. Given time, the community of benthic organisms may be able to return a sediment, made inhospitable by human activities, to its natural state of ecological equilibrium (Rhoads and Boyer, 1982; Pearson and Rosenberg, 1978). Concepts such as 'biological availability' and 'toxicity' of trace metals and other substances, defined by measurements of the chemical composition of the sediment, are thus difficult to apply to sediments unless one has a thorough understanding of how living organisms interact with the sedimentary habitat (Chapters 7 and 8).

REFERENCES

Aller, R.C. (1977) The influence of macrobenthos on chemical diagenesis of marine sediments. PhD thesis, Yale University, New Haven, Connecticut, 600 pp.

Aller, R.C. (1978) The influence of animal–sediment interactions on geochemical processes near the sediment–water interface, in *Estuarine Interactions*, (ed. M.L. Wiley), Academic Press, New York, pp. 157–172.

Aller, R.C. (1980) Quantifying solute distributions in the bioturbated zone of marine sediments by defining an average microenvironment. *Geochimica et Cosmochimica Acta* **44**, 1955–1965.

Aller, R.C. (1982) The effects of macrobenthos on chemical properties of marine sediment and overlying water, in *Animal–Sediment Relations*, (eds P.L. McCall and M.J.S. Tevsz), Plenum Press, New York, pp. 53–102.

Aller, R.C. (1988) Benthic fauna and biogeochemical processes in marine sediments: the role of burrow structures, in *Nitrogen Cycling in Coastal Marine Environments*, (eds T.H. Blackburn and J. Sçrensen), Wiley, New York, pp. 301–338.

Aller, R.C. and Cochran, J.K. (1976) Th-234/U-238 disequilibrium in nearshore sediment: particle reworking and diagenetic time scales. *Earth and Planetary Science Letters* **20**, 37–50.

Aller, R.C. and Rude, P.D. (1988) Complete oxidation of solid phase sulfides by manganese and bacteria in anoxic marine sediments. *Geochimica et Cosmochimica Acta* **52**, 751–765.

Aller, R.C. and Yingst, J.Y. (1978) Biogeochemistry of tube-dwellings: a study of the sedentary polychaete *Amphitrite ornata* (Leidy). *Journal of Marine Research* **36**, 201–254.

Anderson, C.E. (1974) A review of structure in several North Carolina salt marsh plants, in *Ecology of Halophytes*, (eds R.J. Reimold and W.H. Queen), Academic Press, New York, pp. 307–344.

Armstrong, W. (1967) The oxidising activity of roots in waterlogged soils. *Physiology of Plants* **20**, 920–926.

Armstrong, W. (1970) Rhizosphere oxidation in rice and other species: a mathematical model based on the oxygen flux component. *Physiology of Plants* **23**, 623–630.

Armstrong, W. (1978) Root aeration in the wetland condition, in *Plant Life in Anaerobic Environments*, (eds D.D. Hook and R.M.M. Crawforth), Ann Arbor Science Publishers, Ann Arbor, Michigan, pp. 269–298.

Bartlett, R.J. (1961) Iron oxidation proximate to plant roots. *Soil Science* **92**, 372–379.

Berner, R.A. (1980) *Early Diagenesis. A Theoretical Approach*, Princeton University Press, Princeton, NJ.

Caçador, I. (1994) Accumulação e retenção de metais pesados no sapais do estuario do Tejo. PhD thesis, University of Lisbon, Lisbon, 142 pp.

Caçador, I., Vale, C. and Catarino, F. (1996a) Accumulation of Zn, Pb, Cu, Cr and Ni in sediments between roots of the Tagus estuary salt marshes, Portugal. *Estuarine, Coastal and Shelf Science* **42**, 393–403.

Caçador, I., Vale, C. and Catarino, F. (1996b) The influence of plants on concentration and fractionation of Zn, Pb, and Cu in salt marsh sediments (Tagus estuary, Portugal). *Journal of Ecosystem Health* **5**, 193–198.

Di Toro, D.M. (1990) Toxicity of cadmium in sediments: the role of acid volatile sulfide. *Environmental Toxicology and Chemistry* **9**, 1487–1502.

Froelich, P.N., Klinkhammer, G.P., Bender, M.L. *et al.* (1979) Early oxidation of organic matter in pelagic sediments of the eastern equatorial Atlantic: suboxic diagenesis. *Geochimica et Cosmochimica Acta* **43**, 1075–1090.

Gobeil, C., Silverberg, N., Sundby, B. and Cossa, D. (1987) Cadmium diagenesis in Laurentian Trough sediments. *Geochimica et Cosmochimica Acta* **51**, 589–596.

Goldberg, E.D. and Koide, M. (1962) Geochronological studies of deep sea sediments by the ionium/thorium method. *Geochimica et Cosmochimica Acta* **26**, 417–450.

Green, M.S. and Etherington, J.R. (1977) Oxidation of ferrous iron by rice (*Oryza sativa* L.) roots: a mechanism for waterlogging tolerance? *Journal of Experimental Botany* **104**, 678–690.

Luther, G.W. and Church, T.M. (1992) An overview of the environmental chemistry of sulphur in wetland systems, in *Sulphur Cycling on the Continents*, (eds R.W. Howarth, J.W.B. Stewart and M.V. Ivanov), John Wiley, New York, pp. 125–142.

Madureira M.J., Vale, C. and Simões-Gonçalves M.L. (1997) Effect of plants on sulphur geochemistry in the Tagus salt-marsh sediments. *Marine chemistry*, **58**, 27–37.

McCall, P.L. and Tevesz, M.J.S. (1982) *Animal–Sediment Relations. The Biogenic Alteration of Sediments*, Plenum Press, New York.

Nye, P.H. and Tinker, P.B. (1977) *Solute Movement in the Soil–Root System*, University of California Press, Berkeley, CA.

Orson, R.A., Simpson, R.L. and Good, R.E. (1992) A mechanism for the accumulation and retention of heavy metals in tidal freshwater marshes of the upper Delaware River Estuary. *Estuarine, Coastal and Shelf Science* **34**, 171–186.

Otte, M.L., Rozema, J., Koster, L. *et al.* (1989) Iron plaque on roots of @*Aster tripolium* L.: interaction with zinc uptake. *New Phytology* **111**, 309–317.

Pearson, T.H. and Rosenberg, R. (1978) Macrobenthic succession in relation to organic enrichment and pollution of the marine environment. *Oceanography and Marine Biology Annual Review* **16**, 229–311.

Rhoads, D.C. (1974) Organism–sediment relations on the muddy sea floor. *Oceanography and Marine Biology Annual Review* **12**, 263–300.

Rhoads, D.C. and Boyer, L.F. (1982) The effects of marine benthos on physical properties of sediments. A successional perspective, in *Animal–Sediment Relations* (eds P.L. McCall and M.J.S. Tevesz), Plenum Press, New York, pp. 3–52.

Rice, D.L. (1986) Early diagenesis in bioadvective sediments: relationships between the diagenesis of berylium-7, sediment reworking rates, and the abundance of conveyor-belt deposit-feeders. *Journal of Marine Research* **44**, 149–184.

Rice, D.C. and Whitlow, S.I. (1985) Diagenesis of transition metals in bioadvective marine sediments, in *Heavy Metals in the Environment*, (ed. D.T. Lekkas), C.E.C. Consultants, Edinburgh, pp. 353–355.

Schink, D.R. and Guinasso, N.L. (1977) Effects of bioturbation on sediment–seawater interaction. *Marine Geology* **23**, 133–154.

Silverberg, N., Nguyen, H.V., Delibrias, G. *et al.* (1986) Radionuclide profiles, sedimentation rates, and bioturbation in modern sediments of the Laurentian Trough, Gulf of St Lawrence. *Oceanologica Acta* **9**, 285–290.

Sundby, B. and Silverberg, N. (1985) Manganese fluxes in the benthic boundary layer. *Limnology and Oceanography* **30**, 374–382.

Taylor, G.J., Crowder, A.A. and Rodden, R. (1984) Formation and morphology of an iron plaque on the roots of *Typha latifolia* L. grown in solution culture. *American Journal of Botany* **71**, 666–675.

Teal, J.M. and Kanwisher, J.W. (1966) Gas transport in the marsh grass, *Spartina alterniflora*. *Journal of Experimental Botany* **17**, 355–361.

Tenore, K.R. and Coull, B.C. (1980) *Marine Benthic Dynamics*, University of South Carolina Press, Columbia, South Carolina.

Tinker, P.B. and Barraclough, P.B. (1988) Root–soil interactions, in *The Handbook of Environmental Chemistry, Vol. 2, Part D: Reactions and Processes*, (ed. O. Hutzinger), Springer Verlag, New York, pp. 154–171.

Vale, C., Catarino, F., Cortesão, C. and Caçador, I. (1990) Presence of metal-rich rhizoconcretions on the roots of *Spartina maritima* from the salt marshes of the Tagus estuary, Portugal. *Science of the Total Environment* **97/98**, 617–626.

Microorganisms and the biogeochemical cycling of metals in aquatic environments

DAVID BARRIE JOHNSON

3.1 INTRODUCTION

Microorganisms may alter the availability of metals in the environment by a variety of means. Whilst eukaryotic microorganisms can bring about certain of the various metal transformations described below, the major protagonists of metal cycling in aquatic ecosystems are prokaryotes (bacteria and archaea). The cycling of metals generally involves phase changes (usually between soluble and insoluble forms) which have major impacts on the biological availability of metals. These may produce problems of metal deficiency on the one extreme, and metal toxicity on the other. Various reviews on different aspects of this subject are available in the scientific literature. These include a concise review of metal–microbe interactions by Ford *et al.* (1995) and a more detailed account of the same subject matter in a text by Hughes and Poole (1989). Two older review articles remain among the most regularly cited works in this area: one by Summers and Silver (1978), which focused on the role of microorganisms in metal transformations (defined as those changes involving changes in valence or changes from inorganic to covalently linked organic forms), and the second by Gadd and Griffiths (1978), which reviewed the toxicity of heavy metal to microbes. An ecological perspective of metal cycling and metal toxicity was described in the review by Duxbury (1985), while the role of microbially mediated metal transformations in bioremediation of contaminated waters was reviewed by Brierley (1990). Finally, a number of articles have focused on the biosorption of metals by microorganisms; the text edited by Volesky (1990) and the review article by White *et al.* (1995) are recommended reading in this area.

Metal Metabolism in Aquatic Environments. Edited by William J. Langston and Maria João Bebianno. Published in 1998 by Chapman & Hall, London. ISBN 0 412 80370 4

This chapter considers first the various mechanisms by which microorganisms promote the cycling of metals in aquatic environments. In subsequent sections, the biogeochemical transformations of Fe and Mn, which, in both cases, can have profound impact in aquatic ecosystems, are described in greater detail.

3.2 MECHANISMS OF MICROBIALLY MEDIATED TRANSFORMATIONS AND MOBILIZATION/IMMOBILIZATION OF METALS

There are four major processes involved in the cycling of metals by microorganisms in aquatic ecosystems: assimilation/adsorption and mineralization; dissolution and precipitation; oxidation and reduction; and methylation and dealkylation. Compared with the complexities of the biogeochemical cycling of nitrogen and sulphur cycles, both of which elements have multiple oxidation states, those involving metals tend to be far simpler, though somewhat greater complexity of biogeochemical cycling is observed for those metals (such as Fe, Mn and Cr) that can exist in different ionic states under constraints imposed by the environment. Similar microbial transformations occur with some metalloid elements, such as As, Se and Te (Summers and Silver, 1978; White *et al.*, 1995).

3.2.1 ASSIMILATION, ADSORPTION AND MINERALIZATION

From a biological perspective, metals may be divided into those that are essential to living organisms, those that are potentially toxic and relatively available, and those that are potentially toxic but which, due to their general rarity in the environment, are relatively unavailable (Duxbury, 1985). Assimilation of the first group by intracellular accumulation results in these metals being associated with either living or dead biomass. However, non-essential metals may also become associated with microorganisms via either intracellular accumulation, cell surface interactions or adsorption on to extracellular polymeric materials. In all such cases, the net result is immobilization of the metal concerned. Re-mobilization will be a function of the decomposition of the microbial biomass and mineralization of organic materials with which the metals are associated. The abilities of microorganisms to adsorb metals have implications for biotechnology: 'biosorption' (defined as the uptake of metals from solution by living or dead biomass involving active metabolic processes or physicochemical mechanisms such as ion exchange, adsorption or entrapment) has considerable potential for the bioremediation of metal-contaminated waters (Brierley, 1990; White *et al.*, 1995).

Metals that are essential to microorganisms include the s-block elements Na, K, Mg and Ca, and the transition elements V, Cr, Mn, Fe, Co, Ni, Cu, Mo and also Zn (Hughes and Poole, 1989). More recently, a biological role for W in some archaea (e.g. *Pyrococcus furiosus* and *Methanobacterium* spp.) and bacteria (e.g. *Desulfovibrio gigas* and *Clostridium* spp.) has been identified (Kletzin and Adams, 1996). The alkali and alkali earth metals have multiple functions in microbial cells, such as in maintaining osmotic balance, cross-linking negatively charged ligand groups in cell walls, and specific association with biomolecules (such as ATP), while the transition metals and Zn are generally components of certain enzymes and cofactors, though some (notably Fe and Cu) are important catalysts in biological redox processes. Accumulation of metals inside microbial cells usually involves an energy-dependent transport system, whereby they bind first to the extracellular cell surface and are then actively transported across the cell membrane. The same mechanism may result in the intracellular accumulation of toxic metals, if their charge and size are such that they mimic some essential elements. Other factors, such as external pH, and the presence of complexing organic materials and of other metals may also affect intracellular accumulation of toxic metals (Brierley, 1990). Once inside the cell, a metal may undergo enzymic transformation, which may render it less toxic – for example, the reduction of Hg^{2+} to Hg^0 by mercuric reductase, an enzyme that appears to be widely distributed in the bacterial kingdom (Ford *et al.*, 1995).

Metal accumulation by microorganisms may be mediated via interactions with biomolecules which occur on the cell surface. Functional groups in the polymers present in the cell walls of bacteria, fungi and algae include carboxylate and phosphate (which, together, confer a net negative charge on the surface of most microorganisms) amine, imidazole, hydroxide, sulphate and sulphydryl, which may facilitate the uptake of metals – for example, via ion exchange reactions (Brierley, 1990). Since the charge densities on these functional groups are affected by solution pH, this particular parameter has a crucial bearing on the nature and extent of metal sorption; for example, by lowering pH it is possible to promote uptake of anionic metals (such as chromate) rather than cationic forms, as protonation of functional groups on cell surfaces increases. Metal uptake by cell wall constituents may also involve covalent bonding and van der Waals forces. Binding of metals to lipopolysaccharides in Gram-negative bacteria may serve as a mechanism for the accumulation of essential metals when present at low solution concentration; they may promote the structural integrity of bacteria and serve as a mechanism to immobilize toxic metals such as cadmium, thereby preventing their entry into cells (Ford *et al.*, 1995).

Many microorganisms synthesize polymeric materials and secrete them outside cells. Such exopolymers (often referred to as glycocalyx, slime layers and capsules) are variable in their chemical compositions, and may include proteins and nucleic acids, though the principal material present is usually polysaccharidic (EPS). Again, metal uptake is a feature of many microbial exopolymers, though the nature and extent of metal sorption does vary widely between polymers of different origins. As with microbial cell walls, the net charge on EPS, at circum-neutral pH values, is often negative, so that sorption of cationic metal ions is favoured. Metal uptake by such polymers causes the release of protons, which can result in significant lowering of solution pH (Geesey and Jang, 1990). Polysaccharides are linear and highly flexible biopolymers, and binding of polyvalent cations can result in considerable folding of EPS, due to the formation of salt bridges; conformational changes to EPS induced by some metals may be detrimental to the sorption of others (Brierley, 1990).

Release of biologically adsorbed metals into the environment may result, in some instances, from changes in environmental parameters (e.g. in pH, where uptake results from ion exchange to pH-dependent charged sites on macromolecules) but in most cases re-mobilization will depend on the rate of biodegradation of the organic moieties with which the metals are associated. Mineralization of organic matter is a function of various physicochemical (e.g. temperature, pH, oxygen availability) and biological (microorganisms involved in the biodegradation process) parameters; in addition, association with a particular metal may modify the biodegradability of organic materials. Whilst there is a considerable amount of knowledge relating to organic matter mineralization in the environment (e.g. Schlesinger, 1991), there have been relatively few studies on the decomposition of heavy metal-contaminated organic residues. Duxbury (1985) made reference to reports of apparent retarded mineralization of metal-contaminated biomass compared with control materials, and also to research which indicated that some apparent effects were site-related rather than due directly to metal contamination.

3.2.2 DISSOLUTION AND PRECIPITATION

Microorganisms may promote the mobilization or immobilization of metals in ecosystems by causing, either directly or indirectly, their dissolution or precipitation. One way in which this may be achieved is by inducing changes to (micro)environment pH. Most metals are cationic, and the solubility of many is related to pH (greater solubility in acidic solutions). Some activities of microorganisms, such as fermentation, nitrification and sulphur oxidation, generate acidity, whilst others, such as ammonification, denitrification, methanogenesis and sulphate reduction, generate alkalinity. The net effect on environmental pH will depend on a number of factors such as the buffering capacity of the external solution and the scale of microbial activities; however, 'bulk' measurements (e.g. of pH) may disguise the profound variations in environmental parameters that might occur at the microsite level.

Organic acids may be formed as a product of fermentation, though they may also accumulate in culture solutions of some microorganisms when grown aerobically (Berthelin, 1983). Some organic acids are adept at chelating metals, producing highly soluble metal complexes, and inducing further metal solubilisation by causing a shift in equilibrium, as in equation 3.1:

$$Metal_{insoluble} \leftrightarrow Metal_{soluble} \rightarrow Metal_{complexed} \qquad (3.1)$$

Citric acid is produced by many yeasts and fungi and is able to chelate a number of metal cations (Hughes and Poole, 1989). This has been cited as a mechanism whereby these microorganisms obtain protection against Cu poisoning and Pb toxicity (Gadd and Griffiths, 1978); in addition, citric acid production by a *Penicillium* sp. has been used to extract Zn selectively from industrial waste (Schinner and Burgstaller, 1989). Oxalic acid is another metal-complexing organic acid produced by several fungi. Again, soluble metal chelates (which may be less toxic than both the soluble metal and free oxalic acid) may be produced; alternatively, insoluble salts may be precipitated, as in the case of copper oxalate production by wood-rotting fungi (e.g. *Poria monticula*) growing on wood treated with metal-containing preservatives (cited in Hughes and Poole, 1989). A mixture of organic acids produced by *Aspergillus niger* and *Candida lipolytica* have been used in the selective leaching of Al, Sc and Y from red muds (Talasova *et al.*, 1995).

Metal availability may also be enhanced by the production of siderophores. These are organic compounds produced by bacteria and fungi, typically under conditions of limited Fe availability, which have both high affinities and specificities for ferric iron. Other trivalent metals (Al, Ga and Co) may also be strongly complexed by siderophores, as can Pu, suggesting that siderophores could have a potential role in the bioremediation of radionuclide-contaminated sites (White *et al.*, 1995).

Microbially mediated precipitation of metals can result from the formation of extracellular insoluble hydroxides, carbonates, phosphates and sulphides. In addition, intracellular deposition of metal precipitates may occur. These processes are often referred to as 'biomineralization'. *Alcaligenes eutrophus* is able to bring about the crystallization of metal hydroxides and carbonates as a result of pH changes induced in its periplasmic space due to concomitant uptake of hydrogen ions and efflux of metal ions (Springel *et al.*, 1993). Biomineralization of metal phosphates by a *Citrobacter* sp. has been described by Macaskie (1991). In this situation, bacteria generate inorganic phosphate by the cleavage of an organic phosphate (such as glycerol 1-phosphate) using a metal-resistant periplasmic phosphatase. In the presence of metal-contaminated waters, the resultant localized high concentration of phosphate can cause the solubility products of metal phosphates to be exceeded even when metal concentrations are relatively low. This results in the deposition of inorganic metal phosphates, which can accumulate at up to 9 g metal g^{-1} dry biomass.

Sulphate-reducing bacteria (SRB) have a dual role in promoting metal immobilization. Firstly, the reduction of sulphate to sulphide is an alkali-generating reaction:

$$2CH_2O + SO_4^{2-} \rightarrow H_2S + 2HCO_3^- \qquad (3.2)$$

and so a resulting increase in water pH (e.g. in acidic waters) may reduce the solubility of metals such as Al. In addition, many heavy metals form highly insoluble metal sulphides (e.g. the solubility product of copper sulphide is 10^{-45} M) and are readily precipitated in the presence of small concentrations of sulphide:

$$M^{2+} + S^{2-} \rightarrow MS_{solid} \qquad (3.3)$$

Hydrogen sulphide may be produced by yeasts, apparently as a mechanism for reducing metal toxicity; colonies grown on solid media containing copper may become brown or black in colour as a result of CuS deposition around the cell wall (Gadd and Griffiths, 1978). Bacteria, such as *Clostridium cochlearium* and *Klebsiella aerogenes*, can also produce H_2S. The best-known and most globally significant source of biogenic sulphide, however, originates from the activities of SRB, as a result of reactions involved in energy conservation by these bacteria. This dissimilatory reduction of sulphate, a form of anaerobic respiration, is thought to account for the genesis of around 10^9 tonnes sulphide per annum in terrestrial, swampy and coastal areas alone. Most of the bacteria involved are obligate anaerobes, and include bacteria of several different genera (including *Desulfovibrio*, *Desulfobacter* and *Desulfotomaculum*) and the archaean *Archaeoglobus* (Barton, 1995). Bacterial sulphate reduction has traditionally been considered to be one of the key reactions involved in the bioremediation of acid mine drainage waters using constructed wetlands (see below). More recently, a commercial bioreactor of 1800 m³ capacity, capable of treating up to 7000 m³ of metal-contaminated waste water per day utilizing SRB has been established in the Netherlands (White *et al.*, 1995).

Intracellular precipitates of metals may occur when microorganisms are grown in media containing high concentrations of some metals. For example, yeasts may deposit accumulated Tl as thallium oxide within mitochondria, which are then discharged and excreted. There have been various other reports of metal deposition, often localized within cells, particularly with yeasts and fungi (Hughes and Poole, 1989).

3.2.3 OXIDATION AND REDUCTION

Microorganisms are capable of transforming many metals that possess multiple valency states, by oxidation and/or reduction reactions. Whilst these redox transformations generally involve actively metabolizing microbes, they may also occur passively when metals bind to extracellular and intracellular reactive sites on resting cells. In some cases, redox transformations of metals are concerned with energy conservation. The best-known microbially mediated oxido-reduction of metals concerns Fe and Mn; these are dealt with in more detail in sections 3.3 and 3.4, respectively. Other metals for which microbial oxido-reduction has been documented include Cr, Hg, U, Sb, Mo, Cu and Au. Eh–pH stability diagrams of metals (Brookins, 1988) are useful in this context. Whilst these do have limitations (e.g. they tend not to take into account the effects of organic chelates on stabilities of metal species), they are useful in predicting which ionic form of a multivalent metal is most thermodynamically stable in a particular environment.

Chromium is a widespread element in aquatic ecosystems, though in most freshwaters it is present at only around 1 μg l^{-1} (Ehrlich, 1996). There are two important oxidation states of Cr in the environment, +3 (as in the chromic cation) and +6 (as in the chromate and dichromate anions). Chromate (and dichromate) are more toxic and mutagenic than Cr(III) compounds, and also much more soluble at physiological pH. Whilst oxidation of Cr(III) to Cr(VI) appears to be exclusively abiotic (e.g. catalysed by manganese(IV) oxides), chromate-reducing bacteria appear to be widespread in the environment and Cr(VI) reduction can occur under both aerobic and anaerobic conditions (Wang and Shen, 1995). This may be via a direct mechanism (for example, where bacteria use chromate as an electron acceptor for anaerobic growth) or indirect (for example, the reduction of chromate by SRB which is mediated through H$_2$S production).

Mercury is another metal for which microbiological reduction is well known though oxidation has yet to be demonstrated unequivocally. Early reports that some bacteria were able to catalyse elemental mercury (Hg^0) to Hg^{2+} were later challenged when it was noted that various metabolic products, and even yeast extract, were also able to oxidize Hg^0 (Ehrlich, 1996). In contrast, the enzymic nature of mercury (Hg^{2+}) reduction is well documented. The enzyme, mercuric reductase, appears to be widespread among bacteria; the transformation of a soluble, highly toxic form to a volatile is the most common mechanism for microbial resistance to this metal, and is also involved (as a second stage) in the detoxification of methyl mercury by bacteria (see below). Mercuric reductase is also capable of reducing cationic Ag and Au to corresponding native metal colloids (Ehrlich, 1996). The topic of Hg in aquatic environments is described in detail in Chapter 5.

Uranium can exist in two ionic states: U(IV), which is insoluble under physiological conditions, and U(VI), which is soluble. Oxidation of U(IV) coupled to carbon dioxide fixation has been observed for the chemolithotrophic acidophile *Thiobacillus ferrooxidans* (DiSpirito and Tuovinen, 1982). The reverse reaction, in which U(VI) is used as an electron sink, has been reported for the neutrophilic heterotrophs *Shewanella putrefaciens* and *Geobacter metallireducens* (Lovley, 1995; Lovley *et al.*, 1991) and the sulphate-reducer *Desulfovibrio desulfuricans*, though in the latter case U(VI) reduction is not coupled to bacterial growth (Lovley and Phillips, 1992). Bacterial reduction of U has been proposed as a mechanism for remedying uranium-contaminated waters.

There have been reports of microbial oxidation of at least three other metals: Mo, Sb and Cu. Lyalikova and Lebedeva (1984), working with enrichment cultures of molybdenite, reported that some microorganisms were able to use reduced compounds of molybdenum as energy sources. Trivalent antimony can be oxidized to Sb(V) by the bacterium *Stibiobacter senarmontii* (Lyalikova, 1974) and possibly also by *T. ferrooxidans* (Torma and Gabra, 1977). Oxidation of copper from cuprous (Cu^+) to cupric (Cu^{2+}) has also been reported for *T. ferrooxidans* (Lewis and Miller, 1977). Reduction of molybdenum by a strain of *Enterobacter cloacae* (Ghani *et al.*, 1993) and of copper(II) by sulphur-grown *T. ferrooxidans* (Sugio *et al.*, 1990) have also been described, though there have been no reports of reduction of Sb(V) by microorganisms. Finally, passive reduction of gold ($Au^{3+} \rightarrow Au^+ \rightarrow Au^0$) has been observed with the alga *Chlorella vulgaris* (Greene and Darnall, 1990); the reduction of both Au and Ag, mediated by mercuric reductase, has been referred to earlier.

3.2.4 METHYLATION AND DEALKYLATION

Transformation of metals by methylation and dealkylation reactions may be carried out by a number of bacteria and fungi, apparently as detoxification mechanisms. A major contrast to other metal transformations is that methylation usually results in the production of volatile complexes. The best-studied of these is the methylation of Hg. Many bacteria and some fungi are able to methylate Hg, but the most important group appear to be the SRB (Ford *et al.*, 1995). The primary route for Hg methylation is via an extracellular reaction involving the methyl-donor methylcobalamin (Summers and Silver, 1978). Methyl mercury (CH_3Hg^+) may be further methylated to form dimethylmercury (CH_3HgH_3C), though this reaction is about 6000 times slower than that of mercury methylation (Hughes and Poole, 1989). Other mechanisms for methyl mercury formation involve cysteine (or homocysteine) in the fungus *Neurospora crassa*, and possible intracellular formation involving methyl donors other than methylcobalamin in *E. coli* (Hughes and Poole, 1989). Methyl mercury is both water soluble and able to diffuse readily through biological membranes, and is considerably more toxic (50–100 times) to most forms of life (with the possible exception of those microorganisms that methylate mercury) than is Hg^{2+} (Hughes and Poole, 1989). Accumulation of methyl mercury within the food chain has been reported to have caused human poisoning in Japan, Iraq, Pakistan, Guatemala and the United States, (Ehrlich, 1996), the most infamous example of which occurred in Japan at Minamata Bay in the 1950s, providing the name for Minamata disease (Chapter 5).

Microbiological methylation of metals other than Hg has been noted. Formation of mono-, di-, tri- and tetramethyltin by *Pseudomonas* spp. isolated from waters and sediments has been demonstrated in the laboratory, but it is thought that abiotic and anthropogenic sources of methylated tin are far more important than that produced biologically (Hughes and Poole, 1989). A metal-resistant *Pseudomonas* sp. has also been found to form a volatile derivative of Cd, which was considered to be methylcadmium (Summers and

Silver, 1978). The question of whether microorganisms can form methylated lead compounds is more contentious, with evidence both for and against having been presented (described in Hughes and Poole, 1989). Biological methylation of Tl and Sb has also been demonstrated (Ford *et al.*, 1995).

Microbiological formation of, for example, ethyl and phenyl derivatives of metals is generally considered to be insignificant. However, such compounds are synthesized anthropogenically and inevitably find their way into the environment. Formation of the free metal from these and other alkyl derivatives is collectively referred to as dealkylation. Decomposition of organomercurials may be brought about by organomercurial lyases, which cleave the molecules to produce, for example, methane, ethane and benzene, and Hg^{2+}. A second phase in the detoxification process is the reduction of Hg^{2+} by mercuric reductase to produce volatile metallic Hg. Organotin compounds are used as biocides (e.g. triphenyltin acetate, which has use as a molluscicide, algicide and fungicide) and, more importantly in terms of scale, in the production of some plastics. Whilst research has shown that some microorganisms are able to degrade organotins, the rates of decomposition are very slow, and some compounds appear to be highly resistant to microbial attack. This has given rise to a problem of organotin pollution which is escalating in scale, particularly in highly industrialized nations such as the United States (Hughes and Poole, 1989).

3.3 CYCLING OF IRON IN AQUATIC ENVIRONMENTS

Iron is the most abundant element in the planet earth, and the fourth most plentiful in the lithosphere, where O, Si and Al are present at higher concentrations. It occurs in a variety of minerals, including oxides, sulphides, carbonates and silicates. Banded Fe formations (which contain *c.* 28% Fe) are the largest accumulations of Fe in the lithosphere (Nealson, 1983a). These are oxidized deposits of Precambrian age, while more recent accumulations include bog and lake iron ores and ochre deposits (Ehrlich, 1996). In most aquatic ecosystems, however, the concentrations of dissolved Fe are very small, with that in oceanic waters being about 3 μg l^{-1}, and that in fresh waters averaging slightly more (Lundgren and Dean, 1979) though some polluted waters, notably acid mine drainage, may contain much greater concentrations of soluble Fe. Cycling of Fe in both aquatic and terrestrial environments may be mediated through the first three of the mechanisms described above (there are no reports of methylation and dealkylation of this metal). The best-documented mechanism for iron cycling is via oxido-reduction and is most readily observed in environments where oxic and anoxic sites are in close proximity or juxtaposition, such as freshwater and marine sediments, and gley soils. A number of reviews have considered the biogeochemical cycling of Fe in detail, including those by Lundgren and Dean (1979), Nealson (1983a), Johnson (1995a) and Ehrlich (1996).

All micro- and macro-organisms, with the exception of the homolactic fermenting streptococci, have a nutritional requirement for Fe, though in most cases only as a micronutrient. Within cells, Fe is found in the haem prosthetic group (e.g. in cytochromes), in non-haem iron–sulphur proteins (e.g. ferredoxin) and in some superoxide dismutases. In some microorganisms, iron assumes a more significant role, acting as an electron donor or electron sink. The E_0 value of the ferrous/ferric couple (+770 mV) is not far below that of the $\frac{1}{2}O_2/H_2O$ couple, so that, in thermodynamic terms, ferric iron is an attractive alternative electron acceptor to oxygen. On the other hand, the position of the ferrous/ferric redox couple also dictates that oxygen is the only feasible electron acceptor for those bacteria that use ferrous iron oxidation as an energy source, and that the energy available from iron oxidation ($\Delta F = -30$ kJ mol^{-1}, at pH 2.0) is small, even in comparison with other chemolithotrophic reactions.

Iron is a relatively reactive element, and occurs in three oxidation states: metallic iron (Fe^0); ferrous iron, or iron(II) (Fe^{2+}); and ferric iron, or iron(III) (Fe^{3+}). The pH and redox potential is highly important in dictating which ionic form is the more stable in any particular environment (Brookins, 1988). Ferrous iron is thermodynamically unstable in oxidizing environments and spontaneously oxidizes to ferric, though this phenomenon is pH-related, in that the thermodynamic stability of free ferrous iron is far greater in acidic solutions (particularly in those below pH 2) even when Eh values are highly oxidizing (700–750 mV). The half-life of free ferrous iron in aquatic environments is < 1 minute to *c.* 3 minutes, depending on the ionic strength of the water body (increasing salinity slows down abiotic iron oxidation). Whilst ferrous iron is relatively soluble, ferric iron is highly insoluble in all but the most acidic of aquatic ecosystems, and hydrolyses to form a variety of amorphous and crystalline phases, and a solution phase concentration in the order of 10^{-17} M. However, both ferrous and ferric iron may be complexed by a range of organic chelating agents, such as some di- and tricarboxylic acids, siderophores and humic colloids. Complexing of Fe may result in much higher solution phase concentrations of the metal and may stabilize either ionic species in environments in which the corresponding uncomplexed ions are thermodynamically unstable. Because of the great influence of pH on both the stability and the solubility of ionic Fe, the cyclings of Fe in highly acidic and near-neutral environments are often considered separately. A further rationale behind this is that the microflora responsible for oxido-reduction transformations of Fe are quite distinct in such ecosystems. However, natural environments exist as continuums, and pH gradients occur in many situations, such as in natural and constructed wetlands used for ameliorating acid mine drainage waters. It might be anticipated, therefore, that a certain degree of merging and overlap of the microbiological regimes described in the following paragraphs occurs in natural environments.

3.3.1 IRON CYCLING IN CIRCUM-NEUTRAL pH AQUATIC ECOSYSTEMS

Because of the rapid autoxidation of ferrous iron in waters of pH > 4–5, it is often difficult to differentiate between enzymic and abiotic iron oxidation in most aquatic ecosystems. Whilst some neutrophilic bacteria may be responsible for the formation of ferric iron precipitates by oxidizing ferrous iron to ferric directly, iron oxidation and deposition may be brought about by other, indirect mechanisms, including the following.

- Increasing the pH, Eh or dissolved oxygen concentrations (e.g. cyanobacteria) in (micro)environments. The rate of ferrous iron oxidation is given by the equation:

$$dFe(II)/dT = kFe(II)(OH^-)pO_2$$

  where $k = 8.0 \pm 2.5 \times 10^{13}$ min^{-1} atm^{-1} mol^{-2} at 20°C (Nealson, 1983a).
- By producing oxidants that chemically oxidize ferrous iron.
- By 'ligand destruction' of iron chelates, in which microorganisms metabolize the organic moiety of an Fe complex, thereby releasing either ferrous (which rapidly autoxidizes) or ferric iron.

The ferric iron precipitates formed may become intimately associated with certain microorganisms, accumulating on, for example, extracellular sheaths of some filamentous bacteria. Neutrophilic bacteria that have a tendency to be associated with accumulations of ferric iron precipitates have been referred to as 'iron bacteria', a term that has been variously defined; some authors (e.g. Lundgren and Dean, 1979; Johnson, 1995a) have referred to all prokaryotes that either oxidize, reduce or accumulate iron as 'iron bacteria'.

A list of neutrophilic bacteria that either oxidize ferrous iron or deposit ferric iron is given in Table 3.1. Of these microorganisms, there is convincing evidence for energy conservation associated with iron oxidation occurring in only one: *Gallionella ferruginea*. This bacterium is readily identified from the twisted stalk structures (composed of fibrils that accumulate ferric precipitates) produced by the bean-shaped cells. Characteristically, *Gallionella* inhabits waters of low oxygen tensions (typically 0.1–1.0 mg O$_2$ l^{-1}) and low redox potential (typically +200 to +300 mV) such as some well-waters. In these environments, even though the pH may be relatively high (the pH range for *G. ferruginea* is 6.0–7.6), chemical oxidation of ferrous iron is retarded and biological oxidation is therefore a more competitive option. Even though *Gallionella* was one of the first bacteria to be described (by Ehrenberg in 1836), difficulties in obtaining pure cultures and culturing this bacterium in the laboratory have tended to limit progress in understanding its physiology. However, it is now recognized that *G. ferruginea* can grow either autotrophically on ferrous iron, or mixotrophically (obtaining its carbon from both CO$_2$ and organic substrates) in media containing ferrous iron and sugars such as glucose (Hallbeck and Pederson, 1991).

Table 3.1 Neutrophilic iron-oxidizing and iron-depositing bacteria

Morphological/ physiological type	Genera/species	Notes
(a) Bacteria for which there is evidence of direct iron oxidation		
Stalked bacteria	*Gallionella ferruginea*	Energy conserved from Fe^{2+} oxidation; microaerophile
Sheathed bacteria	*Leptothrix* spp. *Sphaerotius natans*	Heterotrophic bacteria; no evidence for energy conservation from Fe^{2+} oxidation
Phototrophic bacteria	*Rhodomicrobium-* and *Thiodyctyon*-like isolates	Fe^{2+} oxidized under anoxic conditions, using photosystem I
(b) Bacteria which deposit ferric iron, but for which there is no firm evidence for Fe^{2+} oxidation		
Sheathed bacteria	*Crenothrix polyspora, Clonothrix, Lieskeela*	Taxonomic position unclear
Budding/prothecate bacteria	*Pedomicrobium ferrugineum, Hyphomicrobium*	Reported pH range 3.5–10
Budding/non-prosthecate bacteria	*Blastocaulis/Planctomyces* group	Heterogeneous group; taxonomic position unclear
Appendaged bacteria	*Metallogenium*-like bacteria	Considerable doubt on the validity of this group
Gliding bacteria	*Toxothrix*	Fe^{3+} sometimes deposited on (U-shaped) filaments

Of the other iron oxidizing/depositing bacteria listed in Table 3.1, the best-studied are those of the *Leptothrix/Sphaerotilus* group. These are sheathed heterotrophic bacteria that synthesize tubular extracellular envelopes, or sheaths, predominantly composed of polysaccharides and proteins (Emerson and Ghiorse, 1993) on which ferric iron precipitates may accumulate. The mechanism involved in iron oxidation and accumulation by filamentous neutrophiles has often been assumed to be that of ligand destruction followed by binding of the released Fe (and autoxidation in the case of ferrous) to negatively charged sites in the sheath polymers. More recently, a strain of *Leptothrix discophora* that did not form a sheath synthesized a soluble macromolecule (probably a protein) that oxidized Fe, and could be separated from another that oxidized Mn (see below; Corstjens *et al.*, 1992). *Leptothrix* spp. tend to populate unpolluted, slow-running waters (often at aerobic/anaerobic interfaces, such as in field drains) whereas *Sphaerotilus natans* (the dominant organism in the mis-named 'sewage fungus' community) tends to thrive in waters that are rich in organic matter. Cell lysates of *S. natans* have also been shown to catalyse ferrous iron oxidation (Corstjens *et al.*, 1992), but there is no evidence that either *Leptothrix* or *Sphaerotilus* conserves energy from iron oxidation.

A novel mechanism for bacterial iron oxidation, operating in the absence of molecular oxygen, was demonstrated by Widdel *et al.* (1993). It was shown that phototrophic bacteria resembling *Rhodomicrobium* and *Thiodictyon* spp., growing under anoxic conditions, coupled the oxidation of Fe to the reduction of CO_2. This observation implies that, since other bacteria may also reduce ferric iron reduction under the same conditions, cycling of Fe can occur in anaerobic environments that are illuminated. It has also been postulated that iron oxidation by phototrophic anaerobes is a means by which large scale Fe oxidation occurred in Archaean times, possibly contributing to the formation of oxidized banded Fe formations.

As with ferrous iron oxidation, neutrophilic bacteria may promote the reduction of ferric iron either directly or indirectly. Spontaneous reduction resulting from bacterially mediated acidification, oxygen depletion and fall in redox potential is generally considered to be of minor importance in natural environments. However, bacteria may synthesize and excrete compounds that act as reductants of ferric iron, such as formic acid (e.g. by *Escherichia coli*) and sulphide (by SRB, though some of these may catalyse Fe reduction directly, as noted below). The ability to reduce ferric iron is widespread among bacteria, though in the majority of cases this is associated solely with the assimilation of Fe into biomass (as noted earlier) and accounts for relatively little turnover of the metal. Most of the bacteria (over 12 genera) that have been reported to reduce ferric iron as part of a dissimilatory process do so during fermentation. In these cases, ferric iron acts as a supplementary and minor sink for electrons (typically < 5% of total reducing equivalents). The metabolic strategy of using a combination of fermentation and ferric iron reduction is of greater advantage, in thermodynamic terms, than using fermentation alone. In contrast to fermentative iron-reducers, bacteria that are able to oxidize organic substrates to carbon dioxide using ferric iron as the exclusive electron acceptor have been described only relatively recently. These include the genus *Geobacter* – obligate anaerobes currently comprising two species (*G. metallireducens* and *G. sulfurreducens*) – and *Shewanella putrefaciens*, a facultative anaerobe. A *Pseudomonas* sp. that coupled the oxidation of hydrogen to the reduction of ferric iron has been described (Balashova and Zavarzin, 1980); and some sulphate-reducing bacteria, such as *Desulfovibrio* spp., can also couple hydrogen oxidation to iron reduction, though they are apparently unable to conserve energy for growth from this metabolism (Lovley, 1995). The attraction of using ferric iron as an electron acceptor (in terms of the relatively high redox potential of the ferric/ferrous redox couple) tends to be offset, at least to some extent, by the insolubility of ferric iron in most environments. For those solid-phase ferric minerals that are reduced by neutrophilic bacteria, relative rates of reduction are correlated with the crystallinity of the mineral, with amorphous iron oxyhydroxides being reduced relatively readily and haematite (Fe_2O_3) very slowly, if at all. Secondary minerals may be produced as a result of dissimilatory Fe reduction – for example, magnetite (Fe_3O_4) formed by *G. metallireducens* when reducing amorphous ferric oxides (Lovley, 1991).

Magnetotactic bacteria are a distinct group of 'iron bacteria' that synthesize single-domain magnetic particles, principally of magnetite, though some of these bacteria have been found to synthesize magnetic iron sulphides. These magnetic minerals are contained within cells in structures called magnetosomes; they cause the bacteria to become orientated within the earth's magnetic field, and thereby to position themselves optimally within their environment (many appear to be microaerophilic sediment-inhabiting bacteria; Blakemore, 1982). The locomotory responses of magnetotactic bacteria in the northern and southern hemispheres are opposite to each other, reflecting the polarity of the earth's magnetic field. Magnetite synthesis in these bacteria involves the reduction and possible partial re-oxidation of the ferric iron that they take up, as magnetite contains both ferrous and ferric iron. Insofar as the magnetite formed is contained within biomass, this form of bio-mineralization may be considered to be an assimilatory process, though it is quite distinct from other such processes in that the end product is a mineral form rather than organically bound Fe. Speculation that magnetotactic bacteria may also catalyse the dissimilatory reduction of ferric iron has yet to be substantiated (Lovley, 1995).

3.3.2 IRON CYCLING IN ACIDIC AQUATIC ENVIRONMENTS

Two major differences in Fe chemistry in acidic compared with circum-neutral pH environments are, firstly, the greater stability of ferrous iron (autoxidation is far slower) and secondly, the greater solubility of ferric iron (particularly in environments of pH < 2.5). This gives rise to greater potential for microbial exploitation both for using the energy available from ferrous iron oxidation (competition from chemical oxidation being far less important) and for utilizing ferric iron as a terminal electron acceptor – Fe(III) is soluble in the most extremely acidic waters, and the solubility of many ferric iron minerals is greater in acidic than in neutral solutions. As in circum-neutral pH environments, iron-cycling in acidic waters is mediated chiefly through oxido-reduction transformations. Ferrous iron oxidation by acidophilic bacteria has received far more attention than ferric iron reduction, primarily because of the importance of this reaction in the processing of sulphide ores ('biomining'; Barrett *et al.*, 1993) and also because of the environmental significance of the same bacteria in the genesis of acid mine drainage pollution (Figure 3.1).

Iron is a chalcophilic element and occurs in a range of sulphide minerals, in which it may be the sole metal component (e.g. pyrite and marcasite, both of which have the mineralogical composition FeS_2 but which differ in crys-

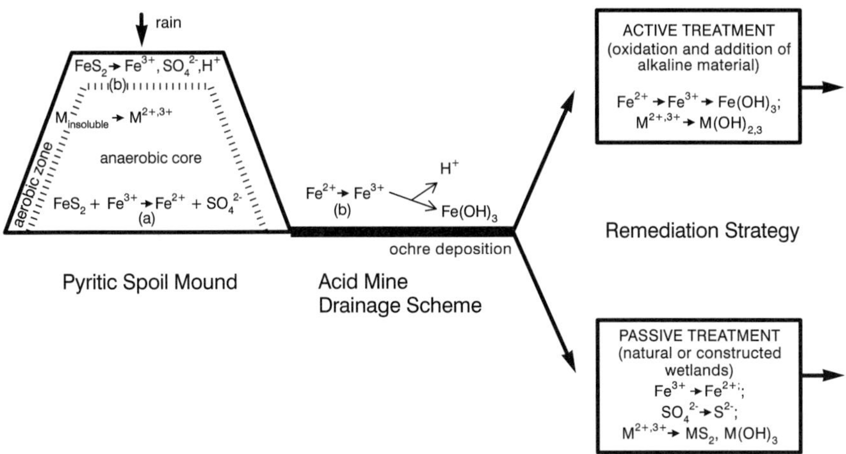

Figure 3.1 Acid mine drainage genesis in a sulphide mineral-rich spoil mound, and subsequent downstream events associated with active and passive remediation strategies. The chemical reactions indicated (a) are primarily abiotic, while those indicated (b) are predominantly biological. $M^{2+,3+}$ represents polyvalent metal cations.

tallinities) or in association with other metals and metalloids (such as in chalcopyrite, $CuFeS_2$, and arsenopyrite, $FeAsS$). Sulphide minerals are the prime ores of many metals of commercial importance, such as Cu, Pb and Zn; in the case of gold, the metal is finely disseminated within sulphide minerals in the increasingly important 'refractory' ores. Coal deposits contain varying amounts of pyrite and/or other sulphide minerals; individual seams may contain up to 15–20% FeS_2, though the figure is generally less than this. Mining of sulphide-containing ores and coals exposes the minerals to moisture and oxygen, both of which are required for biological oxidation of the minerals. For example:

$$FeS_2 + 3.5O_2 + H_2O \rightarrow FeSO_4 + H_2SO_4 \tag{3.5}$$

In equation 3.5, the oxidation state of Fe has not been changed, though ferrous iron is readily oxidized by iron-oxidizing acidophiles:

$$2FeSO_4 + H_2SO_4 + \tfrac{1}{2}O_2 \rightarrow Fe_2(SO_4)_3 + H_2O \tag{3.6}$$

Ferric iron is a powerful oxidizing agent and may oxidize pyrite and other sulphide minerals in a purely abiotic reaction. For example:

$$CuFeS_2 + 2Fe_2(SO_4)_3 \rightarrow CuSO_4 + 5FeSO_4 + 2S \tag{3.7}$$

The ferrous iron generated in reaction 3.7 is re-oxidized by iron-oxidizing acidophiles, thereby causing the mineral oxidation process to continue. The latter process, known as the 'indirect' mechanism, is considered by many researchers to be the major means by which bacteria accelerate the oxidative dissolution of sulphide minerals, though a 'direct' mechanism (as in equation 3.5) may also contribute to the process. Metals present in sulphide ores are made soluble as a result of this bacterial activity and may be readily recovered (e.g. by solvent extraction and electrowinning) in industrial leaching operations that use this biotechnology. The same reactions are responsible for generating extreme acidity and metal pollution in streams and rivers draining mining areas (Figure 3.1). Biological sulphide mineral oxidation may continue in mine spoils and tailings dumps long after commercial mining in an area has ceased, constituting an ongoing major pollution hazard. Ferric iron produced by iron- oxidizing bacteria may hydrolyse, causing the generation of further acidity (equation 3.8) and precipitation of ferric iron-rich deposits (ochre) in stream sediments; the latter are the most obvious manifestations of acid mine drainage pollution.

$$Fe^{3+} + 3H_2O \rightarrow Fe(OH)_3 + 3H^+ \tag{3.8}$$

Because of the increased solubility of many metals in acidic compared with neutral waters, potentially toxic metals (such as Al) may also be present at elevated concentrations in streams and rivers affected by acid mine (or rock) drainage. Although some of these metals may originate in minerals other than sulphides (such as aluminosilicates), these minerals may also undergo accelerated weathering in the highly acidic environments associated with sulphide oxidation.

Microorganisms that are capable of oxidizing ferrous iron in extremely acidic environments (Table 3.2) comprise certain bacteria that are active in low to moderately thermal sites (0 to ~60°C), and archaea that are important in environments of relatively high temperature (> 60°C) (Norris and Johnson, 1998). Of these, the best-studied is the mesophile *Thiobacillus ferrooxidans*, a chemolithotrophic bacterium that typically obtains its energy from the oxidation of ferrous iron (it can also use various reduced forms of inorganic sulphur as substrates), its carbon from fixation of CO_2, and oxygen as electron acceptor. *T. ferrooxidans* is now known to be more versatile in its metabolism than was once thought, also being able, for example, to use formic acid or hydrogen as electron donors, and ferric iron as an electron sink. In contrast, the only metabolic strategy open to the chemolithotroph *Leptospirillum ferrooxidans* appears to be the oxidation of ferrous iron under aerobic conditions, thus delineating it as a unique (and archetypal) 'iron bacterium'. Recent evidence suggests that *L. ferrooxidans*, rather than *T. ferrooxidans*, is the principal metal-mobilizing acidophile in many bioleaching operations and in acid mine drainage streams (Norris and Johnson, 1998). Other chemolithotrophic iron oxidizers have been described, though not all have been ascribed generic or species labels – an exception being *Thiobacillus prosperus*, a halotolerant iron-oxidizing mesophile. Heterotrophic iron-oxidizing acidophiles have more recently been characterized and noted, again, to be widely distributed in

Table 3.2 Acidophilic iron-oxidizing bacteria

Genus/species	Notes
(a) Mesophilic bacteria	
Thiobacillus spp.	Gram-negative, sulphur-oxidizers
T. ferrooxidans	Autotrophic
T. prosperus	Halotolerant
'*Thiobacillus*' m-1	Phylogenetically distinct; S appears not to be oxidized
Leptospirillum ferrooxidans	Vibrioid cells; autotrophic
Heterotrophic isolates	Unicellular and filamentous forms
(b) Moderately thermophilic bacteria	
Sulfobacillus spp.	Gram-positive rods; form endospores
S. thermosulfidooxidans	Versatile nutrition
S. acidophilus	Autotrophic growth on S
Acidimicrobium ferrooxidans	Gram-positive rods, sometimes forming filaments
Leptospirillum thermoferrooxidans	Thermotolerant, though otherwise similar physiology to *L. ferrooxidans*
(c) Extremely thermophilic archaea	
Acidianus brierleyi	Aerobically oxidize or anerobically reduce sulphur; temp. range 45–75°C
Sulfolobus metallicus	Aerobe; temp. range 50–75°C
Metallosphaera sedula	Aerobe; temp. rang 50–80°C
Sulfurococcus yellowstonii	Aerobe; temp. range 40–80°C

acidic industrial and environmental sites (Johnson *et al.*, 1995). Phylogenetic analysis (based on 16S rRNA gene sequencing) have shown that these bacteria are quite distinct from other acidophiles (F. Roberto and B. Johnson, unpublished data), further illustrating the biodiversity of bacteria involved in iron-cycling.

Thermophilic iron-oxidizing bacteria have been isolated from thermal springs in geothermal areas such as the Yellowstone National Park, and the island of Montserrat in the West Indies. Temperatures in leaching heaps and spoil mounds may also become elevated (sulphide mineral oxidation is an exothermic process and may lead to spontaneous ignition of coal-spoil mounds) and thermophilic isolates have been obtained from acidic drainage streams in non-geothermal areas, including temperate zones such as the UK.

These microorganisms may be subdivided into those that grow at *c*. 40–60°C (thermotolerant, or moderately thermophilic bacteria) and those that grow optimally at > 60°C (extremely thermophilic archaea). The former were initially referred to as *Thiobacillus*-like isolates, but are in fact quite different from *T. ferrooxidans*, both physiologically and morphologically. There are now two designated genera of moderately thermophilic iron-oxidizers. *Sulfobacillus* spp. are Gram-positive, spore-forming eubacteria that display highly versatile metabolisms, being able to grow autotrophically, heterotrophically or mixotrophically. *Acidimicrobium ferrooxidans* does not produce spores, and is typically smaller than *Sulfobacillus* spp. (Clark and Norris, 1996); both occur as rod-shaped cells. In contrast, the extremely thermophilic archaea that oxidize ferrous iron are generally coccoid, and include species of the genera *Sulfolobus*, *Acidianus* and *Metallosphaera*; *Sulfurococcus yellowstonii* has also been reported to oxidize Fe (Norris and Johnson, 1996). The highest reported temperature permitting growth of these archaea is about 80°C.

The biological reduction of ferric iron by acidophilic microorganisms was first reported by Brock and Gustafson (1976) who observed that sulphur-grown autotrophs (*Thiobacillus thiooxidans*, a mesophilic bacterium, and *Sulfolobus acidocaldarius*, a thermophilic archaean) both accumulated ferrous iron when grown in liquid media containing Fe(III). In contrast to iron reduction at circum-neutral pH, conditions did not need to be anaerobic for this to occur – probably, again, because of the retarded autoxidation of Fe(II) at low pH. *T. ferrooxidans* can also reduce ferric iron when growing on reduced sulphur, but because of its capacity also to oxidize ferrous iron, net Fe reduction is only observed when grown anaerobically (e.g. on sulphur) or when culture pH is very low (< 1.3), since the Fe oxidation system of *T. ferrooxidans* appears to be more pH-sensitive than its S oxidation system. Many heterotrophic bacteria (e.g. *Acidiphilium* spp.) that inhabit low pH environments also appear to have the ability to use ferric iron as an electron sink and, as with the neutrophiles *Geobacter* and *Shewanella*, this is mediated via anaerobic respiration rather than fermentation (Johnson and McGinness, 1991). Solid-phase ferric iron minerals (such as jarosites, magnetite and haematite) and complex materials such as ochre deposits may be subjected to reductive solubilization by bacteria such as *Acidiphilium* SJH; the rates of Fe reduction estimated for acidophilic heterotrophic bacteria are approximately an order of magnitude greater than those observed for neutrophilic bacteria (T. Bridge and B. Johnson, unpublished data). Ferric iron reduction has also been reported for iron-oxidizing heterotrophic bacteria (Johnson *et al.*, 1995) and iron-oxidizing moderate thermophiles (Johnson *et al.*, 1993), though both groups appear to have a greater propensity for Fe oxidation than Fe reduction and net Fe reduction is generally only observed under conditions of very low oxygen tension. Cycling of Fe between the ferrous and ferric states in laboratory cultures has been demonstrated using mixed cultures of iron-oxidizing and iron-reducing acidophiles (e.g. *T. ferrooxidans* and *Acidiphilium* SJH) and also using pure cultures of bacteria that can reduce and oxidize iron, depending on prevailing aeration conditions (e.g. Johnson *et al.*, 1993).

The reductive dissolution of solid-phase ferric iron compounds generates net alkalinity, as:

$$4Fe(OH)_3 + CH_2O \rightarrow HCO_3^- + 4Fe^{2+} + 7OH^- + 3H_2O \qquad (3.9)$$

though changing environmental conditions (e.g. in water aeration) may cause the ferrous iron produced to be re-oxidized and acidity to be generated from hydrolysis of the ferric iron produced. Removal of the ferrous iron produced by bacterial reduction as an insoluble compound or mineral would preclude this. One possible mechanism for this is via the formation of iron sulphide (initially FeS) resulting from the production of sulphide by sulphate-reducing bacteria, some strains of which are known to be acidophilic or acid tolerant (Johnson *et al.*, 1993). Bioremediation of acid mine drainage waters, using immobilized populations of acidophilic iron- and sulphate-reducing bacteria, has been proposed (Johnson, 1995b) and the same bacteria are probably among the most significant of those involved in the mitigation of acid mine waters using natural and constructed wetlands (Figure 3.1).

3.4 CYCLING OF MANGANESE IN AQUATIC ENVIRONMENTS

After Fe, Mn is the most abundant heavy metal in the lithosphere. Whilst its average concentration is only 1/50 that of Fe, local concentrations of Mn can vary widely (0.002–10%, in terrestrial ecosystems; Ehrlich, 1996). The average concentration of soluble Mn in fresh waters is about 8 μg l^{-1}, whilst in seawater it is somewhat less (0.2 μg l^{-1}), but local concentrations of Mn can be considerably greater than these values. For example, in anoxic hypolimnia of some lakes, soluble Mn levels can exceed 1000 μg l^{-1}, and Mn concentrations around hydrothermal vents in oceanic spreading centres may approach *c.* 20 μg l^{-1} (Ehrlich, 1996).

As with Fe, the major processes involved in Mn cycling in aquatic ecosystems are oxidation and reduction. Manganese can exist in several oxidation states: 0, +2, +3, +4, +6 and +7. Of these, only the +2, +3 and +4 states are important in nature, and Mn(III) is only stable in solution when complexed with suitable ligands. There is some parallel between the behaviour of the manganous and manganic ions and the ferrous/ferric ions. Manganous ions (Mn(II)) can occur as free ions in solution (as Mn^{2+} and $Mn(OH)^+$ in circum-neutral pH waters) but they are more stable than ferrous ions under most pH/Eh regimes (Brookins, 1988). Mn(II) can also occur as a soluble complex with a variety of organic and inorganic ligands. For example, the dominant form of Mn in seawater is Mn^{2+}, which is not as predicted from Eh/pH characteristics of marine environments; the enhanced stability of Mn(II) has been attributed to complexation with chloride, sulphate and bicarbonate ions, and by various organic compounds. Manganic (Mn(IV)) manganese forms highly insoluble oxides; MnO_2 has a point of zero charge of pH 2.8 which, in combination with its large specific surface area, gives rise to a marked capacity to adsorb metal cations, including Mn(II). On thermodynamic grounds, the oxidation of Mn(II) to Mn(IV) is favoured in aerated waters of neutral pH; but the activation energy for Mn(II) oxidation is relatively high, which slows down the rate of spontaneous autoxidation of manganese in such situations.

As with Fe, the major impact of microorganisms on global-scale Mn transformations involves dissimilatory processes. However, Mn is an essential micronutrient, being involved in many cellular reactions which involve oxygen (such as photosynthesis) and as an activator of several enzymes, so that mechanisms for Mn assimilation are widespread amongst microorganisms. The biogeochemical cycling of Mn has been reviewed by a number of authors, including Marshall (1979), Nealson (1983b), Gounot (1994) and Ehrlich (1996).

3.4.1 MANGANESE OXIDATION IN AQUATIC ENVIRONMENTS

Many different microorganisms have been reported to be capable of oxidizing Mn, using either direct or indirect mechanisms. Whilst these include some eukaryotes, bacteria form the dominant group of Mn(II) oxidizers: common Gram-positive (e.g. *Bacillus* and *Arthrobacter* spp.) and Gram-negative (e.g. *Pseudomonas* and *Vibrio* spp.) bacteria, as well as budding/appendaged bacteria (e.g. *Hyphomicrobium* and *Caulobacter*) have been demonstrated to oxidize Mn(II) *in vitro*. Unlike ferrous iron oxidation, biological Mn(II) oxidation in very acidic conditions has not been observed, probably because the free energy available from the reaction (using oxygen as electron acceptor) becomes progressively smaller as solution pH declines; the free energy change assumes a positive value at around pH 1 (Ehrlich, 1996).

A scheme summarizing the various routes whereby Mn(II) may be biologically oxidized is shown in Figure 3.2. In contrast to iron oxidation, there is no incontrovertible proof that any bacterium can grow autotrophically using Mn(II) oxidation as sole energy source; however, it is known that some bacteria are able to incorporate the energy released from Mn(II) oxidation as part of their energy budget, though organic carbon is also required (as a carbon and supplementary energy source). Whilst some bacteria oxidize solution-phase $Mn^{2+}/Mn(OH)^+$, others appear only to utilize Mn(II) that is bound to solid surfaces, such as MnO_2 or clay minerals. In some bacteria (e.g. *Leptothrix pseudoochracea*) Mn oxidation appears to be a mechanism for eliminating excess hydrogen peroxide produced by the cells, using catalase enzymes that function as peroxidases. *Leptothrix discophora*, on the other hand, produces a Mn(II)-oxidizing protein which is associated with acidic polysaccharides in its sheath, in a scenario analogous to the ferrous iron-oxidizing system of this bacterium described earlier. Dormant spores of marine and freshwater *Bacillus* spp. have also been shown to bind and oxidize Mn(II). The various non-enzymic routes for biological Mn(II) oxidation listed in Figure 3.2 appear to be the main mechanisms whereby fungi and algae oxidize the metal, though extracellular peroxidases may also be involved in the case of some white-rot fungi (Ehrlich, 1996).

Bacteria that oxidize Mn(II) may also accumulate the insoluble manganese oxides produced, but in other microorganisms manganese accumulation is a passive rather than an active process. In both cases, deposits of manganese oxides tend to accumulate on extracellular structures, such as sheaths, stalks and slime layers (Ehrlich, 1996).

3.4.2 MANGANESE REDUCTION IN AQUATIC ENVIRONMENTS

Manganese (IV) reduction has been reported to be more widespread among microorganisms than the ability to oxidize Mn(II). A number of phylogenetically unrelated bacteria, and a more limited range of fungi, may catalyse the dissimilatory reduction of Mn(IV) using direct or indirect mechanisms (Figure 3.3). Some bacteria, such as *Athrobacter* sp. and several Gram-negative marine isolates, have been found to be capable of both Mn(II) oxidation

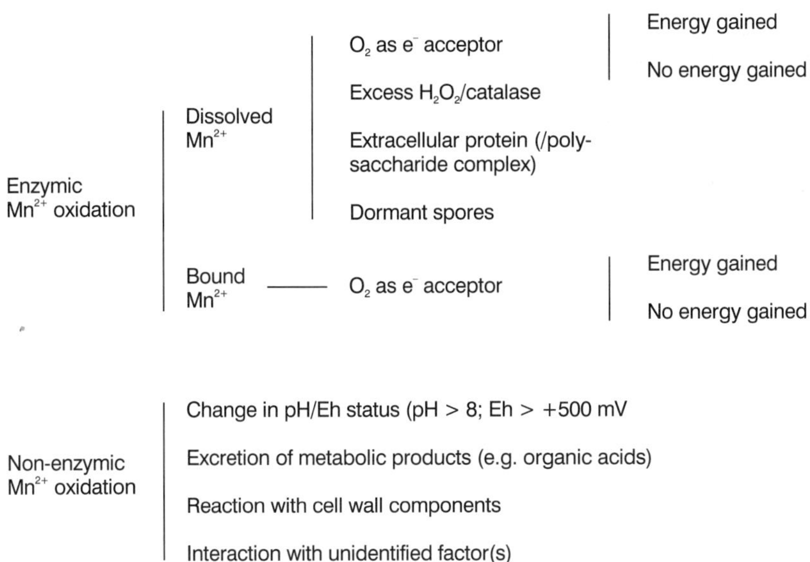

Figure 3.2 Summary of the various routes by which microorganisms may catalyse the oxidation of manganese(II).

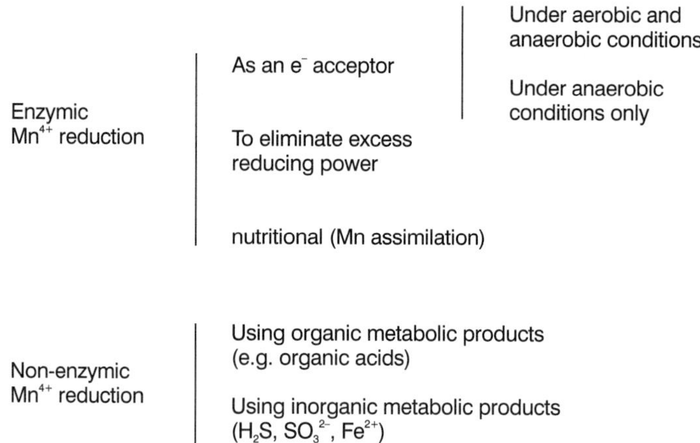

Figure 3.3 Summary of the various routes by which microorganisms may catalyse the reduction of manganese(IV).

and Mn(IV) reduction depending on the aeration status of their growth medium (Ehrlich, 1996). Manganese oxides may differ in their degree of crystallinity and, as with ferric iron minerals, this determines the ease by which they may be reduced by microorganisms, with amorphous and pseudo-amorphous forms being reduced in preference to crystalline MnO_2 (Gounot, 1994). Intimate surface contact, or solubilization of solid manganese (IV) oxides, may be necessary prerequisites for direct reduction to occur.

The most significant process involving enzymic dissimilatory reduction of Mn is that in which Mn(IV) is used as a terminal electron acceptor. For some bacteria (e.g. some *Bacillus* spp.) this can occur under either aerobic or anaerobic conditions, though in others (e.g. *Geobacter metallireducens* and *Shewanella putrefaciens*) it is only observed in anoxic cultures even though *S. putrefaciens* is a facultative anaerobe. Some SRB (e.g. *Desulfovibrio desulfuricans*) can also reduce Mn(IV) directly, and the sulphide produced by all such bacteria may also chemically reduce Mn(IV), as noted below. Most manganese-reducing bacteria are heterotrophic, but some are apparently able to use hydrogen as electron donor (Ehrlich, 1996).

Indirect (non-enzymic) reduction of Mn(IV) may be brought about by the excretion by microorganisms of certain organic acids, such as formic acid (by *E. coli*) and oxalic acid (by a number of fungi), though in some cases reduction of Mn(IV) by organic acids is pH dependent. Some manganese oxides may also be reduced by inorganic metabolites, such as sulphide, sulphite and ferrous iron, thereby implying that all iron-reducing bacteria also have the potential to reduce Mn(IV) indirectly.

3.4.3 MANGANESE TURNOVER IN FRESHWATER AND MARINE ENVIRONMENTS

Microbial oxido-reduction of Mn in aquatic environments is most evident where oxic and anoxic zones are in close proximity, such as areas near sediment–water interfaces. Cycling of Mn between oxidized and reduced forms may be rapid, particularly in disturbed sediments. Manifestations of Mn oxido-reduction may be the production of (ferro)magnesium concretions on the one hand, and of abnormally high concentrations of the soluble metal on the other.

The occurrence of ferromanganese concretions in freshwater lakes and ponds has been recorded on a number of occasions (Ehrlich, 1996). These may take on different forms, including pea-sized nodules, and larger disc-shaped structures or crusts covering rocks within lakes. The compositions of these concretions can vary considerably; Mn contents of between 0.69 and 35.9% (by weight) have been recorded for samples taken from different lakes in North America (Ehrlich, 1996). Ferromanganese nodules and crusts are also found at depth in all oceanic waters, often far removed from the volcanic mid-ocean spreading centres where seawater concentrations of soluble manganese tend to be greatest. Again, marine concretions have variable compositions; an average Mn content of 24.2% (by weight) has been recorded for nodules from the Pacific Ocean. Manganese oxidation in water distribution systems has occasionally produced accumulations of manganese oxides on surfaces within pipelines; these deposits may cause significant lowering of water pressure and, in the case of hydroelectric plants, reduce the efficiency of power generation. Often *Hyphomicrobium*-like bacteria are associated with these Mn deposits in water distribution systems. On the positive side, bacterial manganese oxidation is used in different water treatment processes for removal of both Mn(II) and Fe(II) (Gounot, 1994).

Whilst mobilization of Mn may follow as a consequence of microbial reduction of Mn(IV), the Mn(II) produced may form insoluble compounds (such as $MnCO_3$ and Mn_3O_4) or be adsorbed on to manganic oxides, clays, etc. thereby making the true extent of Mn(IV) reduction more difficult to estimate. Examples of Mn mobilization in lakes have, however, been reported by several groups. Such events are often seasonal, depending on temperature, availability of suitable organic substrates, etc. as well as on water pH and Eh. The major location of Mn(IV) reduction may also change on a seasonal basis; for example, Davison *et al.* (1982) found that Mn(IV) reduction in the English Lake District was most intense in lake sediments in the winter months but that in the summer it occurred mainly within the hypolimnion (i.e. within the water column). Anthropogenic activities may also affect Mn mobilization. Gounot (1994) described how hydrological changes caused by construction of hydroelectric dams in the Rhone Valley resulted in greatly increased Mn(II) concentrations in aquifers that were used as domestic water supplies: Mn levels exceeded 1000 μg l^{-1} on occasions. Soluble Mn is not especially toxic to humans, though dissolution of manganese(IV) oxides results in remobilization of adsorbed metals (including toxic metals) which may constitute a greater danger to health. This is possibly a major reason for the occasional poisonings that have been reported following ingestion of manganese-containing well-waters (Nealson, 1983b).

3.5 OVERVIEW AND BIOTECHNOLOGICAL PERSPECTIVES

The biogeochemical cycling of metals in aquatic ecosystems constitutes a major component of the ongoing transformations of metals on a global scale. As with the cycling of other elements that have biological roles, such as C, N and S, metal transformations are fundamental in maintaining the integrity of the biosphere, though not all metals that are subject to microbial transformations are essential nutrients. Biological activities and physicochemical factors may combine to produce solution-phase concentrations of metals which are growth limiting on the one extreme or toxic on the other, though micro- and macro-biota can vary widely in their sensitivities to, requirements for, and abilities to scavenge metals.

There has been major progress in recent years in elucidating the mechanisms whereby microorganisms bring about transformations of metals. New biotechnologies based on these findings have emerged and continue to develop. Whilst some of these, such as bioleaching of sulphide minerals and ores, are exploitive technologies, others are concerned with environmental protection. Bioremediation of metal-contaminated aquatic and terrestrial environments may involve the use of populations of microorganisms that can be isolated from polluted ecosystems – which is highly pertinent, given the concern that exists about possible accidental or controlled release of genetically modified microorganisms into the environment.

REFERENCES

Balashova, V.V. and Zavarzin, G.A. (1980) Anaerobic reduction of ferric iron by hydrogen bacteria. *Microbiology* **48**, 635–639.

Barrett, J., Hughes, M.N., Karavaiko, G.I. and Spencer, P.A. (1993) *Metal Extraction by Bacterial Oxidation of Minerals*, Ellis Horwood, Chichester, UK.

Barton, L.L. (1995) *Sulfate-Reducing Bacteria*, Plenum Press, New York.

Berthelin, J. (1983) Microbial weathering processes, in *Microbial Geochemistry* (ed. W.E. Krumbein), Blackwell, Oxford.

Blakemore, R.P. (1982) Magnetotactic bacteria. *Annual Review of Microbiology* **36**, 217–238.

Brierley, C.L. (1990) Bioremediation of metal-contaminated surface and groundwaters. *Geomicrobiology Journal* **8**, 201–223.

Brock, T.D. and Gustafson, J. (1976) Ferric iron reduction by sulfur- and iron-oxidizing bacteria. *Applied and Environmental Microbiology* **32**, 567–571.

Brookins, D.G. (1988) *Eh–pH Diagrams for Geochemistry*, Springer-Verlag, Berlin.

Clark, D.A. and Norris, P.R. (1996) *Acidimicrobium ferrooxidans* gen. nov., sp. nov.: mixed-culture ferrous iron oxidation with *Sulfobacillus* species. *Microbiology* **141**, 785–790.

Corstjens, P.L.A.M., de Vrind, J.P.M., Westbroek, P. and de Vrind-de Jong, E.W. (1992) Enzymatic iron oxidation by *Leptothrix discophora*: identification of an iron-oxidizing protein. *Applied and Environmental Microbiology* **58**, 450–454.

Davison, W., Woof, C. and Rigg, E. (1982) The dynamics of iron and manganese in a seasonally anoxic lake. *Limnology and Oceanography* **27**, 987–1003.

DiSpirito, A.A. and Tuovinen, O.H. (1982) Uranous ion oxidation and carbon dioxide fixation by *Thiobacillus ferrooxidans*. *Archives of Microbiology* **133**, 33–37.

Duxbury, T. (1985) Ecological aspects of heavy metal responses in microorganisms, in *Advances in Microbial Ecology* **8** (ed K.C. Marshall), Plenum, New York, pp. 185–235.

Ehrlich, H.L. (1996) *Geomicrobiology*, 3rd edn, Marcel Dekker, New York.

Emerson, D. and Ghiorse, W.C. (1993) Ultrastructure and chemical composition of the sheath of *Leptothrix discophora* SP-6. *Journal of Bacteriology* **175**, 7808–7827.

Ford, T., Maki, J. and Mitchell, R. (1995) Metal–microbe interactions, in *Bioextraction and Biodeterioration of Metals*, (eds C.C. Gaylarde and H.A. Videla), Cambridge University Press, Cambridge, pp. 1–23.

Geesey, G. and Jang, L. (1990) Extracellular polymers for metal binding, in *Microbial Mineral Recovery* (eds H.L.Ehrlich and C.L.Brierley), McGraw-Hill, New York, pp. 223–247.

Gadd, G.M. and Griffiths, A.J. (1978) Microorganisms and heavy metal toxicity. *Microbial Ecology* **4**, 303–317.

Ghani, B., Takai, M., Hishani, N.Z. *et al.* (1993) Isolation and characterisation of a Mo^{6+}-reducing bacterium. *Applied and Environmental Microbiology* **59**, 1176–1180.

Gounot, A.-M. (1994) Microbial oxidation and reduction of manganese: consequences in groundwater and applications. *FEMS Microbiology Reviews* **14**, 339–350.

Greene, B. and Darnall, D.W. (1990) Microbial oxygenic phototrophs (cyanobacteria and algae) for metal-ion binding, in *Microbial Mineral Recovery*, (eds H.L. Ehrlich and C.L. Brierley), McGraw-Hill, New York, pp. 277–302.

Hallbeck, L. and Pederson, K. (1991) Autotrophic and mixotrophic growth of *Gallionella ferruginea*. *Journal of General Microbiology* **137**, 2657–2661.

Hughes, M.N. and Poole, R.K. (1989) *Metals and Micro-organisms*, Chapman & Hall, London.

Johnson, D.B. (1995a) Mineral cycling by microorganisms: iron bacteria, in *Microbial Diversity and Ecosystem Function*, (eds D. Allsop, D.L. Hawksworth and R.R. Colwell) CAB International, Wallingford, UK, pp. 137–160.

Johnson, D.B. (1995b) Acidophilic microbial communities: candidates for bioremediation of acidic mine effluents. *International Biodeterioration & Biodegradation* **35**, 41–58.

Johnson, D.B., Bacelar-Nicolau, P., Bruhn, D.F. and Roberto, F.F. (1995) Iron-oxidising heterotrophic acidophiles: ubiquitous novel bacteria in leaching environments, in *Biohydrometallurgical Processing I*, (eds T. Vargas, C.A. Jerez, J.V. Wiertz, and H. Toledo), University of Chile Press, Santiago, pp. 47–56.

Johnson, D.B., McGinness, S. and Ghauri, M.A. (1993) Biogeochemical cycling of iron and sulfur in leaching environments. *FEMS Microbiology Reviews* **11**, 63–70.

Johnson, D.B. and McGinness, S. (1991) Ferric iron reduction by acidophilic heterotrophic bacteria. *Applied and Environmental Microbiology* **57**, 207–211.

Kletzin, A. and Adams, M.W.W. (1996) Tungsten in biological systems. *FEMS Microbiology Reviews* **18**, 5–63.

Lewis, A.J. and Miller, J.D.A. (1977) Stannous and cuprous ion oxidation by *Thiobacillus ferrooxidans*. *Canadian Journal of Microbiology* **23**, 319–324.

Lovley, D.R. (1991) Dissimilatory Fe(III) and Mn(IV) reduction. *Microbiological Reviews* **55**, 259–287.

Lovley, D.R. (1995) Microbial reduction of iron, manganese, and other metals. *Advances in Agronomy* **54**, 175–231.

Lovley, D.R. and Phillips, E.J.P (1992) Microbial reduction of uranium by *Desulfovibrio desulfuricans*. *Applied and Environmental Microbiology* **58**, 850–856.

Lovley, D.R., Phillips, E.J.P., Gorby, Y.A. and Landa, E.R. (1991) Microbial reduction of uranium. *Nature* **350**, 413–416.

Lundgren, D.G. and Dean, W. (1979) Biogeochemistry of iron, in *Biogeochemical Cycling of Mineral-Forming Elements*, (eds P.A. Trudinger and D.J. Swaine), Elsevier, Amsterdam, pp. 211–251.

Lyalikova, N.N. (1974) *Stibiobacter senarmontii* – a new antimony-oxidizing microorganism. *Mikrobilogiya* **43**, 799–805 (English translation).

Lyalikova, N.N. and Lebedeva, E.V. (1984) Bacterial oxidation of molybdenum in ore deposits. *Geomicrobiology Journal* **3**, 307–318.

Macaskie, L.E. (1991) The application of biotechnology to the treatment of wastes produced from the nuclear fuel cycle: biodegradation and bioaccumulation as a means of treating radionuclide-containing streams. *CRC Critical N/68 Reviews of Biotechnology* **11**, 41–112.

Marshall, K.C. (1979) Biogeochemistry of manganese minerals, in *Biogeochemical Cycling of Mineral-Forming Elements*, (eds P.A. Trudinger and D.J. Swaine), Elsevier, Amsterdam, pp. 253–292.

Nealson, K.H. (1983a) The microbial iron cycle, in *Microbial Geochemistry*, (ed. W.E. Krumbein), Blackwell, Oxford.

Nealson, K.H. (1983b) The microbial manganese cycle, in *Microbial Geochemistry*, (ed. W.E. Krumbein), Blackwell, Oxford.

Norris, P.R. and Johnson, D.B. (1998) Acidophilic microorganisms, in *Extremophiles: Microbial Life in Extreme Environments*, (eds K. Horikoshi and W.D. Grant), Wiley, New York (in press).

Schinner, F. and Burgstaller, W. (1989) Extraction of zinc from industrial waste by a *Penicillium* sp. *Applied and Environmental Microbiology* **55**, 1153–1156.

Schlesinger, W.H. (1991) *Biogeochemistry: an Analysis of Global Change*, Academic Press, San Diego.

Springel, D., Diels, L., Hooyberghs, L. *et al.* (1993) Construction and characterization of heavy metal-resistant haloaromatic-degrading *Alcaligenes eutrophus* strains. *Applied and Environmental Microbiology* **59**, 334–339.

Sugio, T., Tsujita, Y., Inagaki, K. and Tano, T. (1990) Reduction of cupric ions with elemental sulphur by *Thiobacillus ferrooxidans*. *Applied and Environmental Microbiology* **56**, 693–696.

Summers, A.O. and Silver, S. (1978) Microbial transformations of metals. *Annual Review of Microbiology*, **32**, 637–672.

Talasova, I.I., Khavski, N.N., Khairullina, R.T. *et al.* (1995) Red mud leaching with fungal metabolites, in *Biohydrometallurgical Processing I* (eds T. Vargas, C.A. Jerez, J.V. Wiertz and H. Toledo), University of Chile Press, Santiago, pp. 379–384.

Torma, A.E. and Gabra, G.G. (1977) Oxidation of stibnite by *Thiobacillus ferrooxidans*. *Antonie van Leeuwenhoek* **43**, 1–6.

Volesky, B. (ed.) (1990) *Biosorption of Heavy Metals*, CRC Press, Boca Raton, USA.

Wang, Y.T. and Shen, H. (1995) Bacterial reduction of hexavalent chromium. *Journal of Industrial Microbiology* **14**, 159–163.

White, C., Wilkinson, S.C. and Gadd, G.M. (1995) The role of microorganisms in biosorption of toxic metals and radionuclides. *International Biodeterioration and Biodegradation* **35**, 17–40.

Widdel, F., Schnell, S., Heising, S. *et al.* (1993) Ferrous iron oxidation by anoxygenic phototrophic bacteria. *Nature* **362**, 834–836.

4   *Metal accumulation and impacts in phytoplankton*

JAMES G. SANDERS AND GERHARDT F. RIEDEL

4.1 INTRODUCTION

This chapter examines processes that regulate and influence trace element uptake, accumulation, transformation, release and toxicity in natural aquatic systems. Phytoplankton are efficient accumulators of most reactive elements. Many inorganic compounds, including most metals, are involved in the oxidative and reductive reactions that comprise cellular metabolism. Some elements are required in small quantities but are toxic in larger quantities. Although materials associated with phytoplankton behave in a manner somewhat similar to particulates, incorporation of a toxic compound markedly affects the compound's transport through aquatic ecosystems. Toxic compounds may also pass to successively higher trophic levels through interactions between predators and their prey.

Furthermore, participation of toxic substances in biological processes can lead to chemical transformation of the compound into a form quite different from the original. If these transformed chemical species have biological affinities or chemical stabilities that differ from the original compound, this transformation can alter the compound's reactivity and perhaps its transport through the ecosystem. In addition, biological uptake and transformation may enhance or alleviate its toxicity to other organisms, increasing or minimizing the potential for harm.

4.2 METAL ACCUMULATION

The interactions of trace elements with phytoplankton are governed by a complex mixture of physical, chemical and biological constraints. Many elements are extremely particle reactive and sorb readily to both living (phyto-

Metal Metabolism in Aquatic Environments. Edited by William J. Langston and Maria João Bebianno. Published in 1998 by Chapman & Hall, London. ISBN 0 412 80370 4

plankton) and non-living (sediment) particles. Phytoplankton are efficient scavengers of trace elements, accumulating high concentrations from the surrounding medium with concentration factors ranging from 10^3 to 10^6 (Fisher, 1986). Both active and passive mechanisms can be involved in trace element sorption; the relative importance of different sorption pathways varies with the element of interest.

The uptake of most cationic trace elements has been shown to be related to the concentration of free metal ions (e.g. Sunda and Guillard, 1976; Gavis *et al.*, 1981; Sunda, 1988/89, 1994; Morel *et al.*, 1991), though the mechanism that lies behind this apparent relationship is likely to be ligand exchange between the uptake site on the cell membrane and inorganic and organic metal complexes (Morel *et al.*, 1991) and can largely be modelled using saturation uptake kinetics. Exceptions to this general model for uptake appear to be limited to elements associated with neutrally charged complexes such as $HgCl_2$ (Gutknecht, 1981) or AgCl (Sanders and Abbe, 1989), or with lipophilic organic complexes (Phinney and Bruland, 1994). In addition, specialized uptake mechanisms are also present, such as the production of siderophores to enhance iron uptake (see below).

In the uptake of trace elements by phytoplankton, the chemical speciation of the element and the presence of other ions or chelators that can regulate speciation are of utmost importance. Silver uptake, for example, is inversely proportional to salinity due to complexation by chloride (Sanders and Abbe, 1987). For elements generally present as anions, such as arsenic (AsO_4, AsO_3), chromium (CrO_4), selenium (SeO_4, SeO_3), or germanium (GeO_4), uptake is often related to competition for uptake with a similar nutrient ion (e.g. As-P, Cr-S, Se-S, Ge-Si), thus regulating incorporation and controlling cellular content regardless of dissolved concentrations (Blum, 1966; Azam *et al.*, 1973; Planas and Healey, 1978; Sanders and Windom, 1980; Wheeler *et al.*, 1982; Riedel, 1985).

With many of the cationic elements (Ag, Cd, Hg, Zn), sorption is rapid, and generally in proportion to element concentrations within the surrounding medium (Fisher *et al.*, 1984; Sanders and Abbe, 1987). There are differences in accumulation between different phytoplankton species, however, that may be driven by physiological or biochemical differences between cells. Often, cellular content is strongly correlated with cell surface area or cell volume (Fisher *et al.*, 1984; Sanders and Abbe, 1987).

When attempting to compare the accumulation of trace elements by natural phytoplankton of varying species composition, a metal partition coefficient (K_D, the ratio of the amount of metal found in the phytoplankton relative to the dissolved metal available) is often used. While this is a useful computational tool, it must be remembered that many extraneous factors can influence the K_D, such as inorganic complexation, organic complexation, pH, dilution by varying amounts of inorganic and non-living particles; thus, only in particular cases can direct comparisons of K_D be of value. One way to examine the partitioning of different phytoplankton is to sample during bloom events in the natural environment. For example, dinoflagellates (*Gymnodinium, Gyrodinium* species) sampled from a summer bloom in the Patuxent River contained less copper and much less cadmium than did non-bloom assemblages collected from outside of the bloom region at the same time (Figure 4.1), a reflection either of differences between different species or simply of higher biomass in bloom regions. In addition, systems that exhibit elevated levels of trace elements (e.g. Baltimore Harbour in Chesapeake Bay) may have different patterns of partitioning. Baltimore Harbour blooms contain rather high concentrations of copper and cadmium relative to other regions in Chesapeake Bay, and metal K_D is higher, as well (Table 4.1).

Metal partitioning also varies through time, as phytoplankton communities undergo species succession and changes in dominant species. In a series of experiments performed using natural phytoplankton communities maintained under ambient conditions in outdoor mesocosms, interactions with a variety of trace elements have been examined (e.g. Sanders *et al.*, 1981, 1989, 1991; Sanders and Cibik, 1985a; Sanders and Riedel, 1993). In one experiment, mercury partitioning fell from 1.5×10^5 to 6.8×10^4 as dominant species shifted from centric diatoms to an assemblage dominated by the cyanophyte, *Anacystis marina* (Table 4.2). In separate experiments, partitioning of Cadmium and Copper both decreased with a shift in species composition from small flagellates to the centric diatom, *Skeletonema costatum* (Table 4.2). Changes in K_D as shifts in phytoplankton dominance occur could result from differing attractions of trace elements for the several different kinds of cell walls present, with the non-silica surfaces of flagellates having a higher affin-

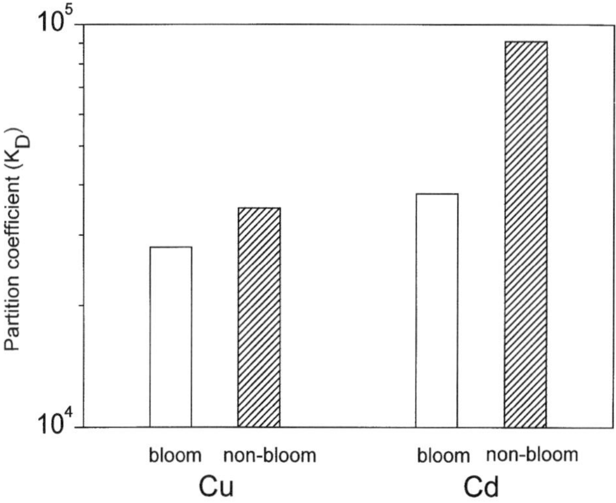

Figure 4.1 Changes in partition coefficients for copper and cadmium inside and outside of blooms of *Gymnodinium/Gyrodinium* spp. in the Patuxent River.

Table 4.1 Metal content and K_D in Baltimore Harbour relative to the Patuxent River

Metal	Baltimore Harbour		Patuxent River	
	Cell content (μg/g)	K_D	Cell content (μg/g)	K_D
Cu	57.2	5.8×10^4	18.3	2.6×10^4
Cd	4.17	7.3×10^4	1.13	3.3×10^4

Table 4.2 Changes in partition coefficient (K_D) relative to changes in dominant species through time in natural phytoplankton communities

Metal	Early		Late	
	K_D	Dominants	K_D	Dominants
Hg	1.5×10^5	Centric diatoms	6.8×10^4	*Anacystis marina*
Cd	6.0×10^4	Flagellates, monads	2.8×10^4	*Skeletonema costatum*
Cu	4.4×10^4	Flagellates, monads	9.0×10^3	*Skeletonema costatum*

ity for some elements than the silica diatom frustules (Sick and Windom, 1975; Brand *et al.*, 1986), from selective releases of metal-binding compounds (see below), or from changes in cell size and cell surface area (Fisher *et al.*, 1984).

The incorporation of trace elements by phytoplankton can regulate trace element form and availability. Most simply, uptake by phytoplankton reduces the amount of dissolved metal, thereby reducing the availability of dissolved species to other phytoplankton or to higher trophic levels (Figure 4.2A). Once associated with a cell, elements can be more or less tightly bound to cellular components, which can influence their availability through feeding to higher trophic levels (Reinfelder and Fisher, 1991, 1994; Connell *et al.*, 1991; Lee and Fisher, 1992). In general, the extent to which phytoplankton accumulate metals depends on the phytoplankton density and the affinity of the phytoplankton for the element. In general, the affinity of the phytoplankton for (and the toxicity of) cationic metals follows a known sequence (Fisher, 1986). At the low end of this series, phytoplankton do not normally remove a substantial fraction of the dissolved metal even when phytoplankton densities are high; at the high end, even low densities of phytoplankton can reduce dissolved concentrations of metal.

As metals are accumulated rapidly by phytoplankton, they can also be released, particularly as cells die and decompose. In studies of trace element release, Fisher and co-workers (Lee and Fisher, 1992, 1994; Fisher and Wente, 1993) demonstrated that elements associated with cellular cytoplasm (Se, Cd) are lost more quickly than particle reactive elements, which would be associated with cell walls (e.g. Sn, Ag). This selective release can result in

76525100-55

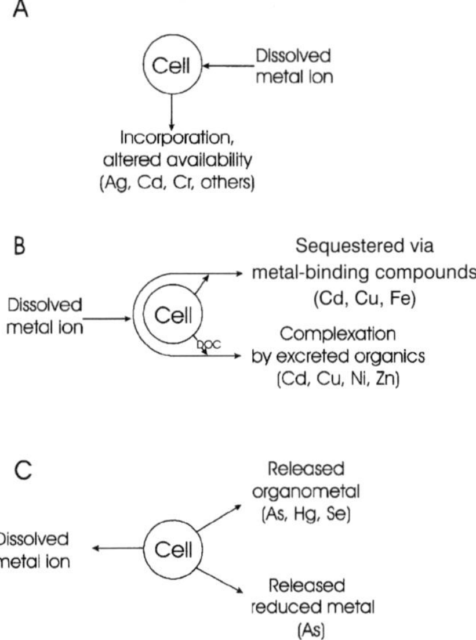

Figure 4.2 Potential reactions of phytoplankton with dissolved metal ions, including changes in (A) partitioning, (B) complexation and sequestration, and (C) transformation. (Adapted from Sanders and Riedel, 1987.)

particle reactive elements being depleted from surface waters in deep aquatic systems.

4.3 METAL COMPLEXATION, BINDING STRATEGIES AND BIOTRANSFORMATION

Phytoplankton have a number of direct and indirect mechanisms available for modifying, capturing and sequestering both necessary and toxic elements. Many species of phytoplankton release compounds capable of complexing some trace elements. In the simplest form, phytoplankton release a variety of dissolved organic compounds (DOC), which can complex many elements (Figure 4.2B). Algal blooms undergo distinct phases in dynamic systems (Margalef, 1962; Davis, 1982), with generally an initial phase when a few species exhibit rapid growth and increases in cell density, a period of species succession, and a later stage of cell density decreases. There are corresponding changes in the chemical composition of the surrounding water, with pH fluctuating through photosynthesis and respiration, and often concentrations of DOC increasing as the bloom senesces (Sharp, 1977; Brockmann and Dahl, 1990). Thus, there are ample reasons why trace element form should be expected to change through time in such systems. In studies performed with coastal phytoplankton assemblages, copper underwent changes in its degree of organic association, with highest levels of complexed copper present as blooms senesced (Sanders and Riedel, 1993). Similar results have been reported from both algal cultures and from natural communities (Wangersky *et al.*, 1989; Zhou *et al.*, 1989; Zhou and Wangersky, 1989).

Relatively little is known about the nature of metal complexing organics in natural waters, but recent work has shown that two classes of ligands – a low concentration, high affinity class, and a more abundant, low affinity class – appear to be responsible for metal complexation (Coale and Bruland, 1988). It is assumed that humic and fulvic acids are important in the abundant, weaker ligand class, producing a host of similar molecules with ligands having a range of affinities. Although this is a complex group of compounds, it seems possible to model the complexing abilities of this suite of compounds as a single ligand with a fixed affinity (Dzombak *et al.*, 1986). In addition to this group of weaker ligands, strong ligands have been observed for copper, cadmium, zinc and nickel (e.g. Bruland, 1989, 1992; Donat *et al.*, 1994). However, the origin and composition of the strong ligands are poorly known. It appears that the high affinity ligands for cadmium and zinc are not identical (Bruland, 1992); thus, these ligands may be highly specific. Gordon *et al.* (1996) characterized high affinity ligands for copper as having relatively low molecular weight, and correlated with the abundance of picoplankton. *Synechococcus* species also have been observed to produce a high affinity ligand in culture (Moffett *et al.*, 1990; Moffett and Brand, 1996).

Phytoplankton are capable of producing specific metal-binding compounds (McKnight and Morel, 1979; Fisher and Fabris, 1982; Moffett and Brand, 1996) such as phytochelatins (Gekeler *et al.*, 1988; Ahner *et al.*, 1994, 1995; Ahner and Morel, 1995; Morelli and Scarano, 1995; Lee *et al.*, 1996) or siderophores (Trick *et al.*, 1983; Wilhelm and Trick, 1994; Wilhelm *et al.*, 1996) in response to elevated levels of metals such as iron, cadmium and copper (Figure 4.2B). Such complexation reactions essentially form a mechanism to regulate the activity and availability of both necessary and toxic metals, presumably at levels acceptable to cellular function.

Another important aspect of biological uptake is the transformation of the chemical form of a contaminant. Aside from the concentration and sequestering of toxic substances, organisms have the capability to alter the chemical form of many inorganic elements and organic compounds (Figure 4.2C). Such shifts alter biological reactivity and toxicity of these compounds and can influence their rate of transport through ecosystems as well as their eventual fate. For example, biological uptake of arsenic leads to the production of reduced and methylated arsenic compounds, some of which are more toxic to higher trophic levels than was the original arsenic compound (Peoples, 1975; Knowles, 1982; Knowles and Benson, 1983; see also Chapter 6). Within many aquatic ecosystems during some periods of the year, large fractions (up to 80%) of the arsenic may be present in these reduced or methylated forms (Sanders, 1980, 1985; Howard *et al.*, 1982, 1984, 1988; Andreae and Froelich, 1984; Froelich *et al.*, 1985; Sanders and Riedel, 1993). The production of such compounds also increases the rate of transport of arsenic through ecosystems by altering its biological and geochemical reactivity (Sanders and Riedel, 1987).

Selenium undergoes a similar, if more complicated series of post-accumulative transformations. Taken up by the cell apparently as an analogue of sulphur, selenium is reduced and incorporated into a wide variety of compounds. Dimethylselenide, dimethylselenone, dimethyldiselenide, and peptides containing the seleno-amino acid are all known to be produced by phytoplankton or bacteria in natural waters (Wrench, 1978; Cooke and Bruland, 1987; Besser *et al.*, 1994).

4.4 ECOTOXICOLOGICAL IMPLICATIONS

Assessing the impact of chronic inputs of contaminants is a difficult task. There are usually no obvious effects from such inputs, but chronic sources, because they are so widespread and occur regularly, may be more damaging on a regional scale and over the long term than are visible events with localized mortalities. There has long been a presumption by both the public and scientific communities that chronic releases of contaminants are responsible for a perceived degradation of the environment, but there is very little scientific evidence for such a conclusion. Conclusively assigning blame for any trend in living resources is difficult. Given the enormous variability of such resources due to climate, weather, fishing pressure, and other unknowns, along with the lack of suitable control systems for comparison, such perceptions must be treated with at least a little scepticism. Generally, the contaminant levels found in many aquatic systems are measurably elevated above background concentrations, but they are not so elevated as to suggest that direct toxicity will occur. Moreover, single species acute and chronic toxicity studies ordinarily suggest that orders of magnitude higher concentrations of contaminants than are found in aquatic systems are necessary to cause significant deleterious effects. However, a perfectly reasonable presumption remains that simply because organisms are not suffering acutely from measurable contaminant loadings does not mean that contaminants are not exerting subtle negative effects on the ecosystem.

While techniques exist to detect trace quantities of a variety of contaminants in natural samples, small alterations in community structure caused by a specific pollutant are not easily assessed. Once accomplished, it is still necessary to evaluate how such a community shift will influence the ecosystem as a whole. In an attempt to study such effects at the community level, attention has been focused on the impact of low-level contamination by trace elements on natural phytoplankton assemblages. Phytoplankton comprise the base of the food web and lend themselves quite well to such studies. Subtle changes in community structure or density can be detected in a matter of days as a result of their rapid growth rates. Thus, it is possible to build upon our

Table 4.3 Sensitivity of phytoplankton species to arsenic and free copper, chromium and cadmium ions; concentrations shown are those necessary to cause a 50% reduction in growth rates (taken from Sanders *et al.*, 1991)

Species	As (μg l^{-1})	Cu (ng l^{-1})	Cd (μg l^{-1})	Cr[1] (μg l^{-1})	References
Rhizosolenia fragilissima	2.2		0.8–1.5		Sanders and Cibik, 1998b; Berland *et al.*, 1977
Skeletonema costatum	5.2	0.2–3.2	0.05		Sanders, 1979; Morel *et al.*, 1978; Brand *et al.*, 1986
Cerataulina pelagica	3				Sanders *et al.*, 1991
Thalassiosira pseudonana	> 25	0.1–2.5	7.5	8.8	Sanders and Riedel, 1987; Sunda and Guillard, 1976; Braek *et al.*, 1980; Riedel, 1988
Cylindrotheca closterium	> 100				Sanders and Vermersch, 1982
Phaeodactylum tricornutum			> 150		Braek *et al.*, 1980
Thalassiosira weissflogii			1	6.8	Riedel, 1988; Foster and Morel, 1982
Amphidinium carterae	9.8				Sanders and Vermersch, 1982
Peridinium trochoidium	90				Sanders and Riedel, 1987
Scrippsiella faeroense			0.4		Kayser, 1982
Prorocentrum spp.			0.5	3.1	Kayser and Sperling, 1980; Riedel, 1988
Gonyaulax tamarensis		0.04			Anderson and Morel, 1978
Isochrysis galbana	2.2		2.4	4.7	Sanders and Vermersch, 1982; Riedel, 1988; Li, 1978
Tetraselmis contracta	> 100				Sanders and Vermersch, 1982
Chlorella sp.				70–210	Riedel, 1988
Monochrysis lutheri		0.1		5.7	Gavis *et al.*, 1981; Riedel, 1988
Monallantus salina		3.2			Gavis *et al.*, 1981

[1]Chromium data are for fresh water; all others are for salt water.

investigations of this important community by assessing the potential impact on trophic relationships and overall effect to the ecosystem.

In many cases, the introduction of low concentrations of trace elements causes shifts in the composition and succession of important dominant species, at concentrations lower than those that will cause overall reduction in phytoplankton biomass. Species exhibit differential sensitivity to a toxic substance (Table 4.3). Individuals or populations can disappear as a result of exposure to chronic loadings of a given contaminant while others are not affected, thus sensitive species are lost from the community and resistant species dominate. Overall, diversity can be lowered, but biomass and primary production rates may not be greatly affected (Schindler, 1987; Winner and Owen, 1991). For example, copper has been shown to shift species composition and lower species diversity in both freshwater (Oliviera, 1985; Winner and Owen, 1991; Gustavsen and Wangberg, 1995) and marine (Harrison *et al.*, 1977; Thomas *et al.*, 1977; Gavis *et al.*, 1981; Sanders *et al.*, 1981; Brand *et al.*, 1986) ecosystems, though the species that proved sensitive have varied considerably. As shown in Table 4.3, and demonstrated by others, similar species may exhibit wide variations in sensitivity to different toxic metals.

In mesocosm experiments, arsenic has also been shown to be quite differentially toxic to phytoplankton. Chronic dosing led to a decline in large, chain-forming centric diatoms such as *Cerataulina pelagica, Rhizosolenia fragilissima* and *Chaetoceros* species (Figure 4.3) and their replacement by smaller, solitary diatom species such as *Thalassiosira* sp. (Sanders and Vermersch, 1982; Sanders and Cibik, 1985a; Sanders *et al.*, 1989, 1991). The result in such shifts was not only a continuation of diatom dominance of the community, but also a dramatic change in the size and shape in the

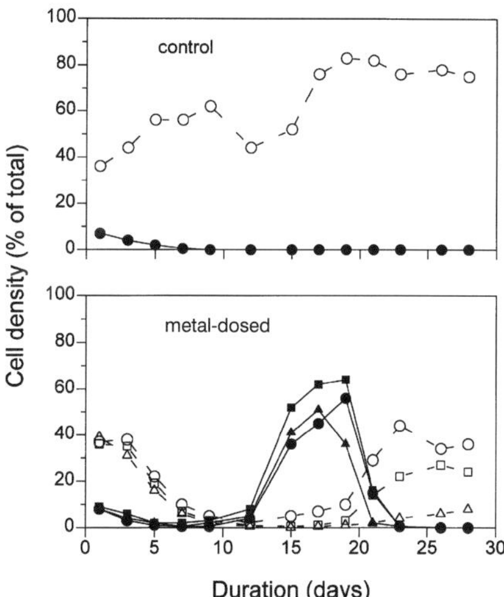

Figure 4.3 Changes (as percentage of total assemblage density) through time of two dominant centric diatoms, *Thalassiosira pseudonana* (solid symbols) and *Cerataulina pelagica* (open symbols), in control (0.3 µg l⁻¹) and arsenic-dosed (3–21 µgl⁻¹) microcosms. At all concentrations above control, *Cerataulina* growth was greatly reduced and *Thalassiosira* was able to dominate. (Adapted from Sanders *et al.*, 1994.)

dominant species – a change that can greatly affect carbon transfer between trophic levels.

Some toxic trace elements have been noted to impact the growth rate and success of nuisance algal species, such as cyanophytes and red tide-forming dinoflagellates. Elevated concentrations of mercury, and in some instances copper, actually enhanced the growth of a colonial cyanophyte, *Anacystis marina*, by 15–40% in estuarine phytoplankton maintained in mesocosms (Sanders *et al.*, 1989).

In natural systems, it is not likely that a single contaminant will be found; rather, it is more likely that mixtures of a variety of contaminants will be present. Chronic dosing of communities with mixtures of toxic elements can also lead to significant changes in phytoplankton species composition. In experiments performed with Patuxent River phytoplankton, chronic doses of a mixture of arsenic, copper and cadmium resulted in a phytoplankton community dominated almost solely by the cyanophyte, *Synechococcus* sp., while control assemblages were dominated by centric and, later in time, pennate diatoms (Figure 4.4). Overall biomass did not change greatly between control and metal-dosed assemblages; however, the extremely small size of the cyanophyte led to very high cell densities in the dosed assemblages relative to controls. Such shifts, from larger diatoms to smaller cyanophytes, would be expected to provide a lower 'quality' food base and lead to reduced success of higher trophic levels. Indeed, when the two phytoplankton assemblages were offered as food to a dominant copepod, *Acartia tonsa*, the result was

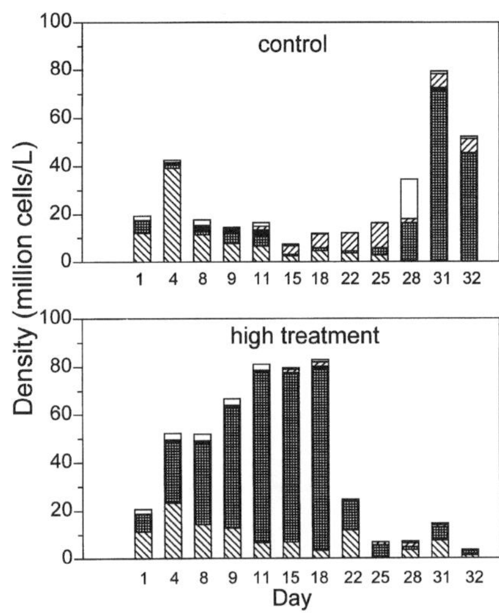

Figure 4.4 Phytoplankton density and importance of different phytoplankton groups through time in natural assemblages held in control and metal-dosed (arsenic 31.8 ± 3.4 µg l⁻¹, copper 15.6 ± 0.6 µg l⁻¹ and cadmium 4.77 ± 0.18 µg l⁻¹) microcosms. ▨ = centric diatoms; ▦ = cyanophytes; ■ = flagellates; ▨ = pennate diatoms; ☐ = other.

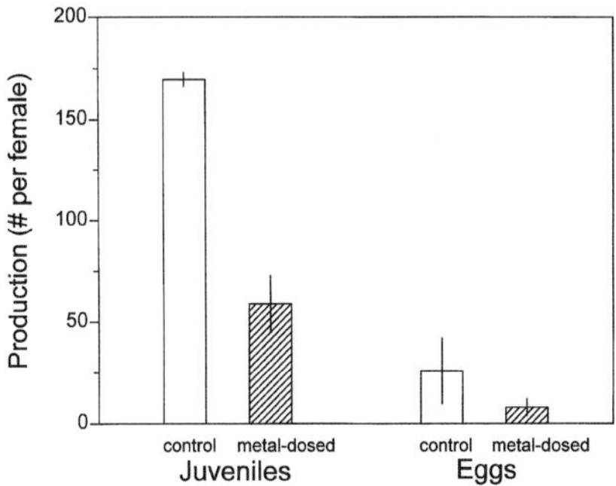

Figure 4.5 Egg and juvenile production (mean ± SD) in female *Acartia tonsa* held for 7 days and fed either control or metal-dosed phytoplankton assemblages shown in Figure 4.4.

reduced adult survival (less than 40% after 4 days and none survived 7 days, relative to 85–90% survival over 7 days in controls) and a greatly decreased egg and juvenile production rate for those individuals feeding on the metal-dosed phytoplankton (Figure 4.5).

Other trace elements, although toxic to higher trophic levels, may have little effect on overall species composition of phytoplankton. Of note is selenium. While selenium can be toxic to phytoplankton, the concentrations required to cause toxicity are extremely high, approaching 1 mg l⁻¹ for most species (Kumar and Prakish, 1971; Wheeler *et al.*, 1982; Foe and Knight, 1986; Kiffney and Knight, 1990). In studies of natural phytoplankton assemblages from a southeastern US lake, selenate was found to reduce cell growth at lower levels, above 10 µg l⁻¹, but did not influence species composition (Riedel *et al.*, 1996). Because selenite and selenate uptake (and presumably toxicity) are inversely proportional to phosphate and sulphate concentrations (Wheeler *et al.*, 1982; Riedel and Sanders, 1996) it may be possible that natural phytoplankton under ambient conditions (i.e. low phosphate, potentially low sulphate in some fresh-water systems) could exhibit higher sensitivities to selenium.

In a similar fashion, chromium uptake is regulated by sulphate concentration. In waters of low sulphate concentration, chromium has been shown to be much more toxic, with sensitivities of some phytoplankton approaching ambient water concentrations (Riedel, 1984; Sanders and Riedel, 1987), a sensitivity that is alleviated with higher sulphate concentrations.

To date, it is not possible to predict accurately biological change to communities that may result from chronic dosing with low levels of toxic trace elements. Further research with natural communities will be required to achieve a predictive capability. While many trace elements can have a profound effect upon species composition and species dominance within phytoplankton communities, others do not. With those elements that affect phytoplankton growth and dominance, the impact on higher trophic levels will vary.

Perhaps the most logical conclusion is that, in systems receiving chronic doses of toxic trace elements, shifts in phytoplankton species composition will occur, and will vary seasonally and depending on the element(s) added. If the system contains herbivores that feed selectively on particular algal species or cell shapes, these herbivores may be subjected to unpredictable shifts in the quantity of available prey. Perhaps most importantly, reduction in cell size appears to be a general result of chronic dosing by many contaminants. If we assume that, given sublethal concentrations of contaminants, cell biomass and to a large extent overall growth rates are regulated by the amount of available nitrogen and phosphorus, the result of such reductions in cell size will be higher densities of smaller phytoplankton. This altered community may not be suitable for some herbivores (e.g. Figure 4.6A) and may select for smaller grazers, diverting carbon from metazoan food chains and increasing the importance of microbial food chains and degradation pathways (Figure 4.6B). Beyond this very general comment, prediction of potential ecosystem effects are difficult to make.

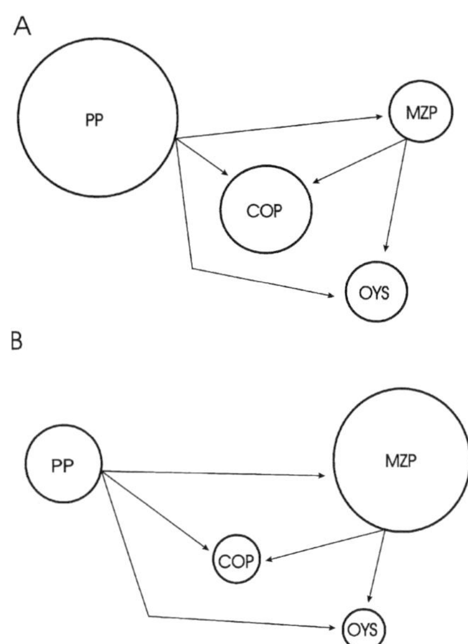

Figure 4.6 Conceptual model for ecosystem-wide effects of trace elements. Size of circles denotes organism size/abundance; arrows denote feeding relationships. PP, phytoplankton; MZP, microzooplankton; COP, copepods; OYS, benthic filter feeders. (A) Normal community with larger phytoplankton species and dominance by copepods and benthic filter feeders. (B) Trace element-stressed community containing smaller phytoplankton species (small diatoms, flagellates, or cyanophytes), reduced feeding by copepods and benthic filter feeders and increased dominance by microzooplankton and microbial heterotrophs. (Adapted from Sanders *et al.*, 1991.)

4.5 SUMMARY

It is crucial that we understand how several processes – physicochemical controls of trace element speciation, algal incorporation, transformation and release, and trace element control of phytoplankton speciation – are coupled in productive ecosystems before we can predict with confidence the transport, availability and impact of any toxic trace element. In some instances the important effect of phytoplankton uptake may be the production of toxic methylated compounds, because of their persistence, stability and toxicity to organisms within the ecosystem. Arsenic, selenium and perhaps tin, mercury, antimony and lead are elements of this type. In other cases, phytoplankton may reduce the toxicity of elements by transforming them into less toxic forms through complexation and sequestration by DOC or by a metal-binding compound. In both instances, however, the modifications mediated by phytoplankton are of significance to the aquatic ecosystem. They must be considered when ecosystem assessments are made.

ACKNOWLEDGEMENTS

This review has benefited from interactions with many colleagues and we are grateful for their interest. We thank D. Connell for assistance with much of the research described here and for critical review. This research has been supported by grants from the National Oceanic and Atmospheric Administration, Coastal Ocean Program, Maryland Sea Grant (R/CBT-13), the Environmental Protection Agency, and the Maryland Department of Natural Resources. Preparation of the manuscript was supported by the Academy of Natural Sciences.

REFERENCES

Ahner, B.A. and Morel, F.M.M. (1995) Phytochelatin production in marine algae. 2. Induction by various metals. *Limnology and Oceanography* **40**, 658–665.

Ahner, B.A., Price, N. and Morel, F.M.M. (1994) Phytochelatin production by marine phytoplankton at low free metal ion concentrations: laboratory studies and field data from Massachusetts Bay. *Proceedings of the National Academy of Sciences, USA* **91**, 8433–8436.

Ahner, B.A., Kong, S. and Morel, F.M.M. (1995) Phytochelatin production in marine algae. 1. An interspecies comparison. *Limnology and Oceanography* **40**, 649–657.

Anderson, D.M. and Morel, F.M.M. (1978) Copper sensitivity of *Gonyaulax tamarensis*. *Limnology and Oceanography* **23**, 283–295.

Andreae, M.O. and Froelich, P.N. (1984) Arsenic, antimony and germanium biogeochemistry in the Baltic Sea. *Tellus* **36B**, 101–117.

Azam, F., Hemmingsne, B.B. and Volcani, B.J. (1973) Germanium incorporation into the silica of diatom cell walls. *Archiv für Mikrobiologie* **92**, 11–20.

Berland, B.R., Bonin, D.J., Guérin-Ancey, O.J. *et al.* (1977) Action de métaux lourds à des doses sublétales sur les caractéristiques de la croissance chez la diatomée *Skeletonema costatum*. *Marine Biology* **42**, 17–30.

Besser, J.M., Huckins, J.N. and Clark, R.C. (1994) Separation of selenium species released from Se-exposed algae. *Chemosphere* **29**, 771–780.

Blum, J.J. (1966) Phosphate uptake by phosphate-starved *Euglena*. *Journal of General Physiology* **49**, 1125–1136.

Braek, G.S., Malnes, D. and Jensen, A. (1980) Heavy metal tolerance of marine phytoplankton. IV. Combined effect of zinc and cadmium on growth and uptake in some marine diatoms. *Journal of Experimental Marine Biology and Ecology* **42**, 39–54.

Brand, L.E., Sunda, W.G. and Guillard, R.R.L. (1986) Reduction of marine phytoplankton reproduction rates by copper and cadmium. *Journal of Experimental Marine Biology and Ecology* **96**, 225–250.

Brockmann, U. and Dahl, E. (1990) Distribution of organic compounds during a bloom of *Chrysochromulina polylepis* in the Skagerrak, in *Toxic Marine Phytoplankton*, (eds E. Granelli, B. Sundstrom, L. Edler, and D.M. Anderson), Elsevier, New York, pp. 104–109.

Bruland, K.W. (1989) Complexation of zinc by natural organic ligands in the central North Pacific. *Limnology and Oceanography* **34**, 269–285.

Bruland, K.W. (1992) Complexation of cadmium by natural organic ligands in the central North Pacific. *Limnology and Oceanography* **37**, 1008–1017.

Coale, K.H. and Bruland, K.W. (1988) Copper complexation in the northeast Pacific. *Limnology and Oceanography* **33**, 1084–1101.

Connell, D.B., Sanders, J.G., Riedel, G.F. and Abbe, G.R. (1991) Pathways of silver uptake and trophic transfer in estuarine organisms. *Environmental Science and Technology* **25**, 921–924.

Cooke, T.D. and Bruland, K.W. (1987) Aquatic chemistry of selenium: evidence of biomethylation. *Environmental Science and Technology* **21**, 1214–1219.

Davis, C.O. (1982) The importance of understanding phytoplankton life strategies in the design of enclosure experiments, in *Marine Mesocosms*, (eds G.D. Grice and M.R. Reeve), Springer-Verlag, New York, pp. 323–332.

Donat, J.R., Lao, K.A. and Bruland, K.W. (1994) Speciation of dissolved copper and nickel in South San Francisco Bay: a multi-method approach. *Analytica Chimica Acta* **284**, 547–571.

Dzombak, D.A., Fish, W. and Morel, F.M.M. (1986) Metal–humate interactions. 1. Discrete ligand and continuous distribution models. *Environmental Science and Technology* **20**, 669–675.

Fisher, N.S. (1986) On the reactivity of metals for marine phytoplankton. *Limnology and Oceanography* **31**, 443–449.

Fisher, N.S. and Fabris, J.G. (1982) Complexation of Cu, Zn, and Cd by metabolites excreted from marine diatoms. *Marine Chemistry* **11**, 245–255.

Fisher, N.S. and Wente, M. (1993) Release of trace elements by dying marine phytoplankton. *Deep-Sea Research* **40**, 671–694.

Fisher, N.S., Bohé, M. and Teyssié, J.-L. (1984) Accumulation and toxicity of Cd, Zn, Ag, and Hg in four marine phytoplankters. *Marine Ecology Progress Series* **18**, 201–213.

Foe, C. and Knight, A.W. (1986) Selenium bioaccumulation, regulation, and toxicity in the green alga, *Selenastrum capricornutum*, and dietary toxicity of contaminated alga to *Daphnia magna*, in *Selenium in the Environment*, (ed. L.J. Slocum), California State University, Fresno, pp. 77–88.

Foster, P.L. and Morel, F.M. (1982) Reversal of cadmium toxicity in a diatom: an interaction between cadmium activity and iron. *Limnology and Oceanography* **27**, 745–751.

Froelich, P.N., Daul, L.W., Byrd, J.T. *et al.* (1985) Arsenic, barium, germanium, tin, dimethylsulfide and nutrient biogeochemistry in Charlotte Harbor, Florida, a phosphorus-enriched estuary. *Estuarine, Coastal and Shelf Science* **20**, 239–264.

Gavis, J., Guillard, R.R.L. and Woodward, B.L. (1981) Cupric ion activity and the growth of phytoplankton clones isolated from different marine environments. *Journal of Marine Research* **39**, 315–333.

Gekeler, W., Grill, E., Winnacker, E.-L. and Zenk, M.H. (1988) Algae sequester heavy metals via synthesis of phytochelatin complexes. *Arch. Microbiol.* **150**, 197–202.

Gordon, A.S., Dyer, B.J., Kango, R.A. and Donat, J.R. (1996) Copper ligands isolated from estuarine water by immobilized metal affinity chromatography: temporal variability and partial characterization. *Marine Chemistry* **53**, 163–172.

Gustavson, K. and Wangberg, S.-A. (1995) Tolerance induction and succession in microalgae communities exposed to copper and atrazine. *Aquatic Toxicology* **32**, 283–302.

Gutknecht, J. (1981) Inorganic mercury (Hg^{2+}) transport through lipid bilayer membranes. *Journal of Membrane Biology* **61**, 61–66.

Harrison, W.G., Eppley, R.W. and Renger, E.H. (1977) Phytoplankton nitrogen metabolism, nitrogen budgets, and observations on copper toxicity: controlled ecosystem pollution experiment. *Bulletin of Marine Science* **27**, 44–57.

Howard, A.G., Arbab-Zavar, M.H. and Apte, S.C. (1982) Seasonal variability of biological arsenic methylation in the estuary of the river Beaulieu. *Marine Chemistry* **11**, 493–498.

Howard, A.G., Arbab-Zavar, M.H. and Apte, S.C. (1984) The behaviour of dissolved arsenic in the estuary of the river Beaulieu. *Estuarine, Coastal and Shelf Science* **19**, 493–504.

Howard, A.G., Apte, S.C., Comber, S.D.W. and Morris, R.J. (1988) Biogeochemical control of the summer distribution and speciation of arsenic in the Tamar estuary. *Estuarine, Coastal and Shelf Science* **27**, 427–444.

Kayser, H. (1982) Cadmium effects in food chain experiments with marine plankton algae (Dinophyta) and benthic filter feeders (Tunicata). *Netherlands Journal of Sea Research* **16**, 444–454.

Kayser, H. and Sperling, K.R. (1980) Cadmium effects and accumulation in cultures of *Prorocentrum micans* (Dinophyta). *Helgolander Meeresuntersuchungen* **33**, 89–102.

Kiffney, P. and Knight, A. (1990) The toxicity and bioaccumulation of selenate, selenite, and seleno-L-*methionine in the cyanobacterium Anabaena flos-aquae*. *Archives of Environmental Contamination and Toxicology* **19**, 488–494.

Knowles, F.C. (1982) The enzyme inhibitory form of inorganic arsenic. *Biochemistry International* **4**, 647–653.

Knowles, F.C. and Benson, A.A. (1983) The biochemistry of arsenic. *Trends in Biochemical Science* **8**, 178–180.

Kumar, H.K. and Prakish, G. (1971) Toxicity of selenium to the blue-green algae *Anacystis nidulans* and *Anabaena variabilis*. *Annals of Botany* **35**, 697–705.

Lee, B.-G. and Fisher, N.S. (1992) Degradation and elemental release rates from phytoplankton debris and their geochemical implications. *Limnology and Oceanography* **37**, 1345–1360.

Lee, B.-G. and Fisher, N.S. (1994) Effects of sinking and zooplankton grazing on release of elements from planktonic debris. *Marine Ecology Progress Series* **110**, 271–281.

Lee, J.G., Ahner, B.A. and Morel, F.M.M. (1996) Export of cadmium and phytochelatin by the marine diatom *Thalassiosira weissflogii*. *Environmental Science and Technology* **30**, 1814–1821.

Li, W.K.W. (1978) Kinetic analysis of interactive effects of cadmium and nitrate on growth of *Thalassiosira fluviatilis* (Bacillariophyceae). *Journal of Phycology* **14**, 454–460.

Margalef, R. (1962) Succession in marine populations. *Advancing Frontiers of Plant Sciences New Delhi*, **2**, 137–188.

McKnight, D.M. and Morel, F.M.M. (1979) Release of weak and strong copper-complexing agents by algae. *Limnology and Oceanography* **24**, 823–837.

Moffett, J.W. and Brand, L.E. (1996) Production of strong, extracellular Cu chelators by marine cyanobacteria in response to Cu stress. *Limnology and Oceanography* **41**, 388–395.

Moffett, J.W., Zika, R.G. and Brand, L.E. (1990) Distribution and potential sources and sinks of copper chelators in the Sargasso Sea. *Deep-Sea Research* **37**, 27–36.

Morel, F.M.M., Hudson, R.J.M. and Price, N.M. (1991) Limitation of productivity by trace metals in the sea. *Limnology and Oceanography* **36**, 1742–1755.

Morel, N.M., Rueter, J.E. and Morel, F.M.M. (1978) Copper toxicity to *Skeletonema costatum*. *Journal of Phycology* **14**, 43–48.

Morrelli, E. and Scarano, G. (1995) Cadmium induced phytochelatins in marine alga *Phaeodactylum tricornutum*: effect of metal speciation. *Chemical Speciation and Bioavailability* **7**, 43–47.

Oliviera, R. (1985) Phytoplankton communities response to a mine effluent rich in copper. *Hydrobiologia* **128**, 61–69.

Peoples, S.A. (1975) Review of arsenical pesticides, in *Arsenical Pesticides*, (ed. E.A. Woolson), American Chemical Society, Washington, DC, pp. 1–12.

Phinney, J.T. and Bruland, K.W. (1994) Uptake of lipophilic organic Cu, Cd, and Pb complexes in the coastal diatom *Thalassiosira weissflogii*. *Environmental Science and Technology* **28**, 1781–1817.

Planas, D. and Healey, F.P. (1978) Effects of arsenate on growth and phosphorus metabolism of phytoplankton. *Journal of Phycology* **14**, 337–341.

Reinfelder, J.R. and Fisher, N.S. (1991) The assimilation of elements ingested by marine copepods. *Science* **251**, 794–796.

Reinfelder, J.R. and Fisher, N.S. (1994) The assimilation of elements ingested by marine planktonic larvae. *Limnology and Oceanography* **39**, 12–20.

Riedel, G.F. (1984) Influence of salinity and sulfate on the toxicity of chromium (VI) to the estuarine diatom *Thalassiosira pseudonana*. *Journal of Phycology* **20**, 496–500.

Riedel, G.F. (1985) The relationship between chromium(VI) uptake, sulfate uptake, and chromium(VI) toxicity in the estuarine diatom, *Thalassiosira pseudonana*. *Aquatic Toxicology* **7**, 191–204.

Riedel, G.F. (1988) Interspecific and geographic variation of the chromium sensitivity of algae, in *Proceedings, 11th Symposium on Aquatic Toxicology and Hazard Assessment,* (eds G.W. Suter and M. Lewis), American Society of Testing and Materials, Philadelphia, pp. 537–548.

Riedel, G.F. and Sanders, J.G. (1996) The influence of pH and media composition on the uptake of inorganic selenium by *Chlamydomonas reinhardtii*. *Environmental Toxicology and Chemistry* **15**, 1577–1583.

Riedel, G.F., Sanders, J.G. and Gilmour, C.C. (1996) Uptake, transformation, and impact of selenium in freshwater phytoplankton and bacterioplankton communities. *Aquatic Microbiology and Ecology* **11**, 43–51.

Sanders, J.G. (1979) Effects of arsenic speciation and phosphate concentration on arsenic inhibition of *Skeletonema costatum* (Bacillariophyceae). *Journal of Phycology* **15**, 424–428.

Sanders, J.G. (1980) Arsenic cycling in marine systems. *Marine Environmental Research* **3**, 257–266.

Sanders, J.G. (1985) Arsenic geochemistry in Chesapeake Bay: dependence upon anthropogenic inputs and phytoplankton species composition. *Marine Chemistry* **17**, 329–340.

Sanders, J.G. and Abbe, G.R. (1987) The role of suspended sediments and phytoplankton in the partitioning and transport of silver in estuaries. *Continental Shelf Research* **7**, 1357–1361.

Sanders, J.G. and Abbe, G.R. (1989) Silver transport and impact in estuarine and marine systems, in *Aquatic Toxicology and Environmental Fate*, Vol. 11, (eds G. Suter and M. Lewis), American Society for Testing and Materials STP 1007, Philadelphia, pp. 5–18.

Sanders, J.G. and Cibik, S.J. (1985a) Adaptive behavior of euryhaline phytoplankton communities to arsenic stress. *Marine Ecology Progress Series* **22**, 199–205.

Sanders, J.G. and Cibik, S.J. (1985b) Reduction of growth rate and resting spore formation in a marine diatom exposed to low levels of cadmium. *Marine Environmental Research* **16**, 165–180.

Sanders, J.G. and Riedel, G.F. (1987) Control of trace element toxicity by phytoplankton, in *Recent Advances in Phytochemistry*, Vol. 21, (eds J.A. Saunders, L. Kosak-Channing and E.E. Conn), Plenum Press, New York, pp. 131–149.

Sanders, J.G. and Riedel, G.F. (1993) Trace element transformation during the development of an estuarine algal bloom. *Estuaries* **16**, 521–532.

Sanders, J.G. and Vermersch, P.S. (1982) Response of marine phytoplankton to low levels of arsenate. *Journal of Plankton Research* **4**, 881–893.

Sanders, J.G. and Windom, H.L. (1980) The uptake and reduction of arsenic species by marine algae. *Estuarine and Coastal Marine Science* **10**, 555–567.

Sanders, J.G., Ryther, J.H. and Batchelder, J.H. (1981) Effects of copper, chlorine, and thermal addition on the species composition of marine phytoplankton. *Journal of Experimental Marine Biology and Ecology* **49**, 81–102.

Sanders, J.G., Osman, R.W. and Riedel, G.F. (1989) Pathways of arsenic uptake and incorporation in estuarine phytoplankton and filter-feeding invertebrates, *Eurytemora affinis*, *Balanus improvisus,* and *Crassostrea virginica. Marine Biology* **103**, 319–325.

Sanders, J.G., Riedel, G.F. and Ferrier, D.P. (1991) Changes in community structure of Chesapeake Bay phytoplankton when exposed to low levels of trace metals: management implications, in *New Perspectives in the Chesapeake System: A Research and Management Partnership*, (eds J.A. Mihursky and A. Chaney), Proceedings of a Conference 4–6 December 1990, Baltimore, MD, Chesapeake Research Consortium Publication No. 137, pp. 451–460.

Sanders, J.G., Riedel, G.F. and Osman, R.W. (1994). Arsenic cycling and impact in estuarine and coastal marine ecosystems, in *Arsenic in the Environment, Part I: Cycling and Characterization*, (ed. J.O. Nriagu), John Wiley and Sons, New York, pp. 289–308.

Schindler, D.W. (1987) Detecting ecosystem response to anthropogenic stress. *Canadian Journal of Fisheries and Aquatic Sciences* **44** (Suppl. 1), 6–25.

Sharp, J.H. (1977) Excretion of organic matter by marine phytoplankton: do healthy cells do it? *Limnology and Oceanography* **22**, 381–399.

Sick, L.V. and Windom, H.L.(1975) Effects of environmental levels of mercury and cadmium on rates of metal uptake and growth physiology of selected genera of marine phytoplankton, in *Mineral Cycling in Southeastern Ecosystems*, (eds F.G. Howell, J.B. Gentry and M.H. Smith), CONF-740513, NTIS, Springfield, VA, pp. 239–249.

Sunda, W.G. (1988/89) Trace metal interactions with marine phytoplankton. *Biological Oceanography* **6**, 411–442.

Sunda, W.G. (1994) Trace metal/phytoplankton interactions in the sea, in *Chemistry of Aquatic Systems: Local and Global Perspectives*, (eds G. Bidoglio and W. Stumm),ECSC, EEC, EAEC, Brussels and Luxembourg, pp. 213–247.

Sunda, W.G. and Guillard, R.R.L. (1976) The relationship between cupric activity and the toxicity of copper to phytoplankton. *Journal of Marine Research* **34**, 511–529.

Thomas, W.H., Holm-Hansen, O., Seibert, D.L.R. *et al.* (1977) Effects of copper on phytoplankton standing crop and productivity: controlled ecosystem pollution experiment. *Bulletin of Marine Science* **27**, 34–43.

Trick, C.G., Andersen, R.J., Price, N.M. *et al.* (1983) Examination of hydroxamate-siderophore production by neritic eukaryotic marine phytoplankton. *Marine Biology* **75**, 9–17.

Wangersky, P.J., Bradley Moran, S., Pett, R.J. *et al.* (1989) Biological control of trace metal residence times: An experimental approach. *Marine Chemistry* **28**, 215–226.

Wheeler, A.E., Zingaro, R.A. and Irgolic, K. (1982) The effect of selenate, selenite and sulfate on six species of unicellular algae. *Journal of Experimental Marine Biology and Ecology* **57**, 181–194.

Wilhelm, S.W. and Trick, C.G. (1994) Iron-limited growth of cyanobacteria: multiple siderophore production is a common response. *Limnology and Oceanography* **39**, 1979–1984.

Wilhelm, S.W., Maxwell, D.P. and Trick, C.G. (1996) Growth, iron requirements, and siderophore production in iron-limited *Synechococcus* PCC 7002. *Limnology and Oceanography* **41**, 89–97.

Winner, R.W. and Owen, H.A. (1991) Seasonal variability in the sensitivity of freshwater phytoplankton communities to a chronic copper stress. *Aquatic Toxicology* **19**, 73–88.

Wrench, J.J. (1978) Selenium metabolism in the marine phytoplankton *Tetraselmis tetrathele* and *Dunaliella minuta*. *Marine Biology* **49**, 231–236.

Zhou, X. and Wangersky, P.J. (1989) Production of copper-complexing organic ligands by the marine diatom *Phaeodactylum tricornutum* in a cage culture turbidostat. *Marine Chemistry* **26**, 239–259.

Zhou, X., Slauenwhite, D.E., Pett, R.J. and Wangersky, P.J. (1989) Production of copper-complexing organic ligands during a diatom bloom: Tower tank and batch culture experiments. *Marine Chemistry* **27**, 19–30.

5   *Mercury in aquatic ecosystems*

TOGWELL A. JACKSON

And loathsome canker lives in sweetest bud.
Shakespeare

5.1 INTRODUCTION

Mercury (Hg) is one of the most toxic heavy metals. From a biological perspective it has no redeeming virtue, for, unlike a number of other heavy metals, it is not known to perform any essential biochemical function (Bowen, 1966). Traces of Hg are ubiquitous in soils, natural waters, sediments, organisms and air (Jonasson and Boyle, 1972), and anomalously high Hg concentrations occur in many ecosystems owing to Hg pollution (a serious, widespread problem), natural Hg enrichment in certain rocks, distinctive properties of Hg such as its tendency to form highly stable complexes and compounds (including species that are easily taken up by organisms but not readily excreted), natural processes (e.g. methylation) which enhance the bioavailability of Hg, increased bioavailability of Hg due to environmental changes caused by human activities, and efficient accumulation of Hg by organisms and certain natural materials, such as soil organic matter and fine-grained sediments. Moreover, Hg is a relatively volatile element, and this accounts, in large part, for its wide distribution.

Of crucial importance in aquatic ecosystems is microbial conversion of inorganic Hg(II) into the readily bioaccumulated toxic organometallic species methyl Hg (CH_3Hg^+). Unlike inorganic Hg(II), CH_3Hg^+ undergoes biomagnification up the aquatic food chain; hence, most of the Hg in fish consists of CH_3Hg^+, though inorganic Hg predominates in associated water and sediments. Elevated CH_3Hg^+ concentrations in fish are found not only in Hg-contaminated habitats but also in virtually unpolluted ones where conditions favouring CH_3Hg^+ production occur, even if environmental Hg levels are low.

Metal Metabolism in Aquatic Environments. Edited by William J. Langston and Maria João Bebianno. Published in 1998 by Chapman & Hall, London. ISBN 0 412 80370 4

CH_3Hg^+ is a health hazard to people who eat contaminated fish, as well as an economic problem for those who fish for a living; and both CH_3Hg^+ and inorganic Hg may poison aquatic organisms. Consumption of fish and shellfish with high CH_3Hg^+ levels has led to outbreaks of Minamata disease, a syndrome of acute or chronic toxic effects, notably irreversible nerve and brain damage leading, in extreme cases, to severe physical and mental disability or death (Takizawa, 1979). Hg has aroused intense and sustained concern because of its toxicity and widespread occurrence as a pollutant.

The purpose of this chapter is to review the biogeochemistry and biological effects of Hg in aquatic ecosystems. But to understand these phenomena, we must know the basic properties and characteristic reactions of the element. Therefore, we begin with a survey of the chemistry of Hg. For brevity, elements and compounds will be represented by their chemical symbols.

5.2 THE CHEMISTRY OF Hg

5.2.1 PROPERTIES AND REACTIONS

Hg, Cd and Zn comprise a family of non-transition heavy metals classed as the Group IIB elements. Heavy metals are defined as metallic elements of specific gravity > 5 g cm^{-3} (Martin and Coughtrey, 1982) and are distinguished by a tendency to form stable complexes (e.g. chelates, sulphides and hydroxylated species) with partly covalent metal–ligand bonds (Douglas and McDaniel, 1965). But Hg differs strikingly from other heavy metals, including Cd and Zn, in a number of respects.

Within Group IIB, Hg has the highest atomic number and weight, and its divalent cation has the largest ionic radius (Cotton and Wilkinson, 1988). The electron configurations of Hg are $[Xe]4f^{14}5d^{10}6s^2$, the two outermost s electrons being the valence electrons. In nature, Hg can exist in any of three interconvertible oxidation states: Hg(0), Hg(I) and Hg(II) (Jonasson and Boyle, 1972). Hg(0) (elemental Hg) has the unique distinction of being a volatile liquid at most earth-surface temperatures (Lide, 1992). It is rather unreactive, and oxidation to HgO by O_2, though thermodynamically favoured, is very slow (Cotton and Wilkinson, 1988); besides, HgO decomposes to Hg(0) + O_2 on exposure to light (Budavari $et\ al.$, 1989). Therefore, Hg(0) can persist for significant lengths of time in contact with air. Hg(I), which consists of the diatomic cation Hg_2^{2+}, is unstable in most natural environments; it forms no stable aqueous complexes and disproportionates spontaneously to Hg(0) + complexed Hg(II) in the presence of ligands that bind Hg(II) (Jonasson and Boyle, 1972; Beneš and Havlík, 1979; Cotton and Wilkinson, 1988; Douglas and McDaniel, 1965). Thus, the main oxidation states of Hg in nature are Hg(0) and Hg(II). Hg(II), or Hg^{2+}, is the principal form of Hg in aquatic environments. It has a strong tendency to form extremely stable coordination complexes and organometallic compounds whose bonds are mainly covalent, the most stable bonds being those linking the Hg directly to S, N, P or C. Being rather inert, Hg species do not react readily with O_2 and water (Cotton and Wilkinson, 1988).

The behaviour of Hg is attributable to the electronic configuration and nuclear charge of the Hg atom or Hg^{2+} ion. The valence electrons of a Group IIB metal comprise two s electrons situated just outside a filled d subshell. Zn has one filled d subshell and Cd has two, but Hg has three of them along with a filled f subshell. As the shielding effect of the inner electrons decreases in the order $s > p > d > f$, the nuclear charge is poorly shielded by the d and f electrons (Sisler, 1963). Therefore, the primary valence electrons and pairs of secondary valence electrons donated by ligands are strongly attracted to the nucleus. Indeed, the primary valence electrons penetrate the inner electron shells, spending part of their time near the nucleus, for such penetration is most readily accomplished by s electrons (Sisler, 1963; Douglas and McDaniel, 1965). Because of these effects, the metals form stable, partly covalent bonds with ligands. This tendency is most pronounced in Hg, because Hg, besides having the most d orbitals, possesses an f orbital and has the highest nuclear charge, exerting the strongest pull on the valence electrons (Douglas and McDaniel, 1965). Thus, Hg has a higher electronegativity and higher stability constant of metal cation hydrolysis than Cd or Zn, has higher first and second ionization potentials, is much less reactive, and displays the 'inert pair' effect (Sillén and Martell, 1964, 1971; Douglas and McDaniel, 1965; Cotton and Wilkinson, 1988; Lide, 1992). Furthermore, Hg^{2+} has a high degree of polarizability (Phillips and Williams, 1965), meaning that its d and f electron subshells are readily deformed by an applied electric field (Douglas and McDaniel, 1965; Cotton and Wilkinson, 1988). Being far from the nucleus and unable to get closer by penetration of inner subshells, d and f electrons are not as securely anchored as s and p ones and are more easily pushed aside by the repulsive force of electrons in an approaching ligand. Polarization of Hg^{2+} by a ligand decreases the shielding of the cation's nuclear charge and shortens the Hg–ligand distance, enhancing the stability and covalent character of the bond.

5.2.2 SPECIATION AND COMPLEXING OF INORGANIC HG(II)

Owing to its marked ability to attract and retain electrons donated by ligands, Hg(II) forms a variety of very stable complexes whose bonds have a high degree of covalent character, and the formation constants of Hg(II) complexes exceed those of Zn(II) and Cd(II) complexes by orders of magnitude (Cotton and Wilkinson, 1988). Among Hg(II) complexes the most usual coordination number is 2, and the molecular structure is linear, as in $HgCl_2$. The bonds in linear Hg complexes have the highest degree of covalent character, and the dipole moment of such a complex is 0 if the two ligands are identical. The second most common coordination number is 4, as in $HgCl_4^{2-}$; more rarely the coordination number is 6, as in $Hg(H_2O)_6^{2+}$, or 3 or 5.

An important consequence of the polarizability of Hg(II) (the 'softness' of the Lewis acid Hg^{2+}) is that Hg(II) has a much stronger affinity for large, highly polarizable ligands ('soft' Lewis bases), such as sulphide species, than for smaller, less polarizable ligands ('hard' Lewis bases) such as O-bearing ones (e.g. -COOH) (Douglas and McDaniel, 1965; Williams, 1971; Schuster, 1991). The preferential binding of Hg(II) by soft Lewis bases is due to stabilization of the resulting complexes by mutual polarization of the cation and its ligands (Douglas and McDaniel, 1965). In theory, the relative affinities of Hg(II) for electron-donating elements within different groups are as follows: Group VIb, O $\ll$ S $<$ Se $\approx$ Te; Group VIIB, F $\ll$ Cl $<$ Br $<$ I (Williams, 1971). The Hg(II)-binding capacities of aliphatic complexing agents of similar size decrease in the order R-SH $\gg$ $R-NH_2$ $>$ R-COOH (Reimers and Krenkel, 1974; Reimers *et al.*, 1975), demonstrating that Hg(II) forms more stable bonds with -SH groups than with the harder N- and O-bearing groups (Williams, 1971). In anoxic natural environments, Hg(II) is preferentially bound by sulphides (Gavis and Ferguson, 1972; Jackson, 1988b).

Dissolved inorganic Hg(II) readily combines with inorganic sulphide (H_2S, HS^-, and S^{2-}) to precipitate black HgS, which is kinetically favoured over the thermodynamically more stable red HgS (cinnabar) (Cotton and Wilkinson, 1988). Black HgS has a very low solubility product (10^{-54}), but some Hg(II) and sulphide remain in solution as hydrolysis products (Cotton and Wilkinson, 1988). Moreover, Hg(II) forms soluble sulphide and mixed-ligand complexes, including HgS_2^{2-}, $Hg(SH)_2$, $Hg(SH)S^-$, $HgSH^+$, $Hg(S_6)_2^{2-}$, HgOHSH and HgClSH as well as solid HgS, the proportions of the products depending on the pH and the concentrations of the reactants (Gavis and Ferguson, 1972; Cotton and Wilkinson, 1988; Dyrssen and Wedborg, 1991; Hudson *et al.*, 1994). If Fe(II) is vastly more abundant than Hg(II), as is usual in nature, the Hg(II) coprecipitates with FeS. Hg(II) also forms highly stable bonds with -SH (thiol) groups of organic sulphides, including proteins, cysteine, and 'thiols' (mercaptans) formed by putrefaction of proteins. Aliphatic thiols bind Hg more effectively than inorganic sulphide does (Douglas and McDaniel, 1965; Dyrssen and Wedborg, 1991), as the alkyl chain releases electrons to the S atom, strengthening the Hg-S bond (Roberts and Caserio, 1965). The stability of sulphide complexes depends on the dissolved O_2 level, because oxidation of the ligand releases the bound Hg (Jernelöv, 1972).

Hg(II) has a greater affinity for Se and Te analogues of sulphides than for sulphides (Rimerman *et al.*, 1977; Dyrssen and Wedborg, 1991), because they are more polarizable (Williams, 1971). The solubility products of Hg sulphide, selenide and telluride decrease in the order HgS $>$ HgSe $>$ HgTe (Björnberg *et al.*, 1988). Hg and Se are closely associated in lake sediments, even at very low or high pH, implying the existence of extremely stable Hg-Se bonds (Jackson, 1991a; Jackson *et al.*, 1980). In nature Se is orders of magnitude less abundant than S; yet both selenides and sulphides play significant parts in the biogeochemistry and toxicology of Hg.

Hg(II) is also complexed by nitrogenous ligands (e.g. $R\text{-}NH_2$) and O-bearing ligands (e.g. $R\text{-}COO^-$, OH^-, and H_2O), including ligands of amino acids and nucleic acids (Douglas and McDaniel, 1965; Gavis and Ferguson, 1972; Ochiai, 1977; Cotton and Wilkinson, 1988). But such ligands, being hard, or intermediate between hard and soft, form less stable bonds with Hg(II) than sulphides do (Williams, 1971; Reimers and Krenkel, 1974).

Organic matter – notably humic substances, which are among the most common, abundant and effective complexing agents in soil, water and sediments (Aiken *et al.*, 1985) – forms very stable complexes with Hg(II) (Trost and Bisque, 1972; Lindberg and Harriss, 1974; Andren and Harriss, 1975; Ramamoorthy and Kushner, 1975b; Reimers *et al.*, 1975; Khalid *et al.*, 1977; Jackson, K.S. *et al.*, 1978; Andersson, 1979; Jackson *et al.*, 1980, 1982; Kerndorff and Schnitzer, 1980; Randle and Hartmann, 1987; Semu *et al.*, 1987; Jackson, 1989; Xu and Allard, 1991; Stordal *et al.*, 1996). It has been shown repeatedly that the complexing capacity of humic matter is far greater for Hg than for other heavy metals such as Cd, Zn, Cu and Pb, apparently because it forms the most highly covalent bonds with ligands (Ramamoorthy and Kushner, 1975b; Kerndorff and Schnitzer, 1980; Jackson *et al.*, 1980). The majority of ligands in humic matter are O-bearing ones (-COOH, phenolic, and enolic(?), -OH, and >C=O), but nitrogenous ligands (e.g. $-NH_2$, $-NH-$, and $-N=$), along with some S-bearing groups (-SH?), are present as well (Aiken *et al.*, 1985). The S- and N-bearing groups (especially -SH), which probably belong to remnants of biochemical compounds (e.g. proteins), may well bind Hg(II) preferentially even though they make up a minority of the ligands (Bidstrup, 1964; Ochiai, 1977; Ramamoorthy and Kushner, 1975a).

The complexing capacity of organic (e.g. humic) matter for metal cations is minimal under extremely acidic conditions but rises to a maximum with increasing pH within a certain pH range that is characteristic of the metal (Kerndorff and Schnitzer, 1980). Hg(II) binding, while conforming to this rule, is anomalous by virtue of its relative insensitivity to pH, remaining close to peak efficiency throughout the pH range of natural waters (85–96% even at pH values as low as $\sim$1.0–2.5), increasing with pH and reaching a maximum of $\sim$94–100% within the pH range $\sim$4–7 (Andersson, 1979; Kerndorff and Schnitzer, 1980; Zvonarev and Zyrin, 1982). Similarly, Hg scavenged by organic (largely humic) lake mud was found to be unextractable with HCl up to a strength of 1 M (Jackson *et al.*, 1980). More efficient binding of Hg at higher pH reflects elimination of competing H^+ ions and creation of negatively charged binding sites by dissociation of acidic ligands (Tipping and Hurley, 1992). The same principle applies to acidic ligands in general, including -SH, H_2S, and the -COOH groups of all organic acids that function as chelators. Neutralization of positively charged, cation-repelling groups like $-NH_3^+$ and $-NH_2^+-$ by removal of protons could be a contributing factor. Metal cation hydrolysis, too, may assist the complexing process, as suggested by a correlation between metal–humic acid affinities under weakly acidic conditions (Kerndorff and Schnitzer, 1980) and the stability constants for metal cation hydrolysis (Sillén and Martell, 1964, 1971). However, as the pH rises into the alkaline range, the efficiency of Hg(II) sorption declines somewhat owing to solubilization of intact Hg–humic complexes – not, as might be supposed, because of the formation of $Hg(OH)_3^-$ anions and negatively charged deprotonated ligands, such as -COO$^-$ and -O$^-$, which repel the anions (Andersson, 1979; Jackson *et al.*, 1980, 1982; Kerndorff and Schnitzer, 1980; Zvonarev and Zyrin, 1982). In brief, Hg–humic complexes have a high degree of stability which is hardly affected by pH – even extremely acidic or alkaline pH values.

Hg(II) shows a preference for larger organic molecules (e.g. humic acids) in some environments but for smaller ones (e.g. fulvic acids) in others (Lindberg and Harriss, 1974; Andren and Harriss, 1975; Ramamoorthy and Kushner, 1975b; Jackson *et al.*, 1980, 1982; Louchouarn *et al.*, 1993), and the pattern of partitioning between large and small size fractions may vary seasonally too (Jackson *et al.*, 1982). The geographical and temporal variations could reflect differences in source material, environment and microbial activity (Semu *et al.*, 1987; Schuster, 1991). Preferential scavenging of Hg by large humic molecules in lake mud relative to small ones has been linked to the high electronegativity of Hg, implying that metal–ligand bonds formed by the larger molecules are more stable because they have a higher degree of covalent character (Jackson *et al.*, 1980).

Under reducing conditions organic complexing agents compete with inorganic sulphide for Hg(II) and may solubilize part of the Hg, even in the presence of H_2S (Lindberg and Harriss, 1974; Timperley and Allan, 1974; Khalid *et al.*, 1977; Hallberg, 1978; Jackson, K.S. *et al.*, 1978; Jackson, 1978, 1979); on the other hand, sulphide may strip Hg(II) from non-thiol organic ligands. In anoxic, H_2S-generating freshwater mud, partitioning of Hg and other metals between sulphide and organic matter may be controlled by the stability of the metal sulphide as represented by either its standard entropy or its standard enthalpy of formation, depending on the prevailing conditions (Jackson, 1978, 1979). However, it is safe to assume that Hg associated with organic matter is bound preferentially by thiol groups and that Hg(II) always has a stronger affinity for both inorganic sulphide and thiols than for non-thiol groups of organic complexing agents (Reimers and Krenkel, 1974; Reimers *et al.*, 1975; Jackson, K.S. *et al.*, 1978).

In natural waters, any dissolved inorganic Hg(II) not bound to organic or sulphide ligands probably consists mainly of hydroxyl and chloride complexes. In the absence of ligands other than H_2O and OH$^-$, aqueous Hg(II), owing to its strong tendency to be hydrolysed, is mostly in the form of hydroxylated species except under extremely acidic conditions (Hahne and Kroontje, 1973; Leckie and James, 1974; MacNaughton and James, 1974; Farrah and Pickering, 1978; Schuster, 1991; Xu and Allard, 1991). The formation of Hg–OH complexes is a function of pH. In theory, the dominant Hg(II) species at pH < ~2 or 3 is the 'free' Hg^{2+} cation, $Hg(H_2O)_6^{2+}$. But the proportion of the Hg in the form of the OH^- complexes $HgOH^+$ and $Hg(OH)_2$ increases sharply to ~100% as the pH rises from ~1 to 4, $HgOH^+$ peaking at pH ~3, while $Hg(OH)_2$ predominates over other species throughout the pH range ~3–14, which brackets the range of most natural waters. As the pH rises through the alkaline range, the complex anion $Hg(OH)_3^-$ becomes increasingly significant, but it never exceeds $Hg(OH)_2$.

Hg(II) forms stable complexes with Cl^-, Br^- and I^- as well as sulphides, organic ligands and OH^-. Hg halide stability in solution increases in the order of increasing polarizability of the ligand, i.e. $Cl^- < Br^- < I^-$ (Douglas and McDaniel, 1965; Williams, 1971). However, in nature most Hg complexes of this class are probably chlorides (Gavis and Ferguson, 1972), as Cl^- is far more abundant than Br^- or I^- and forms the most water-soluble Hg complexes, halide solubility decreasing in the order $HgCl_2 \gg HgBr_2 \gg HgI_2$ (Cotton and Wilkinson, 1988; Budavari *et al.*, 1989; Lide, 1992). Therefore, competition between Cl^- and OH^- ions plays a significant part in the aqueous chemistry of Hg(II), the proportions of the various possible complexes at equilibrium being dependent on the Cl^- and Hg(II) levels and the pH. Assuming values of these variables that might be expected in freshwater environments, and leaving organic and sulphide ligands out of consideration, $HgCl_2$ (with smaller amounts of Hg^{2+}, $HgCl^+$, $HgCl_3^-$, $HgCl_4^{2-}$, and HgOHCl) predominates under acidic to neutral or mildly alkaline conditions, but $HgOH_2$ (with smaller amounts of $Hg(OH)_3^-$ and HgOHCl) is the most abundant species over all or most of the alkaline pH range (Hahne and Kroontje, 1973; Leckie and James, 1974; MacNaughton and James, 1974; Farrah and Pickering, 1978; Schuster, 1991; Hudson *et al.*, 1994). The critical pH separating the field of Hg chloride predominance from that of Hg hydroxide predominance shifts to higher or lower values as the Cl^- level increases or decreases, respectively. The hybrid species HgOHCl peaks at this pH, approaching $HgCl_2$ and $Hg(OH)_2$ in abundance. In short, $HgCl_2$ and $Hg(OH)_2$ are probably the main inorganic Hg species in oxygenated, sulphide-free fresh waters. But $HgCl_3^-$ and $HgCl_4^{2-}$ become increasingly important with rising Cl^- level, attaining predominance in seawater (Andren and Harriss, 1975).

Inorganic Hg(II) ions (e.g. Hg^{2+}, $HgOH^+$, $HgCl_4^{2-}$ and HgS_2^{2-}) are hydrophilic and water soluble; but uncharged linear complexes (e.g. $HgCl_2$, $Hg(OH)_2$, HgOHCl, $Hg(SH)_2$, HgOHSH and HgClSH) are more or less lipophilic owing to their lack of an electric charge and the largely covalent character of their bonds (which limits bond polarity and impedes dissociation in water). In addition, any linear complex whose two ligands are identical (e.g. $HgCl_2$, $Hg(OH)_2$ and $Hg(SH)_2$) has a dipole moment of 0, increasing its lipophilic character (though the individual bonds have polarity). For the same reasons, and also because of its low molecular weight, $HgCl_2$ is slightly volatile at ordinary temperatures (Budavari *et al.*, 1989). Hg salts in general have appreciable volatility (Matheson, 1979).

On the other hand, uncharged Hg(II) species have some degree of hydrophilic as well as lipophilic character and hence are somewhat soluble in water (Budavari *et al.*, 1989). The relative lipophilicity and hydrophilicity of an uncharged species depend on the molecule's dipole moment (if any) and the polarity of its individual bonds, and are critically dependent on the nature of the anions bound to the Hg(II) (Mason *et al.*, 1995b). Thus, $HgCl_2$ is much more lipophilic than $Hg(OH)_2$, its octanol–water partition coefficient (K_{OW}) being two orders of magnitude greater (even though neither species has a dipole moment) (Mason *et al.*, 1995b). This probably indicates that $Hg(OH)_2$, unlike $HgCl_2$, forms H-bonds with water.

In the complexing of Hg(II), Cl^- ions compete with other ligands (even sulphides) and tend to release Hg(II) from them, forming dissolved $Hg–Cl^-$ complexes (Carty and Malone, 1979; Schuster, 1991). Yet different studies have led to quite different conclusions about the effectiveness of Cl^- in releasing Hg from sulphide and organic matter or interfering with the binding of Hg by these substances (Reimers and Krenkel, 1974; Jackson *et al.*, 1980; Randle and Hartmann, 1987; Jackson, 1989; Schuster, 1991), suggesting that reactions between Hg(II) and competing ligands are controlled by an assortment of factors, including environmental variables and the properties and concentrations of the reactants.

Photochemical reactions also play important roles in Hg speciation. Many Hg(II) species are reduced to Hg(I) or Hg(0) by ultraviolet (UV) or visible light, as in Eder's reaction:

$$HgCl_2 \xrightarrow{\text{oxalate, hv}} Hg_2Cl_2$$

(Balzani and Carassiti, 1970; Baughman *et al.*, 1973), owing, perhaps, to the high polarizability of Hg(II). Amyot *et al.* (1994) showed that sunlight can induce production of 'dissolved gaseous mercury' in lake water, suggesting reduction of Hg(II) to Hg(0) (a plausible inference, though the forms of Hg were not identified). Reactions requiring UV light of wavelength (λ) < 290 nm must be negligible at the Earth's surface but could be significant for volatile Hg(II) species which, on evaporating, migrate into the stratosphere (Kondratyev, 1969). Photochemical reactions of Hg may be promoted by photosensitizing agents (e.g. $FeCl_3$, MnO_4^-, Mn^{2+}, Mn and Co oxalates, and fluorescent dyes in Eder's reaction) (Balzani and Carassiti, 1970). These agents absorb light, making its energy available for the reactions, and may enable the reactants to use light of frequencies which could not otherwise be employed: Eder's reaction normally requires UV light of λ < 350 nm but can be driven by visible (blue) light if photoreactive impurities such as Fe oxalate are present. There are also many substances that inhibit photochemical reactions; thus, O_2, O_3, H_2O_2, H^+, CrO_4^{2-} and phenol tend to block Eder's reaction. In nature, Hg photochemistry could be very complex, with promotion and inhibition of different reactions occurring simultaneously. Humic matter, Fe and

Mn oxides, and mineral–humic complexes can promote or suppress a variety of photochemical reactions (Baxter and Carey, 1982, 1983; Choudhry, 1984; Langford and Carey, 1987; Zepp, 1988) and may participate in photochemical reactions of Hg (Allard and Arsenie, 1991).

Inorganic Hg is subject to reduction or oxidation by purely chemical or biochemical mechanisms as well as photochemical ones, or by a combination of these processes. Many abiotic reactions occur in the atmosphere; these include oxidation of Hg(0) to Hg(II) by O_3 and reduction of Hg(II) to Hg(I) by aqueous SO_3^{2-}, followed by release of Hg(0) owing to breakdown of the Hg(I) (Lindqvist *et al.*, 1991; Munthe *et al.*, 1991; Schroeder *et al.*, 1991). In water, humic matter may reduce Hg(II) to Hg(0), possibly through donation of electrons by free radicals (Alberts *et al.*, 1974), and the reaction is promoted by light (Allard and Arsenie, 1991; Xiao *et al.*, 1995). H_2O_2 formed photochemically through the agency of humic matter (Baxter and Carey, 1982, 1983), may oxidize Hg(0) to Hg(II) or reduce Hg(II) to Hg(0), depending on the ambient pH (Schroeder *et al.*, 1991). Finally, Hg(II) is converted to Hg(0) by bacteria (including cyanobacteria) and planktonic algae in lakes and in the ocean (Schottel *et al.*, 1974; De Filippis and Pallaghy, 1975; Summers, 1988; Baldi *et al.*, 1989, 1993a; Nakamura *et al.*, 1990; Barkay *et al.*, 1991; Regnell and Tunlid, 1991; Radosevich and Klein, 1993; Mason *et al.*, 1995a). Microbial reduction of Hg(II) to Hg(0) is catalysed by mercuric reductase (Summers, 1988; Baldi *et al.*, 1989; 1991).

5.2.3 SORPTION AND DESORPTION OF INORGANIC HG(II)

Hg(II) is rapidly and efficiently removed from solution through sorption by fine suspended matter and sediments (Jackson, 1998) and is not readily leached out by mild extractants such as solutions of $CaCl_2$ and the chelator DTPA (Hogg *et al.*, 1978; Jackson, 1984, 1988b, 1998; Jackson and Woychuk, 1980a,b; Jackson *et al.*, 1982). As a rule, the finest particles (colloidal – clay-sized – ones, which are < 2 μm in diameter) have the highest sorption capacities owing to their large specific surfaces (Jackson, 1998), though exceptions occur, as with coarse organic debris (Ramamoorthy and Rust, 1976). The most important sorbents in nature are the following:

- clay minerals;
- 'amorphous' (short-range order) oxides, hydroxides and oxyhydroxides ('hydrated' or 'hydrous' oxides) of Fe, Mn and Al (e.g. FeOOH and MnOOH), all of which will, in this review, be termed 'oxides' for brevity;
- amorphous FeS, which occurs only under reducing conditions;
- particulate humic substances;
- non-humic organic matter, including plankton, biofilms, extracellular biogenic colloids, faecal pellets and moulted exoskeletons of planktonic crustaceans;

- composite particles, such as clay and plankton hard parts with coatings of oxide or humic matter, or both (Jackson, 1978; Jackson *et al.*, 1980; Jackson and Bistricki, 1995).

The scavenging of Hg and other metals by FeS and oxides commonly occurs by coprecipitation, whereby the metals are sorbed by, and simultaneously enclosed in, a growing mass of precipitate (Inoue and Munemori, 1979). Environmental processes (e.g. a drop in pH or solubilization of metals by complexing agents) may result in release (desorption) of sorbed metals; but if the metals are sealed inside a precipitate owing to coprecipitation, they are released only if the precipitate is dissolved or decomposed, as in the reduction and dissolution of $MnOOH$ and $FeOOH$ or the oxidation of FeS. The binding and release of metals by sorbents are controlled by many variables (Jackson, 1998), notably metal properties, characteristics of the sorbent, environmental conditions, competition between different metals for sorption sites, and the concentrations of the reactants. A further complication is that sorption energy varies from one sorption site to another, even within a single particle, and does not bear a consistent relationship with sorption capacity (Jackson, 1998). (Sorption energy doubtless affects the sorption capacity but is not the only factor that does so.)

The mechanisms of metal uptake by particulate matter may be summarized as follows:

1. Rather weak, readily reversible sorption of metal cations occurs at sites of permanent negative charge (cation exchange sites) on the 001 (basal cleavage) faces of clay crystals, where the cations, their hydration spheres intact, are held by ionic bonding reinforced by H-bonding but have no direct contact with the atoms of the mineral surface. But these exchange sites make little contribution to the sorption of heavy metals, as heavy metal cations are easily displaced from them by the far more numerous common cations (Na^+, Ca^{2+}, H^+, etc.) through mass action.

2. Small uncharged metal species are subject to weak, ephemeral sorption to uncharged areas of particle surfaces by van der Waals forces, other dipole interactions and H-bonds. This mechanism, too, is relatively ineffectual.

3. 'Specific sorption' or 'surface complexation', of cationic, anionic and uncharged heavy metal species is the characteristic uptake mechanism of oxides and edge faces of clay. Under suitable conditions (e.g. within a favourable pH range), heavy metals are strongly and preferentially complexed by the various oxygenic ligands (O, OH, O^-, etc.) that make up the mineral surfaces, the result being the formation of stable, partly covalent coordinate bonds linking the metals **directly** to structural units of the mineral. This process involves ligand exchange and some ion exchange (displacement of ligands (e.g. H_2O and OH^-) from dissolved metal species, displacement of H^+ and OH^- ions from the mineral by cations and oxyhydroxyl complex anions, respectively, of the metals, etc.). Heavy metals are

rapidly and easily bound in this way but may be desorbed only slowly and with difficulty. Depending on the ambient pH, a surface ligand of a mineral may have a transitory negative or positive charge owing to deprotonation or protonation, functioning as a cation or anion exchange site, respectively; heavy metal ions may be attracted to these sites by coulombic forces initially but end up being fixed to the mineral by partly covalent coordinate bonds, which are more stable than ionic bonds. The process may be augmented somewhat by H-bonding. Moreover, the sorption of heavy metals by FeS and organic particles is due to complexing by surface ligands and is therefore embraced by the term 'specific sorption'.

The discussion of metal complexing by sulphides and organic matter given in the preceding section is applicable here and need not be repeated. For a detailed review of published information about the binding and release of heavy metals by colloidal minerals, see Jackson (1998).

Specific sorption is the principal mechanism – and by far the most effective one – for the binding of Hg(II) and other heavy metals whose cations are readily hydrolysed. This phenomenon, however, is highly pH-dependent, often reaching maximum efficiency with rising pH over the pH range in which the free cation becomes hydroxylated. In the absence of Cl^- ions the sorption capacity of Mn oxide (Lockwood and Chen, 1973) or quartz (Leckie and James, 1974; MacNaughton and James, 1974; Schuster, 1991) for Hg(II) is lowest at pH $\leq$ ~1.5 or 2.5 (at which Hg is chiefly in the form of Hg^{2+}); but sorption efficiency rises sharply to its maximum as the pH is raised to 4 (i.e. as Hg^{2+} is hydrolysed to $HgOH^+$ and $Hg(OH)_2$). (The curve showing this abrupt increase is called the 'sorption edge'.) Over the pH range ~4 to 6 or 7, in which nearly all the Hg is in the form of $Hg(OH)_2$, the percentage of Hg removed from solution is constant or decreases slightly; and the pH effect is essentially the same for Fe oxide, Mn oxide and quartz (Forbes *et al.*, 1974; Kooner *et al.*, 1995). But from pH ~6 or 7 to pH 11 there is a decline in sorption efficiency, possibly reflecting the predicted increase in $Hg(OH)_3^-$ concentration. In brief, Hg(II) is most efficiently sorbed by minerals at moderately acidic to neutral pH values and is poorly sorbed under extremely acidic or alkaline conditions. Comparable results have been reported for coprecipitation of Hg(II) with FeOOH, the percentage bound being highest at pH ~6-8 (Inoue and Munemori, 1979). Data such as these could mean that $HgOH^+$ and $Hg(OH)_2$ are the most efficiently sorbed Hg species (Hahne and Kroontje, 1973; Schuster, 1991). Little Hg^{2+} is sorbed because of competition with H^+ ions, repulsion by positively charged (protonated) sites on oxide surfaces (as would also be true of the edge faces of clay), and possibly interference by H_2O molecules coordinated to the Hg^{2+} ions (Leckie and James, 1974), while $Hg(OH)_3^-$ is not sorbed because it is repelled by negatively charged sites formed by removal of H^+ ions from OH groups of mineral surfaces (and sorption of OH^- ions?). Nevertheless, the exact mechanisms of specific sorption are open to debate (Jackson, 1998). Moreover, specific properties of the mineral surface may play a part. In experiments on the sorption of $Hg(NO_3)_2$ by clay minerals over the pH range ~3–12, Farrah and Pickering (1978) found that illite, kaolinite and montmorillonite gave qualitatively different results. Only illite approximated the expected pattern of variation with rising pH (i.e. a steady increase in sorption capacity up to a plateau followed by a decline). But the results are mostly inconsistent with data reported by Andersson (1979), who showed qualitatively similar variations for all three minerals. As the clay specimens of Farrah and Pickering had been 'finely ground' and subjected to other treatments before being used, the possible occurrence of artefacts cannot be discounted. Coatings or other surface impurities on the clay could also account for the discrepancy.

The Hg(II) sorption capacities of minerals and organic matter usually show qualitatively similar variations with respect to pH, but there is a major quantitative difference. Under acidic conditions the sorption capacities of mineral particles decline sharply with decreasing pH, whereas organic matter shows only a slight decline (Andersson, 1979; Zvonarev and Zyrin, 1982). Thus, 1 M HCl was unable to leach sorbed Hg from organic lake mud (Jackson *et al.*, 1980), whereas it extracted 10–25% of the Hg sorbed to mineral-rich soils (Hogg *et al.*, 1978). In general, organic matter is the main binding agent for Hg(II) in acidic sulphide-poor environments.

Oxides and humic matter have much higher sorption capacities for Hg(II) than clay minerals do (Andersson, 1979; Schuster, 1991) – even the clay minerals with the highest cation exchange capacities (e.g. montmorillonite). (Indeed, cation exchange capacity and heavy metal sorption capacity vary independently of each other (Jackson, 1998). Thus, Andersson (1979) reported that the amount of Hg(II) sorbed by soils and soil constituents in the pH range 6–9 decreased in the order hydrated Fe oxide > organic soils > illite-rich clay soil > bentonite (i.e. montmorillonite) > kaolinite. The affinity of Hg(II) for thiol groups probably goes far toward explaining the marked ability of organic matter to sorb Hg, and under reducing conditions both FeS and organic particles containing thiol groups are probably the chief sorbents of Hg(II) (Schuster, 1991). Fe, Mn and Al oxides differ among themselves in their sorption capacities for heavy metals but, regardless of composition, freshly precipitated amorphous oxide gels have much higher sorption capacities than the well crystallized phases that eventually form from them during ageing (Jackson, 1998), and crystallization may be accompanied by release of bound Hg(II) (Waslenchuk, 1975). Among common clay minerals, sorption capacity for Hg(II) decreases in the order illite > montmorillonite > kaolinite (Reimers and Krenkel, 1974; Andersson, 1979). This series bears no consistent relation to cation exchange capacity, which decreases in the order montmorillonite >> illite ≥ kaolinite (Jackson, 1998); it probably reflects distinctive properties of the edge faces of the clay. Note that clay minerals, despite their limitations as sorbents, play important roles as carriers of Hg and other metals bound to associated oxides and humic matter. Clay crystals, silt and sand grains, and plankton hard parts commonly have Fe and Mn oxide coatings, and both clay and oxide minerals may be coated with humic matter. Such coatings probably play important roles in the biogeochemistry and transport of metals (Jackson, 1989, 1998; Jackson and Bistricki, 1995).

Competition between Hg(II) and other metals for binding sites must also be considered, but the results of comparative studies have been somewhat inconsistent. Organic matter preferentially binds Hg(II) with respect to other metals at low pH, but colloidal minerals may have relatively low binding capacities for Hg(II). Even in the pH range most favourable for sorption of Hg(II) (pH $\sim$4–7), the percentage sorbed by a mineral (e.g. goethite) is lower for Hg(II) than for other hydrolysable heavy metal cations, such as divalent Cd, Zn, Cu and Pb (Schuster, 1991). This surprising fact suggests that the largely covalent Hg–OH bond diminishes the stability of the bonds linking Hg–OH complexes to mineral surfaces (Schuster, 1991). In a study mentioned by Siegel and Siegel (1979), the affinities of divalent metals for montmorillonite decreased in the order Pb > Cu > Ca > Ba > Mg > Hg. For kaolinite, sorption energy decreased in the order Hg > Cu > Pb, but calcium replacement decreased in the order Pb > Cu > Hg, signifying an inverse relation between sorption strength and sorption capacity: Hg was the most strongly bound metal, but the quantity of Hg sorbed was smallest. However, Ramamoorthy and Rust (1978) claimed that the 'order of cation exchange' in fluvial sediment poor in organic matter was Hg > Pb > Cu > Cd. These results suggest that the outcome of competition for binding sites is hard to predict, as it is the net result of many interacting variables.

A dissolved complexing agent may either hinder or promote the sorption of a heavy metal by particulate matter or have no apparent effect (Jackson, 1998). The net effect depends on the partitioning of 'free' and complexed forms of the metal between the solution phase and solid phase. A complexing agent may interfere with metal sorption or enhance desorption by forming a highly soluble metal complex, but it may promote metal sorption if the complex itself is readily sorbed. Fulvic acid has been found to enhance the sorption of Hg(II) by α-Al_2O_3 over a wide pH range (2.5–9.5) (Xu and Allard, 1991), reflecting the tendency of humic matter to form stable associations with mineral colloids (Schnitzer and Khan, 1972). Yet Cl^- ions commonly interfere with the binding of Hg(II) by natural sorbents and complexing agents, and may release Hg(II) from these binding agents by the formation of soluble Hg–Cl^- complexes which are more weakly sorbed than Hg–OH complexes (Feick *et al.*, 1972; Lockwood and Chen, 1973; Forbes *et al.*, 1974; Leckie and James, 1974; MacNaughton and James, 1974; Reimers and Krenkel, 1974; Newton *et al.*, 1976; Kinniburgh and Jackson, 1978; Ramamoorthy and Rust, 1978; Inoue and Munemori, 1979; Jackson *et al.*, 1982; Wang *et al.*, 1985, 1988, 1991; Schuster, 1991; Chen *et al.*, 1995). As the concentration of dissolved Cl^- increases, the sorption edge of Hg(II) is displaced toward higher pH values, reflecting competition between Cl^- and

OH⁻ for Hg(II) and preferential sorption of hydroxylated Hg species (Leckie and James, 1974; MacNaughton and James, 1974; Andersson, 1979), and the maximum percentage of Hg(II) sorbed is lowered (Leckie and James, 1974; Schuster, 1991). Thus, dissolved Hg–Cl⁻ complexes are not readily sorbed by suspended organic and mineral particles; and if the particles have a preponderance of negative charges, as is usually the case, they repel anionic Cl⁻ complexes. As would be expected, Cl⁻ ions compete more successfully with sorption sites of clay and -COOH groups of organic matter than with -SH and -NH$_2$ groups of organic matter (Reimers and Krenkel, 1974), and solubilize Hg(II) much more readily from clay and silt than from organic matter (Jackson *et al.*, 1982). The Cl⁻ concentration and Cl⁻/Hg ratio, and various environmental factors such as pH, also have important effects (Reimers and Krenkel, 1974). The negative effect of Cl⁻ ions on sorption is particularly strong at lower pH values, as Hg–OH⁻ complexes, which predominate at higher pH values, are more readily sorbed than Hg–Cl⁻ complexes. Occasionally, however, Cl⁻ promotes the sorption of Hg by forming anionic Hg–Cl⁻ complexes which are bound by positively charged sites of oxides or edge faces of clay (Xu and Allard, 1991).

Hg and other heavy metals differ widely in their relative affinities for particular complexing agents and sorbents. For instance, the order of decreasing ability of complexing agents to desorb metals from fluvial sediment poor in organic matter was Hg > Cd ≫ Pb for Cl⁻ but was Cd > Hg > Pb in the case of the chelator NTA (Ramamoorthy and Rust, 1978). Moreover, the partitioning of metals between binding agents of water and sediments varies widely from one environment to another. Thus, when Hg and other metals are introduced into lakes by rivers, their relative tendencies to be trapped in lake sediments may correlate with the standard entropy of metal cation hydration in some river–lake systems and with the standard enthalpy in others (Jackson, 1979). Hg tends to be either the most or the least efficiently trapped metal, depending on whether entropy or enthalpy, respectively, controls metal partitioning between water and sediment.

Distilled water with no added Cl⁻ ions or other ligands is ineffective in desorbing Hg(II) from fluvial sediments (Ramamoorthy and Rust, 1976). Rinsing of Hg(II)-bearing fluvial sediment with Hg-poor river water has been shown to remove at least 50% of the Hg (Bothner and Carpenter, 1973), but the desorption could have been accomplished by complexing agents in the water.

Finally, Hg(II) bound by Fe or Mn oxide is released if the oxide is dissolved by reduction, as happens, for instance, when Hg-contaminated oxides in fluvial suspended matter are deposited in reducing bottom environments of productive riverine lakes (Jackson, 1986, 1993a, 1998; Wang *et al.*, 1989). On the other hand, Hg(II) bound to FeS may be released by oxidation of the sulphide (Jackson, 1998). Hg released by reduction of oxides may be scavenged by sulphides and organic matter, and Hg released by oxidation of FeS may be immobilized by precipitated Fe oxide and organic matter (Jackson, 1998).

Such phenomena are controlled jointly by Eh, pH and other factors, such as complexing agents, and are mediated by microbes.

5.2.4 ORGANOMETALLIC COMPOUNDS OF HG(II)

Because of their highly covalent Hg–C bonds, organometallic Hg(II) compounds are resistant to oxidation and hydrolysis and are quite stable kinetically (though not thermodynamically) in water and O_2 (Cotton and Wilkinson, 1988; Douglas and McDaniel, 1965; Roberts and Caserio, 1965). The naturally occurring organometallic Hg(II) species are methyl Hg and dimethyl Hg (CH_3Hg^+ and $(CH_3)_2Hg$, respectively). The cation CH_3Hg^+ is usually associated with anionic ligands. CH_3Hg^+ species (free CH_3Hg^+ and complexes such as CH_3HgCl, CH_3HgOH, CH_3HgSH, etc.) will be designated collectively as CH_3Hg^+ for convenience when the anion is not specified. The cationic species will be referred to as free CH_3Hg^+.

CH_3Hg^+ and $(CH_3)_2Hg$ are synthesized mostly by microbial methylation of bioavailable inorganic Hg(II) species, a reaction mediated by many different species and strains of free-living bacteria and fungi (methylating microbes, or methylators), ranging from anaerobes to aerobes, under a wide range of environmental conditions (Jensen and Jernelöv, 1969; Landner, 1971; Fagerström and Jernelöv, 1972; Vonk and Sijpesteijn, 1973; Bisogni and Lawrence, 1975; Hamdy and Noyes, 1975; Blum and Bartha, 1980; Pan-Hou and Imura, 1982; Compeau and Bartha, 1985, 1987; Jackson, 1987, 1988b, 1989, 1991a,b, 1993a,b; Kerry *et al.*, 1991; Mason and Fitzgerald, 1991; Matilainen *et al.*, 1991; Regnell and Tunlid, 1991; Choi and Bartha, 1993; Zhang and Planas, 1994; Matilainen, 1995; Watras *et al.*, 1995). In a number of aquatic ecosystems (e.g. estuaries), SO_4^{2-}-reducing bacteria (e.g. *Desulfovibrio desulfuricans*) are the dominant methylators in sediments and water under anoxic conditions, provided that ambient SO_4^{2-} levels are low enough to be limiting, thus compelling the bacteria to live by fermentation rather than anaerobic respiration (Compeau and Bartha, 1985, 1987; Kerry *et al.*, 1991; Choi and Bartha, 1993; Matilainen, 1995; Watras *et al.*, 1995). When SO_4^{2-} levels are high enough to sustain production of H_2S, methylating activity declines owing to interference by sulphide. Production of CH_3Hg^+ or $(CH_3)_2Hg$ by anaerobes may be accompanied by release of CH_4 (Wood, 1971; Jackson, 1987, 1988b, 1991b), and extracts from methane bacteria have been shown to methylate Hg (Wood *et al.*, 1968; Wood, 1971), but independent studies by different workers have established that methane bacteria are unable to synthesize methyl Hg species (McBride and Edwards, 1977; Compeau and Bartha, 1985; Kerry *et al.*, 1991). Microbes, including *Desulfovibrio desulfuricans*, may also convert CH_3Hg^+ to $(CH_3)_2Hg$ and inorganic Hg(II) (Fagerström and Jernelöv, 1972; Baldi *et al.*, 1993b, 1995). Other organometallic Hg compounds, such as phenyl Hg ($C_6H_5Hg^+$), have been introduced into aquatic environments by pollution (Gavis and Ferguson, 1972; Hintelmann *et al.*, 1995a) but are not known to occur naturally. $C_6H_5Hg^+$, being rather unstable, is readily converted to inorganic Hg(II) or transformed into CH_3Hg^+ by microbes (Gavis and Ferguson, 1972).

Microbial production of CH_3Hg^+ and $(CH_3)_2Hg$ is usually accomplished by enzyme-catalysed and non-enzymatic mechanisms whereby methyl groups are transferred to inorganic Hg(II) by methylated cobalamin (vitamin B_{12}), a common coenzyme among both aerobic and anaerobic microbes (Bertilsson and Neujahr, 1971; Wood, 1971; DeSimone *et al.*, 1973; Vonk and Sijpesteijn, 1973; D'Itri, 1991; Choi and Bartha, 1993; Choi *et al.*, 1994). Methylators exploit various sources of methyl groups (Choi *et al.*, 1994). A different mechanism employing the biosynthetic pathway for methionine, with transfer of the methyl group to Hg complexed by homocysteine, has been proposed for the fungus *Neurospora crassa* (Landner, 1971; see also Jackson and Woychuk, 1980a,b); and certain microbes may use other biochemical pathways, including reactions for biosynthesis of unsaturated hydrocarbons such as ethylene and acetylene (De Filippis and Pallaghy, 1975). Another possible mechanism is microbial methylation of Sn(IV) followed by abiotic transfer of the methyl group to inorganic Hg(II), forming CH_3Hg^+ (Huey *et al.*, 1974). CH_3Hg^+ can also be generated abiotically by humic substances (Rogers, 1977; Nagase *et al.*, 1982; Weber *et al.*, 1985; Weber, 1993) and by acetate ions in the presence of sunlight or UV radiation (Akagi *et al.*, 1977), and CH_3Hg^+ can be transformed abiotically into $(CH_3)_2Hg$ by reaction with H_2S (Craig and Bartlett, 1978). The significance of abiotic mechanisms in nature is unknown (Zhang and Planas, 1994), but there are grounds for believing that the microbially mediated pathways for the synthesis of methyl Hg species are the most important ones, especially in sediments (Jensen and Jernelöv, 1969; Jackson, 1987, 1988b, 1989; Korthals and Winfrey, 1987; Zhang and Planas, 1994). Of course, the 'abiotic' mechanisms are indirect consequences of biological activity, as humic matter, acetate, methyl Sn, and H_2S are by-products of the microbial decomposition of the remains of organisms; besides, the apparent role of humic matter could, in fact, be due to extracellular bacterial enzymes associated with the humic matter (Matilainen and Verta, 1995). The effect of humic matter on microbial production of CH_3Hg^+ is equally ambiguous, as both enhancement and inhibition of the process have been reported (Jackson, 1989; Matilainen and Verta, 1995). But there is no contradiction: humic matter may have many different effects, the net effect being either enhancement or inhibition, depending on circumstances.

CH_3Hg^+ is the principal methylated form of Hg in aquatic organisms, but both CH_3Hg^+ and $(CH_3)_2Hg$ occur in natural waters, the proportion of the one species to the other varying with environmental conditions (Fagerström and Jernelöv, 1972; Mason and Fitzgerald, 1991). Moreover, the possibility cannot be ruled out that inadvertent conversion of $(CH_3)_2Hg$ to CH_3Hg^+ by reagents used for extraction of CH_3Hg^+ has led to overestimates of the abundance of CH_3Hg^+ (Gavis and Ferguson, 1972).

CH_3Hg^+ and inorganic Hg(II) have similar chemical affinities and form analogous complexes and species, but systematic quantitative differences exist as well. Like inorganic Hg(II), CH_3Hg^+ has a marked preferential affinity for sulphide and thiols, and is strongly fixed by them (Zepp *et al.*, 1974; Ochiai, 1977; Dyrssen and Wedborg, 1991), the formation constants of thiol complexes of CH_3Hg^+ being 10^8 times higher than those of amino complexes (Cotton and Wilkinson, 1988); but, unlike Hg^{2+}, CH_3Hg^+ easily and rapidly exchanges one thiol ligand for another (Rabenstein and Reid, 1984; Cotton and Wilkinson, 1988). In sediments CH_3Hg^+ is bonded to sulphides (e.g. FeS) and thiols (e.g. proteinaceous residues in organic matter). Nonetheless, H_2S volatilizes CH_3HgCl (Rowland *et al.*, 1977) by converting it to $(CH_3)_2Hg$ (Craig and Bartlett, 1978), although the CH_3Hg^+ may be partially converted to inorganic Hg(II) (e.g. black HgS) + CH_4 (Baldi *et al.*, 1993b). CH_3Hg^+ also reacts with H_2S and thiols to form low molecular weight complexes such as CH_3HgSH and CH_3HgSR, especially at pH values of 7 and slightly higher (Zepp *et al.*, 1974; Dyrssen and Wedborg, 1991); these could include volatile as well as water-soluble species. In natural waters CH_3Hg^+ is thought to be mostly complexed by inorganic and organic sulphide ligands, or, if these are absent, by OH^- or Cl^- ions, although CH_3Hg^+ may be released from sulphide complexes under acidic conditions or at high Cl^- concentrations (Zepp *et al.*,1974). At ambient sulphide levels exceeding a critical value, sulphide inhibits Hg methylation, probably by immobilizing inorganic Hg(II) and by converting CH_3Hg^+ to $(CH_3)_2Hg$ (Bartlett and Craig, 1979; Craig and Moreton, 1986). This explains why SO_4^{2-}-reducing bacteria are major CH_3Hg^+ producers only in SO_4^{2-}-poor environments where they are unable to reduce SO_4^{2-} to H_2S.

In theory (Dyrssen and Wedborg, 1991), the affinities of both free CH_3Hg^+ and Hg^{2+} for different ligands decrease in the order $R\text{-}S^- > SH^- > OH^- > Cl^-$, and CH_3Hg^+ and inorganic Hg(II) have about the same affinity for -SH groups (Gavis and Ferguson, 1972). By the same token, CH_3Hg^+ (Hintelmann *et al.*, 1995b) and inorganic Hg(II) react in very similar ways with humic matter. Hintelmann *et al.* (1995b) reported that CH_3Hg^+ forms extremely stable complexes with humic matter and is probably bound preferentially by thiol groups rather than O- and N-bearing groups (Zepp *et al.*, 1974); and the stability constants of CH_3Hg^+–humic complexes were found to lie within the range of those reported for inorganic Hg(II). CH_3Hg^+, as often observed with inorganic Hg(II), was shown to be accumulated preferentially by the higher molecular weight fractions of humic and fulvic acids. As with inorganic Hg(II), the ability of humic matter to bind CH_3Hg^+ declines, but only moderately, as the pH drops from 7 to 3; and the binding capacity of humic matter for CH_3Hg^+ decreases with increasing CH_3Hg^+ concentration.

Despite the similarities between the chemical reactions of CH_3Hg^+ and inorganic Hg(II), systematic quantitative differences exist. Dyrssen and Wedborg (1991) demonstrated that the stability constants of sulphide, OH^-, and Cl^- complexes of CH_3Hg^+ are consistently lower than those of equivalent inorganic Hg(II) complexes; and Hogg *et al.* (1978) showed that the sorption capacities of two soils differing in clay and organic content were lower for CH_3HgCl than for $HgCl_2$. Similar results have been reported for sediments (Regnell and Tunlid, 1991). These differences, and the fact that CH_3Hg^+ easily exchanges thiol groups, probably reflect the tendency of the methyl group to release electrons to its Hg atom, thereby diminishing the positive charge on the Hg and weakening the Hg–ligand bonds. Besides, CH_3Hg^+, unlike Hg^{2+}, can be bound by only one ligand at a time (Gavis and Ferguson, 1972).

As with inorganic Hg(II), aqueous CH_3Hg^+ speciation is largely a function of pH and the concentrations of anionic ligands. Theoretically, in model freshwater systems consisting of CH_3Hg^+ and Cl^- solutions at pH values in the range 2–10 (Faust, 1992), the fraction of the CH_3Hg^+ in the form of CH_3HgOH increases with rising pH over nearly the entire pH range 2–10, levelling off in the range 8–10, whereas the proportions of both CH_3HgCl and free CH_3Hg^+ are constant from pH 2 to pH $\sim$6.5 and then decline sharply with rising pH. The dominant CH_3Hg^+ species at different pH values are as follows: in the pH range 2–10, CH_3HgCl >> free CH_3Hg^+; at pH 2 to $\sim$4.7, CH_3HgCl >> free CH_3Hg^+ > CH_3HgOH; at pH $\sim$4.7 to $\sim$7.5, CH_3HgCl > CH_3HgOH > free CH_3Hg^+; and at $\sim$7.5–10, CH_3HgOH > CH_3HgCl >> free CH_3Hg^+. The K_{OW} value of CH_3Hg^+ increases with Cl^- and, at constant Cl^- concentration, decreases with rising pH over the range 2–10 (falling most sharply in the range 7–8) (Major *et al.*, 1991), as CH_3HgCl is much more lipophilic than CH_3HgOH (Mason *et al.*, 1995b). Free CH_3Hg^+ is undoubtedly the least lipophilic CH_3Hg^+ species because of its charge, but it makes up no more than a minor fraction of the dissolved CH_3Hg^+ at any given pH. In brief, the lipophilicity of CH_3Hg^+ in freshwater environments is apt to be higher under acidic conditions owing to a preponderance of CH_3HgCl over CH_3HgOH, and lower under alkaline conditions owing to a greater abundance of CH_3HgOH. In sulphide-free marine waters, where Cl^- levels are high, CH_3Hg^+ may well be mainly in the form of CH_3HgCl despite the mild alkalinity of seawater.

A number of CH_3Hg^+ species have a much stronger affinity for lipids than for water owing to their stable, largely covalent bonds and the non-polar character of the methyl group. But CH_3Hg^+ species differ widely among themselves in their degree of lipophilicity, and all of them must have some degree of hydrophilic as well as lipophilic character because of the polarity and dipole moment of the molecule and, when dissociated, its charge. Therefore, CH_3Hg^+ is subject to hydration and is somewhat water soluble (Gavis and Ferguson, 1972; Cotton and Wilkinson, 1988). As with inorganic Hg(II), the relative lipophilicity and hydrophilicity of CH_3Hg^+ depend on its associated anion. CH_3Hg^+ halides (e.g. CH_3HgCl) are among the more lipophilic CH_3Hg^+ species, owing to the covalent character of their bonds and hence their tendency to remain undissociated and uncharged in aqueous solution (Cotton and Wilkinson, 1988); thus, they are two orders of magnitude more soluble in non-polar solvents than in water (Bidstrup, 1964). CH_3Hg^+ sulphate and nitrate have bonds with a greater degree of ionic character (and presumably a greater tendency to dissociate), and are correspondingly more hydrophilic (Cotton and Wilkinson, 1988). Cotton and Wilkinson (1988) imply that CH_3HgOH is predominantly lipophilic, but Mason *et al.* (1995b) have shown that it is, in fact, one of the more hydrophilic species. CH_3HgCl and CH_3HgOH are comparable to their inorganic Hg(II) analogues in their absolute and relative degrees of lipophilic character (Mason *et al.*, 1995b). Thus, their K_{OW} values are of the same order of magnitude as those of the inorganic Hg(II) species, and the K_{OW} value of CH_3HgCl exceeds that of CH_3HgOH by two orders of magnitude. H-bonding between OH groups and H_2O probably enhances the hydrophilicity of CH_3HgOH. However, the K_{OW} value of CH_3HgCl is only about half that of $HgCl_2$, meaning that CH_3HgCl is less lipophilic and more water-soluble – presumably because CH_3HgCl, unlike $HgCl_2$, has a dipole moment. CH_3HgOH has a slightly higher K_{OW} value than $Hg(OH)_2$; apparently H-bonding of H_2O by the pair of OH groups in $Hg(OH)_2$ makes this species more hydrophilic than CH_3HgOH.

Being small, uncharged, largely undissociated and weakly hydrated, species such as CH_3HgCl are somewhat volatile (Matheson, 1979). CH_3Hg^+ species in general are more volatile than inorganic Hg(II) compounds (Ochiai, 1977); indeed, the vapour pressure of CH_3HgCl is two orders of magnitude greater than that of $HgCl_2$ (Phillips *et al.*, 1959).

The properties of $(CH_3)_2Hg$ differ radically from those of CH_3Hg^+. $(CH_3)_2Hg$ is extremely lipophilic and non-polar, as it possesses two non-polar groups, is devoid of a dipole moment, and has stable, largely covalent bonds that do not dissociate in water (Fagerström and Jernelöv, 1972; Gavis and Ferguson, 1972; Ochiai, 1977; Cotton and Wilkinson, 1988). $(CH_3)_2Hg$ is also highly volatile (much more so than CH_3Hg^+) and practically insoluble in water (Fagerström and Jernelöv, 1972; Gavis and Ferguson, 1972; Ochiai, 1977; Cotton and Wilkinson, 1988). Hence, $(CH_3)_2Hg$ readily escapes into the atmosphere, whereas CH_3Hg^+ has a greater tendency to be retained by water.

Turning to the destruction of methyl Hg species, we find that in aquatic environments, such as lake sediments, various free-living demethylating microbes, or demethylators, ranging from anaerobes to aerobes, readily decompose (demethylate) CH_3Hg^+ with release of Hg(0) and in some instances CH_4 or $CH_4 + CO_2$, thereby limiting the net rate of CH_3Hg^+ production (Spangler *et al.*, 1973; Schottel *et al.*, 1974; Mason *et al.*, 1979; Shariat *et al.*, 1979; Compeau and Bartha, 1984; Jackson, 1987, 1988b, 1989, 1991a, 1993b; Korthals and Winfrey, 1987; Baldi *et al.*, 1989, 1991; D'Itri, 1991; Matilainen *et al.*, 1991; Oremland *et al.*, 1991; Regnell and Tunlid, 1991; Baldi *et al.*, 1993b; Mason and Fitzgerald, 1993; Pahan *et al.*, 1994; Zhang and Planas, 1994; Matilainen and Verta, 1995). $(CH_3)_2Hg$ may be converted to CH_3Hg^+, and thence to Hg(0) (Mason and Fitzgerald, 1993). Many bacterial strains capable of volatilizing CH_3Hg^+ (possibly through demethylation) have been isolated from Hg-polluted sediments (Nakamura *et al.*, 1990). Microbes that demethylate CH_3Hg^+ include methane bacteria and SO_4^{2-} reducing bacteria (though the latter also synthesize CH_3Hg^+) (Oremland *et al.*, 1991). The enzyme organomercurial lyase degrades methyl Hg species, yielding inorganic Hg(II), which is then reduced to Hg(0) by mercuric reductase (Baldi *et al.*, 1991; Pahan *et al.*, 1994); demethylation can supposedly be mediated by cobalamin too (DeSimone *et al.*, 1973), even though this coenzyme is the chief catalyst for methylation. Abiotic decomposition of CH_3Hg^+ by Mn oxide has been observed in experimental aquatic systems, though its significance in nature is unknown (Jackson, 1989). Aqueous CH_3Hg^+ complexes may also be broken down abiotically to inorganic Hg(I), Hg(0), and other products by photochemical reactions on exposure to sunlight ($\lambda \geq 290$ nm at the Earth's surface) or UV light (Baughman *et al.*, 1973; Inoko, 1981), and the process has been detected in lake water (Sellers *et al.*, 1996).

Baughman *et al.* (1973) demonstrated that various aqueous organic and inorganic sulphide complexes of CH_3Hg^+ undergo photochemical decomposition, yielding inorganic Hg products, in the presence of sunlight. The rates of photolysis decreased in the order $CH_3HgS^- >> CH_3Hg^+$–thiol complexes, and the quantum yields were highest in the absence of O_2. The rates of photolysis for CH_3Hg^+ halide and OH^- complexes decreased in the order $CH_3HgI >> CH_3HgBr > CH_3HgCl$, CH_3HgOH, and the authors concluded that CH_3HgCl, CH_3HgOH, and free CH_3Hg^+ are not decomposed appreciably by sunlight at the Earth's surface, although volatile species such as CH_3HgCl may be broken down by UV radiation in the stratosphere (Inoko, 1981). These results show that susceptibility to photochemical decomposition increases with the polarizability of the ligand and hence the degree of covalent character of the bond which it forms with the Hg atom. Sellers *et al.* (1996) found that variations in lake water composition (e.g. dissolved organic C content) had no effect on CH_3Hg^+ photolysis in sunlight. As they failed to point out, however, it does not follow that photolysis of CH_3Hg^+ is independent of water chemistry. The work of Baughman *et al.* (1973) and well-known basic principles of photochemistry invalidate any such inference. Possibly the water samples of Sellers *et al.*, even those that were poorest in organic C, had enough thiol groups to scavenge all the CH_3Hg^+, in which case the variations in water chemistry that fell within the scope of their study were irrelevant.

Sunlight has little or no effect on $(CH_3)_2Hg$ at the Earth's surface (Baughman *et al.*, 1973); but UV light of short wavelength (e.g. 254 nm) decomposes $(CH_3)_2Hg$ with production of Hg(0) (Balzani and Carassiti, 1970; Fagerström and Jernelöv, 1972; Inoko, 1981), implying that gaseous $(CH_3)_2Hg$ is destroyed on reaching the upper atmosphere. In addition, $(CH_3)_2Hg$ in water is spontaneously converted to CH_3Hg^+ under acidic conditions (pH < 5.6) (Wood, 1971; Fagerström and Jernelöv, 1972; Gavis and Ferguson, 1972).

5.2.5 ELEMENTAL MERCURY

Although Hg(II) is the most characteristic oxidation state of Hg in aquatic environments, Hg(0) must also be taken into account, as it has been introduced into natural waters by pollution and natural processes, including *in situ* formation of Hg(0) from Hg(I) and Hg(II). We may infer that strong retention of valence electrons explains why Hg(0) is a liquid at room temperature, is readily vaporized, and is relatively inert: Hg atoms have only a weak mutual attraction, which probably owes more to van der Waals forces than to the sharing of delocalized outer electrons (the characteristic mechanism of bonding in metals; Sisler, 1963), and they do not readily combine with other substances such as O_2 by sharing or giving up their valence electrons. Being volatile, rather inert, and only slightly soluble in water, Hg(0) tends to escape into the atmosphere as Hg(0) vapour (Budavari *et al.*, 1989; Bidstrup, 1964; Gavis and Ferguson, 1972; Carty and Malone, 1979). As Hg(0) does not react readily with free O_2, it may persist for significant periods of time in the atmosphere or in O_2-saturated water, but it is subject to eventual oxidation to Hg(II) by various biological and abiotic mechanisms (Jonasson and Boyle, 1972; Carty and Malone, 1979; Schroeder *et al.*, 1991). Hg(0) is lipophilic and hence more soluble in non-polar organic liquids than in water (Bidstrup, 1964; Gavis and Ferguson, 1972; Carty and Malone, 1979).

5.3 BIOAVAILABILITY AND BIOACCUMULATION: CONTROLLING FACTORS

In aquatic environments, Hg concentrations in organisms and rates of Hg bioaccumulation (the net result of uptake and excretion) depend on:

- ambient production rates and concentrations of bioavailable Hg species, which comprise a small, variable proportion of the total Hg content of sediments and water;
- physicochemical variables and biological activities which determine the bioavailability of Hg by controlling the speciation, binding, release, distribution and biogeochemical pathways of Hg in the environment;
- the total supply of Hg, from which are formed the bioavailable species;
- the nature and activities of the biota and individual organisms within it;
- biochemical reactions of Hg inside organisms.

It must be borne in mind that what we see in nature is the net effect of many diverse phenomena operating at once.

5.3.1 HG SPECIATION

The principal bioavailable forms of Hg in aquatic ecosystems are CH_3Hg^+ and inorganic Hg(II) species. $(CH_3)_2Hg$ and Hg(0) are less important insofar as direct effects on aquatic organisms are concerned.

CH_3Hg^+ is created primarily by microbial methylation of bioavailable inorganic $Hg(II)$. Immediately bioavailable inorganic $Hg(II)$ usually makes up only a small percentage of the total inorganic Hg content of sediment or water (Jackson and Woychuk, 1980a,b; Jackson *et al.*, 1982; Jackson, 1988b). Similarly, the CH_3Hg^+ content is generally at least an order (often several orders) of magnitude less than the total Hg inorganic content (e.g. Jackson and Woychuk, 1980a,b, 1981; Jackson *et al.*, 1982; Jackson, 1986, 1988b, 1993a; Bloom and Watras, 1989; Parks *et al.*, 1989; Lee and Iverfeldt, 1991; Wilken and Hintelmann, 1991; Hintelmann and Wilken, 1995). Many aquatic organisms accumulate CH_3Hg^+ preferentially and thus have much higher CH_3Hg^+ / total Hg ratios than do the water and sediments of their habitat, although the ratio varies greatly among different organisms, reaching its maximum at the upper end of the food chain. Hg in fish is mostly in the form of CH_3Hg^+, but Hg in sediments, the chief repositories of Hg in aquatic environments, is mainly inorganic (Watras *et al.*, 1994).

CH_3Hg^+ is considered the most baneful form of Hg in aquatic ecosystems, as it is highly toxic, is readily accumulated by organisms, and becomes increasingly concentrated upward through the food chain. Certain inorganic $Hg(II)$ species, too, are taken up easily, but, unlike CH_3Hg^+, they tend to be immobilized immediately after crossing biological membranes and have a more limited ability to spread to different parts of an organism's body, are more rapidly eliminated and less efficiently accumulated, are not subject to amplification up the food chain, and are less toxic (Knauer and Martin, 1972; Westöö, 1973; Bishop and Neary, 1974; Miettinen, 1975; Mortimer and Kudo, 1975; Wobeser, 1975; Ochiai, 1977; Grieb *et al.*, 1990; Boudou *et al.*., 1991; Wright *et al.*, 1991; Watras and Bloom, 1992; Odin *et al.*, 1994; Mason *et al.*, 1995b; Southworth *et al.*, 1995). However, bioavailable inorganic $Hg(II)$ species are of key significance, primarily because CH_3Hg^+ is synthesized by methylation of these species (Jackson, 1988b, 1991a, 1993b; Farrell *et al.*, 1990), but also because they themselves may contaminate and poison aquatic organisms to some extent (Knauer and Martin, 1972; Nuzzi, 1972; Alexander, 1974; Wobeser, 1975; Hamdy *et al.*, 1977; Ochiai, 1977; Farrell *et al.*, 1990; Liebert *et al.*, 1991).

The bioavailable forms of inorganic $Hg(II)$ consist of $HgOH^+$, $Hg(OH)_2$, $HgClOH$, $HgCl^+$, $HgCl_2$, $HgCl_3^-$, and other low molecular weight complexes (Farrell *et al.*, 1990), including thiol and inorganic sulphide species such as $Hg(SH)_2$ and $Hg(SH)S^-$ (Dyrssen and Wedborg, 1991; Gottofrey and Tjälve, 1991; Hudson *et al.*, 1994). The uncharged $Hg(II)$ species display both lipophilicity and hydrophilicity in varying degrees, and each of these qualities, in its way, promotes biological uptake. Hydrophilicity is conducive to solubility and mobility in the aquatic milieu, release of Hg from binding agents (e.g. sediment particles) that compete with organisms for Hg, and retention of Hg in the water rather than loss by volatilization; thus, it helps to bring the Hg into contact with methylators and other organisms and enhances bioavailability. On the other hand, lipophilicity enables the Hg species to diffuse easily and rapidly through a cell or mucus membrane by dissolving in the membrane's lipid phase, resulting in passive uptake by organisms (Chapter1). Because of its greater lipophilicity, $HgCl_2$ penetrates membranes more readily than does $Hg(OH)_2$ (Farrell *et al.*, 1990; Boudou *et al.*, 1991; Mason *et al.*, 1995b). Ionic species and other strongly hydrophilic species cannot diffuse through membrane lipids but may enter cells by means of active transport and pores in the membrane; however, passive diffusion of uncharged lipophilic species through membranes is by far the principal uptake mechanism (Boudou *et al.*, 1991; Hudson *et al.*, 1994; Mason *et al.*, 1995b). The small dimensions of the bioavailable species also facilitate the penetration of membranes, especially in the case of hydrophilic species.

In oxygenated fresh water the bioavailability of inorganic Hg(II) is enhanced in the range of Cl^- levels and pH values at which dissolved inorganic Hg(II) is mostly in the form of the lipophilic species $HgCl_2$ (Farrell *et al.*, 1990; Boudou *et al.*, 1991; Mason *et al.*, 1995b). By the same token, lipophilic thiol and inorganic sulphide complexes such as $Hg(SH)_2$ probably account for much of the bioavailability of Hg under reducing conditions, especially in freshwater environments (Gottofrey and Tjälve, 1991; Hudson *et al.*, 1994). But the bioavailability of Hg is much lower if the dissolved Hg(II) is mostly in the form of ionic complexes or $Hg(OH)_2$ (Ribeyre and Boudou, 1982; Walczak *et al.*, 1986; Ribo *et al.*, 1989; Boudou *et al.*, 1991; Mason *et al.*, 1995b). In brief, inorganic Hg(II) is probably most bioavailable in acidic fresh waters in which it is mainly in the form of $HgCl_2$ or $Hg(SH)_2$; it should be less bioavailable in weakly acidic or alkaline fresh waters in which $Hg(OH)_2$ exceeds $HgCl_2$ or $Hg(SH)S^-$ exceeds $Hg(SH)_2$, or in seawater or brackish water in which $HgCl_3^-$, $HgCl_4^-$, or $Hg(SH)S^-$ prevails. Yet fish can absorb inorganic Hg(II) to some extent directly from seawater (Windom and Kendall, 1979), suggesting uptake by anion transport mechanisms (Boudou *et al.*, 1991). Sea salt levels near those of seawater inhibit methylation of Hg by SO_4^{2-}-reducing bacteria in anoxic estuarine sediments, but this is not due to conversion of inorganic Hg(II) to anionic Cl^- complexes (Compeau and Bartha, 1987). At sufficiently high pH and low Cl^- levels for $Hg(OH)_2$ to exceed $HgCl_2$, the bioavailability of inorganic Hg(II) should be relatively low for two reasons: $Hg(OH)_2$ penetrates membranes less easily than $HgCl_2$, and it is more readily sorbed by suspended matter (see above).

Inorganic Hg(II) absorbed by an organism is excreted slowly, as it is strongly retained by -SH groups of proteins (unlike more typical lipophilic substances, which are excreted slowly because they are selectively accumulated by fat) (Miettinen, 1975; Ochiai, 1977; Mason *et al.*, 1995b). On entering an organism by penetration of cell or mucous membranes, inorganic Hg(II) tends to be fixed to proteinaceous components of the membranes, though it is partly distributed elsewhere in the organism's body (Boudou *et al.*, 1991). After exposure of trout to water or food containing $HgCl_2$, the highest inorganic Hg concentrations in the tissues of the fish were found to be in either the gut or the gills, depending on whether the Hg came from food or water, respectively, the next highest levels being in the kidney and spleen (Boudou *et al.*, 1991). Following ingestion of Hg-contaminated food, Hg was at least an order of magnitude more abundant in the intestine than in any other organ. Mayfly nymphs yielded comparable results. Similarly, on exposure of a rooted aquatic plant to sediment containing $HgCl_2$, most of the Hg taken up was concentrated in the roots, though Hg was also detected in leaves and stems (Boudou *et al.*, 1991). Thus, in both plants and animals inorganic Hg(II) tends to remain fixed at its points of entry, resisting assimilation into the inner tissues.

CH_3Hg^+ species differ strikingly from analogous inorganic Hg(II) species in their interactions with organisms, though there are basic similarities too. As a rule, organisms absorb CH_3Hg^+ rapidly, both directly from ambient water and from food, and retain it tenaciously, excreting it only very slowly (Miettinen, 1975). CH_3Hg^+ easily penetrates cell and gill membranes (if absorbed from water) and the mucous membrane of the gut (if ingested with food), and is quickly assimilated by cytoplasm and internal tissues. CH_3Hg^+ species do not necessarily differ significantly from their inorganic Hg(II) analogues in membrane penetration kinetics, but they behave quite differently once they get inside (Boudou *et al.*, 1991; Mason *et al.*, 1995b).

Comparison of the membrane penetration kinetics of different Hg species using diatom cells has shown that analogous CH_3Hg^+ and inorganic Hg(II) species cross membranes at comparable rates (Mason *et al.*, 1995b). CH_3HgCl crossed the membranes at virtually the same rate as $HgCl_2$ despite its lower degree of lipophilicity, whilst CH_3HgOH was taken up 2.6 times as fast as $Hg(OH)_2$, although it is only 1.4 times as lipophilic. Apparently the methyl group facilitates membrane penetration irrespective of the lipophilicity of the molecule as a whole. But at the most, the results for analogous CH_3Hg^+ inorganic Hg(II) species differed by far less than an order of magnitude. An experiment on the diffusion of CH_3Hg^+ and inorganic Hg(II) species through model membranes revealed the same broad tendencies, except that CH_3HgCl penetrated the membrane more rapidly than $HgCl_2$ (Boudou *et al.*, 1991). In short, CH_3Hg^+ penetrates membranes at the same rate as its inorganic Hg(II) analogues or somewhat more rapidly. Note, however, that inorganic Hg(II) alters membrane permeability (Ochiai, 1977; Boudou *et al.*, 1991), introducing a degree of ambiguity into membrane penetration data.

Another important observation is that CH_3HgCl passes through membranes much more easily and rapidly than CH_3HgOH. Therefore, the kinetics of CH_3Hg^+ uptake by organisms must be a function of the Cl^- content and pH of the ambient solution, CH_3HgCl predominating over CH_3HgOH in strongly to moderately acidic fresh waters, whereas the reverse is true under weakly acidic to alkaline conditions. The same reasoning applies to sulphide complexes; thus, the bioavailable species CH_3HgSH must be prevalent at low pH, changing to the less bioavailable species CH_3HgS^- at higher pH values. In acidic fresh water at least two factors favour bioaccumulation of CH_3Hg^+: greater availability of inorganic Hg(II) to methylators and more rapid biological uptake of the CH_3Hg^+, because both are largely in the form of uncharged lipophilic complexes. On the other hand, CH_3Hg^+ could, conceivably, be more prone to demethylation under these conditions because of enhanced availability to demethylators. In seawater, CH_3Hg^+ is probably taken up rapidly by organisms, as it is mainly in the form of CH_3HgCl, but production of CH_3Hg^+ could be retarded somewhat by the prevalence of $HgCl_3^-$ and $HgCl_4^-$ over $HgCl_2$.

On entering an organism by passage through a membrane, CH_3Hg^+, like inorganic Hg(II), is preferentially complexed by thiol groups of proteins (and peptides and amino acids) and accumulates in proteinaceous material such as the edible muscle tissues of fish (Ochiai, 1977; Boudou *et al.*, 1991; Mason *et al.*, 1995b); it is also bound by nucleic acids, nucleotides, pyrimidines, etc. (Cotton and Wilkinson, 1988). But CH_3Hg^+ spreads much more readily through the internal tissues of both plants and animals than inorganic Hg(II) does, and it shows a far weaker tendency to be retained at the points of entry (Boudou *et al.*, 1991). CH_3Hg^+ is also eliminated far more slowly than inorganic Hg (Ochiai, 1977); its half-time in mussels, for instance, is 100 times greater than that of inorganic Hg (Miettinen, 1975). Comparing effects of $HgCl_2$ and CH_3HgCl on trout and aquatic plants, Boudou *et al.* (1991) showed that the concentrations in most tissues were much higher for CH_3Hg^+ than for inorganic Hg(II) after exposure; only in the intestines of the fish and the roots of the plant did inorganic Hg(II) predominate. But Mortimer and Kudo (1975) concluded that aquatic plants accumulate $HgCl_2$ and CH_3HgCl with equal ease. Be that as it may, CH_3Hg^+ is much more efficiently accumulated (more rapidly assimilated and more slowly excreted) by many organisms, notably those at higher trophic levels (fish and fish-eating animals) (Westöö, 1973; Bishop and Neary, 1974; Ochiai, 1977; Huckabee *et al.*, 1979; Windom and Kendall, 1979; Grieb *et al.*, 1990; Wright *et al.*, 1991; Watras and Bloom, 1992; Odin *et al.*, 1994; Watras *et al.*, 1994; Southworth *et al.*, 1995). It is reported that organisms at and near the lower end of the aquatic food chain (plankton and benthic invertebrates) have lower CH_3Hg^+/inorganic Hg ratios and more inorganic Hg than CH_3Hg^+ (Jernelöv and Lann, 1971; Koeman *et al.*, 1975; Huckabee *et al.*, 1979; Windom and Kendall, 1979; May *et al.*, 1987; Jackson, 1988a, 1991a), but such observations may be biased by errors due to inorganic Hg in gut contents of invertebrates, non-living particles associated with plankton, and coatings (e.g. oxide deposits) on hard parts of organisms (Jackson and Bistricki, 1995). As a rule, CH_3Hg^+ undergoes amplification up the food chain (although at least one exception is on record; Knauer and Martin, 1972), but inorganic Hg does not (D'Itri, 1972; Huckabee *et al.*, 1979; Windom and Kendall, 1979; May *et al.*, 1987; Jackson, 1991a). Thus, the proportion of CH_3Hg^+ in the total Hg body burden is typically very high (as high as 99%) in fish, especially fish-eating predators.

The contrast between CH_3Hg^+ and inorganic $Hg(II)$ probably reflects the fact that CH_3Hg^+ readily exchanges one thiol group for another, thereby spreading rapidly through the bodies of organisms (Cotton and Wilkinson, 1988), whereas inorganic $Hg(II)$ may be strongly fixed by the first proteins that it encounters, the protein components of membranes. Another factor contributing to the more efficient bioaccumulation of CH_3Hg^+ is that it is less strongly bound by non-living matter in the environment. Furthermore, the relative stability of CH_3Hg^+ in the presence of water and dissolved O_2 allows the compound to remain in the environment long enough to have a fair probability of coming into contact with organisms, although demethylation limits the net production rates and concentrations of CH_3Hg^+.

As with other lipophilic Hg species, aquatic organisms easily absorb $(CH_3)_2Hg$ and $Hg(0)$ from their environment by passive diffusion through membranes. Being more lipophilic than CH_3Hg^+ and unable to bind thiol groups, $(CH_3)_2Hg$ accumulates in fat (Ochiai, 1977) unless it is transformed into CH_3Hg^+ or inorganic Hg inside the organisms (Bidstrup, 1964; Wood, 1971; Gavis and Ferguson, 1972). Because of their great volatility and low solubility in water, however, $(CH_3)_2Hg$ and $Hg(0)$ are not retained effectively by aquatic environments and are readily lost to the atmosphere; thus, they are of secondary importance to the biota despite their bioavailability (Fagerström and Jernelöv, 1972; Baldi *et al.*, 1995).

Owing to the overriding importance of speciation (along with the environmental and biological factors that control it) as the basis of bioavailability, the total Hg content of water or sediment is an unreliable and generally poor guide to the biological effects of Hg – far too crude a parameter except where the grossest comparisons are concerned (e.g. between heavily polluted and pristine environments) (Langley, 1973; Jackson and Woychuk, 1980a,b, 1981; Jackson *et al.*, 1982, 1993; Jackson, 1986, 1988a,b, 1991a,b, 1993a,b; Kelly *et al.*, 1995). Only a small fraction of the total inorganic Hg is readily available for methylation, and its magnitude is a function of physicochemical and biological variables, as are the production and decomposition rates and concentrations of CH_3Hg^+ in ecosystems. CH_3Hg^+ and total Hg in water and sediments are positively correlated in some instances (Jackson *et al.*, 1982; Jackson, 1988b, 1993a; Parks *et al.*, 1989; Hudson *et al.*, 1994; Watras *et al.*, 1995); but within a wide range of total Hg concentrations the ambient CH_3Hg^+ levels and production rates and biological Hg accumulations commonly vary independently of the total supply of inorganic Hg, or correlate much more weakly with it than with environmental and biological factors or, in some cases, are inversely related to it. In river systems polluted with Hg from point sources, total Hg levels decrease sharply (as expected) with distance downstream from the source of pollution, whereas CH_3Hg^+ levels may decrease much more gradually, or show no consistent trend, or even increase over a considerable stretch of the river's course (Langley, 1973; Jackson and Woychuk, 1980a,b, 1981; Jackson *et al.*, 1982; Jackson, 1986, 1988b, 1993a,b; Parks and Hamilton, 1987; Parks *et al.*,

1991b). In a long, narrow, highly productive riverine lake characterized by physicochemical and biological gradients extending from a shallow deltaic environment at the inflow end to a deep basin at the outflow end, the total Hg content of the water was found to decrease from the inflow to the outflow as a result of the settling out and dilution of Hg-contaminated fluvial detritus, but the CH_3Hg^+ content increased because environmental conditions became increasingly favourable for the activities of methylating microbes owing to a greater abundance of organic nutrients derived from phytoplankton blooms (Jackson, 1986, 1993a). CH_3Hg^+ and total Hg in the water of a river system may also display inversely related seasonal trends because conditions that maximize total Hg content (e.g. high flow rates resulting in erosion and resuspension of Hg-contaminated sediment during the spring flood) are not favourable for CH_3Hg^+ production, whereas conditions favourable for CH_3Hg^+ production (e.g. low flow rates in the summer and autumn) are not conducive to high total Hg levels (Jackson, 1986; Parks *et al.*, 1986, 1989). Cores of organic sediment taken from a severely Hg-polluted river, several years after the discharge of Hg into the river had largely ceased, showed that CH_3Hg^+ was most abundant at the sediment-water interface, because microbial activity was most intense there, whereas the total Hg maximum occurred below the interface, reflecting a gradual decline in Hg loading since the cessation of Hg discharges (Jackson and Woychuk, 1980a, 1980b). In general, rates of CH_3Hg^+ production and bioaccumulation may be high where total Hg levels are low if the prevailing conditions enhance the activities of methylating microbes or the bioavailability of inorganic Hg, whilst the rates may be low where total Hg levels are high if the conditions are unfavourable for methylating activity or the release of bioavailable inorganic Hg(II) (Langley, 1973; Jackson and Woychuk, 1980a, 1980b, 1981; Jackson *et al.*, 1982; Jackson, 1986, 1993a, b; Kelly *et al.*, 1995). Even in a virtually unpolluted system with no more than low background Hg levels, the Hg content of fish may increase greatly in response to environmental changes that foster microbial production of CH_3Hg^+, as usually occurs when reservoirs are created by the impoundment of river systems and flooding of adjacent land (Bodaly *et al.*, 1984; Jackson, 1987, 1988b, 1991a; Messier and Roy, 1987; Verdon *et al.*, 1991; Morrison and Thérien, 1995).

At extremely low concentrations, total inorganic Hg(II) may, of course, be the limiting factor in CH_3Hg^+ production: CH_3Hg^+ concentrations and production rates increase markedly when a virtually pristine aquatic environment with low background Hg levels is polluted with inorganic Hg. At the opposite extreme, exceedingly high concentrations of inorganic Hg(II) could suppress CH_3Hg^+ production by poisoning the methylators (Jensen and Jernelöv, 1969; Jackson *et al.*, 1982; Jackson, 1991b), and a large enough build-up of CH_3Hg^+ itself could have this effect (D'Itri, 1991). But over a wide range of ambient total inorganic Hg levels, as in different regions within Hg-polluted river systems, CH_3Hg^+ production varies independently of the total inorganic Hg supply and is controlled, instead, by environmental conditions and biological activities.

The production, decomposition and bioaccumulation of CH_3Hg^+ are functions of numerous environmental and biological factors, all acting and interacting at once, directly and indirectly. Some factors reinforce each other whilst others tend to cancel each other's effects, and one factor may have multiple effects; moreover, different combinations of environmental variables (e.g. low Eh and high salinity, low pH and high dissolved O_2 concentration, etc.) may lead to quite different results (e.g. Compeau and Bartha, 1984; Jackson, 1987, 1989, 1993b). As the overall result is the net effect of many diverse phenomena and conditions, it is hard to interpret in detail or to predict accurately, and the various cause-and-effect relations contributing to it are difficult to disentangle. The complexity of the situation is compounded by the fact that both methylation and demethylation are mediated by many different kinds of bacteria and fungi that differ from one another widely in their ecological requirements and limits of tolerance, besides interacting (it may be assumed) with other microbes in multifarious direct and indirect ways ranging from symbiosis and mutualism to competition and antagonism (e.g. Compeau and Bartha, 1985, 1987; Jackson, 1989, 1991b, 1995a, 1998; Oremland *et al.*, 1991). Besides affecting the bioavailability of inorganic Hg(II) and the overall activity of the microbial community, a change in environmental conditions is apt to modify the species composition of the active part of the microbial community by initiating ecological succession, thereby altering the rates of methylation and demethylation; the result may be a marked increase or decrease, or hardly any change, in the net rate of CH_3Hg^+ production (Jackson, 1984, 1989, 1991b). Microbial synthesis and decomposition of CH_3Hg^+ occur under a wide variety of conditions (ranging, for instance, from O_2-rich to highly reducing, from oligotrophic to eutrophic, and from acidic to alkaline) (e.g. Bisogni and Lawrence, 1975; Jackson, 1987, 1989, 1991b); but the net rate of CH_3Hg^+ production is subject to large spatial and temporal (e.g. seasonal) variations related to environmental variations (e.g. Jackson, 1986, 1987, 1988b, 1993a,b). The efficiency of CH_3Hg^+ production depends both on microbial growth and activities (which are regulated by factors such as nutrient supply, pH, temperature, and oxidation-reduction conditions) and on the bioavailability of inorganic Hg(II) (which is limited by inorganic Hg speciation and the sorption and complexing of Hg by various metal-binding agents – especially sulphides – that compete with methylators for inorganic Hg species), as well as loss of Hg through volatilization. Ideal conditions for the growth and activities of methylating microbes do not necessarily favour maximum availability of inorganic Hg(II) for methylation, and vice versa. Thus, different tendencies may offset each other's effects, the net rate of CH_3Hg^+ production representing a compromise between high intensity of methylating activity and low availability of inorganic Hg(II) or low intensity of methylating activity and high availability of inorganic Hg(II) (Jackson, 1988b; 1993a,b; Mason and Fitzgerald, 1991).

Given such a complex web of effects and relationships, it is difficult to predict or explain pathways of bioavailable Hg in detail in any particular ecosystem, each system being a unique and ever-changing complex of physicochemical and biological characteristics that affect the behaviour of Hg in countless direct and indirect ways. Each ecosystem in which Hg is a problem must be investigated comprehensively and in depth with a view to achieving an interdisciplinary synthesis of a wide range of information if the biogeochemistry and environmental impact of Hg in that system are to be adequately understood and dealt with. Nevertheless, from existing knowledge it is possible to formulate a number of widely applicable generalizations about the factors that control the speciation, bioavailability, bioaccumulation and biogeochemical cycling of Hg, and sophisticated models combining empirical data and theoretical calculations – for example, the Mercury Cycling Model for lakes in northern Wisconsin (United States) (Hudson *et al.*, 1994) – have been developed. Despite their inevitable limitations (oversimplification, questionable assumptions, lack of sufficient information about natural processes, omission of relevant phenomena that may be important, limited applicability to specific situations, correlations that do not necessarily indicate cause and effect, etc.), such models may assist our interpretation and prediction of phenomena occurring in natural waters and help to narrow down the number of possible explanations for observed phenomena. But a model should never be accepted uncritically at face value, and it is necessary to keep clearly in mind the limits of its usefulness – for example, the limited ability of theoretical calculations based on the chemistry of simple aqueous solutions at equilibrium to predict Hg speciation in complex natural environments dominated by innumerable microbes, metal-binding agents, and processes subject to kinetic rather than thermodynamic control.

The rest of this section will be devoted to effects of specific environmental factors on the bioavailability and bioaccumulation of Hg.

5.3.2 ORGANIC NUTRIENTS, OXYGEN, SULPHIDES AND SELENIDES

Optimal conditions for the production and bioaccumulation of CH_3Hg^+ in aquatic ecosystems include an ample supply of biodegradable organic substances (such as remains of dead algae or plants), anoxic or O_2-poor (low Eh) conditions, and absence or paucity of sulphides (Fagerström and Jernelöv, 1972; Langley, 1973; Olson and Cooper, 1976; Shin and Krenkel, 1976; Bisogni, 1979; Wright and Hamilton, 1982; Bodaly *et al.*, 1984; Jackson, 1986, 1987, 1988b, 1991a, 1993a,b; Korthals and Winfrey, 1987; Björnberg *et al.*, 1988; Matilainen *et al.*, 1991; Mason and Fitzgerald, 1991, 1993; Regnell and Tunlid, 1991; Regnell, 1994, 1995; Slotton *et al.*, 1995; Watras *et al.*, 1995; Gagnon *et al.*, 1996). These factors are linked, as microbes utilizing labile organic matter consume O_2 and generate sulphides.

Labile organic matter fosters production of CH_3Hg^+ by furnishing nutrient substrates for direct utilization by methylators and by promoting heterotrophic microbial growth in general, resulting in anoxic conditions; thus, methylating activity tends to correlate with heterotrophic microbial activity as a whole. Enrichment of lake or river water with labile organic matter, whether it be autochthonous (e.g. the remains of a plankton bloom in a eutrophic lake) or allochthonous (e.g. remains of land plants and soil humus submerged by the waters of a newly formed reservoir), is typically followed by a marked upsurge in the microbial production of CH_3Hg^+ (Bodaly *et al.*, 1984; Jackson, 1986, 1987, 1988b, 1991a, 1993a,b; Scruton *et al.*, 1994; Anderson *et al.*, 1995; Morrison and Thérien, 1995). But decomposition of organic matter also produces sulphides, which interfere with methylation because they bind inorganic Hg(II), rendering it less bioavailable (although certain soluble sulphide complexes of inorganic Hg(II) and CH_3Hg^+ are bioavailable); production of selenides and organic complexing agents (especially compounds bearing thiol groups) probably have this effect as well. Dissolved O_2 tends to inhibit microbial methylating activity and may deplete the pool of inorganic Hg potentially available for methylation by fostering microbial reduction of inorganic Hg(II) to Hg(0) in well aerated mixed-layer (surface water) environments (Mason *et al.*, 1995a); but O_2 also enhances the release of bioavailable inorganic Hg(II) into solution by oxidising sulphides, selenides, and organic matter, thereby compensating somewhat for the negative effect of O_2 on methylation (Jernelöv, 1972; Jackson, 1988b, 1993b).

Microbial CH_3Hg^+ production is commonly concentrated in surface sediments (at the sediment–water interface) because microbial activity is most intense there; but in certain regions of the water column, labile organic matter (e.g. remains of plankton) accompanied by local O_2 depletion may support a level of CH_3Hg^+ producing activity comparable to that found in surficial sediments, as may be observed in the hypolimnion of a lake just below the thermocline or below the thermocline of the ocean, and even in the epilimnion of an extremely eutrophic lake following a phytoplankton bloom (Fagerström and Jernelöv, 1972; Jackson, 1986, 1993a; Mason and Fitzgerald, 1991, 1993; Matilainen, 1995; Watras *et al.*, 1995). Results of a field experiment suggest that the quantities of CH_3Hg^+ generated each year in the water column in some regions of the sea are comparable to those produced in associated bottom sediments (Topping and Davies, 1981). Methylating activity is likely to be concentrated at the boundary between anoxic and oxygenated zones, e.g. the sediment–water interface where O_2-rich water overlies anoxic sediment (Jackson and Woychuk, 1980a, b; Korthals and Winfrey, 1987; Watras *et al.*, 1995) and the top of an anoxic hypolimnion in a stratified lake (in which case methylating activity in the bottom sediments of the hypolimnion may be relatively insignificant) (Watras *et al.*, 1995). This may reflect the fact that SO_4^{2-}-reducing bacteria methylate Hg mainly under conditions favouring fermentation rather than SO_4^{2-} reduction; it is also consistent with a role for microaerophilic microbes in

CH_3Hg^+ production. As labile organic substances and sulphides in sediments decrease in relation to mineral detritus, the proportion of inorganic Hg(II) weakly sorbed to mineral particles (e.g. clay and oxides) increases; hence, the Hg becomes correspondingly more exchangeable and more bioavailable (Jackson and Woychuk, 1980a,b, 1981; Jackson *et al.*, 1982; Jackson, 1993b). Thus, the bioavailability of inorganic Hg(II) is greatest in detrital mineral sediment situated in a well aerated environment. Possible exceptions, however, have been noted. Comparing vigorously mixed, well aerated riverine lake environments with relatively stagnant, poorly aerated ones, Jackson (1988a) found that aeration enhances the availability of Hg to plankton by promoting decomposition of organic matter and sulphides, leading to release of their bound Hg, but decreases the availability of Hg to benthic invertebrates by causing precipitation of Hg-immobilizing Fe and Mn oxides.

As a rule, the rate of Hg methylation or rate of methylation per unit rate of demethylation (the M/D ratio) in sediments and water is higher under anoxic conditions than in the presence of O_2 (Olson and Cooper, 1976; Bisogni, 1979; Windom and Kendall, 1979; Jackson, 1987, 1988b; Korthals and Winfrey, 1987; Parks *et al.*, 1989), although at least one study showed the reverse of this tendency (Bisogni and Lawrence, 1975), and certain bacterial species methylate Hg more effectively in oxygenated environments than in anoxic ones (Vonk and Sijpesteijn, 1973). Besides lowering the rate of methylation, dissolved O_2 commonly increases the rate of demethylation, helping to account for the relatively low rate of CH_3Hg^+ production in well aerated environments, whereas anoxic conditions promote methylation at the expense of demethylation (Windom and Kendall, 1979; Compeau and Bartha, 1984; Jackson, 1987). CH_3Hg^+ production is likely to be greatest in anoxic sediments at and near the sediment–water interface, because bacterial activity is normally concentrated in that region (Fagerström and Jernelöv, 1972; Jackson and Woychuk, 1980a,b; Korthals and Winfrey, 1987), or a few centimetres below the interface, if the surface sediment has been oxidized (Gagnon *et al.*, 1996). Korthals and Winfrey (1987) reported that the M/D ratio was highest at the sediment–water interface, even though the rate of demethylation was also maximal there. Matilainen *et al.* (1991) found that methylation rates in lake sediments were highest under anoxic conditions but that demethylation rates in anoxic and oxygenated environments were similar.

The apparent inconsistencies in the effects of dissolved O_2 may reflect the fact that methylation and demethylation are not simple functions of O_2; the role of O_2 is modified by other variables, such as pH and salinity. Jackson (1987) showed that rates of methylation and demethylation in lake sediment amended with organic nutrients were independent of dissolved O_2 levels if the pH was acidic ($\sim$4.5–6.0); but near pH 7.0 the methylation rate was higher and the demethylation rate lower under anoxic conditions (under N_2), whereas the methylation rate was lower and the demethylation rate higher in the presence of O_2 (under air). Without nutrient enrichment, methylation was inhibited under both N_2 and air (equally so in both cases), whereas the rate of demethylation was high under N_2 but low under air. In brief, given an adequate supply of labile organic matter, pH values near 7, and other conditions favourable for heterotrophic microbial activity, lack of O_2 fosters CH_3Hg^+ production (within certain bounds, at least) by promoting methylation and inhibiting demethylation, whereas O_2 depresses CH_3Hg^+ production by inhibiting methylation and enhancing demethylation. But the effects of O_2 depend on other factors, such as nutrient levels, pH and salinity, which affect microbial activities and the availability of inorganic Hg. Regarding effects of dissolved salts, Compeau and Bartha (1984) observed that, at low Eh, methylation in estuarine sediments was fostered by low salinity but inhibited by high salinity, whereas methylation was less sensitive to variations in salinity at high Eh. In contrast, demethylation was suppressed by a combination of low Eh and low salinity, whereas the inhibition was reversed by raising the salinity; but high Eh values promoted demethylation irrespective of the salinity. In estuaries, then, CH_3Hg^+ production as determined by the balance between methylating and demethylating activities is a complex function of at least two independently varying factors, Eh and salinity, and many other variables could be involved as well. In any aquatic environment the rate of CH_3Hg^+ production is the net effect of many variables acting and interacting in different ways.

Owing to their strong preferential binding of Hg, sulphides are of paramount importance as limiting factors in the production and bioaccumulation of CH_3Hg^+ and the biological uptake of inorganic Hg(II). Thus, Hg concentrations in freshwater plankton and at least two species of freshwater fish (lake whitefish and white sucker) that feed on benthic animals have been found to correlate inversely with the sulphide content of associated sediments, suggesting that sulphides interfere with Hg uptake (Jackson, 1988a; Jackson *et al.*, 1993). Sulfides, including thiols (e.g. cysteine) as well as H_2S, tend to reduce the rates of Hg methylation and bioaccumulation by decreasing the bioavailability of inorganic Hg(II) (Fagerström and Jernelöv, 1972; Jernelöv, 1972; Blum and Bartha, 1980; Björnberg *et al.*, 1988; Jackson, 1984, 1988a,b, 1991a; Farrell *et al.*, 1990; Jackson *et al.*, 1993). Thus, in sediments from several lakes polluted with heavy metals (including Hg), the lowest CH_3Hg^+ / total Hg ratios were associated with the highest free sulphide concentrations (Jackson, 1984). Sulphides may also immobilize CH_3Hg^+ itself, interfering with biological uptake. If, however, other conditions, such as nutrient supply, are highly favourable, rates of CH_3Hg^+ production and bioaccumulation may be relatively high even in the presence of sulphide, as in the anoxic hypolimnion of a lake (Jackson, 1984, 1993ab; Matilainen, 1995; Watras *et al.*, 1995), up to a point, at least (Craig and Moreton, 1986). In such cases, the tendency of sulphides to suppress CH_3Hg^+ production is probably offset somewhat by the presence of bioavailable inorganic Hg(II) in the form of sulphide and thiol complexes and hydrolysis products, as well as by the intense microbial methylating activity. Hudson *et al.* (1994) have postulated that passive uptake of the uncharged lipophilic Hg sulphide complexes such as $Hg(SH)_2$ by methylating microbes explains the occurrence of high methylation rates in anoxic environments. As nutrient enrichment and sulphide production go together, a trade-off of positive and negative effects on CH_3Hg^+ production and bioaccumulation can be expected. For these reasons, perhaps, CH_3Hg^+ production in estuarine sediments increases with rising sulphide concentration up to a critical level and then declines with further increases in sulphide (Craig and Moreton, 1986).

Generally speaking, sulphides (along with the analogous but much less abundant selenides and tellurides) are the principal metal-binding agents that limit the bioavailability of Hg(II) in natural waters (Björnberg *et al.*, 1988). Thus, Jackson (1987, 1991a, 1993b) found that the total Hg/sulphide or Hg/Se ratio of lake sediment is a reliable parameter for estimating the bioavailability of inorganic Hg in the sediment. Jackson calculated the relative rates of CH_3Hg^+ production in sediments from different lakes by measuring microbial methylating activity using a special laboratory assay (Jackson, 1987, 1988b, 1989), determining the Hg/sulphide or Hg/Se ratio of the sediment, and then combining the two sets of data by multiplication. The resulting empirical compound variable (but neither of its two component variables alone) gave a very significant positive correlation with mean Hg levels in populations of walleye (predatory fish at the upper end of the aquatic food chain) inhabiting the lakes. Using the concentrations of $0.5\,\mathrm{M}$ $CaCl_2$-extractable Hg in the sediment in place of the Hg/sulphide or Hg/Se ratio to quantify the bioavailable inorganic Hg(II) fraction yielded nearly the same results (Jackson, 1987, 1991a), and the same method accurately predicted CH_3Hg^+ concentrations in sediments and water (Jackson, 1987, 1988b). (Unlike $CaCl_2$, neither dilute acetic acid nor the chelator DTPA proved to be a satisfactory extractant for bioavailable inorganic Hg(II), suggesting that the formation of soluble $Hg–Cl^-$ complexes accounts for the success of the $CaCl_2$ extraction method; Jackson, 1988b, 1991a). The results of this research constitute strong evidence for the following conclusions:

- It is mainly sulphides and selenides that limit the availability of inorganic Hg(II) for methylation.
- Meaningful estimates of relative rates of CH_3Hg^+ production and bioaccumulation can be obtained only if the microbial CH_3Hg^+ generating activity and the supply of inorganic Hg(II) available for methylation are both taken into account (neither variable alone being sufficient).
- Walleye (unlike certain other fish species that were tested) are well suited for whole-lake bioassays of CH_3Hg^+ production.
- The methods employed for quantification of CH_3Hg^+ production in sediments are valid and potentially useful for purposes of research and monitoring, as demonstrated by the strong correlations between the sediment data and independent data for fish.

Owing to the extremely stable bonds that Hg forms with selenides, Se either interferes with the accumulation of Hg in aquatic organisms or is closely associated with Hg in their tissues (probably in the form of Hg–Se or CH_3Hg–Se complexes bound to the -SH groups of proteins; Koeman *et al.*, 1973), depending on the nature of the organisms. Research by several workers has established that Se (probably in the form of selenides, even if initially in the form of selenite in the environment) can be remarkably effective in blocking the accumulation of Hg by fish and other aquatic animals and may further the elimination of Hg from Hg-contaminated fish (Rudd *et al.*, 1980; Turner and Rudd, 1983; Turner and Swick, 1983; Björnberg *et al.*, 1988; Lindqvist *et al.*, 1991; Paulsson and Lundbergh, 1991). Field experiments demonstrated that Se in lake water did not affect the uptake of waterborne Hg by fish, whereas Se ingested with food decreased the Hg content of the fish (Turner and Swick, 1983). However, a number of marine and freshwater animals accumulate Hg and Se together, so that an increase in the one element is accompanied by an increase in the other. Thus, there is a highly significant positive correlation between Hg and Se in the tissues of marine mammals (e.g. dolphins, porpoises and seals), the Hg/Se mole ratio in the tissues being 1 : 1, and the Hg was found to be tightly bound, strongly suggesting fixation of CH_3Hg^+ by the formation of selenide complexes bound to -SH groups of proteins (Koeman *et al.*, 1973, 1975). Similarly, there is a marked positive correlation between Hg and Se in the flesh of tuna fish, which is also attributed to CH_3Hg–Se–S–protein complexes (Ganther *et al.*, 1972). One study showed that Se increased the Hg body burden of goldfish exposed to inorganic Hg(II) in the form of $HgCl_2$, but it ameliorated the toxicity of the Hg, implying strong fixation of Hg, as in marine mammals (Heisinger *et al.*, 1979). Evidently selenides intervene in the uptake, assimilation and excretion of CH_3Hg^+, the principal form of Hg in fish and aquatic mammals. Besides affecting Hg accumulation by animals, Se may have complex effects on microbial production of CH_3Hg^+ and other microbial activities in sediments, as demonstrated experimentally using sediment from a lake polluted with smelter fallout (Jackson, 1991b). With increasing Se concentration, CH_3Hg^+ production alternately increased and decreased, forming a series of CH_3Hg^+ maxima (coinciding with CO_2 maxima) superimposed on a downward trend ending in total inhibition at the highest Se levels (≥ 50 μmol l^{-1} in the aqueous phase). This complex zigzag pattern of variation was ascribed to systematic changes in the species composition of the microflora (ecological succession) owing to suppression of Se-sensitive species accompanied by opportunistic flourishing of Se-tolerant species, which were themselves inhibited as the Se continued to rise, the result being alternate upswings and downswings in the net rate of CH_3Hg^+ production.

Finally, it is important to be aware that organic nutrients, O_2 and sulphides have multiple effects, some reinforcing each other whilst others tend to cancel each other. The net effect of any one of these three factors, or a combination of them, acting in conjunction with a host of other variables, such as pH, may be to stimulate or depress CH_3Hg^+ production, depending on the outcome of a complex trade-off. High primary productivity may promote CH_3Hg^+ production by providing organic nutrient substrates and creating anoxic conditions, while tending to inhibit CH_3Hg^+ production through the formation of sulphide and organic complexing agents (e.g. thiols); and organic matter dispersed in the water may scavenge dissolved CH_3Hg^+, making it less available for uptake by fish (Hudson *et al.*, 1994). In contrast, a well aerated, highly oxidizing environment poor in labile organic matter is less favourable for microbial methylating activity but makes inorganic Hg(II) more available for methylation (Jackson, 1988b, 1991a, 1993a,b). Consequently, in a eutrophic lake the net rate of CH_3Hg^+ production may be high in a zone of O_2 depletion because stimulation of the activities of methylating microbes compensates for immobilization of inorganic Hg(II) by sulphide; but in a well aerated region of the lake the rate could be equally high because enhanced bioavailability of inorganic Hg(II) compensates for a lower level of methylating activity.

These principles are strikingly illustrated by phenomena observed in a chain of extremely eutrophic, Hg-polluted lakes linked by the Qu'Appelle River in the semi-arid prairie region of Saskatchewan, Canada (Jackson, 1986, 1993a,b). In the deepest basins of the lakes the bottom water was poorest in dissolved O_2 and the sediments had the lowest Eh values, highest organic content, highest level of heterotrophic (CO_2-generating) microbial activity and most intense microbial Hg methylating activity, but had the lowest degree of inorganic Hg bioavailability (as estimated by the total Hg/sulphide ratio of the sediment). An analogous trade-off exists in an O_2-depleted zone in ocean water, where a methyl Hg maximum coincides with a minimum in bioavailable inorganic Hg (Mason and Fitzgerald, 1991). At the opposite extreme, the most shallow basins of the lakes were richest in dissolved O_2 and had the lowest levels of heterotrophic and methylating activities (despite a high level of primary productivity) but were characterized by the highest degree of inorganic Hg bioavailability. The observed relationship between CH_3Hg^+ production in the sediments and mean Hg concentrations in walleye populations in the lakes revealed that the Hg content of the fish is determined by the combined effect of microbial methylating activity and inorganic Hg(II) availability, not either factor alone. The highest Hg levels in the fish occurred in lakes with deep anoxic basins (or a deep anoxic basin at one end and a shallow, well aerated basin at the other), because they supported the greatest intensity of methylating and general heterotrophic activity. But in one shallow, well aerated lake (the one situated furthest downstream from the source of Hg pollution), the Hg content of walleye was much higher than might have been expected. The reason for this anomaly is that inorganic Hg(II) in that lake had a high degree of bioavailability, compensating somewhat for a low level of methylating activity. A comparable trade-off was brought to light by comparison of two constrasting basins – one relatively stagnant and poorly aerated, and the other well flushed and aerated by fluvial currents – in Notigi Lake (northern Manitoba, Canada), a natural boreal forest lake artificially expanded to form a reservoir (Jackson, 1988b, 1991a).

One other example of apparent effects of variations in the abundance of labile organic matter is instructive. Huge quantities of both Hg and organic particles (wood chips) have been discharged into the Wabigoon River in Northern Ontario (Canada) from a chlor-alkali plant and pulp-and-paper mill (Jackson, 1980; Jackson and Woychuk, 1980a,b, 1981; Jackson *et al.*, 1982; Parks and Hamilton, 1987). Dispersal and dilution of the pollutants in the downstream direction created an environmental gradient characterized by a sharp decrease in the total Hg and organic C content and a corresponding increase in the natural clay-silt content of the sediment with distance downstream from the source of pollution. But CH_3Hg^+ levels remained high throughout the river, even increasing in the downstream direction over part of the river's course, despite the drop in total Hg (Jackson and Woychuk, 1980a,b, 1981; Jackson *et al.*, 1982; Parks and Hamilton, 1987; Parks *et al.*, 1991b). The bioavailability of the inorganic Hg (based on the percentage extracted with 0.5 M $CaCl_2$ or Ca acetate solution or the amount solubilized per unit concentration of Cl^- ions present as pollutants in the river water) increased in the downstream direction, signifying that Hg sorbed to fine mineral particles was more weakly bound than Hg sorbed to the wood chips (Jackson and Woychuk, 1980a,b, 1981). The results suggest that conditions for CH_3Hg^+ production improved in the downstream direction because the inorganic Hg(II) became more bioavailable (and possibly owing to attenuation of toxic pollutants that inhibited bacteria), though this advantage may have been partially offset by the decrease in the abundance of organic matter and inorganic Hg.

The nature of the labile organic matter as well as its gross abundance may have a bearing on CH_3Hg^+ production. Data for the Qu'Appelle River lakes suggest that the species composition of the phytoplankton, which is the main source of organic nutrients for Hg-transforming microbes, affects the rates of Hg methylation and demethylation. Thus, certain species of cyanobacteria appear to promote methylation, but diatoms and chlorophytes apparently foster demethylation (Jackson, 1993b).

As these observations and others (e.g. Jackson, 1987, 1991a, 1993a) illustrate, organic nutrient substrates are required for the growth of demethylators as well as methylators. But the net effect of adding organic matter to an environment relatively poor in organic matter is an upsurge in CH_3Hg^+ production (Jackson, 1988b, 1991a). Thus, a marked increase in the rate of CH_3Hg^+ production leading to a rise in the Hg content of fish is a usual – perhaps universal – side-effect of the creation of reservoirs owing to the introduction of organic matter into the aquatic environment by the flooding of land (Bodaly *et al.*, 1984; Jackson, 1987, 1988b, 1991a); this phenomenon has occurred repeatedly in newly formed reservoirs located in widely separated, environmentally diverse geographical regions where no local point sources of Hg contamination are known to exist.

In reservoirs of northern Manitoba, Canada, which have been studied in detail, the main reason for this effect is that the labile terrestrial organic matter of recently submerged land areas stimulated the growth and activities of methylating microbes by providing them with nutrient substrates and, at the same time, creating anoxic conditions (Jackson, 1987, 1988b). However, as the organic content and heterotrophic microbial activity of the sediments increased, production of CH_3Hg^+ in the sediments and concentrations of Hg (mostly CH_3Hg^+) in fish inhabiting the overlying water rose to a maximum and then declined (Jackson, 1991a). By the same token, experimental enrichment of lake and reservoir sediments with organic nutrients accelerated methylation only if the sediment was poor in organic matter to begin with; if it was already rich in organic matter, the nutrient amendment depressed methylation somewhat (Jackson, 1991a). The reduced production and bioaccumulation of CH_3Hg^+ following excessive enrichment in organic nutrients was probably caused primarily by the production of sulphide and organic complexing agents, especially thiols, which lowered the bioavailability of inorganic Hg(II) and perhaps the bioavailability of CH_3Hg^+ itself (D'Itri, 1971; Jackson, 1991a). Other possible causes include stimulation of demethylating microbes at the expense of methylators (Jackson, 1991a), conversion of CH_3Hg^+ to volatile derivatives such as $(CH_3)_2Hg$ which were then lost to the atmosphere (Rowland *et al.*, 1977; Craig and Bartlett, 1978), and microbial reduction of inorganic Hg(II) to Hg(0) – mainly in the epilimnion – leading to loss through evaporation and depletion of the pool of inorganic Hg(II) available for methylation (Mason *et al.*, 1995a). These conflicting effects of nutrient enrichment may help to reconcile the documented stimulation of microbial CH_3Hg^+ production and bioaccumulation by labile organic matter in eutrophic lakes with the seemingly paradoxical fact that fish in eutrophic lakes commonly have lower Hg levels than fish in oligotrophic lakes with a comparable degree of Hg contamination (Jernelöv *et al.*, 1975; Björnberg *et al.*, 1988). The systematic difference between Hg concentrations in the biota of eutrophic and oligotrophic lakes has been ascribed to greater 'biodilution' and 'growth dilution' in the more productive lakes – a consequence of their larger biomass and the higher growth rates of their inhabitants (Jernelöv *et al.*, 1975; Björnberg *et al.*, 1988; Meili, 1991). This interpretation is plausible, but Rudd and Turner (1983) and Jackson (1988a, 1991a) found no evidence that biodilution and growth dilution played a significant role in nutrient-enriched systems that they studied, probably because the increase in biological Hg concentrations due to acceleration of CH_3Hg^+ production by the nutrients far outweighed any opposing effect of biodilution and growth dilution.

In the reservoirs of Manitoba and other regions, an increase in the supply of Hg owing to the incorporation of Hg-bearing terrestrial organic matter into the aquatic environment has probably contributed to the increase in the Hg content of the biota following impoundment (Jackson, 1987, 1988b; Louchouarn *et al.*, 1993; Mucci *et al.*, 1995; Rodgers *et al.*, 1995). Jackson (1987, 1988b) considered this to be of secondary importance compared with the large post-impoundment upsurge in the growth of methylating microbes, but it could be significant, as suggested by the next section.

5.3.3 HUMIC MATTER

Allochthonous organic matter (notably humic matter) dispersed in runoff water from soils, peat bogs and wetlands is a vehicle for transporting inorganic Hg(II) and CH_3Hg^+ into lakes in the boreal forest zone (Lee and Hultberg, 1990; Meili, 1991; Meili *et al.*, 1991; Haines *et al.*, 1994; St Louis *et al.*, 1994; Rudd, 1995; Branfireun *et al.*, 1996). High concentrations of humic matter in forest lakes are associated with high Hg levels in fish such as the piscivorous species northern pike (*Esox lucius*) and lake trout (*Salvelinus namaycush*) (Håkanson *et al.*, 1988; McMurty *et al.*, 1989; Rask and Metsälä, 1991; Haines *et al.*, 1994). Haines *et al.* (1994) postulated that humic matter promotes bioaccumulation of Hg by furthering the transport of Hg into lakes and retaining Hg in the water column. Humic matter may also stimulate the activities of heterotrophic microbes that generate CH_3Hg^+ (Jackson, 1989, 1995a), while depressing primary production (Jackson and Hecky, 1980), thereby reducing the biodilution of Hg. But Jackson (1989) observed both positive and negative effects of humic matter on CH_3Hg^+ production in forest lake sediments; and Matilainen and Verta (1995) concluded that methylation in forest lakes was suppressed by humic matter. Moreover, 'dissolved organic matter' (possibly composed largely of humic matter) scavenges dissolved CH_3Hg^+, making it less available to fish (Hudson *et al.*, 1994). Evidently humic matter performs diverse functions, and the overall effect may be the net result of opposing tendencies.

Humic matter may affect the bioavailability and bioaccumulation of Hg by complexing Hg(II) species and influencing microbial activities (Jackson, 1989, 1995a), and by mediating 'abiotic' methylation. Though largely resistant to microbial decomposition, humic substances have profound and complex effects, both harmful and beneficial, on the growth and activities of aquatic organisms, including phytoplankton and benthic bacteria (Jackson, 1995). The experiments of Jackson (1989) revealed that soil humic acid extracts had a number of different effects on microbial transformations of Hg in lake sediments. Humic acid (stripped of associated biochemical compounds by acid hydrolysis and dialysis) caused no immediate change in the rate of microbial Hg methylation and CO_2 production in nutrient-amended, $CaCO_3$-buffered sediment, but after incubation for a few days it increased both CH_3Hg^+ and CO_2 production appreciably. After several more days, however, the humic acid caused a decline in CH_3Hg^+ and CO_2 levels accompanied by an upsurge of CH_4-generating activity. The humic acid also complexed inorganic Hg(II) (HgCl$_2$) strongly, rendering it practically non-dialysable. The results indicate that humic acid stimulated methylating microbes, causing a net increase in the CH$_3$Hg$^+$ production rate even though it probably lowered the rate somewhat as well by decreasing the bioavailability of the inorganic Hg(II); but subsequently the humic acid brought the CH$_3$Hg$^+$ level down by stimulating the activities of demethylating microbes. The data suggest that the humic matter altered the course of ecological succession in the microbial community, thereby altering the balance between methylation and demethylation, first in favour of CH$_3$Hg$^+$ production and then in favour of CH$_3$Hg$^+$ decomposition. In another experiment, in which humic acid was present as a coating on kaolinite, the humic acid inhibited Hg methylation, possibly by masking Fe oxide coatings on the clay (see below), though it enhanced demethylation as before.

These results demonstrate that humic substances interacting with different microbial species and different environmental variables may have complex, variable and not altogether predictable effects on the microbial production and decomposition of CH$_3$Hg$^+$. Considering the wide variety of humic substances, microbial species and strains, and combinations of environmental conditions that may interact in nature, the possible effects could vary enormously, both qualitatively and quantitatively. Humic matter probably has important indirect effects as well (Jackson, 1995). Depending on the nature of the humic matter (e.g. its molecular size), its abundance and the ambient conditions, humic matter may either suppress primary production, as in boreal forest lakes, or promote it, as in some coastal marine waters, and this must influence CH$_3$Hg$^+$ production and bioaccumulation in different ways; besides, humic substances dispersed in water could benefit bacteria by serving as carriers of biodegradable organic matter and extracellular enzymes, and by exerting favourable effects on cell physiology. A general review of such phenomena has been given elsewhere (Jackson, 1995).

5.3.4 PH AND BUFFERING

The pH of natural water, together with related parameters such as alkalinity, hardness and buffering capacity, is of key importance in the production and bioaccumulation of CH$_3$Hg$^+$. The role of pH has attracted much attention in recent years owing to the influence of acid precipitation on the bioaccumulation of Hg in lakes. Widespread acidification of ill-buffered lakes in remote regions, such as the Precambrian shields of Canada and Scandinavia (Schindler, 1988), has been accompanied by a pronounced rise in the Hg content of fish, even though Hg pollution in these lakes has been limited to deposition of trace quantities of Hg transported from distant sources (e.g. coal-burning power plants) by atmospheric circulation (see below). A large body of empirical data produced by extensive field studies and experiments has established the important generalization that Hg concentrations in freshwater fish and other aquatic organisms (and fish-eating animals) tend to increase as the pH, alkalinity, hardness, conductivity and acid-neutralizing capacity of the water decrease (Jernelöv, 1972; Jernelöv *et al.*, 1975; Brouzes *et al.*, 1977; Scheider *et al.*, 1979; Håkanson, 1980; Wren and MacCrimmon, 1983; Håkanson *et al.*, 1988; Richman *et al.*, 1988; Lathrop *et al.*, 1989, 1991; McMurty *et al.*, 1989; Cope *et al.*, 1990; Grieb *et al.*, 1990; Wiener *et al.*, 1990; Winfrey and Rudd, 1990; Ponce and Bloom, 1991; Rask and Metsälä, 1991; Wren *et al.*, 1991; Watras and Bloom, 1992; Hudson *et al.*, 1994; Simonin *et al.*, 1994; Anderson *et al.*, 1995; Meyer *et al.*, 1995).

The relationship between ambient pH and Hg in fish is neither simple nor completely consistent, and its underlying causes are not well understood. Several possible explanations merit serious consideration, but apparently no one of them, by itself, is sufficient to cover all the facts, suggesting that the observed effect is the net result of a complex interplay of many different physicochemical and biological phenomena involving different effects of pH and interactions of pH with other variables (Wood, 1980; Jackson, 1987; Richman *et al.*, 1988; Winfrey and Rudd, 1990; Ponce and Bloom, 1991; Haines *et al.*, 1994). Two examples of empirical observations should suffice to make the point that a number of different factors must be taken into account:

- Andersson *et al.* (1995) reported that Hg levels in fish inhabiting acidified lakes peaked at a pH slightly higher than 5.0, decreasing above and below that value, demonstrating that bioaccumulation of Hg was not a simple function of pH.
- Haines *et al.* (1994) found that the effect of pH on the Hg content of freshwater fish depended on the humic content ('colour') of the water; fish from lakes of high pH had low Hg levels regardless of humic content, but Hg levels in fish from lakes of low pH were quite variable, being highest in lakes of high humic content.

A number of hypothetical explanations have been invoked to explain the effect of pH and related factors on Hg levels in fish. One possible contributing factor is the fact that the acids of acid precipitation are accompanied by airborne anthropogenic Hg, since both volatile Hg and acid-generating S and N oxides are released into the atmosphere by the combustion of fossil fuels (see below). Another likely consequence of acidification is that the proportions of lipophilic, highly bioavailable aqueous Hg species (e.g. $HgCl_2$ and CH_3HgCl) increase at the expense of less bioavailable species (e.g. $Hg(OH)_2$ and CH_3HgOH), with the two-fold result that inorganic Hg(II) is more available for methylation and the CH_3Hg^+ produced is more rapidly and efficiently taken up by aquatic organisms (see above). Lowering the pH also furthers the release of sorbed and complexed inorganic Hg(II) into solution, making it more available for methylation; at higher pH values, aqueous Hg(II) is more efficiently sorbed by particulate matter because a higher proportion of it is in the form of

$Hg(OH)_2$. (This would apply mainly to Hg(II) sorbed by colloidal minerals; organic (and probably sulphide) complexes of Hg are relatively insensitive to pH except under extremely acidic conditions – see above). Furthermore, acidic conditions may promote production of CH_3Hg^+ at the expense of $(CH_3)_2Hg$, both by causing spontaneous conversion of $(CH_3)_2Hg$ to CH_3Hg^+ and by promoting the activities of microbes that produce CH_3Hg^+ rather than $(CH_3)_2Hg$ (Wood, 1971; Fagerström and Jernelöv, 1972; Gavis and Ferguson, 1972). An additional factor linked to acid precipitation is increased SO_4^{2-} loading from the air; in anoxic environments this could foster the growth of SO_4^{2-}-reducing bacteria, which, under anoxic, low-SO_4^{2-} conditions, are important CH_3Hg^+ producers, but the effect of this on methylation is uncertain, as SO_4^{2-} concentrations above a certain critical level block the methylating activities of these bacteria (see above). A direct effect of pH on the bioaccumulation of CH_3Hg^+ has been postulated as well, but it has received only limited support from the available evidence and seems to be of no more than minor significance (Bloom *et al.*, 1991; Ponce and Bloom, 1991).

The arguments relating the increased Hg content of fish in acidified lakes to more efficient production of CH_3Hg^+ by microbes are plausible, but research in this area has yielded seemingly contradictory results. Some publications claim that acidification increases the net rate of CH_3Hg^+ production (Jernelöv, 1972; Fagerström and Jernelöv, 1972; Beijer and Jernelöv, 1979; Jackson and Woychuk, 1980a, b, 1981; Miskimmin *et al.*, 1992; Wood, 1980; Xun *et al.*, 1987; Winfrey and Rudd, 1990; Bloom *et al.*, 1991; Matilainen *et al.*, 1991), but others report evidence that acidification tends to inhibit CH_3Hg^+ production (Shin and Krenkel, 1976; Ramlal *et al.*, 1985; Jackson, 1987; Steffan *et al.*, 1988). This paradox probably indicates that the role of pH is complex and can be understood only by examining the combined effects of pH and other factors. Experimental data reported by Jackson (1987) suggest that combined effects of pH and dissolved O_2 may be involved, and observations reported by Jackson and Woychuk (1980a,b, 1981) and Matilainen *et al.* (1991), as well as literature surveyed by Winfrey and Rudd (1990), are consistent with this possibility. Investigation of Hg transformations in lake sediment over a range of ambient pH values under atmospheres of air and N_2 (Jackson, 1987) revealed that within the pH range ~4.5–8.6 the rates of both methylation and demethylation peaked at pH values close to 7.0 (in the range ~6.0–7.5) under both air and N_2, declining as the pH rose or fell. These results are in agreement with those of Shin and Krenkel (1976). At pH ~7.0, methylating activity was more intense under N_2 than under air, but demethylating activity was stronger under air. Thus, at pH 7.0, anoxic conditions were optimal for CH_3Hg^+ production owing to high rates of methylation combined with low rates of demethylation, whereas exposure to well oxygenated water depressed CH_3Hg^+ production by simultaneously lowering the methylation rate and raising the rate of demethylation. At pH ~4.5–6.0, however, the rates of both methylation and demethylation were the same under air as under N_2. In a well oxygenated lake, therefore, acidification might well increase the net rate of CH_3Hg^+ production. These results are consistent with the possibility that the increase in the Hg content of fish in poorly buffered lakes following acidification may be caused, at least in part, by a combined effect of pH and O_2 which alters the balance between methylation and demethylation, causing a rise in the annual net rates of CH_3Hg^+ production in the lakes.

5.3.5 MISCELLANEOUS PHYSICAL VARIABLES

Various physical characteristics of the aquatic environment are of key importance in CH_3Hg^+ production and the bioaccumulation of Hg because of their effects on biological activities, oxidation–reduction conditions and water chemistry. Depth of water, thermal stratification and water dynamics (flushing, turbulent mixing, etc.) are of major significance; and lake volume and maximum depth have an important bearing on Hg levels in fish, probably owing to their influence on water quality and biological productivity (Wren and MacCrimmon, 1983).

Summer stratification isolates the hypolimnion from the well aerated epilimnion, often leading to anoxia and build-up of CH_3Hg^+ in eutrophic lakes. During the autumn turnover, CH_3Hg^+ concentrated in the hypolimnion is brought to the surface and may then be taken up by organisms in surface waters or (in the case of riverine lakes and lakes drained by streams) flushed out through the outflow, contaminating aquatic environments downstream from the lake (Parks *et al.*, 1989). During the spring turnover, algal blooms formed as a result of nutrient recycling may stimulate CH_3Hg^+ production. During the growing season in the temperate zone, the waters of a broad, shallow lake whose well mixed water column is not subject to stable thermal stratification are apt to be warm, well aerated and rich in nutrients (and possibly resuspended sediment) released from the bottom and continually circulated throughout the water column. The warmth and nutrients probably favour CH_3Hg^+ production, whilst aeration and suspended particles (which limit light penetration besides scavenging Hg) tend to depress it. In river systems, flow rate and discharge are critical, as the biological activity necessary for CH_3Hg^+ production is likely to be greatest in times of slack water, when the flushing and dilution of nutrients and resuspension of bed sediments are minimal (Jackson, 1986; Parks *et al.*, 1989); in the temperate zone, this typically occurs during the summer, when low flow rates and high temperatures maximize microbial activity (Parks and Hamilton, 1987; Parks *et al.*, 1989). A study of seasonal variations in the Hg speciation in the Moose Jaw River, a tributary of the Qu'Appelle River (see above), showed that at the time of minimal discharge (in November) there was a phytoplankton bloom resulting in a major upsurge in CH_3Hg^+ levels in the water even though total Hg levels were lowest, whereas during peak discharge (during the spring flood) CH_3Hg^+

levels were nil even though total Hg concentrations were maximal owing to bottom scour and resuspension of particulate Hg (Jackson, 1986). Unfavourable conditions for CH_3Hg^+ production in a river during the spring flood may result not only from the flushing and dilution of nutrients but also from the binding of Hg by sediments and bank material eroded and brought into suspension by fluvial currents (Jackson *et al.*, 1982; Jackson, 1986; Parks *et al.*, 1986).

Water temperature is of central importance, because it affects the production and bioaccumulation of CH_3Hg^+ by regulating the metabolic rates of organisms. Temperature is a function of climate, season and factors such as the mean depth, surface area and morphology of a body of water. Thermal stratification is dependent on depth, and its stability depends, too, on factors such as wind-driven circulation and fluvial currents flowing through riverine lakes. With rising temperature (all other things being equal) CH_3Hg^+ production rates and concentrations in aquatic environments increase, and the bioavailability of inorganic Hg in sediments and the concentration of inorganic Hg in the overlying water may increase as well, suggesting mobilization of inorganic Hg(II) by microbial activities such as the decomposition of Hg-bearing organic matter (Shin and Krenkel, 1976; Jackson *et al.*, 1982; Parks *et al.*, 1986, 1989; Parks and Hamilton, 1987; Jackson, 1988b, 1991a; Bodaly *et al.*, 1993). Thus, Jackson (1988b) found that in lake and reservoir environments of Manitoba the concentrations of both CH_3Hg^+ in water and bioavailable inorganic Hg in sediments were higher in August than in June. As a rule, the production and bioaccumulation of CH_3Hg^+ in a lake of the temperate zone are probably maximized in midsummer, the height of the growing season (Jackson *et al.*, 1982; Parks and Hamilton, 1987; Jackson, 1988b; Parks *et al.*, 1989); and in a river the combined effects of high temperature and low flow rate would be expected to enhance CH_3Hg^+-generating biological activity in midsummer.

In a study of lakes differing in surface area but otherwise similar, Bodaly *et al.* (1993) found that both CH_3Hg^+ production in epilimnetic bottom sediments and the Hg content of several fish species increased with the mean temperature of epilimnetic water, which, in turn, was inversely related to the 'size' (i.e. surface area) of the lake. The methylation rate increased with rising temperature, but the demethylation rate decreased; hence, the M/D ratio correlated positively with temperature. The correlation between Hg in fish and epilimnion temperature did not apply to benthivorous fish, probably because they frequent cool bottom waters. The authors ascribed these results to direct effects of temperature on both bacterial activities and the metabolic rates of the fish. Their interpretation, though reasonable as far as it goes, is an oversimplification which ignores alternatives that merit consideration. Possible effects of temperature and lake size are more complex than Bodaly *et al.* have indicated: they include indirect effects involving related variables, which were not considered (e.g. enhancement of primary production by heat and solar radiation, leading to intensified activity of methylators). Although temperature is known to affect the bioaccumulation of Hg, the relationships described by the authors may be due to variables that correlate with temperature or lake area, or temperature together with other factors, rather than temperature as such or temperature alone (Jackson, 1988a).

Generally speaking, the rise in temperature from winter to summer, or from cold, deep water to warm, shallow water, probably increases the growth rates and metabolic rates of methylating microbes in the epilimnetic zone, primary and secondary producers that provide these microbes with organic nutrients, and fish and other organisms that accumulate CH_3Hg^+. At higher temperatures, aquatic organisms accumulate both CH_3Hg^+ and inorganic Hg(II) more efficiently and transfer CH_3Hg^+ more readily from one trophic level to another (e.g. from phytoplankton to zooplankton) because the metabolic rates of the organisms are elevated (Reinert *et al.*, 1974; Boudou and Ribeyre, 1981; Ribeyre and Boudou, 1982; Jackson, 1988a, 1991a; Bodaly *et al.*, 1993). But the higher rate of CH_3Hg^+ uptake by fish at higher temperatures is offset somewhat by accelerated excretion of CH_3Hg^+ (Ruohtula and Miettinen, 1975). Another complication arises from the fact that rising water temperature may cause the rate of CH_3Hg^+ accumulation by fish to increase until a critical temperature is reached and then decline (Burkett, 1974). Furthermore, the waters of small boreal forest lakes are commonly enriched in allochthonous humic matter, which affects the production and bioaccumulation of CH_3Hg^+ (see above), whereas in the larger lakes the humic matter is more dilute. Also note that warm conditions do not favour the growth of all bacteria; some species are adapted to cold conditions, as is doubtless the case with bacteria in the sediments and hypolimnetic waters of deep lake basins where the temperature is perpetually low, though rising temperatures and other favourable conditions in the epilimnion may stimulate these microbes indirectly by increasing the supply of plankton-generated organic nutrient substrates that sink to the bottom. It is necessary to add that the effect of temperature may be obscured by that of some other environmental factor; thus, the above-mentioned seasonal maximum in phytoplankton biomass and CH_3Hg^+ content coinciding with minimum discharge in Moose Jaw River water occurred in November, when the temperature of the water (1.0–3.3°C) was close to freezing point, whereas the concentrations in July, when the temperature was high (18.5°C), were much lower (comparable to the levels observed during the spring flood) (Jackson, 1986). This implies that variation in methylating activity during the growing season was controlled by the supply of organic nutrient substrates, and therefore by the flow rate, not by water temperature.

Exposure of surface waters to solar radiation (a function of latitude, time of year and climate) must also be considered because of its relation to temperature and primary production, and because Hg is subject to photochemical alteration. Effects of sunlight depend on characteristics of individual lakes, such as humic content, suspended matter, surface area, mean depth, trophic status and the nature of the biota.

5.3.6 DISSOLVED SALTS

Salinity is of paramount importance, especially where Cl^-, SO_4^{2-}, and possibly Ca^{2+} are concerned, and it has multiple effects. The results of experiments on the effects of variations in salinity on Hg methylation in $HgCl_2$-spiked slurries composed of anoxic estuarine sediments mixed with sea salt solutions at different concentrations (0.3–24 and 1–30‰) show a decline in methylating activity with rising salinity (Blum and Bartha, 1980; Compeau and Bartha, 1987). Compeau and Bartha (1985) concluded that SO_4^{2-} ions in the sea salt prevented SO_4^{2-}-reducing bacteria from synthesizing CH_3Hg^+ by causing the bacteria to switch from fermentation to SO_4^{2-} reduction, leading to immobilization of inorganic Hg(II) by H_2S. Evidently the observed effect of salinity was not caused by the formation of anionic Hg–Cl^- complexes (Compeau and Bartha, 1987). Nevertheless, Cl^- ions at concentrations of 0.2–20‰ have been found to inhibit microbial CH_3Hg^+ production in soil–water slurries (Shin and Krenkel, 1976), proving that suppression of the methylating activities of SO_4^{2-}-reducing bacteria by SO_4^{2-} ions is not the only possible effect of elevated salinity. Hg concentrations in fish from Canadian Shield lakes have been shown to correlate inversely with the conductivity (i.e. salinity) of the water, but this may reflect the influence of variables controlling the pH (e.g. water hardness and alkalinity) rather than salinity as such (Wren and MacCrimmon, 1983; Björnberg *et al.*, 1988). Interactive effects of salinity and Eh on Hg methylation and demethylation (Compeau and Bartha, 1984) have already been discussed.

As explained above, Cl^- ions play an important part in determining the bioavailability of Hg. Owing to the formation of water-soluble, weakly sorbed Hg–Cl^- complexes, Cl^- ions tend to prevent or reverse the complexing or sorption of inorganic Hg(II) by other binding agents, thereby making it more bioavailable and more readily methylated. Accordingly, inorganic Hg(II) sorbed to suspended particles transported to the sea by rivers is largely desorbed by Cl^- ions on coming into contact with the salty waters of estuaries (de Groot *et al.*, 1971; de Groot and Allersma, 1975; Newton *et al.*, 1976; van der Weijden, 1990; Chen *et al.*, 1995); by the same token, contamination of Hg-polluted freshwater environments with Cl^- ions (e.g. from road de-icing salt and effluents from chlor-alkali plants) results in desorption of sediment-bound Hg (Feick *et al.*, 1972; Jackson *et al.*, 1982). As we have seen, moreover, dissolved Cl^- concentrations and pH values favouring the formation of the uncharged, lipophilic species $HgCl_2$ and CH_3HgCl enhance the biological uptake of both inorganic and methyl Hg owing to the ease with which they pass through membranes, and this may assist methylation and demethylation of Hg.

Finally, there is evidence consistent with the possibility that dissolved Ca^{2+} ions interfere with the uptake of Hg by fish. Thus, Wren and MacCrimmon (1983) observed an inverse correlation between the Hg content of fish and Ca levels in the surrounding water. As they themselves pointed out, this may simply be an indirect indication of the effect of pH, as water quality variables (e.g. hardness) that determine the water's buffering capacity are linked to ambient Ca levels; but the relationship is also consistent with the possibility that Ca interferes directly in the biological uptake of CH_3Hg^+, possibly by decreasing gill membrane permeability or by competing with Hg species for cellular binding sites (Wren and MacCrimmon, 1983; Hudson *et al.*, 1994). In any event, the evidence available thus far consists solely of inverse correlations; a cause-and-effect relationship has not been established.

5.3.7 CLAY- AND SILT-SIZED MINERAL PARTICLES

The binding and release of Hg by fine-grained mineral particles are among the major determinants of the bioavailability of the metal (Jackson, 1995, 1998). Thus, the sorption or coprecipitation of inorganic Hg(II) by Fe and Mn oxides (and sorption by organic coatings; Chapter 8) may interfere with the uptake of Hg by aquatic organisms, such as benthic invertebrates, which have a high proportion of inorganic Hg to CH_3Hg^+ (Jackson, 1988a). There is indirect evidence, too, that Fe and Mn oxides have selective effects on Hg uptake by specific kinds of organisms. The results of research on benthic invertebrates in lake and reservoir sediments suggest that the uptake of inorganic Hg by chironomid larvae is limited by FeOOH, whilst uptake by oligochaetes, nematodes and pelecypods at the same sampling sites is controlled by MnOOH, implying discrimination between different forms of inorganic Hg(II) both by the oxides and by the animals themselves (Jackson, 1988a).

Clay- and silt-sized mineral particles strongly influence the microbial production of CH_3Hg^+, but the effects are complex, variable, poorly understood and hard to predict. Inorganic Hg(II) in sediments is less strongly sorbed and hence more available for methylation if it is sorbed to clay, oxides and silt than if it is bound to organic matter or sulphide (Jackson and Woychuk, 1980a,b, 1981; Jackson *et al.*, 1982; Schuster, 1991; Jackson, 1993b). But clay and silt introduced into lakes by fluvial transport or the erosion of shoreline material inhibits CH_3Hg^+ production in sediments, and both methylation and demethylation may be adversely affected (Jackson, 1987, 1988a,b, 1989, 1991a, 1993a,b, 1995). Possible reasons include: reduced light penetration owing to turbidity, limiting primary production of the organic nutrients needed by microbes; rapid burial and dilution of organic matter and other nutrients; the smothering of microbes by prevention of the exchange of dissolved nutrients and wastes between the sediment and water; scavenging and immobilization of inorganic Hg(II); and selective inhibitory or stimulatory effects of the minerals on specific kinds of microbes.

Regarding selective effects, experiments performed by Jackson (1987, 1989, 1995) showed that clay had no effect, or a slightly inhibitory effect, on Hg methylation in sediments from a boreal forest lake but strongly enhanced subsequent demethylation. In contrast, clay strongly inhibited Hg methylation in sediment from a prairie lake but did not stimulate subsequent demethylation. Either of these radically different effects could lower the net rate of CH_3Hg^+ production in an aquatic ecosystem. The disparity between the two sets of observations probably reflects a major difference between the microbial communities of the two unlike lakes. In general, experiments on effects of clay minerals and oxides on microbial Hg transformations in sediments yielded complex and variable results, suggesting selective effects on particular microbes, and therefore involvement of ecological succession in the microbial community, rather than general effects such as suppression of microbial activity as a whole or decreased bioavailability of inorganic Hg(II) owing to sorption. Depending on the nature, abundance and surface chemistry of the mineral colloid, the experimental conditions and the source of the sediment (i.e. the species composition of the microflora), the minerals either strongly inhibited or promoted Hg transformations or had little net effect, or exerted a succession of different effects as the experimental conditions (e.g. the abundance of the added colloid) changed or as the incubation time increased (Jackson, 1989, 1995). Coatings on the mineral particles were of decisive importance. FeOOH tended to enhance methylation, and FeOOH coatings on clay crystals greatly mitigated the inhibitory effect of the clay. Indeed, the extent to which a specimen of clay inhibited methylation was attributable entirely to the amount of FeOOH on the surfaces of the clay crystals, not to the nature of the clay mineral itself (Jackson, 1989, 1995). Removal of oxide coatings from clay depressed both methylation and demethylation. Furthermore, environmental changes, such as nutrient enrichment, altered the effects of the coatings on the clay: on addition of organic nutrients, oxide coatings promoted methylation and impeded demethylation, but without nutrient enrichment the reverse tended to occur.

The importance of particle coatings and other impurities associated with clay cannot be overemphasized, and failure to recognize it may lead to serious errors. An object lesson recorded in the literature will serve to drive this point home. Specimens of natural silty clay used in some of the experiments of Jackson (1987, 1989, 1995) contain calcite and dolomite; pH buffering by these carbonates stimulated the activities of Hg methylating microbes in sediments amended with the clay, offsetting the negative effect of the non-carbonate minerals and causing a net rise in the rate of CH_3Hg^+ production. To measure the impact of the non-carbonate minerals, it was necessary to eliminate this pH effect by buffering all experimental and control systems with added $CaCO_3$. Meanwhile, Hecky *et al.* (1987, 1991) independently carried out a comparable experiment employing the same clay but neglected to take pH buffering into account (though this phenomenon was obvious from their own raw data). As in the initial experiments of Jackson, their data showed a favourable effect of the clay on methylation; but, despite the published findings of Jackson, they failed to realize that their results were attributable solely to buffering by carbonates. Thus, their inferences about the effects of clay are unfounded.

5.3.8 POLLUTANTS OTHER THAN MERCURY

Hg in aquatic environments is commonly accompanied by other pollutants which may either exacerbate or ameliorate the undesirable biological effects of Hg. A common and notorious instance of this is the association between volatile Hg and strong acids introduced into the air by combustion of coal and transported over great distances by winds. As we have seen, the acids aggravate the adverse effects of Hg, apparently promoting the accumulation of CH_3Hg^+ by fish.

Synergistic and antagonistic effects of toxic pollutants such as heavy metals other than Hg may occur as well (Chapter 11). A study of sediments from lakes polluted with heavy metals showed a strong inverse correlation between Hg methylating activity and the abundance of bioavailable (DTPA-extractable) Cd, probably owing to inhibition of methylators by the Cd (Jackson *et al.*, 1993). In contrast, total Cd content, as well as other variables (such as the Eh, pH and organic content of the sediment) that might be expected to control CH_3Hg^+ production, did not correlate significantly with methylating activity. Hence, bioavailable Cd exerted a controlling influence on the Hg content of northern pike in the lakes. Cu and Zn gave comparable results, but Cd, though much less abundant, had by far the strongest inhibitory effect. The order of decreasing inhibition was Cd > Cu > Zn, which is the order of decreasing metal sulphide stability as represented by standard entropy, suggesting that inhibition of methylators resulted from the binding of Cd by -SH groups of enzymes. Also note that low Hg levels have been found in fish inhabiting lakes polluted with Zn (Björnberg *et al.*, 1988). Experiments, too, have shown inhibition of methylating activity by added Cd, Cu and Zn in lake sediments (Jackson, 1991b). In sediment from an essentially pristine lake, CH_3Hg^+ production simply declined with increasing Cd, Cu or Zn concentration, revealing an absence of metal-tolerant methylating microbes; but in sediments from metal-polluted lakes methylating activity was alternately inhibited and enhanced, yielding a zigzag pattern of variation, with increasing Cd, Cu or Zn. Only at the highest metal concentrations was methylation completely suppressed. Surprisingly, within certain ranges of metal concentrations the CH_3Hg^+ yield was higher than in control systems containing no added Cd, Cu or Zn. These complex results imply ecological succession based on competition between metal-sensitive and metal-tolerant microbes, with tolerant species supplanting sensitive ones as the metal concentration increased, the result being large upswings and downswings in the net rate of CH_3Hg^+ production. As would be expected, metal-tolerant microbes were detected only in metal-contaminated sediments. Moreover, the relative toxicities of different metals apparently depended on the nature of the microbes. For example, Cd had either greater or less toxicity than Cu, or the same toxicity as Cu, depending on the circumstances.

The effects of organic wastes are complex and variable and some of them are mutually antagonistic, the net result being either enhancement or suppression of CH_3Hg^+ production. Thus, wood chip deposits in the Wabigoon River (see above) provide nutrients for methylators but restrict the availability of inorganic Hg(II) (Jackson and Woychuk, 1980a,b, 1981). Sewage, too, may play many parts. A lake near a base metal mine at Flin Flon, Manitoba, is the receiving basin for both municipal sewage effluent and tailings pond effluent contaminated with heavy metals (including Cu, Zn, Cd and some Hg), along with SO_4^{2-} (Jackson, 1978, 1979, 1984). Nutrients in the sewage support the growth of algal blooms, resulting in rapid immobilization of the metals owing to scavenging by plankton and non-living suspended particles (aided by alkaline conditions due to photosynthesis) followed by sedimentation of the particulate metals and precipitation of metal sulphides by H_2S (generated by decomposition of dead algae accompanied by SO_4^{2-} reduction). Hg is more efficiently trapped in the sediments than other metals because the stability of black HgS, as represented by its standard entropy, is greater than that of any of the other metal sulphides (Jackson, 1978, 1979). Compared with sediments in sulphide-poor lakes in the vicinity, the sediments of this lake were found to have a low CH_3Hg^+/total Hg ratio (Jackson, 1984), indicating net inhibition of CH_3Hg^+ production owing to immobilization of inorganic Hg(II) by sulphide, despite presumed stimulation of methylating activity by the high primary productivity of the lake. An altogether different effect of sewage was seen in lakes of the Qu'Appelle River system (see above). Jackson (1993b) found evidence that the discharge of nutrients from sewage effluents into one of the lakes favours development of plankton blooms with anomalously high proportions of diatoms and chlorophytes, which apparently foster the growth of demethylators at the expense of methylators (Jackson, 1993b), resulting in relatively weak CH_3Hg^+ producing activity.

5.3.9 BIOLOGICAL FACTORS

The bioaccumulation of Hg is a function of the characteristics and activities of the microbial community that controls the speciation and bioavailability of the Hg and of the aquatic food-chain organisms that accumulate the Hg. An extremely complex, ever-shifting interplay of biological and physicochemical factors is involved in this process, and a list of possible direct and indirect effects, feedback mechanisms and interactions between different factors would be endless.

As discussed above, the activities of microbial communities in sediments and water are multifarious. Natural microfloras contain many diverse species and strains of both methylating and demethylating microbes and microbes that convert inorganic Hg(II) to Hg(0), together with other kinds of microbes that do not mediate Hg speciation reactions but may affect these processes indirectly by influencing the growth and activities of Hg-transforming species (through mutualism, competition, antagonism, etc.) (Jackson, 1995). These microbes differ widely in their ecological requirements and limits of tolerance to different conditions, and they interact with each other in various complex ways. In any given environment at any point in time, the microflora is composed of an assemblage of those species that are best adapted to the prevailing conditions. A shift in environmental conditions, either imposed by external events (e.g. an influx of suspended silt or dead algae from a plankton bloom) or caused by the microbes themselves (e.g. when aerobes deplete the local supply of dissolved O_2 – creating a favourable environment for anaerobes – or when anaerobes start generating H_2S), leads to a shift in the species composition of the active portion of the microflora (i.e. ecological succession) whereby microbes ill adapted to the new conditions die or become inactive whilst well adapted species replace them. There may also be important changes in overall microbial biomass or activity; thus, enrichment in organic nutrients increases the microbial biomass and the general level of heterotrophic microbial activity besides changing the species composition of the microflora. Not surprisingly, the nature and activities of the microflora are subject to major spatial and temporal (e.g. seasonal) variations. As we have seen, these variations commonly result in large changes in the rates of Hg methylation and demethylation and in the balance between the two processes (the M/D ratio). An essential role of food-chain organisms, chiefly phytoplankton, is to produce the labile organic matter utilized as nutrient substrates by the microflora. Irrespective of their influence on Hg-transforming microbes, however, the food-chain organisms moderate their uptake, retention and excretion of Hg through an assortment of vital functions that differ from species to species, vary over the lifetime of an individual organism, and are modified by the environment; and different combinations of these biological variables produce different net effects (Jernelöv and Lann, 1971; Scott and Armstrong, 1972; Burrows and Krenkel, 1973; Scott, 1974; de Freitas and Hart, 1975; Miettinen, 1975; Norstrom *et al.*, 1976; de Freitas *et al.*, 1977; Huckabee *et al.*, 1979; Windom and Kendall, 1979; Jackson and Woychuk, 1980a,b; Jackson, 1988a, 1991a; Nicoletto and Hendricks, 1988; Richman *et al.*, 1988; Grieb *et al.*, 1990; Cabana *et al.*, 1994; Rodgers, 1994).

A most important general principle of Hg bioaccumulation is the marked tendency of Hg concentrations in the muscle tissue of fish (mature fish, at least) to increase with size (fork length or weight; Scott, 1974; Jackson, 1991a). In some instances, as in newly formed reservoirs following a sudden increase in CH_3Hg^+ production, Hg content and size are poorly correlated and the correlations may even be negative in the case of small fish with high growth rates (Jackson, 1991a), but these cases are exceptional. The positive correlation between Hg content and size in fish is thought to be primarily an effect of age, which, of course, correlates with size: the older the fish, the greater its cumulative body burden of CH_3Hg^+ because it has been exposed to environmental CH_3Hg^+ for a longer time and because CH_3Hg^+ is usually taken up rapidly but excreted slowly (Huckabee *et al.*, 1979; Windom and Kendall, 1979). However, Scott (1974) inferred that the correlation between size and Hg content reflects interactions between age, growth rate and condition. Indeed, we can go further and draw the conclusion that CH_3Hg^+ (or total Hg) levels in fish and other organisms are complex functions of many biological variables, including metabolic rate, growth rate, diet, excretory pathways, population biomass and habitat preference, as well as age and size (Jackson, 1991a; see also Chapter 8). Smaller, younger fish take up CH_3Hg^+ more rapidly than larger, older ones because of their higher rate of metabolism (de Freitas and Hart, 1975), but they also excrete it more rapidly because of their small body size, and growth dilution tends to keep the concentrations low. As with growth dilution, biomass dilution may result in lower Hg concentrations in the tissues of individual fish in the relatively large populations of the more productive lakes (Jernelöv *et al.*, 1975; Björnberg *et al.*, 1988; Rask and Metsälä, 1991), though in some environments its quantitative importance relative to other factors appears to be small (Rudd and Turner, 1983; Jackson, 1988a, 1991a).

Retention of CH_3Hg^+ by fish is a function of body size, diet and species and is not affected by growth rate (de Freitas *et al.*, 1977). Miettinen (1975) maintains that excretion of CH_3Hg^+ and other forms of Hg becomes more rapid with rising temperature, but de Freitas *et al.* (1977) deny that temperature affects whole-body retention of CH_3Hg^+ by fish. Another consideration is that juvenile fish have different feeding habits than adults of the same species; thus, among piscivorous predators, immature fish feed on small invertebrates, which are likely to have low CH_3Hg^+ concentrations, whereas adults prey on other fish, which are enriched in CH_3Hg^+. Fish that feed on invertebrates generally have lower Hg concentrations than fish that feed on other fish (Jackson, 1991a; Brouard *et al.*, 1994; Rodgers, 1994). In brief, there are many different tendencies controlled by different physiological phenomena, some of them tending to offset each other, but the net result is that larger fish tend to be richer in Hg than smaller ones even though smaller fish, paradoxically, accumulate Hg faster.

Another generalization of major importance is that CH_3Hg^+ concentrations and CH_3Hg^+/inorganic Hg ratios in aquatic organisms usually increase progressively up the food chain, the result being that the lower organisms (plankton and benthos) have relatively high, though variable, proportions of inorganic Hg, whilst the Hg in fish, especially piscivorous predators, is mostly in the form of CH_3Hg^+ (Jernelöv and Lann, 1971; D'Itri, 1972; Bishop and Neary, 1974; Koeman *et al.*, 1975; Huckabee *et al.*, 1979; Windom and Kendall, 1979; May *et al.*, 1987; Jackson, 1988a, 1991a; Grieb *et al.*, 1990). The longer the food chain, the higher is the Hg content of the fish at the upper end of it (Cabana *et al.*, 1994); this may explain the absence of biomagnification in the short food chain studied by Knauer and Martin (1972). There is some doubt, however, about the mechanism of biomagnification, and the very concept of it is thought to be misleading: biomagnification may be no more than an expression of the longer life spans and lower growth rates of animals at the upper end of the food chain (de Freitas *et al.*, 1974; Huckabee *et al.*, 1979). Nevertheless, the fact that animals at higher trophic levels consume food that is richer in CH_3Hg^+ than do animals at lower levels is undoubtedly a contributory factor (Jackson, 1991a; Brouard *et al.*, 1994; Rodgers, 1994). Supporting evidence includes: an observed rise in the Hg content of a whitefish population owing to the inclusion of more fish in their diet (Brouard *et al.*, 1994); higher Hg levels in pike populations that preyed on fish species of higher Hg content (Rask and Metsälä, 1991); and the occurrence of two separate but parallel food chain segments, one linking benthic invertebrates to whitefish and the other linking zooplankton to spottail shiner, implying that anomalously low CH_3Hg^+ levels in the benthos resulted in correspondingly low CH_3Hg^+ levels in the whitefish (Jackson, 1991a).

Whatever the underlying causes of biomagnification, its effects are subject to modification by many biological and environmental factors. For instance, even animals at the same trophic level may differ considerably in Hg content, Hg-fork length relationships, and spatial and temporal variations in Hg content (Jackson, 1991a). Walleye and northern pike populations coexisting in reservoirs and a riverine lake were found to differ appreciably in this regard, probably because they differ in their habitat preferences and spatial distributions within the bodies of water that they occupy (Jackson, 1991a). Owing to their lower metabolic rate, pike prefer shallow, weedy waters near the shore, whilst walleye have a greater tendency to venture into open water. Consequently, they differ in their degree of exposure to the regions of most intense Hg methylating activity, which, in recently formed reservoirs, are the near-shore zones of flooded land. Systematic differences in Hg concentrations in two other coexisting fish species at the same trophic level were tentatively ascribed to different spatial variations in diet, rate of food intake, growth rate, metabolic rate and biomass, resulting in different rates of Hg uptake, elimination, and biodilution (Jackson, 1991a). To take another example from lakes of the Wabigoon River system (see above), Hg in walleye decreased with distance downstream from the source of Hg pollution owing to attenuation of fluvially transported CH_3Hg^+ in surface water, whereas Hg in white sucker increased, reflecting a trend in the CH_3Hg^+ content of sediments and bottom water (Jackson and Woychuk, 1980a,b). Even within a single species, individuals differ among themselves: females commonly have higher Hg levels than males, a difference that is not related to body size (Nicoletto and Hendricks, 1988), and, as already discussed, smaller, younger fish have lower Hg concentrations than larger, older ones. The mechanism of Hg uptake must also be considered. Animals take up CH_3Hg^+ in two ways: by ingestion with food and direct absorption from water (through the gill membrane in the case of fish) (de Freitas *et al.*, 1974; Norstrom *et al.*, 1976; Gottofrey and Tjälve, 1991; Jackson, 1991a). The proportions of CH_3Hg^+ assimilated by these two pathways appear to vary from one species to another (Jackson, 1991a). According to D'Itri (1991), fish generally take up most of their CH_3Hg^+ with food, but absorption through the gill membrane and body surface is also a significant pathway. Diet is the main source of CH_3Hg^+ for certain predatory fish, and the nature of the diet affects the transfer of CH_3Hg^+ to the fish (see above): differences in feeding strategy cause differences in the Hg content of fish (Richman *et al.*, 1988).

Some aquatic animals appear to have evolved biochemical mechanisms for transforming and eliminating Hg. In reservoirs where the Hg content of fish was found to be abnormally high owing to effects of impoundment, Hg levels in whitefish peaked early and then declined over time, whilst levels in walleye, pike and trout remained high (Bodaly *et al.*, 1984; Jackson, 1991a; Anderson *et al.*, 1995; Morrison and Thérien, 1995), suggesting that the biochemical and physiological pathways of Hg in whitefish are fundamentally different from those of the other species, and that they include a special mechanism for excretion of Hg. Compared with other fish species studied, whitefish in northern Manitoba lake and reservoir waters have a weak tendency to accumulate CH_3Hg^+ (although most of the Hg in their flesh is in the form of CH_3Hg^+) and, compared with walleye and pike, they have an anomalously high ratio of liver CH_3Hg^+ to muscle CH_3Hg^+, suggesting, again, a special mechanism of excretion, such as the binding of CH_3Hg^+ by carrier molecules (possibly proteins with -SH groups, or other thiols), creating hydrophilic complexes that are readily eliminated (Jackson, 1991a). The fact that whitefish feed on benthic invertebrates which, themselves, have a low affinity for CH_3Hg^+ suggests that the hypothetical complexes are formed by the benthos consumed by the fish and are subsequently excreted by the fish. This theory is in agreement with evidence that the CH_3Hg^+ which fish ingest with food is less toxic than CH_3Hg^+ absorbed directly from water (Wobeser, 1974). Experimental results showing that CH_3Hg^+ administered to rainbow trout as a protein complex had a shorter half-time in the fish than CH_3Hg^+ administered as a nitrate (Ruohtula and Miettinen, 1975) are also consistent with the idea of a molecular carrier originating in food consumed by the fish. In any case, the low CH_3Hg^+ diet of whitefish is surely linked to the low CH_3Hg^+ content of the fish. Habitat preference may be a contributing factor, as whitefish frequent cold bottom waters, where rates of CH_3Hg^+ are relatively low – partly because of the low temperatures and, in reservoirs, partly because organic nutrients are concentrated in the nearshore zone of flooded land (Jackson, 1991a). Demethylation of CH_3Hg^+ in liver or kidney tissues also occurs in certain fish species, including whitefish (Burrows and Krenkel, 1973; Windom and Kendall, 1979).

5.4 TOXICITY AND DETOXIFICATION

The chief cause of Hg toxicity is inhibition of enzymes owing to strong binding of their -SH groups by inorganic Hg(II) and CH_3Hg^+ (although Hg(II) may also activate certain enzymes; Bidstrup, 1964; Ochiai, 1977). Hg(0) and $(CH_3)_2Hg$, too, are toxic, probably as a result of being transformed within the body into inorganic Hg(II) and CH_3Hg^+, respectively (Bidstrup, 1964; Wood, 1971; Gavis and Ferguson, 1972; Carty and Malone, 1979). However, because of their high degree of volatility, they are more fugitive in the environment, and hence have a lower probability of being taken up by the aquatic biota. Besides combining with -SH groups of enzymes and other proteins, Hg(II) forms complexes with the -COOH and $-NH_2$ groups of amino acids and proteins and with the nitrogenous bases and phosphate groups of nucleic acids, altering the conformations of the molecules (Ochiai, 1977); it also interferes with the normal functions of biological membranes, including cell membranes of phytoplankton (Ochiai, 1977) and gill membranes of fish (Walczak *et al.*, 1986). Moreover, CH_3Hg^+ denatures nucleic acids (Ochiai, 1977; Cotton and Wilkinson, 1988).

Both inorganic Hg(II) and CH_3Hg^+ in aqueous solution are very toxic to a wide range of aquatic organisms, including bacteria (Hamdy *et al.*, 1977; Hamdy and Wheeler, 1978; Ribo *et al.*, 1989; Farrell *et al.*, 1990), yeast (Kidby, 1974), planktonic marine and freshwater algae (Knauer and Martin, 1972; Nuzzi, 1972; Röderer, 1983) and fish (Alexander, 1974; Ruohtula and Miettinen, 1975; Wobeser, 1975; Walczak *et al.*, 1986). Experiments, however, have shown CH_3Hg^+ to be more effective than inorganic Hg(II) in suppressing the photosynthetic activities of marine phytoplankton (Knauer and Martin, 1972). Similarly, CH_3HgCl was found to be almost an order of magnitude more toxic than $HgCl_2$ to rainbow trout fingerlings (Wobeser, 1975); as $HgCl_2$ and CH_3HgCl penetrate membranes with equal ease (Mason *et al.*, 1995b), the difference in toxicity probably reflects differences in the fate of the two species inside the fish. The greater toxicity of CH_3Hg^+ may arise from the fact that this species more readily exchanges one thiol group for another, rapidly spreading through the contaminated organism's body, impairing the functions of many enzymes in succession (Cotton and Wilkinson, 1988). CH_3Hg^+ and inorganic Hg(II) have qualitatively as well as quantitatively different toxic effects on aquatic organisms (Röderer, 1983).

Wobeser (1974) claimed that certain fish (which he did not identify) are more sensitive to CH_3Hg^+ taken up directly from water than to CH_3Hg^+ ingested with food. Even after long-term consumption of food rich in CH_3Hg^+, there is little evidence of toxicity. Therefore, fish can accumulate large amounts of Hg without suffering detectable adverse effects, whereas predators that prey on the fish are poisoned by the contaminated flesh.

Environmental factors (e.g. Cl^-, sulphides and pH) that determine the bioavailability of Hg also affect toxicity correspondingly. Experiments with model aqueous systems showed that addition of Cl^- ions increased the toxicity of inorganic Hg(II) to bacteria (Farrell *et al.*, 1990), probably by forming the lipophilic species $HgCl_2$, thereby facilitating uptake by bacterial cells. In contrast, addition of Cl^- to rainbow trout in water initially poor in Cl^- protected the fish against otherwise fatal concentrations of inorganic Hg(II) (Walczak *et al.*, 1986). No interpretation of these results was offered, but conversion of part of the Hg to anionic chloride species that were not readily taken up by the fish is one possible explanation. By the same token, the fact that CH_3Hg^+ is more toxic than inorganic Hg(II) to marine phytoplankton (Knauer and Martin, 1972) could be explained, in part at least, by the fact that the lipophilic uncharged species CH_3HgCl crosses cell membranes more easily and rapidly than the hydrophilic anionic species $HgCl_3^-$ and $HgCl_4^{2-}$.

The strong tendency of sulphides and thiols to bind inorganic Hg(II) and CH_3Hg^+ probably has a detoxifying effect except in the case of uncharged, lipophilic, low molecular weight sulphide and thiol complexes, which are bioavailable and therefore presumably toxic (see above). Experiments have demonstrated amelioration of the toxicity of inorganic Hg(II) to bacteria in the presence of cysteine (Ribo *et al.*, 1989; Farrell *et al.*, 1990), suggesting strong binding of Hg(II) by the -SH group of the molecule and inability of the cysteine–Hg complex to pass easily through cell membranes because of the hydrophilic character of the amino acid's -COOH and $-NH_2$ groups (especially when they are ionized).

On exposure to bioavailable, potentially poisonous Hg species, microbial populations undergo natural selection in favour of Hg-tolerant or Hg-resistant strains (Liebert *et al.*, 1991), including ones that protect themselves by means of biochemical mechanisms for converting the toxic Hg species to less harmful forms (Schottel *et al.*, 1974; Hamdy *et al.*, 1977; Pan-Hou and Imura, 1982; Compeau and Bartha, 1985; Summers, 1988; Baldi *et al.*, 1991, 1993b). Both methylation and demethylation, as well as the other microbially mediated Hg speciation and Hg-binding reactions described above, are widely regarded as mechanisms of detoxification. The detoxification strategies of microbes include volatilization of Hg by conversion of inorganic Hg(II) or CH_3Hg^+ to Hg(0), and by transformation of inorganic Hg(II) or CH_3Hg^+ to $(CH_3)_2Hg$, resulting in removal of the Hg from the microbe's immediate vicinity (Schottel *et al.*, 1974; Summers, 1988; Baldi *et al.*, 1991, 1993b). Conversion of inorganic Hg(II) to CH_3Hg^+ is also seen as a detoxification reaction (Hamdy and Noyes, 1975; Hamdy *et al.*, 1977; Pan-Hou and Imura, 1982; Compeau and Bartha, 1985; D'Itri, 1991). This may seem illogical, since many microbes find it necessary to get rid of CH_3Hg^+ by changing it into more volatile, less water-soluble products, but the fact that CH_3HgCl is more hydrophilic than $HgCl_2$ may help to explain it. Besides, CH_3Hg^+ is more volatile, as well as more water-soluble, and it forms less stable bonds than equivalent inorganic Hg(II) species; hence, it has a greater tendency to diffuse away from the microbes (D'Itri, 1991). Moreover, methylation may meet the detoxification requirements of some microbial species, whereas demethylation is more satisfactory for others, or the choice of a detoxification strategy may depend on external conditions. The volatility and relative lipophilicity and hydrophilicity depend on the nature of the ligands bound to the CH_3Hg^+ and Hg^{2+} ions, and on other factors, such as pH.

Immobilization of Hg by production of H_2S or thiols is another possible strategy for preventing Hg toxicity. Accordingly, SO_4^{2-}-reducing bacteria have two protective mechanisms which they use under different conditions: methylation (conversion of inorganic Hg(II) to CH_3Hg^+, and CH_3Hg^+ to $(CH_3)_2Hg$) and production of H_2S (Compeau and Bartha, 1985; Baldi *et al.*, 1993b). The pH of water also has an important bearing on Hg toxicity. The toxicity of inorganic Hg(II) to bacteria was found to be lower at pH 9 than at pH 5 or 6 (Ribo *et al.*, 1989); no doubt the reason for this is that the proportion of $Hg(OH)_2$ to $HgCl_2$ is higher at pH 9 (see above). The same principles apply to the analogous species CH_3HgOH and CH_3HgCl.

There are grounds for suspecting that fish and other aquatic animals, as with microbes, have evolved special mechanisms for protecting themselves against Hg poisoning. The evidence suggesting excretion of CH_3Hg^+ complexed with hydrophilic carrier molecules, the ability of certain fish species to demethylate CH_3Hg^+ in their tissues, and the fact that many fish habitually consume food heavily contaminated with CH_3Hg^+ without suffering untoward consequences all point to the possible existence of specific adaptations designed to protect fish from harmful effects of Hg (see above, and Chapter 10).

An agent of CH_3Hg^+ and inorganic Hg(II) detoxification which may be of great significance to aquatic organisms is Se in the form of selenides (see above). It has long been known that Se is effective in ameliorating the effects heavy metals, including Hg and Cd, leading to the hypothesis that it detoxifies the Hg accumulated by marine mammals. The highly significant positive correlation between Hg and Se, the 1 : 1 Hg/Se mole ratio and the apparently strong binding of Hg by Se in liver and brain tissues of marine mammals are consistent with this possibility (Koeman *et al.*, 1973, 1975). Se has also been shown to abate the acute toxicity of $HgCl_2$ in freshwater fish; interestingly, Se increased the total Hg content of the fish while decreasing its toxicity, strongly suggesting immobilization of Hg inside the animal by formation of stable Hg–Se complexes (Heisinger *et al.*, 1979). Se in marine fish containing high CH_3Hg^+ concentrations may also protect animals and humans that eat the flesh of the fish (Ganther *et al.*, 1972).

The ability of Se to prevent or abate the toxicity of CH_3Hg^+ and inorganic Hg(II) has inspired the idea of deliberately adding Se to Hg-polluted lakes for purposes of remediation (Rudd *et al.*, 1980; Turner and Rudd, 1983; Turner and Swick, 1983). Field experiments designed to test this concept succeeded in demonstrating that Se was effective in reducing Hg concentrations in freshwater fish (Rudd *et al.*, 1980; Turner and Rudd, 1983; Turner and Swick, 1983; Paulsson and Lundbergh, 1991; Lindqvist *et al.*, 1991). Unfortunately,

Se itself is a potentially toxic element (Lemly and Smith, 1987; Magos, 1991). In natural waters, Se levels as low as 10 μg l^{-1} (the legal upper limit for drinking water in Canada; Turner and Rudd, 1983) can be harmful to fish, and Se at concentrations exceeding 2–5 μg l^{-1} may undergo biomagnification in food chains, resulting in toxic effects (Lemly and Smith, 1987). The perils of using Se to combat Hg were revealed dramatically by a field experiment in Sweden: in four of 11 lakes treated with Se for detoxification of Hg, the reproduction of fish declined catastrophically (Lindqvist *et al.*, 1991). Thus, although the method has shown some efficacy, further work is needed, and application of such a method cannot be recommended without a guarantee that its potentially disastrous side-effects can be prevented.

5.5 THE BIOGEOCHEMICAL CYCLE OF MERCURY

5.5.1 SOURCES OF MERCURY

Hg is introduced into air, water and soil by various natural processes and human activities (Nriagu and Pacyna, 1988; Nriagu, 1989; D'Itri, 1991; Lindqvist *et al.*, 1991; Hudson *et al.*, 1995; Pacyna and Keeler, 1995; Jackson, 1997). The Hg may be derived mainly from local inputs or imported from distant sources by atmospheric or fluvial transport. It may arise from point discharges or diffuse inputs, or both, and can originate from secondary sinks (e.g. soil organic matter) as well as primary sources (e.g. volcanoes or coal-burning power plants); and it is cycled between different compartments, undergoing chemical transformations by various biological and abiotic pathways. There are continual interchanges between the Hg pools in the lithosphere, hydrosphere and atmosphere.

Largely owing to the volatility of Hg, major quantities of natural and anthropogenic Hg are released into the atmosphere every year and are conveyed to aquatic and terrestrial ecosystems near and far by atmospheric circulation. Hg is a ubiquitous trace constituent of the atmosphere. Airborne Hg is mostly in the form of gaseous Hg(0), but includes volatile Hg(II) species along with Hg(II) (and traces of Hg(0)) sorbed to dust or dissolved in water droplets (Lindberg, 1987; Brosset and Lord, 1991; Lindqvist *et al.*, 1991). Hg(0) and Hg(II) are transported over thousands of kilometres by global atmospheric circulation, but dissolved and particulate Hg(II) have a far stronger tendency than Hg(0) to be returned to the Earth's surface by wet and dry deposition, much of which occurs within 100 km of the source (Lindberg, 1987; Lindqvist *et al.*, 1991); at low temperatures Hg(0) too may undergo appreciable dry deposition (Steinnes and Andersson, 1991). The separation of these pathways is never absolute, as Hg(0) and Hg(II) are interconvertible (Lindqvist *et al.*, 1991; Munthe *et al.*, 1991; Schroeder *et al.*, 1991). On a global scale, the natural and anthropogenic contributions to the atmospheric Hg burden are of comparable magnitude, each amounting to thousands of tonnes per year; but there is considerable geographic and temporal variation (Nriagu and Pacyna, 1988; Nriagu, 1989; D'Itri, 1991; Lindqvist *et al.*, 1991; Hudson *et al.*, 1995; Jackson, 1997) and most estimates of the natural flux may be too high because of failure to take secondary anthropogenic emissions into account (Hudson *et al.*, 1995). Anthropogenic emissions have increased over time (largely owing to combustion of fossil fuels) since the onset of the Industrial Revolution, accelerating in the mid-twentieth century, and then declining somewhat in some regions whilst continuing unabated elsewhere. The evidence for a temporal increase in the rate of Hg deposition from the atmosphere includes profiles of Hg (and associated pollutants and micro-fossils indicative of fuel combustion) in dated cores of fine-grained sediments representing continuous, undisturbed stratigraphic sequences from remote lakes with no history of local pollution (e.g. Heit *et al.*, 1981; Ouellet and Jones, 1983; Evans, 1986; Johnson *et al.*, 1986; Lockhart *et al.*, 1993; Louchouarn *et al.*, 1993; Engstrom *et al.*, 1994). The cores typically show Hg enrichment in the uppermost (youngest) horizons.

Rasmussen (1994) has questioned the quantitative significance of long-range atmospheric transport of anthropogenic Hg, asserting that insufficient attention has been paid to natural sources of Hg and maintaining that the distribution of Hg in cores may be due to post-depositional remobilization of Hg rather than temporal variation in loading. However, a comprehensive review and synthesis of the literature (Jackson, 1997) has refuted Rasmussen's arguments. The case against Rasmussen's views is especially convincing because it is based on a large, diverse body of evidence characterized by agreement between different kinds of information amassed by many investigators studying various natural systems in widely separated regions.

The ultimate natural sources of Hg are in the Earth's crust and upper mantle. Volatile Hg is vented into the atmosphere by volcanic activity and other degassing processes, whilst Hg in rocks exposed at the surface is released by weathering (Jonasson and Boyle, 1972; Nriagu, 1989, 1992; D'Itri, 1991; Painter *et al.*, 1994; Friske and Coker, 1995). Hg is widely dispersed as a trace element in various rocks, but anomalously high concentrations of Hg, including Hg ores, are found in mineralized bedrock and at the Earth's surface in zones of vulcanism, faulting, fracturing, and hot spring activity, mainly in mobile belts along plate boundaries; coal and fine-grained sedimentary rocks (e.g. shales, especially those of high organic and sulphide content), are also somewhat enriched in Hg (Gavis and Ferguson, 1972; Jonasson and Boyle, 1972; Painter *et al.*, 1994; Friske and Coker, 1995). The Hg deposited in mineralized zones is mainly in the form of cinnabar but may include minor amounts of metacinnabar, calomel, Hg(0) and assorted rare Hg minerals, and occurs as an impurity in other minerals. Primary Hg deposits usually precipitate in faults, fractures and pores in the host rock (commonly sedimentary rock), or replace the rock, and are deposited in hot springs, through the agency of hydrothermal solutions emanating from subsurface magma. Other phenomena that transfer Hg, directly or indirectly, from natural sources to aquatic ecosystems include soil erosion, forest fires, the formation of aerosols from sea spray, and various microbially mediated and abiotic reactions whereby Hg(II) is converted to volatile species in soil and water (Nriagu, 1989, 1992). Owing to the patchy distribution of Hg-rich materials and the episodic nature of phenomena such as volcanic eruptions, natural Hg levels in Earth surface environments show considerable spatial and temporal variation. Geographic variations in the composition of lake and stream sediments reflect variations in the Hg content of rock formations in their catchment basins (Evans, 1986; Rognerud and Fjeld, 1993; Painter *et al.*, 1994; Friske and Coker, 1995). Some regional Hg anomalies in soil organic matter and marine animals may be of geological origin (Steinnes, 1994; Wagemann *et al.*, 1995).

The natural background Hg has been enormously augmented by Hg pollution resulting from a variety of human (especially industrial) activities (Gavis and Ferguson, 1972; D'Itri *et al.*, 1978; Lindberg, 1987; Nriagu and Pacyna, 1988; D'Itri, 1991; Lindqvist *et al.*, 1991; Nriagu, 1992). Combustion of coal has been the greatest single cause of atmospheric Hg pollution, with incineration of solid refuse being a close second (Nriagu and Pacyna, 1988; Nriagu, 1992; Pacyna and Keeler, 1995). Air pollution due to the burning of coal and other fossil fuels is responsible for Hg contamination and acidification in many lakes hundreds or thousands of kilometres from the sources of pollution (Jackson, 1997). At many localities industrial and municipal wastewaters containing Hg in the form of Hg(0), inorganic Hg(II) or, in some cases, organometallic Hg(II) compounds have been discharged directly into aquatic environments, usually rivers, which may then transport the Hg (mostly as particulate Hg(II)) over great distances, causing widespread contamination of aquatic organisms and habitats (e.g. Armstrong and Hamilton, 1973; Fimreite and Reynolds, 1973; Bishop and Neary, 1976; Parks, 1976; Parks and Hamilton, 1987; Parks *et al.*, 1991b). Chlor-alkali plants employing Hg(0) electrodes for electrolysis of NaCl have been especially notorious point sources of Hg (Gavis and Ferguson, 1972) and have been sources of airborne Hg too (Lindqvist *et al.*, 1991). Technological changes involving replacement of Hg(0) electrodes with diaphragm cells have abated the problem, resulting in gradual lowering of Hg levels in fish, but harmful effects of past pollution persist for many years (e.g. Armstrong and Scott, 1979; Jackson, 1980; Jackson *et al.*, 1982; Jackson and Woychuk, 1980a,b; Parks and Hamilton, 1987; Parks *et al.*, 1986, 1989, 1991a,b). Many other human activities contribute to Hg pollution. The Hg pollution and consequent mass poisonings by consumption of CH_3Hg^+-contaminated fish and shellfish at Minamata and Niigata, Japan, resulted from discharge of Hg-contaminated effluents from acetaldehyde plants into natural waters (Takizawa, 1979). Recently the use of Hg(0) to extract gold from ore has led to severe pollution of rivers and air in the Amazon region of Brazil (Aula *et al.*, 1994; Palheta and Taylor, 1995).

5.5.2 BIOGEOCHEMICAL PATHWAYS OF MERCURY IN AQUATIC ECOSYSTEMS

Inorganic Hg(II) is the dominant form of Hg in aquatic environments. When introduced into natural waters, the Hg is rapidly and efficiently scavenged by fine-grained suspended particles and most of it accumulates in clay- and silt-sized bottom sediments, although a small and highly variable proportion of it remains in solution, or is resolubilized, in the form of different ionized and uncharged aqueous species and complexes. The partitioning of Hg between the solid and aqueous phases and the nature and quantities of the different forms of Hg are subject to continual change in response to varying environmental conditions and microbial activities. As a rule, Hg(II) is strongly and preferentially bound by inorganic sulphides and the thiol groups of humic and non-humic organic substances (and their much rarer selenide and telluride analogues); if these binding agents are scarce (as in a highly oxidizing environment where most of the particulate matter consists of mineral detritus) the Hg(II) is sorbed by mineral particles, principally by Fe and Mn oxyhydroxides. Because most of the Hg is bound to sediments and suspended particles, the dynamics and distribution of Hg in natural waters are largely determined by the same forces that control the erosion, suspension, transport and deposition of its carrier particles. In river systems, particulate Hg is resuspended from bed and bank deposits during times of high flow rates and bottom scour (e.g. during the spring flood in the temperate zone), is transported in suspension by fluvial currents and is deposited in basins of deposition such as riverine lakes and marine basins. Fairly deep productive basins characterized by reducing conditions and the formation of H_2S on the bottom tend to trap and accumulate Hg, and a series of riverine lakes or settling ponds (especially ones with reducing conditions at the bottom) helps to purify a river and ameliorate the effects of heavy metal pollution through removal, immobilization and burial of the metals (Jackson, 1978, 1979, 1993a,b); wetlands, too, can be effective in trapping riverborne metals (Sinicrope *et al.*, 1992). However, the efficiency of Hg immobilization, and even the relative tendencies of Hg and other heavy metals to be immobilized, varies greatly from one basin of deposition to another (Jackson, 1979). At best, the process of entrapment is far from absolute, and the Hg deposited and immobilized in sediments is, to a greater or lesser extent, subject to remobilization and further transport (Jackson and Woychuk, 1980a,b, 1981; Jackson, 1993a); and Hg buried in the sediments of a riverbed or floodplain can be returned to the aquatic environment through erosion and resuspension (brought on by short-term environmental changes, such as storms, or long-term changes, such as a drop in base level causing the river to shift from aggradation to downward cutting).

Even in a river with a succession of efficient Hg-trapping basins of deposition, Hg pursues its gradual migration in the downstream direction. For example, Hg sorbed to FeOOH coatings on suspended clay and silt may be transported by fluvial currents and deposited in an anoxic lake basin rich in sulphides and organic matter, where the FeOOH is reduced and solubilized, liberating the Hg, which is mostly immobilized by sulphide and thiols but is, to some extent, mobilized as dissolved sulphide and thiol complexes, hydrolysis products, etc., and recycled into the water (Jackson, 1993a). Oxidation of the reduced sediments through contact with water containing dissolved O_2 would liberate sulphide-bound Hg by oxidizing the sulphide, although part of the Hg might immediately be immobilized again owing to coprecipitation with FeOOH or MnOOH and binding by humic matter. Changes of this kind may occur seasonally. During times of slack flow in a river, there is minimal resuspension of Hg bound to mineral detritus (except during episodic events like storms), but there may be an increase in the concentration of Hg dissolved or associated with dispersed colloidal organic matter in the water, as may occur during the summer (Jackson *et al.*, 1982). When Hg sorbed to fluvial particulate matter is transported into an estuary, a number of important changes occur owing to the gradational shift from freshwater to marine conditions (Jackson, 1998). The changes include partial desorption and solubilization of the bound Hg by the formation of water-soluble $Hg–Cl^-$ complexes, as well as flocculation of Hg-bearing clay and humic matter owing to the increase in salinity.

Throughout these processes, Hg is continually in a state of flux owing to biological activities and physicochemical processes controlling speciation reactions and binding and release by various sorbents, complexing agents, and organisms – crucially important reactions because of their relevance to the biological effects of the Hg, even though aqueous Hg species usually comprise only a small percentage of the total Hg supply. In any aquatic ecosystem, a small and highly variable fraction of the total inorganic Hg(II) in the environment consists of bioavailable inorganic Hg(II) species in solution or loosely sorbed by particles. It is from this pool of reactive inorganic Hg(II) species that the methylated species CH_3Hg^+ and $(CH_3)_2Hg$ are synthesized by various free-living microorganisms in the surficial sediments and water column. The $(CH_3)_2Hg$ is largely volatilized, but dissolved CH_3Hg^+ gradually diffuses through the water column, where it is readily accumulated by organisms along with some bioavailable inorganic Hg(II); however, CH_3Hg^+ production is kept in check by the process of demethylation, whereby various microorganisms decompose CH_3Hg^+, reducing the Hg to Hg(0). Inorganic Hg(II) may also be reduced to Hg(0) by microbes. To some extent, such processes are abiotic (as in photochemical production of Hg(0) from CH_3Hg^+ and inorganic Hg(II) in sunlight), but they are mostly mediated by microbes. In contrast to inorganic Hg(II), the CH_3Hg^+ undergoes biomagnification up the food chain; therefore, it is by far the most dominant form of Hg in fish and in piscivorous animals such as marine mammals and aquatic birds; at high concentrations in fish CH_3Hg^+ is extremely hazardous to consumers, including humans. Hg(0) formed in the aquatic environment tends to be lost through volatilization, though Hg(0) gas (along with inorganic Hg(II) bound to dust, in the atmosphere) is also introduced into bodies of water by wet and dry fallout, and some of it may be oxidized biochemically or abiotically to Hg(II).

The net rate of CH_3Hg^+ production depends on the abundance of bioavailable inorganic Hg(II) species and the activities of methylating and demethylating microbes. Both the percentage of the total inorganic Hg(II) pool that is bioavailable (susceptible to methylation) and the activities of the microbes that produce and destroy CH_3Hg^+ are controlled by a wide range of environmental variables; moreover, the kinetics of CH_3Hg^+ accumulation by the aquatic biota depend not only on the conditions controlling the supply of CH_3Hg^+ but also on the inherent characteristics, behaviour, activities and stage of development of the organisms themselves. Consequently, net rates of CH_3Hg^+ production and bioaccumulation show enormous variation in response to spatial and temporal variations in environmental conditions. In general, anoxic but not excessively reducing (sulphide-rich) environments of weakly acidic to neutral pH, and a generous supply of labile organic matter that can provide nutrient substrates for microbes, are optimal for CH_3Hg^+ production, but the process occurs to a greater or lesser extent under a wide variety of conditions. Above all, the formation and bioaccumulation of bioavailable Hg species such as CH_3Hg^+ are controlled by an infinitely complex interplay of biological and physicochemical processes. The biologically reactive unmethylated and methylated species make up only a minor proportion of the total Hg supply, which is generally a very poor and unreliable measure of ecological impact. The net result of Hg contamination in natural environments represents the combined effect of all these multifarious processes acting and interacting at the same time, or in succession, both directly and indirectly. Some processes are synergistic whilst others tend to counteract each other.

To understand the biological effects of Hg in nature, we must take all these different environmental and biological aspects of the problem into account and achieve an interdisciplinary synthesis. One of the greatest challenges in future biogeochemical research will be to attain an in-depth understanding of this most intricate and multidimensional of subjects – the biogeochemical cycle of mercury.

ACKNOWLEDGMENTS

Financial support was provided by the Government of Canada (Department of the Environment).

REFERENCES

Aiken, G.R., McKnight, D.M., Wershaw, R.L. and MacCarthy, P. (1985) *Humic Substances in Soil, Sediment, and Water*, John Wiley & Sons (Wiley Interscience), New York, Toronto, Chichester, Brisbane, Singapore.

Akagi, H., Miller, D.R. and Kudo, A. (1977) Photochemical transformation of mercury, in *Distribution and transport of pollutants in flowing water ecosystems*: Ottawa River project, final report, Vol. 1, Chapter 16, National Research Council of Canada, Ottawa.

Alberts, J.J., Schindler, J.E., Miller, R.W. and Nutter, D.E. Jr (1974) Elemental mercury evolution mediated by humic acid, *Science* **184**, 895–897.

Alexander, D.G. (1974) Mercury effects on swimming and metabolism of trout. *Proc. Int. Conf. on Transport of Persistent Chemicals in Aquatic Ecosystems (Ottawa, Canada, 1–3 May, 1974)*, Sect. III, p. 65-69.

Allard, B. and Arsenie, I. (1991) Abiotic reduction of mercury by humic substances in aquatic system – an important process for the mercury cycle, *Water Air Soil Polln* **56**, 457–464.

Amyot, M., Mierle, G., Lean, D.R.S. and McQueen, D.J. (1994) Sunlight-induced formation of dissolved gaseous mercury in lake waters, *Environ. Sci. Technol.* **28**, 2366–2371.

Anderson, M.R., Scruton, D.A., Williams, U.P. and Payne, J.F. (1995) Mercury in fish in the Smallwood Reservoir, Labrador, twenty one years after impoundment, *Water Air Soil Polln* **80**, 927–930.

Andersson, A. (1979) Mercury in soils, in *The Biogeochemistry of Mercury in the Environment*, (ed. J.O. Nriagu), Elsevier/North-Holland Biomedical Press, Amsterdam, New York, Oxford, pp. 79–112.

Andersson, P., Borg, H. and Kärrhage, P. (1995) Mercury in fish muscle in acidified and limed lakes. *Water Air Soil Polln* **80**, 889–892.

Andren, A.W. and Harriss, R.C. (1975) Observations on the association between mercury and organic matter dissolved in natural waters, *Geochim. Cosmochim. Acta* **39**, 1253–1257.

Armstrong, F.A.J. and Hamilton, A.L. (1973) Pathways of mercury in a polluted Northwestern Ontario lake, in *Trace Metals and Metal–Organic Interactions in Natural Waters*, (ed. P.C. Singer), Ann Arbor Science Publishers, Ann Arbor, pp. 131–156

Armstrong, F.A.J. and Scott, D.P. (1979) Decrease in mercury content of fishes in Ball Lake, Ontario, since imposition of controls on mercury discharges. *J. Fish. Res. Board Can.* **36**, 670–672.

Aula, I., Braunschweiler, H., Leino, T. *et al.* (1994) Levels of mercury in the Tucuruí Reservoir and its surrounding area in Pará, Brazil, in *Mercury Pollution*, (eds C.J. Watras and J.W. Huckabee), Lewis Publishers, Boca Raton, Ann Arbor, London, Tokyo, pp. 21–40.

Baldi, F., Filipelli, M. and Olson, G.J. (1989) Biotransformation of mercury by bacteria isolated from a river collecting cinnabar mine waters, *Microbiol. Ecol.* **17**, 263–274.

Baldi, F., Semplici, F. and Filippelli, M. (1991) Environmental applications of mercury resistant bacteria, *Water Air Soil Polln* **56**, 465–475.

Baldi, F., Parati, F., Semplici, F. and Tandoi, V. (1993a) Biological removal of inorganic Hg(II) as gaseous elemental Hg(0) by continuous culture of a Hg-resistant *Pseudomonas putida* strain FB-1,. *World J. Microbiol. Biotechnol.* **9**, 275–279.

Baldi, F., Pepi, M. and Filippelli, M. (1993b) Methylmercury resistance in *Desulfovibrio desulfuricans* strains in relation to methylmercury degradation, *Appl. Environ. Microbiol.* **59**, 2479–2485.

Baldi, F., Parati, F. and Filippelli, M. (1995) Dimethylmercury and dimethylmercury-sulfide of microbial origin in the biogeochemical cycle of Hg. *Water Air Soil Polln* **80**, 805–815.

Balzani, V. and Carassiti, V. (1970) *Photochemistry of Coordination Compounds*, Academic Press, New York, London.

Barkay, T., Turner, R.R., VandenBrook, A. and Liebert, C. (1991) The relationships of Hg(II) volatilization from a freshwater pond to the abundance of *mer* genes in the gene pool of the indigenous microbial community, *Microbiol. Ecol.* **21**, 151–161.

Bartlett, P.D. and Craig, P.J. (1979) Methylation processes for mercury in estuarine sediments, in *Heavy Metals in the Environment* (Proc. Int. Conf. on Management and Control of Heavy Metals in the Environment, London, Sept., 1979), CEP Consultants, Edinburgh, pp. 354–355.

Baughman, G.L., Gordon, J.A., Wolfe, N.L. and Zepp, R.G. (1973) *Chemistry of Organomercurials in Aquatic Systems*, Ecological Research Series, EPA-660/3-73-012, National Environmental Research Center, Office of Research and Development, US Environmental Protection Agency, Corvallis.

Baxter, R.M. and Carey, J.H. (1982) Reactions of singlet oxygen in humic waters, *Freshwater Biol.* **12**, 285–292.

Baxter, R.M. and Carey, J.H. (1983) Evidence for photochemical generation of superoxide ion in humic waters. *Nature* **306**, 575–576.

Beijer, K. and Jernelöv, A. (1979) Methylation of mercury in aquatic environments, in *The Biogeochemistry of Mercury in the Environment*, (ed. J.O. Nriagu), Elsevier/North-Holland Biomedical Press, Amsterdam, Oxford, New York, pp. 203–210.

Beneš, P. and Havlík, B. (1979) Speciation of mercury in natural waters, in *The Biogeochemistry of Mercury in the Environment*, (ed. J.O. Nriagu), Elsevier/North-Holland Biomedical Press, Amsterdam, Oxford, New York, pp. 175–202.

Bertilsson, L. and Neujahr, H.Y. (1971) Methylation of mercury compounds by cobalamin, *Biochem.* **10**, 2805–2808.

Bidstrup, P.L. (1964) *Toxicity of Mercury and its Compounds*, Elsevier, Amsterdam, London, New York.

Bishop, J.N. and Neary, B.P. (1974) The form of mercury in freshwater fish. *Proc. Int. Conf. on Transport of Persistent Chemicals in Aquatic Ecosystems (Ottawa, Canada, 1–3 May, 1974)*, Sect. III, pp. 25–29.

Bishop, J.N. and Neary, B.P. (1976) *Mercury Levels in Fish from Northwestern Ontario, 1970–1975*, Ministry of the Environment, Ontario, Canada.

Bisogni, J.J. (1979) Kinetics of methylmercury formation and decompositon in aquatic environments, in *The Biogeochemistry of Mercury in the Environment*, (ed. J.O. Nriagu), Elsevier/North-Holland Biomedical Press, Amsterdam, Oxford, New York, pp. 211–230.

Bisogni, J.J. Jr and Lawrence, A.W. (1975) Kinetics of mercury methylation in aerobic and anaerobic aquatic environments, *J. Water Polln Control Fed.* **47**, 135–152.

Björnberg, A., Håkanson, L. and Lundbergh, K. (1988) A theory on the mechanisms regulating the bioavailability of mercury in natural waters, *Environ. Polln* **49**, 53–61.

Bloom, N.S. and Watras, C.J. (1989) Observations of methylmercury in precipitation, *Sci. Total Environ.* **87/88**, 199–207.

Bloom, N.S., Watras, C.J. and Hurley, J.P. (1991) Impact of acidification on the methylmercury cycle of remote seepage lakes, *Water Air Soil Polln* **56**, 477–491.

Blum, J.E. and Bartha, R. (1980) Effect of salinity on methylation of mercury, *Bull. Environ. Contam. Toxicol.* **25**, 404–408.

Bodaly, R.A., Hecky, R.E. and Fudge, R.J.P. (1984) Increases in fish mercury levels in lakes flooded by the Churchill River diversion, northern Manitoba, *Can. J. Fish. Aquat. Sci.* **41**, 682–691.

Bodaly, R.A., Rudd, J.W.M., Fudge, R.J.P. and Kelly, C.A. (1993) Mercury concentrations in fish related to size of remote Canadian Shield lakes, *Can. J. Fish. Aquat. Sci.* **50**, 980–987.

Bothner, M.H. and Carpenter, R. (1973) Sorption–desorption reactions of mercury with suspended matter in the Columbia River. *Proc. Symp. Radioactive Contamination of the Marine Environment (Vienna, Austria, 1972)*, IAEA, pp. 73–87.

Boudou, A. and Ribeyre, F. (1981) Comparative study of the trophic transfer of two mercury compounds – $HgCl_2$ and CH_3HgCl – between *Chlorella vulgaris* and *Daphnia magna*. Influence of temperature. *Bull. Environ. Contam. Toxicol.* **27**, 624–629.

Boudou, A., Delnomdedieu, M., Georgescauld, D. *et al.* (1991) Fundamental roles of biological barriers in mercury accumulation and transfer in freshwater ecosystems, *Water Air Soil Polln* **56**, 807–821.

Bowen, H.J.M. (1966) *Trace Elements in Biochemistry*, Academic Press, London, New York.

Branfireun, B.A., Heyes, A. and Roulet, N.T. (1996) The hydrology and methylmercury dynamics of a Precambrian Shield headwater peatland, *Water Resources Res.* **32**, 1785–1794.

Brosset, C. and Lord, E. (1991) Mercury in precipitation and ambient air – a new scenario. *Water Air Soil Polln* **56**, 493–506.

Brouard, D., Doyon, J.-F. and Schetagne, R. (1994) Amplification of mercury concentrations in lake whitefish (*Coregonus clupeaformis*) downstream from the La Grande 2 Reservoir, James Bay, Québec, in *Mercury Pollution*, (eds C.J. Watras and J.W. Huckabee), Lewis Publishers, Boca Raton, Ann Arbor, London, Tokyo, pp. 369–379.

Brouzes, R.J.P., McLean, R.A.N. and Tomlinson, G.H. (1977) The link between pH of natural waters and the mercury content of fish. Report presented at meeting of US National Academy of Sciences (National Research Council Panel on Mercury), 3 May, 1977, Washington, DC.

Budavari, S., O'Neil, M.J., Smith, A. and Heckelman, P.E. (eds) (1989) *The Merck Index*, 11th edn, Merck & Co., Rahway.

Burkett, R.D. (1974) The influence of temperature on uptake of methylmercury-203 by bluntnose minnows, *Pimephales notatus* (Rafinesque), *Bull. Environ. Contam. Toxicol.* **12**, 703–709.

Burrows, W.D. and Krenkel, P.A. (1973) Studies on uptake and loss of methylmercury-203 by bluegills (*Lepomis macrochiros* Raf.), *Environ. Sci. Technol.* **7**, 1127–1130.

Cabana, G., Tremblay, A., Kalff, J. and Rasmussen, J.B. (1994) Pelagic food chain structure in Ontario lakes: a determinant of mercury levels in lake trout (*Salvelinus namaycush*), *Can. J. Fish. Aquat. Sci.* **51**, 381–389.

Carty, A.J. and Malone, S.F. (1979) The chemistry of mercury in biological systems, in *The Biogeochemistry of Mercury in the Environment*, (ed. J.O. Nriagu), Elsevier/North-Holland Biomedical Press, Amsterdam, Oxford, New York, pp. 433–479.

Chen, J., Tang, F. and Wang, F. (1995) Mobilization of mercury from estuarine suspended particulate matter: a case study in the Yalujiang estuary, northeast China, *Water Qual. Res. J. Can.* **30**, 25–32.

Choi, S.-C. and Bartha, R. (1993) Cobalamin-mediated mercury methylation by *Desulfovibrio desulfuricans* LS. *Appl. Environ. Microbiol.* **59**, 290–295.

Choi, S.-C., Chase, T. Jr and Bartha, R. (1994) Metabolic pathways leading to mercury methylation in *Desulfovibrio desulfuricans* LS. *Appl. Environ. Microbiol.* **60**, 4072–4077.

Choudhry, G.G. (1984) *Humic Substances*, Gordon and Breach Science Publishers, New York, London, Paris, Montreux, Tokyo.

Christman, R.F. and Gjessing, E.T. (eds) (1983) *Aquatic and Terrestrial Humic Materials*, Ann Arbor Science (Butterworth Group), Ann Arbor.

Compeau, G. and Bartha, R. (1984) Methylation and demethylation of mercury under controlled redox, pH, and salinity conditions, *Appl. Environ. Microbiol.* **48**, 1203–1207.

Compeau, G. and Bartha, R. (1985) Sulfate-reducing bacteria: principal methylators of mercury in anoxic estuarine sediment, *Appl. Environ. Microbiol.* **50**, 498–502.

Compeau, G. and Bartha, R. (1987) Effect of salinity on mercury-methylating activity of sulfate-reducing bacteria in estuarine sediments, *Appl. Environ. Microbiol.* **53**, 261–265.

Cope, W.G., Wiener, J.G. and Rada, R.G. (1990) Mercury accumulation in yellow perch in Wisconsin seepage lakes: relation to lake characteristics, *Environ. Toxicol. Chem.* **9**, 931–940.

Cotton, F.A. and Wilkinson, G. (1988) *Advanced Inorganic Chemistry*, 5th edn, Wiley-Interscience (John Wiley & Sons), New York, Toronto, Chichester, Brisbane, Singapore.

Craig, P.J. and Bartlett, P.D. (1978) The role of hydrogen sulphide in environmental transport of mercury. *Nature* **275**, 635–637.

Craig, P.J. and Moreton, P.A. (1986) Total mercury, methyl mercury and sulphide levels in British estuarine sediments, III. *Water Res.* **20**, 1111–1118.

De Filippis, L.F. and Pallaghy, C.K. (1975) A simple model for the non-enzymatic reduction and alkylation of mercuric salts in biological systems. *Bull. Environ. Contam. Toxicol.* **14**, 32–37.

de Freitas, A.S.W. and Hart, J.S. (1975) Effect of body weight on uptake of methyl mercury by fish, *Water Quality Parameters, ASTM STP 573*, American Society for Testing and Materials, pp. 356–363.

de Freitas, A.S.W., Qadri, S.U. and Case, B.E. (1974) Origins and fate of mercury compounds in fish, in *Proc. Int. Conf. on Transport of Persistent Chemicals in Aquatic Ecosystems (1–3 May, 1974, Ottawa, Canada)*, Sect. III, pp. 31–36.

de Freitas, A.S.W., Gidney, M.A.J., McKinnon, A.E. and Norstrom, R.J. (1977) Factors affecting whole-body retention of methyl mercury in fish, in *Biological Implications of Metals in the Environment*, (eds H. Drucker and R.E. Wildung,), Proceedings 15th Annual Hanford Life Sciences Symposium, 29 September–1 October, 1975, Richland, Washington, Technical Information Center, Energy Research and Development Administration, US Department of Commerce, Springfield, Virginia, pp. 441–451.

de Groot, A.J. and Allersma, E. (1975) Field observations on the transport of heavy metals in sediments, in *Heavy Metals in the Aquatic Environment*, (ed. P. A. Krenkel), Pergamon Press, Oxford, New York, Toronto, Braunschweig, Sydney, pp. 85–101.

de Groot, A.J., de Goeij, J.J.M. and Zegers, C. (1971) Contents and behaviour of mercury as compared with other heavy metals in sediments from the rivers Rhine and Ems. *Geologie en Mijnbouw* **50**, 393–398.

DeSimone, R.E., Penley, M.W., Charbonneau, L. *et al.* (1973) The kinetics and mechanism of cobalamin-dependent methyl and ethyl transfer to mercuric ion. *Biochim. Biophys. Acta* **304**, 851–863.

D'Itri, F. (1971) Comparison of mercury levels in an oligotrophic and a eutrophic lake, *Marine Technol. Soc. J.* **5**, 10–14.

D'Itri, F. (1972) *The Environmental Mercury Problem*, Chemical Rubber Co. Press, Cleveland.

D'Itri, F. (1991) Mercury contamination – what we have learned since Minamata, *Environ. Monitoring Assess.* **19**, 165–182.

D'Itri, F.M., Andren, A.W., Doherty, R.A. *et al.* (1978) *An Assessment of Mercury in the Environment*, National Academy of Sciences, Washington, DC.

Douglas, B.E. and McDaniel, D.H. (1965) *Concepts and Models of Inorganic Chemistry*, Blaisdell, Waltham (Mass.), Toronto, London.

Dyrssen, D. and Wedborg, M. (1991) The sulphur–mercury(II) system in natural waters, *Water Air Soil Polln* **56**, 507–519.

Engstrom, D.R., Swain, E.B., Henning, T.A. *et al.* (1994) Atmospheric mercury deposition to lakes and watersheds, in *Environmental Chemistry of Lakes and Reservoirs*, (ed. L.A. Baker), pp. 33–66.

Evans, R.D. (1986) Sources of mercury contamination in the sediments of small headwater lakes in south-central Ontario, Canada. *Arch. Environ. Contam. Toxicol.* **15**, 505–512.

Fagerström, T. and Jernelöv, A. (1972) Some aspects of the quantitative ecology of mercury. *Water Res.* **6**, 1193–1202.

Farrah, H. and Pickering, W.F. (1978) The sorption of mercury species by clay minerals. *Water, Air, Soil Polln* **9**, 23–31.

Farrell, R.E., Germida, J.J. and Huang, P.M. (1990) Biotoxicity of mercury as influenced by mercury(II) speciation, *Appl. Environ. Microbiol.* **56**, 3006–3016.

Faust, B.C. (1992) The octanol/water distribution coefficients of methylmercuric species: the role of aqueous-phase chemical speciation, *Environ. Toxicol. Chem.* **11**, 1373–1376.

Feick, G., Horne, R.A. and Yeapple, D. (1972) Release of mercury from contaminated freshwater sediments by the runoff of road deicing salt, *Science* **175**, 1142–1143.

Fimreite, N. and Reynolds, L.M. (1973) Mercury contamination of fish in Northwestern Ontario,. *J. Wildlife Manage.* **37**, 62–68.

Fitzgerald, W.F., Mason, R.P. and Vandal, G.M. (1992) Atmospheric cycling and air–water exchange of mercury over mid-continent lakes, in *The Deposition and Fate of Trace Metals in Our Environment*, (eds E.S. Verry and S.J. Vermette), Forest Service (US Department of Agriculture), North Central Forest Experiment Station, pp. 139–156.

Forbes, E.A., Posner, A.M. and Quirk, J.P. (1974) The specific adsorption of inorganic Hg(II) species and Co(III) complex ions on goethite, *J. Colloid Interface Sci.* **49**, 403–409.

Friske, P.W.B. and Coker, W.B. (1995) The importance of geological controls on the natural distribution of mercury in lake and stream sediments across Canada, *Water Air Soil Polln* **80**, 1047–1051.

Gagnon, C., Pelletier, É., Mucci, A. and Fitzgerald, W.F. (1996) Diagenetic behavior of methylmercury in organic-rich coastal sediments, *Limnol. Oceanogr.* **41**, 428–434.

Ganther, H.E., Goudie, C., Sunde, M.L. *et al.* (1972) Selenium: relation to decreased toxicity of methylmercury added to diets containing tuna, *Science* **175**, 1122–1124.

Gavis, J. and Ferguson, J.F. (1972) The cycling of mercury through the environment. *Water Res.* **6**, 989–1008.

Gottofrey, J. and Tjälve, H. (1991) Effect of lipophilic complex formation on the uptake and distribution of Hg^{2+} and CH_3-Hg^+ in brown trouts (*Salmo trutta*):

studies with some compounds containing sulphur ligands. *Water Air Soil Polln* **56**, 521–532.

Grieb, T.M., Driscoll, C.T., Gloss, S.P. *et al.* (1990) Factors affecting mercury accumulation in fish in the upper Michigan peninsula, *Environ. Toxicol. Chem.* **9**, 919–930.

Hahne, H.C.H. and Kroontje, W. (1973) Significance of pH and chloride concentration on behavior of heavy metal pollutants: mercury(II), zinc(II), and lead(II), *J. Environ. Qual.* **2**, 444–450.

Haines, T.A., Komov, V.T. and Jagoe, C.H. (1994) Mercury concentration in perch (*Perca fluviatilis*) as influenced by lacustrine physical and chemical factors in two regions of Russia, in *Mercury Pollution*, (eds C.J. Watras and J.W. Huckabee), Lewis Publishers, Boca Raton, Ann Arbor, London, Tokyo, pp. 397–407.

Håkanson, L. (1980) The quantitative impact of pH, bioproduction and Hg-contamination on the Hg-content of fish (pike), *Environ. Polln* (Series B) **1**, 285–304.

Håkanson, L., Nilsson, Å. and Andersson, T. (1988) Mercury in fish in Swedish lakes. *Environ. Polln* **49**, 145–162.

Hallberg, R. (1978) Metal–organic interaction at the redoxcline, in *Environmental Biogeochemistry and Geomicrobiology*, Vol. 3, (ed. W.E. Krumbein), Ann Arbor Science Publishers, Ann Arbor (Michigan), pp. 947–953.

Hamdy, M.K. and Noyes, O.R. (1975) Formation of methyl mercury by bacteria, *Appl. Microbiol.* **30**, 424–432.

Hamdy, M.K. and Wheeler, S.R. (1978) Inhibition of bacterial growth by mercury and the effects of protective agents, *Bull. Environ. Contam. Toxicol.* **20**, 378–386.

Hamdy, M.K., Noyes, O.R. and Wheeler, S.R. (1977) Effect of mercury on bacteria: protection and transmethylation, in *Biological Implications of Metals in the Environment* (Proceedings 15th Annual Hanford Life Sciences Symposium, Richland, Washington, 29th September–1st October, 1975), (eds H. Drucker and R.E. Wildung,), Technical Information Center, Energy Research and Development Administration, US Department of Commerce, Springfield, Virginia, pp. 20–35.

Hecky, R.E., Bodaly, R.A., Strange, N.E. *et al.* (1987) Mercury bioaccumulation in yellow perch in limnocorrals simulating the effects of reservoir formation, in *Technical Appendices to the Summary Report, Canada–Manitoba Agreement on the Study and Monitoring of Mercury in the Churchill River Diversion*, Vol. 2, Chapter 7, Governments of Canada and Manitoba.

Hecky, R.E., Ramsey, D.J., Bodaly, R.A. and Strange, N.E. (1991) Increased methylmercury contamination in fish in newly formed freshwater reservoirs, in *Advances in Mercury Toxicology*, (eds T. Suzuki, N. Imura and T.W. Clarkson), Plenum Press, New York, London, pp. 33–52.

Heisinger, J.F., Hansen, C.D. and Kim, J.H. (1979) Effect of selenium dioxide on the accumulation and acute toxicity of mercuric chloride to goldfish. *Arch. Environ. Contam. Toxicol.* **8**, 279–283.

Heit, M., Tan, Y., Klusek, C. and Burke, J.C. (1981) Anthropogenic trace elements and polycyclic aromatic hydrocarbon levels in sediment cores from two lakes in the Adirondack acid lake region, *Water Air Soil Polln* **15**, 441–464.

Hintelmann, H. and Wilken, R.-D. (1995) Levels of total mercury and methylmercury compounds in sediments of the polluted Elbe River: influences of seasonally and spatially varying environmental factors, *Sci. Total Environ.* **166**, 1–10.

Hintelmann, H., Hempel, M. and Wilken, R.D. (1995a) Observation of unusual organic mercury species in soils and sediments of industrially contaminated sites. *Environ. Sci. Technol.* **29**, 1845–1850.

Hintelmann, H., Welbourn, P.M. and Evans, R.D. (1995b) Binding of methylmercury compounds by humic and fulvic acids. *Water Air Soil Polln* **80**, 1031–1034.

Hogg, T.J., Stewart, J.W.B. and Bettany, J.R. (1978) Influence of the chemical form of mercury on its adsorption and ability to leach through soils, *J. Environ. Qual.* **7**, 440–445.

Huckabee, J.W., Elwood, J.W. and Hildebrand, S.G. (1979) Accumulation of mercury in freshwater biota, in *The Biogeochemistry of Mercury in the Environment*, (ed. J.O. Nriagu), Elsevier/North-Holland Biomedical Press, Amsterdam, Oxford, New York, pp. 277–302.

Hudson, R.J.M., Gherini, S.A., Watras, C.J. and Porcella, D.B. (1994) Modeling the biogeochemical cycle of mercury in lakes: the mercury cycling model (MCM) and its application to the MTL Study lakes, in *Mercury Pollution* (eds C.J. Watras and J.W. Huckabee), Lewis Publishers, Boca Raton, Ann Arbor, London, Tokyo, pp. 473–523.

Hudson, R.J.M., Gherini, S.A., Fitzgerald, W.F. and Porcella, D.B. (1995) Anthropogenic influences on the global mercury cycle: a model-based analysis, *Water Air Soil Polln* **80**, 265–272.

Huey, C., Brinckman, F.E., Grim, S. and Iverson, W.P. (1974) The role of tin in bacterial methylation of mercury. *Proc. Int. Conf. on Transport of Persistent Chemicals in Aquatic Ecosystems (Ottawa, Canada, 1–3 May, 1974)*, Sect. II, pp. 73–78.

Hultberg, H., Parkman, H. and Renberg, I. (1994) Recent decrease in atmospheric deposition of total mercury as reflected by total and methyl mercury profiles in profundal sediments in one acid and one limed lake on the Swedish west coast [abstract], in *Abstracts, International Conference on Mercury as a Global Pollutant (Whistler, British Columbia (Canada), July, 1994)*.

Inoko, M. (1981) Studies on the photochemical decomposition of organomercurials – methylmercury(II) chloride. *Environ. Polln* (Ser. B) **2**, 3–10.

Inoue, Y. and Munemori, M. (1979) Coprecipitation of mercury(II) with iron(III) hydroxide, *Environ. Sci. Technol.* **13**, 443–445.

Jackson, K.S., Jonasson, I.R. and Skippen, G.B. (1978) The nature of metals–sediment–water interactions in freshwater bodies, with emphasis on the role of organic matter, *Earth-Science Rev.* **14**, 97–146.

Jackson, T.A. (1978) The biogeochemistry of heavy metals in polluted lakes and streams at Flin Flon, Canada, and a proposed method for limiting heavy-metal pollution of natural waters, *Environ. Geol.* **2**, 173–189.

Jackson, T.A. (1979) Relationships between the properties of heavy metals and their biogeochemical behaviour in lakes and river–lake systems, in *Heavy Metals in the Environment* (Proceedings International Conference on Management and Control of Heavy Metals in the Environment, September, 1979, London), CEP Consultants, Edinburgh, pp. 457–460.

Jackson, T.A. (ed.) (1980) *Mercury Pollution in the Wabigoon-English River System of Northwestern Ontario, and Possible Remedial Measures: a Progress Report*, Government of Canada (Department of the Environment) and Government of Ontario (Ministry of the Environment).

Jackson, T.A. (1984) Effects of inorganic cadmium, zinc, copper, and mercury on methyl mercury production in polluted lake sediments, in *Environmental Impacts of Smelters*, (ed J.O. Nriagu), John Wiley & Sons (Wiley-Interscience), New York, Toronto, Chichester, Brisbane, Singapore, pp. 551–578.

Jackson, T.A. (1986) Methyl mercury levels in a polluted prairie river–lake system: seasonal and site-specific variations, and the dominant influence of trophic conditions, *Can. J. Fish. Aquat. Sci.* **43**, 1873–1887.

Jackson, T.A. (1987) Methylation, demethylation, and bioaccumulation of mercury in lakes and reservoirs of northern Manitoba, with particular reference to effects of environmental changes caused by the Churchill-Nelson River diversion, in *Technical Appendices to the Summary Report, Canada – Manitoba Agreement on the Study and Monitoring of Mercury in the Churchill River Diversion*, Vol. 2, Chapter 8, Governments of Canada and Manitoba.

Jackson, T.A. (1988a) Accumulation of mercury by plankton and benthic invertebrates in riverine lakes of northern Manitoba (Canada): importance of regionally and seasonally varying environmental factors, *Can. J. Fish. Aquat. Sci.* **45**, 1744–1757.

Jackson, T.A. (1988b) The mercury problem in recently formed reservoirs of northern Manitoba (Canada): effects of impoundment and other factors on the production of methyl mercury by microorganisms in sediments, *Can. J. Fish. Aquat. Sci.* **45**, 97–121.

Jackson, T.A. (1989) The influence of clay minerals, oxides, and humic matter on the methylation and demethylation of mercury by micro-organisms in freshwater sediments, *Appl. Organometal. Chem.* **3**, 1–30.

Jackson, T.A. (1991a) Biological and environmental control of mercury accumulation by fish in lakes and reservoirs of northern Manitoba, Canada, *Can. J. Fish. Aquat. Sci.* **48**, 2449–2470.

Jackson, T.A. (1991b) Effects of heavy metals and selenium on mercury methylation and other microbial activities in freshwater sediments, in *Heavy Metals in the Environment*, (ed. J.P. Vernet), Elsevier, Amsterdam, London, New York, Tokyo, pp. 191–217.

Jackson, T.A. (1993a) Effects of environmental factors and primary production on the distribution and methylation of mercury in a chain of highly eutrophic riverine lakes, *Water Polln Res. J. Can.* **28**, 177–216. (Also see 'Erratum', *Water Polln Res. J. Can.* **28**, after p. 512.)

Jackson, T.A. (1993b) The influence of phytoplankton blooms and environmental variables on the methylation, demethylation, and bio-accumulation of mercury (Hg) in a chain of eutrophic mercury- polluted riverine lakes in Saskatchewan, Canada, in *Heavy Metals in the Environment* (eds R.J. Allan and J.O. Nriagu), Proceedings International Conference, Toronto, September 1993, Vol. 2, CEP Consultants Ltd, Edinburgh, pp. 301–304.

Jackson, T.A. (1995) Effects of clay minerals, oxyhydroxides, and humic matter on microbial communities of soil, sediment, and water, in *Environmental Impact of Soil Component Interactions*, (eds P.M. Huang, J. Berthelin, J.M. Bollag *et al.*), Lewis Publishers (CRC Press), Boca Raton, London, Tokyo, pp. 165–200.

Jackson, T.A. (1997) Long-range atmospheric transport of mercury to ecosystems, and the importance of anthropogenic emissions – a critical review and evaluation of the published evidence. *Environ. Revs* **5**, 99–120.

Jackson, T.A. (1998) The biogeochemical and ecological significance of interactions between colloidal minerals and trace elements, in *Environmental Interactions of Clay Minerals*, (eds J.E. Rae and A. Parker), Springer-Verlag, Berlin, Heidelberg, London, Paris, Barcelona, New York, Tokyo, Hong Kong, (in press).

Jackson, T.A. and Bistricki, T. (1995) Selective scavenging of copper, zinc, lead, and arsenic by iron and manganese oxyhydroxide coatings on plankton in lakes polluted with mine and smelter wastes: results of energy dispersive X-ray microanalysis, *J. Geochem. Explor.* **52**, 97–125.

Jackson, T.A. and Hecky, R.E. (1980) Depression of primary productivity by humic matter in lake and reservoir waters of the Boreal forest zone, *Can. J. Fish. Aquat. Sci.* **37**, 2300–2317.

Jackson, T.A. and Woychuk, R.N. (1980a) Mercury speciation and distribution in a polluted river–lake system as related to the problem of lake restoration, in *Restoration of Lakes and Inland Waters*, Proceedings International Symposium on Inland Waters and Lake Restoration, 8–12 September 1980, Portland, Maine, EPA 440/5-81-010, US Environmental Protection Agency, Office of Water Regulations and Standards, Washington, pp. 93–101.

Jackson, T.A. and Woychuk, R.N. (1980b) The geochemistry and distribution of mercury in the Wabigoon River system, in *Mercury Pollution in the Wabigoon-English River System of Northwestern Ontario, and Possible Remedial Measures – a Progress Report*, (ed. T.A. Jackson), Government of Canada (Department of the Environment) and Government of Ontario (Ministry of the Environment).

Jackson, T.A. and Woychuk, R.N. (1981) Methyl mercury formation and distribution in a polluted river–lake system: the effect of environmental variables, and implications for biological uptake and lake restoration. *Verh. Internat. Verein. Limnol.* **21**, 1114–1115 (abstract).

Jackson, T.A., Kipphut, G., Hesslein, R.H. and Schindler, D.W. (1980) Experimental study of trace metal chemistry in soft-water lakes at different pH levels, *Can. J. Fish. Aquat. Sci.* **37**, 387–402.

Jackson, T.A., Parks, J.W., Jones, P.D. *et al.* (1982) Dissolved and suspended mercury species in the Wabigoon River (Ontario, Canada): seasonal and regional variatons, *Hydrobiol.* **92**, 473–487.

Jackson, T.A., Klaverkamp, J.F. and Dutton, M.D. (1993) Heavy metal speciation and its biological consequences in a group of lakes polluted by a smelter, Flin Flon, Manitoba, Canada, *Appl. Geochem.* (Suppl.) **2**, 285–289.

Jensen, S. and Jernelöv, A. (1969) Biological methylation of mercury in aquatic organisms, *Nature* **223**, 753–754.

Jernelöv, Å. (1972) Factors in the transformation of mercury to methylmercury, in *Environmental Mercury Contamination*, (eds R. Hartung and B.D. Dinman), Ann Arbor Science Publishers, Ann Arbor, pp. 167–172.

Jernelöv, A, Landner, L. and Larsson, T. (1975) Swedish perspectives on mercury pollution. *J. Water Polln Control Fed.* **47**, 810–822.

Jernelöv, A. and Lann, H. (1971) Mercury accumulation in food chains. *Oikos* **22**, 403–406.

Johnson, M.G., Culp, L.R. and George, S.E. (1986) Temporal and spatial trends in metal loadings to sediments of the Turkey Lakes, Ontario, *Can. J. Fish. Aquat. Sci.* **43**, 754–762.

Jonasson, I.R. and Boyle, R.W. (1972) Geochemistry of mercury and origins of natural contamination of the environment, *Can. Mining Metallurg. (CIM) Bull.* **65**, 32–39.

Kelly, C.A., Rudd, J.W.M., St Louis, V.L. and Heyes, A. (1995) Is total mercury a good predictor of methyl mercury concentration in aquatic systems? *Water Air Soil Polln* **80**, 715–724.

Kerndorff, H. and Schnitzer, M. (1980) Sorption of metals on humic acid. *Geochim. Cosmochim. Acta* **44**, 1701–1708.

Kerry, A., Welbourn, P.M., Prucha, B. and Mierle, G. (1991) Mercury methylation by sulphate-reducing bacteria from sediments of an acid stressed lake, *Water Air Soil Polln* **56**, 565–575.

Khalid, R.A., Gambrell, R.P. and Patrick, W.H. Jr (1977) Sorption and release of mercury by Mississippi River sediment as affected by pH and redox potential, in *Biological Implications of Metals in the Environment*, (eds H. Drucker and R.E. Wildung), Proceedings 15th Annual Hanford Life Sciences Symposium, Richland, Washington, 29 September–1 October, 1975), Technical Information Center, Energy Research & Development Administration, US Department of Commerce, Springfield, Virginia, pp. 297–314.

Kidby, D.K. (1974) On the nature and significance of mercury inhibition of invertase from *Saccharomyces cerevisiae. J. Gen. Microbiol.* **84**, 343–349.

Kinniburgh, D.G. and Jackson, M.L. (1978) Adsorption of mercury(II) by iron hydrous oxide gel. *Soil Sci. Soc. Amer. J.* **42**, 45–47.

Knauer, G.A. and Martin, J.H. (1972) Mercury in a marine pelagic food chain. *Limnol. Oceanogr.* **17**, 868–876.

Koeman, J.H., Peeters, W.H.M., Koudstaal-Hol, C.H.M. *et al.* (1973) Mercury–selenium correlations in marine mammals. *Nature* **245**, 385–386.

Koeman, J.H., van de Ven, W.S.M., de Goeij, J.J.M. *et al.* (1975) Mercury and selenium in marine mammals and birds. *Sci. Total Environ.* **3**, 279–287.

Kondratyev, K.Ya. (1969) *Radiation in the Atmosphere*, Academic Press, New York, London.

Kooner, Z.S., Cox, C.D. and Smoot, J.L. (1995) Prediction of adsorption of divalent heavy metals at the goethite/water interface by surface complexation modeling. *Environ. Toxicol. Chem.* **14**, 2077–2083.

Korthals, E.T. and Winfrey, M.R. (1987) Seasonal and spatial variations in mercury methylation and demethylation in an oligotrophic lake. *Appl. Environ. Microbiol.* **53**, 2397–2404.

Landner, L. (1971) Biochemical model for the biological methylation of mercury suggested from methylation from methylation studies *in vivo* with *Neurospora crassa. Nature* **230**, 452–454.

Langford, C.H. and Carey, J.H. (1987) Photocatalysis by inorganic components of natural water systems, in *Photochemistry of Environmental Aquatic Systems*, (eds W.J. Cooper and R.G. Zika), ACS Symposium Series no. 327, American Chemical Society.

Langley, D.G. (1973) Mercury methylation in an aquatic environment, *J. Water Polln Control Fed.* **45**, 44–51.

Lathrop, R.C., Noonan, K.C., Guenther, P.M. *et al.* (1989) *Mercury Levels in Walleye from Wisconsin Lakes of Different Water and Sediment Chemistry Characteristics*, Tech. Bull. 163, Department of Natural Resources, Madison, Wisconsin.

Lathrop, R.C., Rasmussen, P.W. and Knauer, D.R. (1991) Mercury concentrations in walleyes from Wisconsin (USA) lakes, *Water Air Soil Polln* **56**, 295–307.

Leckie, J.O. and James, R.O. (1974) Control mechanisms for trace metals in natural waters, in *Aqueous-environmental Chemistry of Metals*, (ed. A.J. Rubin), Ann Arbor Science Publishers, Ann Arbor (Michigan), pp. 1-76.

Lee, Y.-H. and Hultberg, H. (1990) Methylmercury in some Swedish surface waters. *Environ. Toxicol. Chem.* **9**, 833–841.

Lee, Y.-H. and Iverfeldt, Å. (1991) Measurement of methylmercury and mercury in run-off, lake, and rain waters, *Water Air Soil Polln* **56**, 309–321.

Lemly, A.D. and Smith, G.J. (1987) *Aquatic Cycling of Selenium: Implications for Fish and Wildlife*, Fish and Wildlife Service Leaflet 12, US Department of the Interior (Fish and Wildlife Service), Washington, DC.

Lide, D.R. (ed.) (1992) *CRC Handbook of Physics and Chemistry*, 73rd edn, CRC Press, Boca Raton, Ann Arbor, London, Tokyo.

Liebert, C.A., Barkay, T. and Turner, R.R. (1991) Acclimation of aquatic microbial communities to Hg(II) and CH_3Hg^+ in polluted freshwater ponds. *Microb. Ecol.* **21**, 139–149.

Lindberg, S.E. (1987) Emission and deposition of atmospheric mercury vapor, in *Lead, Mercury, Cadmium and Arsenic in the Environment*, (eds T.C. Hutchinson and K.M. Meema), John Wiley & Sons, Chichester, New York, Toronto, Brisbane, Singapore, pp. 89–106.

Lindberg, S.E. and Harriss, R.C. (1974) Mercury–organic matter associations in estuarine sediments and interstitial water. *Environ. Sci. Technol.* **8**, 459–462.

Lindqvist, O., Johansson, K., Aastrup, M. *et al.* (1991) Mercury in the Swedish environment – recent research on causes, consequences and corrective methods, *Water Air Soil Polln* **55**, 1–261.

Lockhart, W.L., Wilkinson, P., Billeck, B.N. *et al.* (1993) Polycyclic aromatic hydrocarbons and mercury in sediments from two isolated lakes in central and northern Canada. *Water Sci. Technol.* **28**, 43–52.

Lockwood, R.A. and Chen, K.Y. (1973) Adsorption of Hg(II) by hydrous manganese oxides. *Environ. Sci. Technol.* **7**, 1028–1034.

Louchouarn, P., Lucotte, M., Mucci, A. and Pichet, P. (1993) Geochemistry of mercury in two hydroelectric reservoirs in Quebec, Canada. *Can. J. Fish. Aquat. Sci.* **50**, 269–281.

MacNaughton, M.G. and James, R.O. (1974) Adsorption of aqueous mercury(II) complexes at the oxide/water interface. *J. Colloid Interface Sci.* **47**, 431–440.

Magos, L. (1991) Overview on the protection given by selenium against mercurials, in *Advances in Mercury Toxicology*, (eds T. Suzuki, N. Imura and T.W. Clarkson), Plenum Press, New York and London, pp. 289–298.

Major, M.A., Rosenblatt, and Bostian, K.A. (1991) The octanol/water partition coefficient of methylmercuric chloride and methylmercuric hydroxide in pure water and salt solutions. *Environ. Toxicol. Chem.* **10**, 5–8.

Martin, M.H. and Coughtrey, P.J. (1982) *Biological Monitoring of Heavy Metal Pollution*, Applied Science Publishers, London, New York.

Mason, J.W., Anderson, A.C. and Shariat, M. (1979) Rate of demethylation of methylmercuric chloride by *Enterobacter aerogenes* and *Serratia marcescens*. *Bull. Environ. Contam. Toxicol.* **21**, 262–268.

Mason, R.P. and Fitzgerald, W.F. (1991) Mercury speciation in open ocean waters. *Water Air Soil Polln* **56**, 779–789.

Mason, R.P. and Fitzgerald, W.F. (1993) The distribution and biogeochemical cycling of mercury in the equatorial Pacific Ocean. *Deep-Sea Res. J.* **40**, 1897–1924.

Mason, R.P., Morel, F.M.M. and Hemond, H.F. (1995a) The role of microorganisms in elemental mercury formation in natural waters. *Water Air Soil Polln* **80**, 775–787.

Mason, R.P., Reinfelder, J.R. and Morel, F.M.M. (1995b) Bioaccumulation of mercury and methylmercury. *Water Air Soil Polln* **80**, 915–921.

Matheson, D.H. (1979) Mercury in the atmosphere and in precipitation, in *The Biogeochemistry of Mercury in the Environment*, (ed. J.O. Nriagu),

Elsevier/North Holland Biomedical Press, Amsterdam, Oxford, New York, pp. 113–129.

Matilainen, T. (1995) Involvement of bacteria in methylmercury formation in anaerobic lake waters. *Water Air Soil Polln* **80**, 757–764.

Matilainen, T. and Verta, M. (1995) Mercury methylation and demethylation in aerobic surface waters. *Can. J. Fish. Aquat. Sci.* **52**, 1597–1608.

Matilainen, T., Verta, M., Niemi, M. and Uusi-Rauva, A. (1991) Specific rates of net methylmercury production in lake sediments, *Water Air Soil Polln* **56**, 595–605.

May, K., Stoeppler, M. and Reisinger, K. (1987) Studies in the ratio total mercury/methylmercury in the aquatic food chain. *Toxicol. Environ. Chem.* **13**, 153–159.

McBride, B.C. and Edwards, T.L. (1977) Role of the methanogenic bacteria in the alkylation of arsenic and mercury, in *Biological Implications of Metals in the Environment*, (eds H. Drucker and R.E. Wildung), Proceedings 15th Annual Hanford Life Sciences Symposium, Richland, Washington, 29th September–1st October 1975, Technical Information Center, Energy Research and Development Administration, US Department of Commerce, Springfield, Virginia, pp. 1–19.

McMurty, M.J., Wales, D.L., Scheider, W.A. *et al.* (1989) Relationship of mercury concentrations in lake trout (*Salvelinus namaycush*) and smallmouth bass (*Micropterus dolomieui*) to the physical and chemical characteristics of Ontario lakes. *Can. J. Fish. Aquat. Sci.* **46**, 426–434.

Meili, M. (1991) The coupling of mercury and organic matter in the biogeochemical cycle – towards a mechanistic model for the Boreal forest zone, *Water Air Soil Polln* **56**, 333–347.

Meili, M., Iverfeldt, Å. and Håkanson, L. (1991) Mercury in the surface water of Swedish forest lakes – concentrations, speciation and controlling factors. *Water Air Soil Polln* **56**, 439–453.

Messier, D. and Roy, D. (1987) Concentrations en mercure chez les poissons au complexe hydroélectrique de La Grande Rivière (Québec). *Naturaliste Can. (Rev. Écol. Syst.)* **114**, 357–368.

Meyer, M.W., Evers, D.C., Daulton, T. and Braselton, W.E. (1995) Common loons (*Gavia immer*) nesting on low pH lakes in northern Wisconsion have elevated blood mercury content. *Water Air Soil Polln* **80**, 871–880.

Miettinen, J.K. (1975) The accumulation and excretion of heavy metals in organisms, in *Heavy Metals in the Aquatic Environment*, (ed. P.A. Krenkel), Pergamon Press, Oxford, New York, Toronto, Sydney, Braunschweig, pp. 155–166.

Miskimmin, B.M., Rudd, J.W.M. and Kelly, C.A. (1992) Influence of dissolved organic carbon, pH, and microbial respiration rates on mercury methylation and demethylation in lake water, *Can. J. Fish. Aquat. Sci.* **49**, 17–22.

Morris, C. (1992) *Academic Press Dictionary of Science and Technology*, Academic Press, San Diego, New York, Boston, London, Toronto, Sydney, Tokyo.

Morrison, K.A. and Thérien, N. (1995) Changes in mercury levels in lake whitefish (*Coregonus clupeaformis*) and northern pike (*Esox lucius*) in the LG-2 reservoir since flooding. *Water Air Soil Polln* **80**, 819–828.

Mortimer, D.C. and Kudo, A. (1975) Interaction between aquatic plants and bed sediments in mercury uptake from flowing water. *J. Environ. Qual.* **4**, 491–495.

Mucci, A., Lucotte, M., Montgomery, S. *et al.* (1995) Mercury remobilization from flooded soils in a hydroelectric reservoir of northern Quebec, La Grande-2: results of a soil resuspension experiment. *Can. J. Fish. Aquat. Sci.* **52**, 2507–2517.

Munthe, J., Xiao, Z.F. and Lindqvist, O. (1991) The aqueous reduction of divalent mercury by sulfite. *Water Air Soil Polln* **56**, 621–630.

Nagase, H., Ose, Y., Sato, T. and Ishikawa, T. (1982) Methylation of mercury by humic substances in an aquatic environment. *Sci. Total Environ.* **25**, 133–142.

Nakamura, K., Sakamoto, M., Uchiyama, H. and Yagi, O. (1990) Organomercurial-volatilizing bacteria in the mercury-polluted sediment of Minamata Bay, Japan. *Appl. Environ. Microbiol.* **56**, 304–305.

Newton, D.W., Ellis, R. Jr and Paulsen, G.M. (1976) Effect of pH and complex formation on mercury(II) adsorption by bentonite. *J. Environ. Qual.* **5**, 251–254.

Nicoletto, P.F. and Hendricks, A.C. (1988) Sexual differences in accumulation of mercury in four species of centrarchid fishes., *Can. J. Zool.* **66**, 944–949.

Norstrom, R.J., McKinnon, A.E. and deFreitas, A.S.W. (1976) A bioenergetics-based model for pollutant accumulation by fish. Simulation of PCB and methylmercury residue levels in Ottawa River yellow perch (*Perca flavescens*). *J. Fish. Res. Board Can.* **33**, 248–267.

Nriagu, J.O. (1989) A global assessment of natural sources of atmospheric trace metals, *Nature* **338**, 47–49.

Nriagu, J.O. (1992) Worldwide contamination of the atmosphere with toxic metals, in *The Deposition and Fate of Trace Metals in Our Environment*, (eds E.S. Verry. and S.J. Vermette), Forest Service (US Department of Agriculture), North Central Forest Experiment Station, pp. 9–21.

Nriagu, J.O. and Pacyna, J.M. (1988) Quantitative assessment of worldwide contamination of air, water and soils by trace metals. *Nature* **333**, 134–139.

Nuzzi, R. (1972) Toxicity of mercury to phytoplankton. *Nature* **237**, 38–40.

Ochiai, E.-I. (1977) *Bioinorganic Chemistry*, Allyn and Bacon, Boston, Toronto, London, Sydney.

Odin, M., Feurtet-Mazel, A., Ribeyre, F. and Boudou, A. (1994) Actions and interactions of temperature, pH and photoperiod on mercury bioaccumulation by nymphs of the burrowing mayfly *Hexagenia rigida*, from the sediment contamination source. *Environ. Toxicol. Chem.* **13**, 1291– 1302.

Olson, B.H. and Cooper, R.C. (1976) Comparison of aerobic and anaerobic methylation of mercuric chloride by San Francisco Bay sediments *Water Res.* **10**, 113–116.

Oremland, R.S., Culbertson, C.W. and Winfrey, M.R. (1991) Methylmercury decomposition in sediments and bacterial cultures: involvement of methanogens and sulfate reducers in oxidative demethylation *Appl. Environ. Microbiol.* **57**, 130–137.

Ouellet, M. and Jones, H.G. (1983) Historical changes in acid precipitation and heavy metal deposition originating from fossil fuel combustion in eastern North America as revealed by lake sediment geochemistry. *Water Sci. Technol.* **15**, 115–130.

Pacyna, J.M. and Keeler, G.J. (1995) Sources of mercury in the Arctic. *Water Air Soil Polln* **80**, 621–632.

Pahan, K., Ghosh, D.K., Ray, S. *et al.* (1994) Mercury and organomercurial degrading enzymes in a broad-spectrum Hg-resistant strain of *Bacillus pasteurii. Bull. Environ. Contam. Toxicol.* **52**, 582–589.

Painter, S., Cameron, E.M., Allan, R. and Rouse, J. (1994) Reconnaissance geochemistry and its environmental relevance. *J. Geochem. Explor.* **51**, 213–246.

Palheta, D. and Taylor, A. (1995) Mercury in environmental and biological samples from a gold mining area in the Amazon region of Brazil. *Sci. Total Environ.* **168**, 63–69.

Pan-Hou and Imura, N. (1982) Physiological role of mercury-methylation in *Clostridium cochlearium* T-2C. *Bull. Environ. Contam. Toxicol.* **29**, 290–297.

Parks, J.W. (1976) *Mercury in Sediment and Water in the Wabigoon–English River System, 1970–1975*, Ministry of the Environment, Ontario, Canada.

Parks, J.W. and Hamilton, A.L. (1987) Accelerating recovery of the mercury-contaminated Wabigoon/English River system, *Hydrobiology* **149**, 159–188.

Parks, J.W., Sutton, J.A. and Lutz, A. (1986) Effect of point and diffuse source loadings on mercury concentrations in the Wabigoon River: evidence of a seasonally varying sediment–water partition. *Can. J. Fish. Aquat. Sci.* **43**, 1426–1444.

Parks, J.W., Lutz, A. and Sutton, J.A. (1989) Water column methylmercury in the Wabigoon/English river–lake system: factors controlling concentrations, speciation, and net production. *Can. J. Fish. Aquat. Sci.* **46**, 2184–2202.

Parks, J.W., Craig, P.J., Neary, B.P. *et al.* (1991a) Biomonitoring in the mercury-contaminated Wabigoon–English–Winnipeg River (Canada) system: selecting the best available bioindicator. *Appl. Organometal. Chem.* **5**, 487–495.

Parks, J.W., Curry, C., Romani, D. and Russell, D.D. (1991b) Young northern pike, yellow perch and crayfish as bioindicators in a mercury contaminated watercourse. *Environ. Monit. Assess.* **16**, 39–73.

Paulsson, K. and Lundbergh, K. (1991) Treatment of mercury contaminated fish by selenium addition. *Water Air Soil Polln* **56**, 833–841.

Phillips, C.S.G. and Williams, R.J.P. (1965) *Inorganic Chemistry*, Vol. I, Oxford University Press, Oxford, New York.

Phillips, G.F., Dixon, B.E. and Lidzey, R.G. (1959) The volatility of organo-mercury compounds. *J. Sci. Food Agric.* **10**, 604–610.

Ponce, R.A. and Bloom, N.S. (1991) Effect of pH on the bioaccumulation of low level, dissolved methylmercury by rainbow trout (*Oncorhyncus mykiss*). *Water Air Soil Polln* **56**, 631–640.

Porcella, D.B., Huckabee, J.W. and Wheatley, B. (eds) (1995) *Mercury as a Global Pollutant*, Kluwer Academic Publishers, Dordrecht, London, Boston.

Rabenstein, D.L. and Reid, R.S. (1984) Nuclear magnetic resonance studies of the solution chemistry of metal complexes. 20. Ligand-exchange kinetics of methylmercury(II)–thiol complexes. *Inorg. Chem.* **23**, 1246–1250.

Radosevich, M. and Klein, D.A. (1993) Bacterial enumeration and mercury volatilization in deep subsurface sediment samples. *Bull. Environ. Contam. Toxicol.* **51**, 226–233.

Ramamoorthy, S. and Kushner, D.J. (1975a) Binding of mercuric and other heavy metal ions by microbial growth media. *Microbiol. Ecology* **2**, 162–176.

Ramamoorthy, S. and Kushner, D.J. (1975b) Heavy metal binding components of river water. *J. Fish. Res. Board Can.* **32**, 1755–1766.

Ramamoorthy, S. and Rust, B.R. (1976) Mercury sorption and desorption characteristics of some Ottawa River sediments. *Can. J. Earth Sci.* **13**, 530–536.

Ramamoorthy, S. and Rust, B.R. (1978) Heavy metal exchange processes in sediment–water systems. *Environ. Geol.* **2**, 165–172.

Ramlal, P.S., Rudd, J.W.M., Furutani, A. and Xun, L. (1985) The effect of pH on methyl mercury production and decomposition in lake sediments. *Can. J. Fish. Aquat. Sci.* **42**, 685–692.

Randle, K. and Hartmann, E.H. (1987) Applications of the continuous flow stirred cell (CFSC) technique to adsorption of zinc, cadmium and mercury on humic acids. *Geoderma* **40**, 281–296.

Rask, M. and Metsälä, T.-R. (1991) Mercury concentrations in northern pike, *Esox lucius* L., in small lakes of Evo area, southern Finland. *Water Air Soil Polln* **56**, 369–378.

Rasmussen, P.E. (1994) Current methods of estimating atmospheric mercury fluxes in remote areas. *Environ. Sci. Technol.* **28**, 2233–2241.

Regnell, O. (1994) The effect of pH and dissolved oxygen levels on methylation and partitioning of mercury in freshwater model systems. *Environ. Polln* **84**, 7–13.

Regnell, O. (1995) Methyl mercury in lakes: factors affecting its production and partitioning between water and sediment. PhD dissertation, Lund University (Department of Ecology – Chemical Ecology and Ecotoxicology), Sweden.

Regnell, O. and Tunlid, A. (1991) Laboratory study of chemical speciation of mercury in lake sediment and water under aerobic and anaerobic conditions. *Appl. Environ. Microbiol.* **57**, 789–795.

Reimers, R.S. and Krenkel, P.A. (1974) Kinetics of mercury adsorption and desorption in sediments. *Water Polln Control Fed.* **46**, 352–365.

Reimers, R.S., Krenkel, P.A., Eagle, M. and Tragitt, G. (1975) Sorption phenomenon in the organics of bottom sediments, in *Heavy Metals in the Aquatic Environment*, (ed. P.A. Krenkel), Pergamon Press, Oxford, New York, Toronto, Sydney, Braunschweig, pp. 117–129.

Reinert, R.E., Stone, L.J. and Willford, W.A. (1974) Effect of temperature on accumulation of methylmercuric chloride and *P,P'* DDT by rainbow trout (*Salmo gairdneri*). *J. Fish. Res. Board Can.* **31**, 1649–1652.

Ribeyre, F. and Boudou, A. (1982) Study of the dynamics of the accumulation of two mercury compounds – $HgCl_2$ and CH_3HgCl – by *Chlorella vulgaris*: effect of temperature and pH factor of the environment. *Int. J. Environ. Studies* **20**, 35–40.

Ribo, J.M., Yang, J.E. and Huang, P.M. (1989) Luminescent bacteria toxicity assay in the study of mercury speciation. *Hydrobiology* **188/189**, 155–162.

Rich, D. (1965) *Periodic Correlations*, W.A. Benjamin, New York, Amsterdam.

Richman, L.A., Wren, C.D. and Stokes, P.M. (1988) Facts and fallacies concerning mercury uptake by fish in acid stressed lakes. *Water Air Soil Polln* **37**, 465–473.

Rimerman, R.A., Buhler, D.R. and Whanger, P.D. (1977) Metabolic interactions of selenium with heavy metals, in *Biochemical Effects of Environmental Pollutants*, (ed. S.D. Lee), Ann Arbor Science Publishers, Ann Arbor, pp. 377–396.

Roberts, J.D. and Caserio, M.C. (1965) *Basic Principles of Organic Chemistry*, W.A. Benjamin, New York, Amsterdam.

Röderer, G. (1983) Differential toxic effects of mercuric chloride and methylmercuric chloride on the freshwater alga *Poterioochromonas malhamensis*. *Aquat. Toxicol.* **3**, 23–34.

Rodgers, D.W. (1994) You are what you eat and a little bit more: bioenergetics-based models of methylmercury accumulation in fish revisited, in *Mercury Pollution*, (eds C.J. Watras and J.W. Huckabee), Lewis Publishers, Boca Raton, Ann Arbor, London, Tokyo, pp. 427–439.

Rodgers, D.W., Dickman, M. and Han, X. (1995) Stories from old reservoirs: sediment Hg and Hg methylation in Ontario hydroelectric developments. *Water Air Soil Polln* **80**, 829–839.

Rogers, R.D. (1977) *Abiological Methylation of Mercury in Soil*, Report no. EPA-600/3-77-007, Environmental Monitoring and Support Laboratory, Office of Research and Development, US Environmental Protection Agency, Las Vegas, Nevada.

Rognerud, S. and Fjeld, E. (1993) Regional survey of heavy metals in lake sediments in Norway. *Ambio* **22**, 206–212.

Rowland, I.R., Davies, M.J. and Grasso, P. (1977) Volatilisation of methylmercuric chloride by hydrogen sulphide. *Nature* **265**, 718–719.

Rudd, J.W.M. (1995) Sources of methyl mercury to freshwater ecosystems: a review. *Water Air Soil Polln* **80**, 697–713.

Rudd, J.W.M. and Turner, M.A. (1983) The English–Wabigoon River system: V. Mercury and selenium bioaccumulation as a function of aquatic primary productivity. *Can. J. Fish. Aquat. Sci.* **40**, 2251–2259.

Rudd, J.W.M., Turner, M.A., Townsend, B.E. *et al.* (1980) Dynamics of selenium in mercury-contaminated experimental freshwater ecosystems. *Can. J. Fish. Aquat. Sci.* **37**, 848–857.

Ruohtula, M. and Miettinen, J.K. (1975) Retention and excretion of ^{203}Hg-labelled methylmercury in rainbow trout. *Oikos* **26**, 385–390.

Scheider, W.A., Jeffries, D.S. and Dillon, P.J. (1979) Effects of acidic precipitation on Precambrian freshwaters in southern Ontario. *J. Great Lakes Res.* **5**, 45–51.

Schindler, D.W. (1988) Effects of acid rain on freshwater ecosystems. *Science* **239**, 149–157.

Schindler, P.W. and Stumm, W. (1987) The surface chemistry of oxides, hydroxides, and oxide minerals, in *Aquatic Surface Chemistry*, (ed. W. Stumm.), John Wiley & Sons, New York, Toronto, Chichester, Brisbane, Singapore, pp. 83–110.

Schnitzer, M. and Khan, S.U. (1972) *Humic Substances in the Environment*, Marcel Dekker, New York.

Schottel, J., Mandal, A. and Toth, K. (1974) Mercury and mercurial resistance determined by plasmids in *Escherichia coli* and *Pseudomonas aeruginosa*, in *Proc. Int. Conf. on Transport of Persistent Chemicals in Aquatic Ecosystems (Ottawa, Canada, 1–3 May, 1974)*, Sect. II, pp. 65–71.

Schroeder, W.H., Yarwood, G. and Niki, H. (1991) Transformation processes involving mercury species in the atmosphere – results from a literature survey. *Water Air Soil Polln* **56**, 653–666.

Schuster, E. (1991) The behavior of mercury in the soil with special emphasis on complexation and adsorption processes – a review of the literature, *Water, Air, Soil Pollut.*, **56**, 667–680.

Scott, D.P. (1974) Mercury concentration of white muscle in relation to age, growth, and condition in four species of fishes from Clay Lake, Ontario. *J. Fish. Res. Board Can.* **31**, 1723–1729.

Scott, D.P. and Armstrong, F.A.J. (1972) Mercury concentration in relation to size in several species of freshwater fishes from Manitoba and Northwestern Ontario. *J. Fish. Res. Board Can.* **29**, 1685–1690.

Scruton, D.A., Petticrew, E.L., LeDrew, L.J. *et al.* (1994) Methylmercury levels in fish tissue from three reservoir systems in insular Newfoundland, Canada, in *Mercury Pollution*, (eds C.J. Watras and J.W. Huckabee), Lewis Publisheers, Boca Raton, Ann Arbor, London, Tokyo, pp. 441–455.

Sellers, P., Kelly, C.A., Rudd, J.W.M. and MacHutchon, A.R. (1996) Photodegradation of methylmercury in lakes. *Nature* **380**, 694–697.

Semu, E., Singh, B.R. and Selmer-Olsen, A.R. (1987) Adsorption of mercury compounds by tropical soils. II. Effect of soil:solution ratio, ionic strength, pH, and organic matter. *Water Air Soil Polln* **32**, 1–10.

Shariat, M., Anderson, A.C. and Mason, J.W. (1979) Screening of common bacteria capable of demethylation of methylmercuric chloride. *Bull. Environ. Contam. Toxicol.* **21**, 255–261.

Shin, E.-B. and Krenkel, P.A. (1976) Mercury uptake by fish and biomethylation mechanisms. *J. Water Polln Control Fed.* **48**, 473–501.

Siegel, B.Z. and Siegel, S.M. (1979) Biological indicators of atmospheric mercury, in *The Biogeochemistry of Mercury in the Environment*, (ed. J.O. Nriagu), Elsevier/North Holland Biomedical Press, Amsterdam, Oxford, New York, pp. 131–159.

Sillén, L.G. and Martell, A.E. (1964) *Stability Constants of Metal-ion Complexes*, Special Publication no. 17, The Chemical Society, London.

Sillén, L.G. and Martell, A.E. (1971) *Stability Constants of Metal-ion Complexes*, Supplement no. 1, Special Publication no. 25, The Chemical Society, London.

Simonin, H.A., Gloss, S.P., Driscoll, C.T. *et al.* (1994) Mercury in yellow perch from Adirondack drainage lakes (New York, US), in *Mercury Pollution*, (eds C.J. Watras and J.W. Huckabee), Lewis Publishers, Boca Raton, Ann Arbor, London, Tokyo, pp. 457–469.

Sinicrope, T.L., Langis, R., Gersberg, R.M. *et al.* (1992) Metal removal by wetland mesocosms subjected to different hydroperiods. *Ecol. Eng.* **1**, 309–322.

Sisler, H. (1963) *Electronic Structure, Properties, and the Periodic Law*, Reinhold, New York; Chapman & Hall, London.

Slotton, D.G., Reuter, J.E. and Goldman, C.R. (1995) Mercury uptake patterns of biota in a seasonally anoxic northern California reservoir. *Water Air Soil Polln* **80**, 841–850.

Southworth, G.R., Turner, R.R., Peterson, M.J. and Bogle, M.A. (1995) Form of mercury in stream fish exposed to high concentrations of dissolved inorganic mercury. *Chemosphere* **30**, 779–787.

Spangler, W.J., Spigarelli, J.L., Rose, J.M. and Miller, H.M. (1973) Methylmercury: bacterial degradation in lake sediments. *Science* **180**, 192–193.

St Louis, V.L., Rudd, J.W.M., Kelly, C.A. *et al.* (1994) Importance of wetlands as sources of methyl mercury to Boreal forest ecosystems. *Can. J. Fish. Aquat. Sci.* **51**, 1065–1076.

Steffan, R.J., Korthals, E.T. and Winfrey, M.R. (1988) Effects of acidification on mercury methylation, demethylation, and volatilization in sediments from an acid-susceptible lake. *Appl. Environ. Microbiol.* **54**, 2003–2009.

Steinnes, E. (1994) Is mercury affected by the 'global fractionation' process? in *Abstract Book of International Conference on Mercury as a Global Pollutant (Whistler, British Columbia, July, 1994)* (abstract).

Steinnes, E. and Andersson, E.M. (1991) Atmospheric deposition of mercury in Norway: temporal and spatial trends. *Water Air Soil Polln* **56**, 391–404.

Stordal, M.C., Gill, G.A., Wen, L.-S. and Santschi, P.H. (1996) Mercury phase speciation in the surface waters of three Texas estuaries: importance of colloidal forms. *Limnol. Oceanogr.* **41**, 52–61.

Stumm, W., Hohl, H. and Dalang, F. (1976) Interaction of metal ions with hydrous oxide surfaces. *Croatica Chemica Acta* **48**, 491–504.

Summers, A.O. (1988) Biotransformations of mercury compounds, in *Environmental Biotechnology*, (ed. G.S. Omenn), Plenum Press, New York, London, pp. 105–109.

Takizawa, Y. (1979) Epidemiology of mercury poisoning, in *The Biogeochemistry of Mercury in the Environment*, (ed. J.O. Nriagu), Elsevier/North-Holland Biomedical Press, Amsterdam, Oxford, New York, pp. 325–365.

Timperley, M.H. and Allan, R.J. (1974) The formation and detection of metal dispersion halos in organic lake sediments. *J. Geochem. Explor.* **3**, 167–190.

Tipping, E. and Hurley, M.A. (1992) A unifying model of cation binding by humic substances. *Geochim. Cosmochim. Acta* **56**, 3627–3641.

Topping, G. and Davies, I.M. (1981) Methylmercury production in the marine water column. *Nature* **290**, 243–244.

Trost, P.B. and Bisque, R.E. (1972) Distribution of mercury in residual soils, in *Environmental Mercury Contamination*, (eds R. Hartung and B.D. Dinman), Ann Arbor Science Publishers, Ann Arbor, pp. 178–196.

Turner, M.A. and Rudd, J.W.M. (1983) The English–Wabigoon River system: III. Selenium in lake enclosures: its geochemistry, bioaccumulation, and ability to reduce mercury bioaccumulation. *Can. J. Fish. Aquat. Sci.* **40**, 2228–2240.

Turner, M.A. and Swick, A.L. (1983) The English–Wabigoon River system: IV. Interaction between mercury and selenium accumulated from waterborne and dietary sources by northern pike (*Esox lucius*). *Can. J. Fish. Aquat. Sci.* **40**, 2241–2250.

Verdon, R., Brouard, D., Demers, C. *et al.* (1991) Mercury evolution (1978–1988) in fishes of the La Grande hydroelectric complex, Québec, Canada. *Water Air Soil Polln* **56**, 405–417.

Vonk, J.W. and Sijpesteijn, A.K. (1973) Studies on the methylation of mercuric chloride by pure cultures of bacteria and fungi. *Antonie van Leeuwenhoek* **39**, 505–513.

Wagemann, R., Lockhart, W.L., Welch, H. and Innes, S. (1995) Arctic marine mammals as integrators and indicators of mercury in the Arctic. *Water Air Soil Polln* **80**, 683–693.

Walczak, B.Z., Hammer, U.T. and Huang, P.M. (1986) Ecophysiology and mercury accumulation of rainbow trout (*Salmo gairdneri*) when exposed to mercury in various concentrations of chloride. *Can. J. Fish. Aquat. Sci.* **43**, 710–714.

Wang, J.S., Huang, P.M., Hammer, U.T. and Liaw, W.K. (1985) Influence of selected cation and anion species on the adsorption of mercury(II) by montmorillonite. *Appl. Clay Sci.* **1**, 125–132.

Wang, J.S., Huang, P.M., Hammer, U.T. and Liaw, W.K. (1988) Influence of chloride/mercury molar ratio and pH on the adsorption of mercury by poorly crystalline oxides of Al, Fe, Mn, and Si. *Verh. Internat. Verein. Limnol.* **23**, 1594–1600.

Wang, J.S., Huang, P.M., Hammer, U.T. and Liaw, W.K. (1989) Role of dissolved oxygen in the desorption of mercury from freshwater sediment, in *Aquatic Toxicology and Water Quality Management*, (ed. J.O. Nriagu.), John Wiley & Sons, New York, Toronto, Chichester, Brisbane, Singapore, pp. 153–159.

Wang, J.S., Huang, P.M., Liaw, W.K. and Hammer, U.T. (1991) Kinetics of the desorption of mercury from selected freshwater sediments as influenced by chloride. *Water Air Soil Polln* **56**, 533–542.

Waslenchuk, D.G. (1975) Mercury in fluvial bed sediments subsequent to contamination. *Environ. Geol.* **1**, 131–136.

Watras, C.J. and Bloom, N.S. (1992) Mercury and methylmercury in individual zooplankton: implications for bioaccumulation. *Limnol. Oceanogr.* **37**, 1313–1318.

Watras, C.J., Bloom, N.S., Hudson, R.J.M. *et al.* (1994) Sources and fates of mercury and methylmercury in Wisconsin lakes, in *Mercury Pollution*, (eds C.J Watras and J.W. Huckabee), Lewis Publishers, Boca Raton, Ann Arbor, London, Tokyo, pp. 153–177.

Watras, C.J., Bloom, N.S., Claas, S.A. *et al.* (1995) Methylmercury production in the anoxic hypolimnion of a dimictic seepage lake. *Water Air Soil Polln* **80**, 735–745.

Watras, C.J., Morrison, K.A. and Bloom, N.S. (1995) Chemical correlates of Hg and methyl-Hg in northern Wisconsin lake waters under ice-cover. *Water Air Soil Polln* **84**, 253–267.

Weber, J.H. (1993) Review of possible paths for abiotic methylation of mercury(II) in the aquatic environment. *Chemosphere* **26**, 2063–2077.

Weber, J.H., Reisinger, K. and Stoeppler, M. (1985) Methylation of mercury(II) by fulvic acid. *Environ. Technol. Letts* **6**, 203–208.

Westöö, G. (1973) Methylmercury as percentage of total mercury in flesh and viscera of salmon and sea trout of various ages. *Science* **181**, 567–568.

van der Weijden, C.H. (1990) Behaviour of heavy metals upon transition from riverine to marine environment, in *Program and Abstracts, V.M. Goldschmidt Conference (2–4 May, 1990, Baltimore, Maryland)*, pp. 88 (abstract).

Wiener, J.G., Fitzgerald, W.F., Watras, C.J. and Rada, R.G. (1990) Partitioning and bioavailability of mercury in an experimentally acidified Wisconsin lake. *Environ. Toxicol. Chem.* **9**, 909–918.

Wilken, R.-D. and Hintelmann, H. (1991) Mercury and methylmercury in sediments and suspended particles from the River Elbe, North Germany. *Water Air Soil Polln* **56**, 427–437.

Williams, D.R. (1971) *The Metals of Life*, Van Nostrand Reinhold, London, New York, Cincinnati, Toronto, Melbourne.

Windom, H.L. and Kendall, D.R. (1979) Accumulation and biotransformation of mercury in coastal and marine biota, in *The Biogeochemistry of Mercury in the Environment*, (ed. J.O. Nriagu), Elsevier/North-Holland Biomedical Press, Amsterdam, Oxford, New York, pp. 303–323.

Winfrey, M.R. and Rudd, J.W.M. (1990) Environmental factors affecting the formation of methylmercury in low pH lakes. *Environ. Toxicol. Chem.* **9**, 853–869.

Wobeser, G. (1974) Toxicity of methyl mercury for fish and mink, in *Proceedings International Conference on Transport of Persistent Chemicals in Aquatic Ecosystems (1–3 May, 1974, Ottawa, Canada)*, Sect. III, p. 71 (abstract).

Wobeser, G. (1975) Acute toxicity of methyl mercury chloride and mercuric chloride for rainbow trout (*Salmo gairdneri*) fry and fingerlings. *J. Fish. Res. Board Can.* **32**, 2005–2013.

Wood, J.M. (1971) Environmental pollution by mercury, in *Advances in Environmental Science and Technology*, Vol. 2 (eds J.N. Pitts Jr and R.L. Metcalf), Wiley-Interscience (John Wiley & Sons), New York, Toronto, London, Sydney, pp. 39–56.

Wood, J.M. (1980) The role of pH and oxidation–reduction potentials in the mobilization of heavy metals, in *Polluted Rain*, (eds T.Y Toribara, M.W. Miller and P.E. Morrow), Plenum Press, New York, London, pp. 223–232.

Wood, J.M., Kennedy, F.S. and Rosen, C.G. (1968) Synthesis of methyl mercury compounds by extracts of a methanogenic bacterium. *Nature* **220**, 173–174.

Wren, C.D. and MacCrimmon, H.R. (1983) Mercury levels in the sunfish, *Lepomis gibbosus*, relative to pH and other environmental variables of Precambrian shield lakes. *Can. J. Fish. Aquat. Sci.* **40**, 1737–1744.

Wren, C.D., Scheider, W.A., Wales, D.L. *et al.* (1991) Relation between mercury concentrations in walleye (*Stizostedion vitreum*) and northern pike (*Esox lucius*) in Ontario lakes and influence of environmental factors. *Can. J. Fish. Aquat. Sci.* **48**, 132–139.

Wright, D.A., Welbourn, P.M. and Martin, A.V.M. (1991) Inorganic and organic mercury uptake and loss by the crayfish *Orconectes propinquus*. *Water Air Soil Polln* **56**, 697–707.

Wright, D.R. and Hamilton, R.D. (1982) Release of methyl mercury from sediments: effects of mercury concentration, low temperature, and nutrient addition. *Can. J. Fish. Aquat. Sci.* **39**, 1459– 1466.

Xiao, Z.F., Strömberg, D. and Lindqvist, O. (1995) Influence of humic substances on photolysis of divalent mercury in aqueous solution. *Water Air Soil Polln* **80**, 789–798.

Xu, H. and Allard, B. (1991) Effects of a fulvic acid on the speciation and mobility of mercury in aqueous solutions. *Water Air Soil Polln* **56**, 709–717.

Xun, L., Campbell, N.E.R. and Rudd, J.W.M. (1987) Measurements of specific rates of net methyl mercury production in the water column and surface sediments of acidified and circumneutral lakes. *Can. J. Fish. Aquat. Sci.* **44**, 750–757.

Zepp, R.G. (1988) Environmental photoprocesses involving natural organic matter, in *Humic Substances and their Role in the Environment*, (eds F.H. Frimmel and R.F. Christman), Wiley- Interscience (John Wiley & Sons), Chichester, New York, Toronto, Brisbane, Singapore, pp. 193–214.

Zepp, R.G., Baughman, G.L., Wolfe, N.L. and Cline, D.M. (1974) Methylmercuric complexes in aquatic systems. *Environ. Letts* **6**, 117–127.

Zhang, L. and Planas, D. (1994) Biotic and abiotic mercury methylation and demethylation in sediments. *Bull. Environ. Contam. Toxicol.* **52**, 691–698.

Zvonarev, B.A. and Zyrin, N.G. (1982) Patterns of mercury sorption in soils. I. Effect of pH on mercury sorption by soils. *Vestn. Mosk. Univ. Ser. 17: Pochvoved.* **4**, 43–48 (in Russian).

6   *Arsenic metabolism in aquatic ecosystems*

JOHN S. EDMONDS AND KEVIN A. FRANCESCONI

6.1 INTRODUCTION

The concentration of arsenic is higher in marine organisms (Tables 6.1 and 6.2) than in freshwater organisms (Cullen and Reimer, 1989) and therefore arsenic in the sea and in marine organisms has received much greater attention than arsenic in freshwater environments; higher concentrations generally mean greater ease of study and greater fears of toxicity to be stilled. The contents of this chapter dealing with marine and freshwater environments will reflect this disproportionate attention.

Edmonds and Francesconi (1987a) published a scheme (Figure 6.1) for the transformations of arsenic in the marine environment that was based largely on the identification of compounds in the various compartments; arsenate (1) in seawater, arsenic-containing carbohydrates (14) in algae and arsenobetaine (5) in animals. (**Note: throughout this chapter, numbers in parentheses following a chemical name refer to structures in Figure 6.2.**) Arsenobetaine was identified as the major form of arsenic in the western rock lobster (Edmonds *et al.*, 1977), then in two species of shark (Kurosawa *et al.*, 1980; Cannon *et al.*, 1981). Subsequently its virtual ubiquity in marine animals was established (Table 6.3) (Francesconi and Edmonds, 1993).

We knew that arsenic in seawater was largely in the form of arsenate (1) (Andreae, 1979) and assumed that arsenic was being diverted into the food-web by algae. When we examined a species of brown alga (*Ecklonia radiata*, containing about 10 mg arsenic kg^{-1} on a wet weight basis) that contributed at least some of the base of the food-web supporting the western rock lobster, we expected to find some rather obvious precursor to arsenobetaine, if not arsenobetaine itself. This simple notion of food-chain transformations received a setback with the finding that *Ecklonia* contained the bulk of its

Metal Metabolism in Aquatic Environments. Edited by William J. Langston and Maria João Bebianno. Published in 1998 by Chapman & Hall, London. ISBN 0 412 80370 4

Table 6.1 Concentrations of arsenic in some marine animals

Type of animal (No. of species)	Location	Arsenic concentration (mg kg^{-1}, wet weight basis)		References
		Range	Mean	
Teleost fish (4 spp.)	Canada	0.2–11.7	–	Kennedy, 1976
Teleost fish (9 spp.)	Goa	0.3–12.6	–	Zingde *et al.*, 1976
Teleost fish (9 spp.)	Australia (SE)	0.1–4.4	1.0	Bebbington *et al.*, 1977
Teleost fish (8 spp.)	Norway	0.6–7.8	2.7	Egaas & Brækkan, 1977
Teleost fish (many spp.)	USA (all areas)	Range of means 1–7		Hall *et al.*, 1978
Teleost fish (plaice – *Pleuronectes platessa*)	Netherlands (N. Sea)	3–166	23	Luten *et al.*, 1982
Teleost fish (15 spp.)	Scotland	0.2–89.9	–	Falconer *et al.*, 1983
Elasmobranch fish (*Carcharias sorrakowah*)	Goa	6.3–10.8	8.5	Zingde *et al.*, 1976
Elasmobranch fish (several spp.)	USA (all areas)	Range of means 7–20		Hall *et al.*, 1978
Elasmobranch fish (2 spp.)	Australia (SE)	5–30	15	Glover, 1979
Elasmobranch fish (3 spp.)	Papua New Guinea	0.2–7.5	2.5	Powell *et al.*, 1981
Crustaceans (16 spp.)	USA (all areas)	Range of means 3–50		Hall *et al.*, 1978
Crustaceans (krill, whole animals)	Antarctic–Scotia Sea	1.2–3.5	2.0	Stoeppler & Brandt, 1979
Crustaceans (3 spp.)	Scotland	3.3–38.2	11.5	Falconer *et al*, 1983
Bivalve molluscs (12 spp.)	USA (all areas)	Range of means 2–20		Hall *et al.*, 1978
Bivalve molluscs (5 spp.)	Japan	1.0–10.4	3.6	Shiomi *et al.*, 1984
Gastropod molluscs (limpet, *Patella vulgata*)	UK	1.3–4.7	2.5	Peden *et al.*, 1973
Gastropod molluscs (abalone, 2 spp.)	USA	3.0–27	14	Hall *et al.*, 1978
Gastropod molluscs (10 spp.)	Japan	1.6–107	26	Shiomi *et al.*, 1984

Table 6.2 Concentrations of arsenic in some marine algae

Type of alga (No. of species)	Location	Arsenic concentration (mg kg^{-1}, dry weight basis)		References
		Range	Mean	
Brown (8 spp.)	India	8–68	30	Dhandhukia and Seshadri, 1969
Red (5 spp.)		0–5	1.5	
Green (5 spp.)		0.1–6.3	2.2	
Brown (7 spp.)	Norway	15–109	44	Lunde, 1970
Red (2 spp.)		10–13	12	
Brown (3 spp.)	UK (inshore waters)	26–47	39	Leatherland and Burton, 1974
Red (2 spp.)		11–39	25	
Brown (15 spp.)	Japan	< 1–230	46	Tagawa and Kojima, 1976
Red (8 spp.)		< 1–12	4.4	
Green (3 spp.)		< 1–8	3.8	
Brown (24 spp.)	USA	1.1–31.6	10.3	Sanders, 1979
Red (15 spp.)		04–8.2	1.4	
Green (16 spp.)		0.2–23.3	1.5	
Brown (14 spp.)	British Columbia	40.8–92.4	57	Whyte and Englar, 1983
Brown (14 spp.)	Australia	21.3–179	62	Maher and Clarke, 1984
Red (10 spp.)		12.5–31.3	19.2	
Green (9 spp.)		6.3–16.3	10.7	
Brown (7 spp.)	Goa	4.5–20.9	8.1	Rao *et al.*, 1991

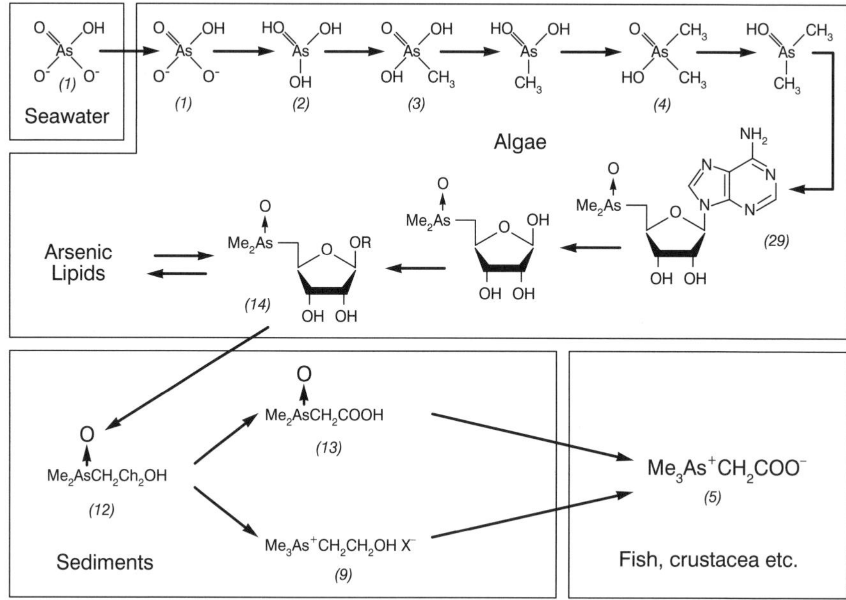

Figure 6.1 Scheme for the biogenesis of arsenobetaine (redrawn from Edmonds and Francesconi, 1987a).

Table 6.3 Arsenobetaine in marine animals (data taken from Francesconi and Edmonds, 1993)

Animal (No. of species)	Arsenic concentration (mg kg^{-1}, wet weight)	Arsenic (%) present as arsenobetaine
FISH		
Elasmobranchs (7 spp.)	3.1–44.3	94–95+
Teleosts (17 spp.)	0.1–166	48–95+
CRUSTACEANS		
Lobsters (4 spp.)	4.7–26	77–95+
Prawns/Shrimps (5 spp.)	5.5–20.8	55–95+
Crabs (6 spp.)	3.5–8.6	79–95+
MOLLUSCS		
Bivalves (11 spp.)	0.7–2.8	12–88
Gastropods (6 spp.)	3.1–116.5	58–95+
Cephalopods (3 spp.)	49	72–95+
ECHINODERMS (1 sp.)	12.4	60
COELENTERATES (1 sp.)	7.5	15
PORIFERA (2 spp.)	3.2–6.8	13–15

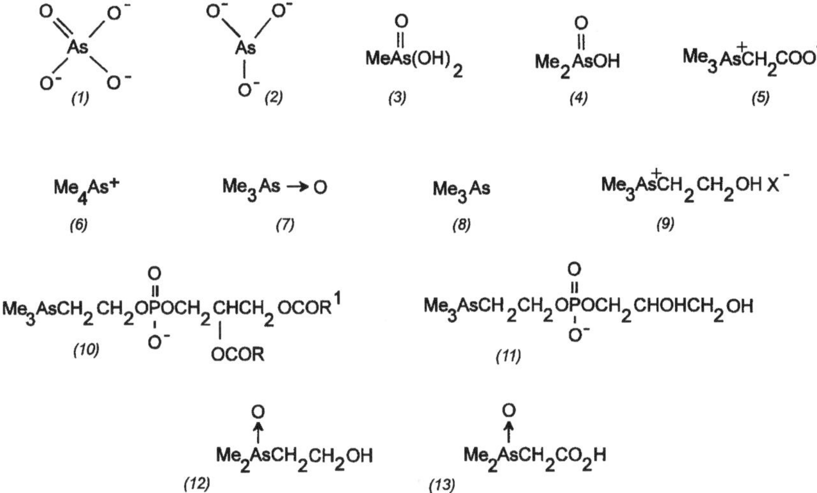

Figure 6.2 Arsenic compounds found in the marine environment and/or referred to in the text (continued on pages 164 and 165):

(1) arsenate
(2) arsenite
(3) methylarsonic acid
(4) dimethylarsinic acid
(5) arsenobetaine
(6) tetramethylarsonium ion
(7) trimethylarsine oxide
(8) trimethylarsine
(9) arsenocholine
(10) phosphatidylarsenocholine
(11) glycerophosphorylarsenocholine
(12) dimethylarsinoylethanol
(13) dimethylarsinoylacetic acid
(14)–(24) dimethylarsinoylribosides
(25),(26) trimethylarsonioribosides
(27),(28) see text
(29) dimethylarsinoyladenosine

Information on the stereochemistry of the dimethylarsinoylribosides (15–20) is given in Edmonds *et al.* (1993).

arsenic as relatively complex carbohydrate derivatives (15, 16) apparently unrelated to arsenobetaine (Edmonds and Francesconi, 1981, 1983). However the balance was re-established in favour of such a route when it was found that allowing decomposition of algal fragments under conditions promoting anaerobic microbial activity caused all arsenic-containing carbohydrates therein to be converted to dimethylarsinoylethanol (12), a compound that

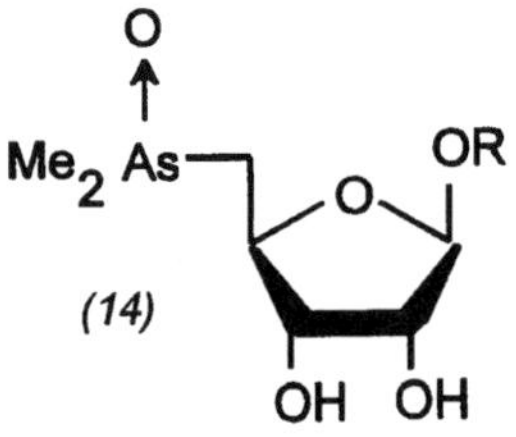

(14)

R = 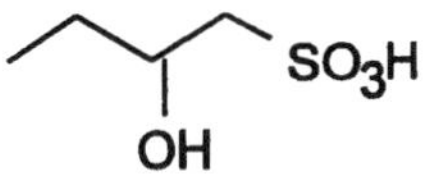
(15)

R = 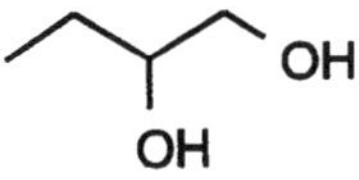
(16)

R = 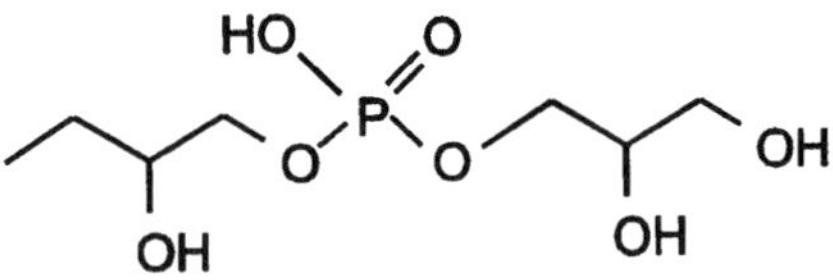
(17)

R = 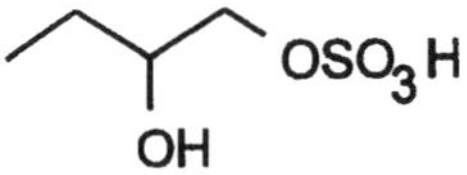
(18)

R = 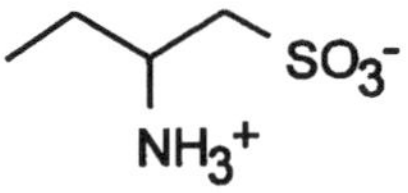
(19)

R = Me
(21)

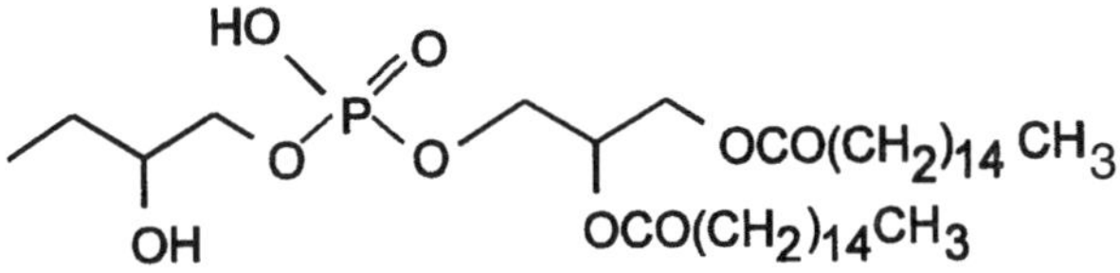

R = 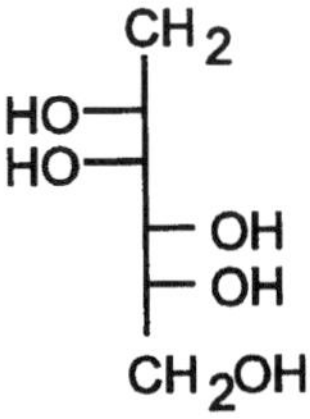
(22)

R = 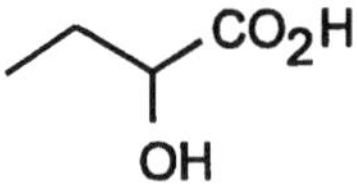
(23)

R = 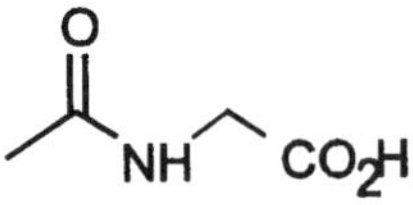
(24)

R =
(20)

Figure 6.2 continued; for key, see page 163.

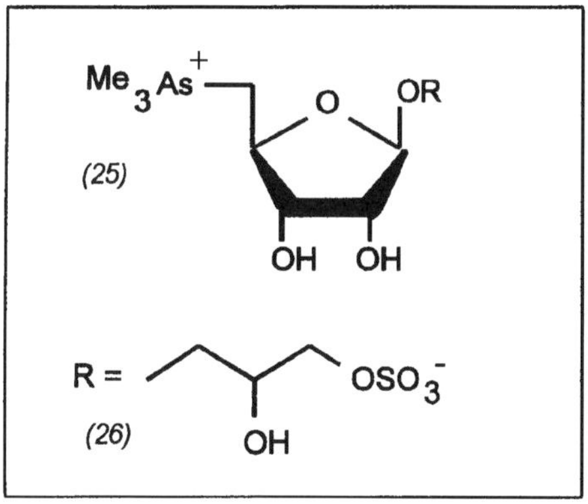

(25)

(26)

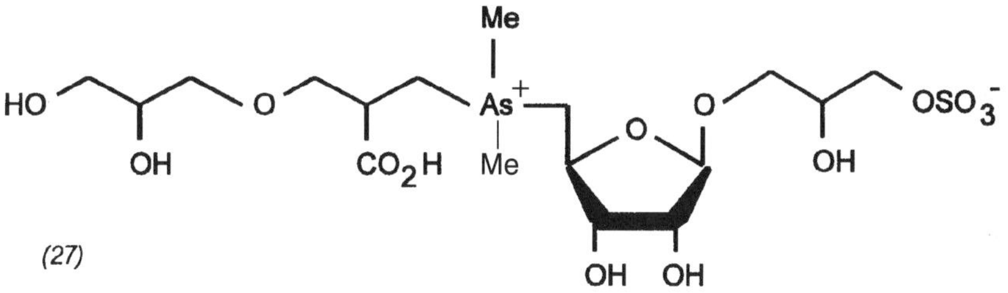

(27)

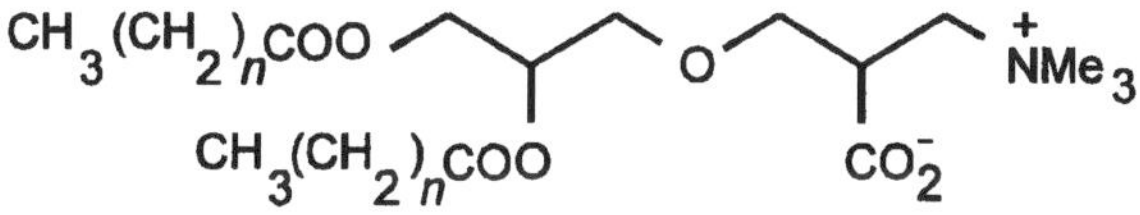

(28)

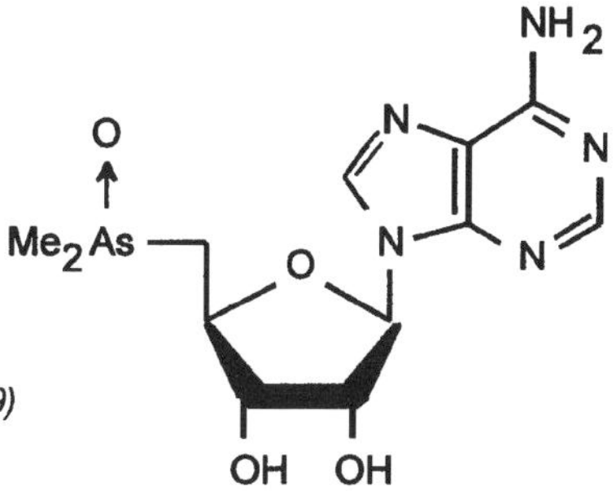

(29)

Figure 6.2 continued; for key, see page 163.

could easily be envisaged as a precursor of arsenobetaine (Edmonds *et al.*,1982).

At this stage the scheme shown in Figure 6.1 was drawn up (Edmonds and Francesconi, 1987a). Subsequently the discovery (Francesconi *et al.*, 1991b, 1992a) of dimethylarsinoyladenosine (29) in the kidney of the giant clam *Tridacna maxima* (arsenic compounds in clam tissues are assumed to result from the presence of symbiotic algae or from phytoplankton ingested as food) added some stiffening to the scheme, at least as far as offering support for the involvement of *S*-adenosylmethionine as the adenosylating as well as the methylating agent (Edmonds and Francesconi, 1983). However, despite dimethylarsinoylethanol (12) providing an apparent link between algal arsenic compounds and arsenobetaine in animals and an idea of the latter compound becoming available to animals through detritivores (because anaerobic processes in sediments had been invoked), details of the processes remained vague.

We would like to take this opportunity to examine this scheme critically, stage by stage, taking into account information that has become available since its publication, and thereby discuss arsenic metabolism in the marine environment – the major part, as we see it, of the purpose of this chapter.

6.2 ARSENIC IN SEAWATER

Arsenic is present in seawater at a concentration of about 2 μg l^{-1} (Armstrong and Harvey, 1950; Smales and Pate, 1952; Portmann and Riley, 1964; Johnson and Pilson, 1972) and there is no substantial evidence of variation with body of water although Johnson and Pilson (1972) found that deep ocean water contained higher concentrations of arsenic as arsenate (1) than surface water samples.

The chemical form of arsenic varies with biological activity (and therefore with depth) and with the degree of oxygenation of the water. In oxic waters arsenate (1) is the major form of arsenic, but because of biological activity arsenite (2) is always present at greater than thermodynamic equilibrium concentrations (Sugawara and Kanamori, 1964; Johnson, 1972; Gohda, 1975; Johnson and Burke, 1978). In the photic zone simple methylated arsenic compounds, methylarsonate (3) and dimethylarsinate (4) have been detected as well as the predominating inorganic species (Braman and Foreback, 1973; Andreae, 1977, 1978, 1979). Probably these compounds are present as a result of the methylating capability of algae (Andreae, 1978, 1979), phytoplankton in open waters and possibly macroalgae adjacent to the coast. Methylarsonic (3) and dimethylarsinic acids (4) are probable intermediates in the production of dimethylarsinoylribosides (14) in algae (Edmonds and Francesconi, 1987a) (though not all algae produce the latter compounds) but whether their presence in seawater results from decomposition of released ribosides or from a release of the simple compounds themselves is not currently clear. Some algae have been shown to contain dimethylarsinic acid in addition to dimethylarsinoylribosides (Jin *et al.*, 1988), but it has not been detected in the majority of macroalgae that have been studied. The possibility remains that phytoplankton rather than macroalgae contribute to the oceanic burden of these simple methylated arsenic compounds. Certainly two species of marine unicellular algae have been shown to contain their arsenic only in the form of methylarsonic (3) and dimethylarsinic acids and dimethylarsinoylribosides were absent (Cullen *et al.*, 1994; Shibata *et al.*, in press). Nevertheless, the production of dimethylarsinoylribosides by phytoplankton appears to be the norm rather than the exception (Shibata *et al.*, in press; Edmonds *et al.*, unpublished results).

Seawater is a complex matrix and the determination of the arsenic species that it contains can be difficult. This problem is usually overcome by using hydride generation techniques that separate the arsenic species from the seawater (Morita and Edmonds, 1992). The released hydrides (arsines) can be cold-trapped and sequentially distilled into a suitable element-specific detector (traditionally an atomic absorption spectrophotometer but recently ICP-mass spectrometers have been used) or separated by GC etc. The problem with this method – and this is only now receiving adequate attention – is that only those arsenic species producing volatile arsines upon treatment with a suitable reductant (usually sodium borohydride) are amenable to analysis (Morita and Edmonds, 1992). All other compounds that might be present, including arsenobetaine, dimethylarsinoylethanol (12) and the dimethylarsinoylribosides, would not be detected.

That seawater contained a substantial component of 'refractory' or 'hidden' arsenic has been suspected for some time (Howard and Comber, 1989; Bettencourt and Andreae, 1991). Elucidation of the nature of this hidden arsenic remains a major problem confronting those concerned with the natural chemistry of arsenic in seawater, and until its form or forms are known the detailed study of the algal and microbial transformations of arsenic compounds in seawater will of necessity remain incomplete. Part of this study must also be concerned with the nature of the compounds released by algae and the rates with which they are produced.

Seawater contains methyl iodide at concentrations up to a few nanograms per litre, with higher concentrations being found in the vicinity of algal (kelp) beds (Lovelock, 1975). Methyl iodide is one of a range of halogenated alkanes released by marine algae (Gschwend *et al.*, 1985) and its possible role, and that of other components of seawater that are released by algae, such as dimethyl propiothetin (Ishida and Kadota, 1967), in the chemical methylation of arsenic species is also an area that requires study. The topic is referred to briefly below.

A scheme for the cycling of arsenic in seawater and algae (and sediments) that involves the production of dimethylarsinoylribosides from absorbed arsenate in algae and their microbial conversion back to arsenate after their

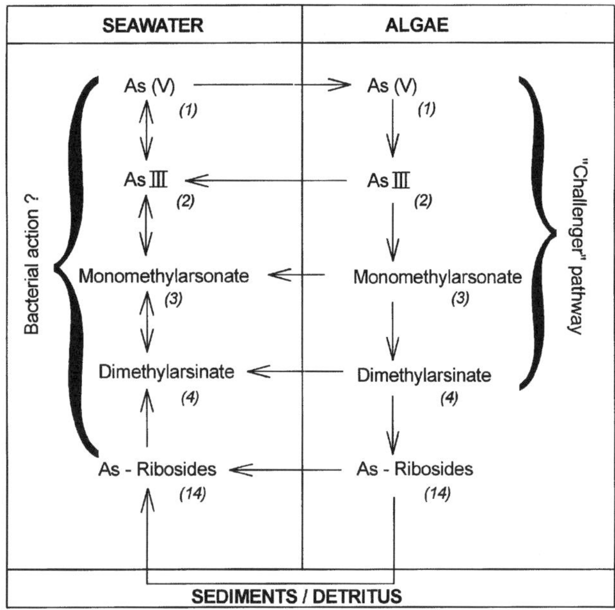

Figure 6.3 Simple scheme for the cycling of arsenic in the sea involving the production of arsenic-containing carbohydrates from arsenate by algae, and their subsequent degradation back to arsenate by bacterial action in the water.

release back into the seawater is shown in Figure 6.3. Such a scheme of course says nothing about arsenobetaine, but the arsenic compounds at all stages would be available for absorption and transformation by marine animals. However, to adhere to our scheme we must first consider the uptake by algae of arsenic in seawater.

6.3 UPTAKE OF ARSENATE BY ALGAE

In many seawaters, particularly those of tropical regions, the concentration of arsenate (1) – the major form of arsenic in seawater – exceeds that of the essential phosphate. Possibly the similarity of arsenate to phosphate results in arsenate being absorbed by algae, which must then undergo a detoxification procedure to prevent it entering and decoupling oxidative phosphorylation processes (Maugh, 1979). This implies of course that arsenate and phosphate are absorbed by algae by the same mechanism but it is far from clear that this is the case. Some workers have reported competitive uptake (Sanders and Windom, 1980) while others have found separate mechanisms for the absorption of the two anions (Andreae and Klumpp, 1979; Klumpp, 1980). While these two options could be equally valid, the difference from species to species that is implied suggests that the situation is not simple and more work is necessary for clarification. If, in some cases at least, arsenate is taken up by a mechanism independent of that used for phosphate absorption, the question as to why must be addressed. Is arsenic an essential element for algal growth?

A possible consequence of competitive uptake is that arsenate, at elevated concentrations, might have a toxic effect on algae, particularly phytoplankton, quite different to that of arsenite (2). Artificial increases in arsenate levels could arise from industrial discharge or from accidental spillage, and are probably more likely to occur in estuarine or coastal water than in the deep ocean. The effect of arsenate, where presumably it is inhibiting phosphate uptake rather than interfering at the biochemical level, on phytoplankton species composition and succession has been studied (Sanders and Cibik, 1985; Chapter 4).

6.4 ARSENIC METABOLISM IN ALGAE

The process by which absorbed inorganic arsenate (1) is sequentially reduced and oxidatively methylated in marine algae was assumed to follow the pathway first described by Challenger for the methylation of arsenic by the bread mould *Scopulariopsis brevicaulus* (Challenger, 1945, 1951) and involving *S*-adenosylmethionine (Adomet) as the methylating agent (Cantoni, 1952). The assumption that Adomet was involved in algae seemed particularly reasonable as it could be easily envisaged as the source of the ribose moiety that was attached to arsenic in the overwhelming majority of arsenic compounds in those algae that have been studied. Thus adomet was seen as first providing two methyl groups and then, instead of a third methyl group, the adenosyl group would be attached to arsenic, giving rise to dimethylarsinoyladenosine (29), a hypothetical intermediate when the overall scheme was first published (Edmonds and Francesconi, 1987a) but subsequently discovered in and isolated from extracts of the kidney of the giant clam *Tridacna maxima* (Francesconi *et al.*, 1991b, 1992a). Loss of adenine and glycosidation with algal metabolites would then produce the range of dimethylarsinoylribosides (14) that have been isolated from marine algae or from the kidney of the giant clam (where algae are assumed to be the source) (Francesconi *et al.*, 1992a).

The metabolism of arsenic compounds within algae is the part of the overall transformations of arsenic in the sea that is most clear; probably only inorganic arsenic can be taken in and, in the majority of cases, this is converted into carbohydrate derivatives. Details of the process remain to be sorted out: the enzymes involved, the order of methylation versus adenosylation, the release or otherwise of the compounds to seawater and whether they contribute to the so-called hidden arsenic.

Although the presence of arsenic in algal cells might result from an inability of the algae to distinguish arsenate from phosphate, the range and complexity of the compounds formed from that arsenate require consideration.

Probably the process is essentially one of detoxification, with the dimethyl-arsinoylribosides formed merely contributing harmless compounds to the osmotic pool. Compound (17), containing a phosphoric acid diester of glycerol as its aglycone, has been found (Morita and Shibata, 1990; Edmonds *et al.*, 1993) in all algae examined – except those in which dimethylarsinoylribosides were totally absent (Cullen *et al.*, 1994; Shibata *et al.*, in press) – and has been isolated as the di-palmitoyl derivative (20) from the Japanese edible alga *Undaria pinnatifida* (Morita and Shibata, 1988). In such a form it is probably much better able to cross membranes and might be instrumental in removing arsenic from cells and thus be a key compound in a detoxification process.

The possibility remains that this compound, or any other of the arsenic compounds found in algae, might have a more fundemental role in cellular metabolism and biochemistry than detoxification, but this has yet to be demonstrated. Compounds of the complexity of (27) can give rise to interesting speculations (Francesconi *et al.*, 1991a). However, this compound, when acylated with long-chain fatty acids, is analogous to a polar lipid containing a quaternary nitrogen atom (28) important as a membrane lipid in certain algae (Vogel *et al.*, 1990), and might also, therefore, be involved in the cellular excretion of unwanted arsenic.

6.5 THE BIOGENESIS OF ARSENOBETAINE

Dimethylarsinoylethanol (12), produced in laboratory experiments by the microbial decomposition, under anaerobic conditions, of dimethylarsinoylribosides (14) contained in algal fragments (or of the pure compounds) (Edmonds *et al.*, 1982) obviously played a key role in the formulation of the scheme (Figure 6.1) because it provided an apparent link between algal compounds and arsenobetaine (5). However, dimethylarsinoylethanol has still not been detected as a natural product and its conversion to arsenobetaine (further methylation at arsenic and oxidation of the side-chain) in the laboratory under conditions designed to replicate those found in the natural environment has not been accomplished. Of the two steps involved – further methylation at arsenic and oxidation of the side-chain – only the methylation step is of significance, the oxidation of (for example) the analogous arsenocholine (9) occurring readily enough upon its administration to fish as part of their diet (Francesconi *et al.*, 1990). The discovery of dimethylarsinoylethanol in the natural environment and its conversion to arsenobetaine under 'natural' conditions would add considerable weight to the proposed scheme. Chemical conversion of dimethylarsinoylethanol to arsenobetaine is easily done (Francesconi and Edmonds, unpublished results).

Possibly the methylation of dimethylarsinoylethanol by 'natural' processes, if it is indeed an important precursor of arsenobetaine in the environment, is too slow for easy detection in laboratory experiments, and highly selective absorption and retention higher in the food-chain accounts for the overwhelming predominance of arsenobetaine in marine animals. Nevertheless the failure to confirm dimethylarsinoylethanol as an integral component of our scheme means that other possible origins of arsenobetaine must be considered. Other observations must also be taken into account such as the other methylated arsenic compounds that have been found in marine organisms and their relationship to arsenobetaine. First, though, a general consideration of the biological methylation of arsenic may help to place the biogenesis of arsenobetaine into context.

Most organisms, from bacteria (McBride and Wolfe, 1971) and fungi (Challenger *et al.*, 1933) to mammals (Vahter, 1994), methylate the inorganic arsenic to which they are exposed, though there are some surprising exceptions – the marmoset monkey (Vahter *et al.*, 1982) and the chimpanzee (Vahter *et al.*, 1995) for example. Almost always, however, methylation does not proceed beyond the production of a trimethylated compound – trimethylarsine (8) or trimethylarsine oxide (7) – and often not beyond the dimethylated stage. Also, Challenger's classical work made no mention of tetramethylarsonium ion (6), the endpoint of methylation of arsenate, being produced by *Scopulariopsis brevicaulis*. Tetramethylarsonium ion is rather more difficult to detect than other simple arsenic compounds by the usual methods of analysis; it does not of course provide a volatile arsine upon treatment with sodium borohydride and is tightly held to the type of cation exchange resin that is customarily employed in the analysis of arsenical metabolites (Morita and Edmonds, 1992). This greater analytical difficulty does provide a complicating factor but to assume that all studies of arsenic methylation have failed to detect tetramethylarsonium ion or other tetra-alkylated species because of it would be unrealistic.

Perhaps naively, we have assumed that this failure of methylation to proceed beyond three methyl groups results from the inevitable release of the arsenic compound, trimethylarsine (8), from a hypothetical substrate that would follow reduction after the third methylation. In simplistic terms, all available valences of the reduced arsenic atom would be occupied by methyl groups (Edmonds and Francesconi, 1988). Such a release, making the arsenic atom unavailable for further methylation, might well be accelerated, in certain circumstances, by the volatility of trimethylarsine. Chemical methylation of trialkylarsine oxides to the arsonium species is very simple, usually only requiring treatment with methyl iodide in the presence of a reductant: British anti-Lewisite (BAL) (Francesconi *et al.*, 1994) and sodium borohydride (Shibata and Morita, 1988) have been used.

The arsenic compounds in marine algae are consistent with those produced from inorganic arsenic by most other organisms inasmuch as most is in the form of dimethylated (trialkylated) arsine oxides and only a very small proportion (< 1% total arsenic in those two organisms where it has been detected) is present as arsonium compounds (Shibata and Morita, 1988; Francesconi *et al.*, 1991a). Possibly such low levels of arsonium compounds result from chemical methylation of dimethylarsinoylribosides (14) rather than a methylation under enzymatic control.

All of this makes the indisputable predominance of the arsonium compound trimethylarsonioacetate, arsenobetaine, in marine animals interesting and surprising. Another factor that must be taken into account in any scheme for the biogenesis of arsenobetaine is its coexistence with other arsenic compounds. Probably the second most abundant organic arsenic species that has been identified in marine animals is tetramethylarsonium ion (6) (Shiomi *et al.*, 1987; Morita and Shibata, 1987; Francesconi *et al.*, 1988; Cullen and Dodd, 1989). This species has been found in some gastropod and bivalve molluscs and, although it is the major arsenic compound in some tissues (gills), arsenobetaine is always present also. Arsenobetaine coexists with other arsenic compounds in some animals: with trimethylarsine oxide (7) in some teleost fish (Edmonds and Francesconi, 1987b), with dimethylarsinoylribosides (14) in molluscs (Morita and Shibata, 1987) and in the digestive gland of the western rock lobster (Edmonds *et al.*, 1992). These compounds are either being accumulated independently and have separate origins, or their joint presences are to be accounted for by interconversions. Any theory of arsenobetaine biogenesis must take account of this.

In addition, arsenobetaine occurs as the overwhelmingly major arsenic compound, often at high concentrations, in situations where the involvement of microbial activity in sediments must, at best, be highly indirect. For example, the presettlement puerulus larvae of the western rock lobster, strictly pelagic in nature, contain arsenic at a concentration of greater than 30 mg kg^{-1} (wet weight basis) and more than 99% of this is arsenobetaine (Francesconi and Edmonds, unpublished results).

Obviously any scheme for the origin of arsenobetaine must satisfactorily account for the presence of the carboxymethyl group. Our original scheme (Edmonds and Francesconi, 1987a) derived this two-carbon moiety from the breakdown of the pentose ring of the algal dimethylarsinoylribosides (14). This and other possibilities for production of the carboxymethyl group are considered briefly at a chemical or biochemical level below. Later the location of these possible mechanisms, together with other stages in the biogenesis of arsenobetaine, is considered in more detail at an organism or environmental component level.

6.5.1 ORIGIN OF THE CARBOXYMETHYL GROUP IN ARSENOBETAINE

(a) From the breakdown of a larger molecule

This process, an integral part of our original scheme, has been considered already. Only the algal dimethylarsinoylribosides (14) currently serve as candidates for the larger precursor. In summary, this step in any overall scheme is supported by the ready generation of a two-carbon side-chain upon anaerobic microbial decomposition of the dimethylarsinoylribosides (14), but opposed by our failure to detect the product – dimethyarsinoylethanol (12) – in the natural environment or to convert it to arsenobetaine under conditions designed to replicate those that might be found naturally, and also, it might be thought, by the presence of arsenobetaine in open-ocean pelagic animals.

However, a variation on this theme (Figure 6.4) in which the precursor is a trimethylarsonioriboside (25) offers a possibility for the production of arsenobetaine without the need for the final methylation. Allowing the trimethylarsonioriboside (26) to decompose in anaerobic sediments (contained in a separating funnel) resulted in its quantitative conversion to arsenocholine (9) (Francesconi *et al.*, 1992b). Oxidation of arsenocholine to arsenobetaine occurs readily and rapidly in those organisms (fish and mussels) to which it has been fed as part of their diet (Francesconi *et al.*, 1989; Gailer *et al.*, 1995). Indeed the speed of the conversion might well account for the absence of arsenocholine in virtually all animals that have been examined. Where it has been reliably reported, highly sensitive HPLC ICP-MS techniques have been used to detect its presence at usually less than 1% of total arsenic (Larsen *et al.*, 1993). Other reports of higher concentrations of arsenocholine (Norin *et al.*, 1983; Lawrence *et al.*, 1986) in some shrimps have been questioned (Cullen and Reimer, 1989) and can probably be discounted.

The problem with the scheme for the production of arsenobetaine involving the decomposition of trimethylarsonioribosides (25) is the small amount of the latter that is likely to be available: in the two organisms where it has been detected, compound (26) has accounted for less than 1% of total arsenic (Shibata and Morita, 1988; Francesconi *et al.*, 1991a). Nevertheless it is possible that arsenobetaine is produced in this way (Figure 6.4) but the apparent necessity for the involvement of anaerobic sediments might again be considered difficult to accommodate in all cases.

(b) By addition of CO_2 to tetramethylarsonium ion

The presence of both arsenobetaine and tetramethylarsonium ion (6) in some molluscs requires either that they are accumulated (or biosynthesized) independently or that one is converted into the other. The predominance of tetra-

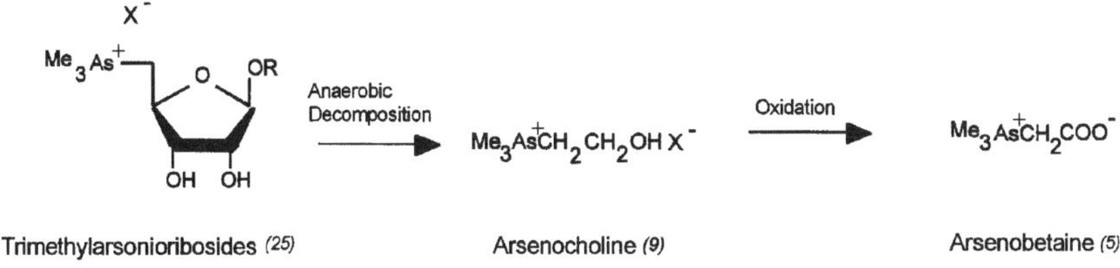

Figure 6.4 Possible scheme for the production of arsenobetaine from trimethylarsonioribosides.

methylarsonium ion in the gills of some molluscs (Shiomi *et al.*, 1987; Francesconi *et al.*, 1988) might suggest that it is absorbed from seawater (where it could contribute to the 'hidden' arsenic) and then, possibly by some enzymic process, CO_2 is added to produce arsenobetaine via the ylid (Figure 6.5). This perhaps fanciful notion would have arsenobetaine being produced in molluscs and thus becoming available to higher components of the food-chain. On the other hand. the presence of tetramethylarsonium ion in the gills of molluscs might be taken as evidence of a procedure for expelling arsenic from the molluscan body. It is also possible that tetramethylarsonium ion is produced by molluscs by some process analogous to that for the biosynthesis of tetramethylammonium ion. The latter compound has been shown to be a component of some molluscs (Asano and Itoh, 1960; Fänge, 1960). Possible parallels between arsenic and nitrogen metabolism are further discussed briefly below.

(c) Direct addition of the carboxymethyl group to trimethylated arsenic

Chemical synthesis of arsenobetaine involves attack of trimethylarsine (8) on ethylbromoacetate, with the former serving as a nucleophile (Edmonds *et al.*, 1977; Cannon *et al.*, 1981). However, the usual source of acetate in biochemical processes, acetyl coenzyme A, normally requires an electrophilic acceptor. It is thus difficult to envisage a process for the carboxymethylation of trimethylarsine that involves acetyl CoA. Although trimethylarsine oxide (7) may provide a more appropriate recipient for the carboxymethyl group, other mechanistic problems would need to be overcome. Nevertheless, the direct carboxymethylation of a trimethyl arsenic species, probably by bacteria in the water column, is a possible source of arsenobetaine that cannot be discounted.

(d) Processes that parallel nitrogen metabolism

The obvious similarity of arsenobetaine and glycine betaine immediately suggests that they might be derived by analogous pathways. Such a notion has, at least in part, fuelled discussion of biogenetic pathways for arsenic in marine organisms (Phillips and Depledge, 1985). However, a complete parallel would involve arsenic analogues of all the precursors of glycine betaine,

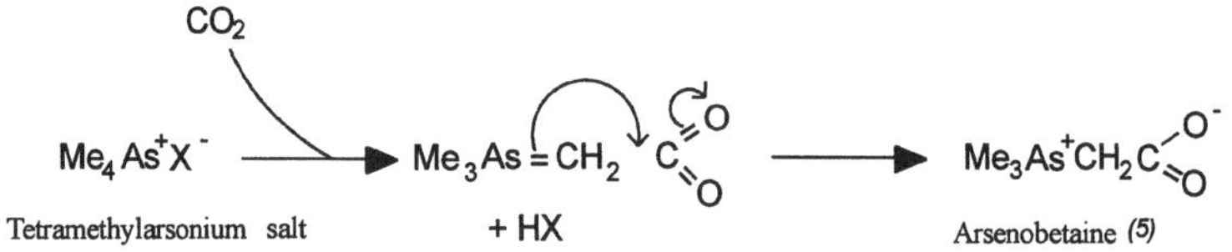

Figure 6.5 Speculative scheme for the production of arsenobetaine from trimethylarsonium ion.

including ethanolamine and serine. None of these has ever been detected in any organism and the possible toxic hazard that their presence might entail ('arsenoethanolamine', in particular, would appear to be a potentially toxic species) might be considered to render unlikely a pathway for arsenic metabolism that parallels that of nitrogen.

6.5.2 THE SITE OF ARSENOBETAINE FORMATION

Marine algae do not contain arsenobetaine, yet almost all marine animals do. Despite the brief speculation above on possible conversion of tetramethylarsonium ion (6) to arsenobetaine, there is currently no evidence at all to support the idea that marine animals have the ability to synthesize arsenobetaine from any precursor except arsenocholine (9) and this conversion is likely to be insignificant in the total scheme. Certainly no marine animal has been shown to biosynthesize arsenobetaine *de novo* from ingested arsenate (1), and no animal has been shown to possess the ability to convert ingested dimethylarsinoylribosides (14) to arsenobetaine. Indeed it has been shown that the lobster *Homarus americanus* is unable to make this conversion (Cooney and Benson, 1980) and preliminary feeding experiments with a teleost fish, *Rhabdosargus sarba*, also failed to support such a route (Francesconi and Edmonds, unpublished results). For the purposes of this discussion we shall assume that marine animals receive arsenobetaine – or possibly arsenocholine, though this is very much less likely – ready made.

It has been clearly demonstrated that the mussel *Mytilus edulis* rapidly absorbs arsenobetaine from ambient water (Francesconi and Edmonds, 1987; Gailer *et al*, 1995) and that the fish *Aldrichetta forsteri* is equally retentive of arsenobetaine presented in its food (Francesconi *et al.*, 1989). Although these animals also retain the synthetic arsenocholine (9) (rapidly converting it to arsenobetaine) and, to a lesser extent, tetramethylarsonium ion (6) that they are offered, they do not accumulate other arsenic compounds to which they are subjected (Francesconi *et al.*, 1989). The selectivity in favour of arsenobetaine in particular may be important in explaining the high proportion of the compound in the arsenic burden of almost all marine animals, and a comparison with the highly selective accumulation of methylmercury by fish has been made (Edmonds and Francesconi, 1987a; Francesconi and Edmonds, in press).

For completeness we should mention here that a total about 4% of the arsenocholine administered to fish as part of their diet was converted to, and accumulated as, phosphatidylarsenocholine (10) and glycerophosphorylarsenocholine (11), the rest of course being retained as arsenobetaine (Francesconi *et al.*, 1990). It is difficult to assess the significance of the small quantity of phosphatidylarsenocholine occurring naturally in the digestive gland of the western rock lobster (Edmonds *et al.*, 1992). It was possibly derived from a small amount of arsenocholine produced by reduction of a small fraction of the relatively large concentration of arsenobetaine present. On the other hand it might suggest that the western rock lobster is indeed exposed to a source of arsenocholine.

Marine algae do not make arsenobetaine, then, and neither do marine animals. It is made in sediments or in the water column by microbial activity supplemented possibly by chemical methylation by naturally occurring agents such as methyl iodide and dimethyl propiothetin. It seems unlikely that its production can be restricted to sediments, anaerobic or otherwise, because its presence in pelagic organisms would suggest a possibly unreasonable rate of release and distribution. On the other hand benthic fish do tend to contain more arsenic (and therefore arsenobetaine) than pelagic fish . If arsenobetaine is being derived from dimethylarsinoylribosides (14), as in our original scheme, we have to allow the possibility that the conversion to arsenobetaine can occur in aerobic conditions. A transformation under such conditions that retains a two-carbon side-chain has not yet been demonstrated, and in any case the difficulties and provisos that apply to the necessary additional methylation would still apply.

Another indigestible aspect of schemes involving largely microbial transformations in sediments or water of compounds excreted by or lost from algae on senescence is the double accumulation that is needed. Such schemes require that, first, arsenate is absorbed from water by algae and converted to dimethylarsinoylribosides (14) which are then dispersed in water, and that the arsenic, now present as arsenobetaine, is absorbed by animals, or at least diverted into the food-chain by animals, such as bivalve molluscs, that possess a remarkable facility to scavenge arsenobetaine from ambient water. A factor militating against the absorbance of arsenobetaine from ambient water is the noticeable decomposition of arsenobetaine (presumably again under microbial influence), first to trimethylarsine oxide (7) and, ultimately, to arsenate (1) (Hanaoka *et al.*, 1992). Rates of decomposition are uncertain and it is possible that the rate of absorbance by biota of arsenobetaine formed in seawater may be significantly greater than its microbially mediated decomposition.

Arsenobetaine may be produced in seawater from arsenate entirely by the action of bacteria. As discussed above, only the carboxymethylation step offers something new; that bacteria can generate trimethylarsine (8) or its oxide (7) from arsenate is well known. There is a further variation: carboxymethylation occurs after dimethylation – dimethylarsinic acid (4) being carboxymethylated to dimethylarsinoyl acetic acid (13), which again would require the final methylation. However, the carboxymethylation of dimethylarsinic acid may not pose the same mechanistic impediments to the involvement of acetyl coenzyme A as do arsenic species already possessing three methyl groups. The indigestible aspect of this scheme is that absorption and transformation of arsenate by algae would be entirely independent of the process giving rise to arsenobetaine in marine animals. The two schemes would not be related.

Although our apparent failure to bring about the methylation of dimethylarsinoylethanol (12) under conditions that would be expected in the natural environment might suggest that this step cannot be involved in the biogenesis of arsenobetaine, it is possible (as implied above) that a chemical methylation depending, for example, on naturally occurring methyl iodide might occur but at a rate that would make detection in laboratory experiments difficult. This is an area requiring further study. In other words, the slow production of arsenobetaine partly by biological processes – to give dimethylarsinoylribosides (14) and dimethylarsinoylethanol (12) – and partly by chemical processes (final methylation) would occur at a rate less than that of its absorbance by biota. There would thus be a rapid flux of arsenobetaine through seawater but its concentration at any time would be very low. The same considerations would apply if the entire process (or most of it) was under microbial rather than algal control.

Arsenobetaine was known as a synthetic compound long before its discovery as a natural component of the western rock lobster. It was first synthesized for pharmacological studies examining the possible physiological role of arsenic analogues of some simple nitrogen metabolites (Welch and Landau, 1942). The large number of studies demonstrating its presence, often in surprisingly large quantities, in marine animals have resulted in it being regarded solely in a marine context. Two recent reports have considerably widened our knowledge of its distribution and must be accommodated in any biogenetic theories. It has been identified in two species of Japanese freshwater fishes (Shiomi *et al.*, 1995). One of these (*Salmo gairdneri* containing 1.08 mg arsenic kg^{-1} on a wet-weight basis) was cultured and there was a possibility that the source of arsenobetaine was marine-derived material used as a feedstuff. The other (*Hypomesus nipponensis*; 1.46mg arsenic kg^{-1}) was a wild river fish and the assumption was that the arsenobetaine it contained was there without the intervention of the sea. The authors point out that their results are at variance with those of other workers (Lawrence *et al.*, 1986) who failed to find arsenobetaine in Canadian freshwater species. In terms of concentrations their results also differed from those of New Zealand workers (Robinson *et al.*, 1995) who found only 0.02 to 0.05 mg arsenic kg^{-1} on a wet-weight basis in trout *Oncorhyncus mykiss* and *Salmo trutta* from the Taupo volcanic zone and Waikato river – areas that would be expected to contain elevated arsenic concentrations (Lancaster *et al.*, 1971). The arsenic concentrations in fresh water can vary very widely and in volcanically active countries with an abundance of hot-springs and volcanic lakes, such as Japan (Noguchi and Nakagawa, 1970) and New Zealand (Anonymous, 1977), the levels can be high. In other regions of the world the concentration of arsenic in rivers and lakes is often lower than in seawater, though there is much variation (Anonymous, 1977). A start has been made (Maeda, 1994) but the metabolism of arsenic in freshwater systems is evidently an area for further study. Do freshwater algae biosynthesize dimethylarsinoylribosides (14), for example, when the concentration of arsenic (arsenate?) to which they are exposed is high enough? If they do not, the discovery of arsenobetaine in freshwater fish might be taken to support the notion that such compounds are not its origin.

The second recent report is more surprising and must cause substantial reassessment of all theories of the origin of arsenobetaine. It has been identified in terrestrial mushrooms (Byrne *et al.*, 1995). Whether the mushrooms are making it from absorbed arsenate (1) (or other precursor) or whether they are absorbing it (together with other simple arsenic compounds such as the methylated arsenic acids) ready made from soil, where it had been made by other organisms (bacteria?), is not clear at present. However these fungi might well provide a more convenient model system for studying the biosynthesis of arsenobetaine than any marine organism. We can only assume at this time that the results of such an investigation would be pertinent to the origins of arsenobetaine in marine organisms. However, methylation of arsenic, wherever it occurs, has been shown (or assumed) to involve Adomet as the methylating agent. In all cases, then, there may be potential for the transfer of an adenosyl as well as methyl groups to the arsenic atom. While the dimethylarsinoylribosides (14) would seem to be typical algal products insofar as the aglycone in most cases is an obvious algal compound (mannitol and cysteinolic acid, for example), their precursor and a key intermediate in the algal metabolism of arsenic, dimethylarsinoyladenosine (29), might well have a more universal presence. Possibly the transfer of the adenosyl group, and thus the ribose moiety and the potential for its degradation, is not confined to the marine environment and parallels in arsenic metabolism between marine and terrestrial organisms may be closer than they appear at present. More work is clearly necessary.

REFERENCES

Andreae, M.O. (1977) Determination of arsenic species in natural waters. *Anal. Chem.* **49**, 820–823.

Andreae, M.O. (1978) Distribution and speciation of arsenic in natural waters and some marine algae. *Deep-Sea Res.* **25**, 391–402.

Andreae, M.O. (1979) Arsenic speciation in seawater and interstitial waters: the influence of biological-chemical interactions on the chemistry of a trace element. *Limn. Oceanog.* **24**, 440–452.

Andreae, M.O. and Klumpp, D. (1979) Biosynthesis and release of organo-arsenic compounds by marine algae. *Environ. Sci. Technol.* **13**, 738–741.

Anonymous (1977) *Medical and Biological Effects of Environmental Pollutants: Arsenic*, National Academy of Sciences, Washington, DC.

Armstrong, F.A.J. and Harvey, H.W. (1950) The cycle of phosphorus in the waters of the English Channel. *J. Mar. Biol. Assoc. UK* **29**, 145–162.

Asano, M. and Itoh, M. (1960) Salivary poison of a marine gastropod *Neptunea arthritica* (Bernardi), and the seasonal variation of its toxicity. *Ann. NY Acad. Sci.* **90**, 674–688.

Bebbington, G.N., Mackay, N.J., Chvojka, R. *et al.* (1977) Heavy metals, selenium and arsenic in nine species of Australian commercial fish. *Aust. J. Mar. Freshw. Res.* **28**, 277–286.

Bettencourt, A.M.M. de and Andreae, M.O. (1991) Refractory arsenic species in estuarine waters. *Appl. Organometal. Chem.* **5**, 111–116.

Braman, R.S. and Foreback, C.C. (1973) Methylated forms of arsenic in the environment. *Science* **182**, 1247–1249.

Byrne, A.R., Slejkovec, Z., Stijve, K. *et al.* (1995) Arsenobetaine and other arsenic species in mushrooms. *Appl. Organometal. Chem.* **9**, 305–313.

Cannon, J.R., Edmonds, J.S., Francesconi, K.A. *et al.* (1981) Isolation, crystal structure and synthesis of arsenobetaine, a constituent of the western rock lobster, *Panulirus cygnus*, the dusky shark, *Carcharhinus obscurus*, and some samples of human urine. *Aust. J. Chem.* **34**, 787–798.

Cantoni, G.L. (1952) The nature of the active methyl donor formed enzymatically from L-methionine and adenosinetriphosphate. *J. Am. Chem. Soc.* **74**, 2942–2943.

Challenger, F. (1945) Biological methylation. *Chem. Revs* **36**, 315–361.

Challenger, F. (1951) Biological methylation. *Adv. Enzymol.* **12**, 429–491.

Challenger, F., Higginbottom, C. and Ellis, L. (1933) The formation of organo-metalloid compounds by microorganisms. Part 1. Trimethylarsine and dimethylethylarsine. *J. Chem. Soc.* 95–101.

Cooney, R.V. and Benson, A.A. (1980) Arsenic metabolism in *Homarus americanus*. *Chemosphere* **9**, 335–341.

Cullen, W.R. and Dodd, M. (1989) Arsenic speciation in clams of British Columbia. *Appl. Organometal. Chem.* **3**, 79–88.

Cullen, W.R. and Reimer, K.J. (1989) Arsenic speciation in the environment. *Chem. Revs* **89**, 713–764.

Cullen, W.R., Harrison, L.G., Li, H. and Hewitt, G. (1994) Bioaccumulation and excretion of arsenic compounds by a marine unicellular alga, *Polyphysa peniculus*. *Appl. Organometal. Chem.* **8**, 313–324.

Dhandhukia, M.M. and Seshadri, K. (1969) Arsenic content in marine algae. *Phykos* **8**, 108–111.

Edmonds, J.S. and Francesconi, K.A. (1981) Arseno-sugars from brown kelp (*Ecklonia radiata*) as intermediates in cycling of arsenic in a marine ecosystem. *Nature* **289**, 602–604.

Edmonds, J.S. and Francesconi, K.A. (1983) Arsenic-containing ribofuranosides: isolation from brown kelp *Ecklonia radiata* and NMR spectra. *J. Chem. Soc., Perkin Transactions* **1**, 2375–2382.

Edmonds, J.S. and Francesconi, K.A. (1987a) Transformations of arsenic in the marine environment. *Experientia* **43**, 553–557.

Edmonds, J.S. and Francesconi, K.A. (1987b) Trimethylarsine oxide in estuary catfish (*Cnidoglanis macrocephalus*) and school whiting (*Sillago bassensis*) after oral administration of sodium arsenate; and as a natural component of estuary catfish. *Sci. Total Environ.* **64**, 317–323.

Edmonds, J.S. and Francesconi, K.A. (1988) The origin of arsenobetaine in marine animals. *Appl. Organometal. Chem.* **2**, 297–302.

Edmonds, J.S., Francesconi, K.A., Cannon, J.R. *et al.* (1977) Isolation, crystal structure and synthesis of arsenobetaine, the arsenical constituent of the western rock lobster *Panulirus longipes cygnus* (George). *Tetrahedron Letts* **18**, 1543–1546.

Edmonds, J.S., Francesconi, K.A. and Hansen, J.A. (1982) Dimethyloxarsylethanol from anaerobic decomposition of brown kelp *Ecklonia radiata*: a likely precursor of arsenobetaine in marine fauna. *Experientia* **38**, 643–644.

Edmonds, J.S., Shibata, Y., Francesconi, K.A. *et al.* (1992) Arsenic lipids in the digestive gland of the western rock lobster *Panulirus cygnus*: an investigation by HPLC ICP-MS. *Sci. Total Environ.* **122**, 321–335.

Edmonds, J.S., Francesconi, K.A. and Stick, R.V. (1993) Arsenic compounds from marine organisms. *Nat. Prod. Reps* **10**, 421–428.

Egaas, E. and Brækkan, O.R. (1977) The arsenic content of some Norwegian fish products. *Fiskeridirektoratets Skrifter Serie Ernæring* **1**, 93–98.

Falconer, C.R., Shepherd, R.J., Pirie, J.M. and Topping, G. (1983) Arsenic levels in fish and shellfish from the North sea. *J. Exper. Mar. Biol. Ecol.* **71**, 193–203.

Fänge, R. (1960) The salivary gland of *Neptunea antiqua*. *Anns NY Acad. Sci.* **90**, 689–694.

Francesconi, K.A. and Edmonds, J.S. (1987) Accumulation of arsenobetaine from seawater by the mussel (*Mytilus edulis*), in *Heavy Metals in the Environment*, Vol. 2, (eds S.E. Lindberg and T.C. Hutchinson), CEP Consultants, Edinburgh, pp. 71–73.

Francesconi, K.A. and Edmonds, J.S. (1993) Arsenic in the sea. *Oceanog. Mar. Biol. Ann. Rev.* **31**, 111–151.

Francesconi, K.A. and Edmonds, J.S. (in press) Arsenic and marine organisms. *Adv. Inorg. Chem.*

Francesconi, K.A., Edmonds, J.S. and Hatcher, B.G. (1988) Examination of the arsenic constituents of the herbivorous marine gastropod *Tectus pyramis*: isolation of tetramethylarsonium ion. *Comp. Biochem. Physiol.* **90C**, 313–316.

Francesconi, K.A., Edmonds, J.S. and Stick, R.V. (1989) Accumulation of arsenic in yelloweye mullet (*Aldrichetta forsteri*) following oral administration of organoarsenic compounds and arsenate. *Sci. Total Environ.* **79**, 59–67.

Francesconi, K.A., Stick, R.V. and Edmonds J.S. (1990) Glycerylphosphorylarsenocholine and phosphatidylarsenocholine in yelloweye mullet (*Aldrichetta forsteri*) following oral administration of arsenocholine. *Experientia* **46**, 464–466.

Francesconi, K.A., Edmonds, J.S., Stick, R.V. *et al.* (1991a) Arsenic-containing ribosides from the brown alga *Sargassum lacerifolium*: X-ray molecular structure of 2-amino-3-[5′-deoxy-5′-(dimethylarsinoyl)ribosyloxy]propane-1-sulphonic acid. *J. Chem. Soc., Perkin Transactions* **1**, 2707–2716.

Francesconi, K.A., Stick, R.V. and Edmonds, J.S. (1991b) An arsenic-containing nucleoside from the kidney of the giant clam, *Tridacna maxima*. *J. Chem. Soc., Chem. Commun.* 928–929.

Francesconi, K.A., Edmonds, J.S. and Stick, R.V. (1992a) Arsenic compounds from the kidney of the giant clam *Tridacna maxima*: Isolation and identification of an arsenic-containing nucleoside. *J. Chem. Soc., Perkin Transactions* **1**, 1349–1357.

Francesconi, K.A., Edmonds, J.S. and Stick, R.V. (1992b) Arsenocholine from anaerobic decomposition of a trimethylarsonioriboside. *Appl. Organometal. Chem.* **6**, 247–249.

Francesconi, K.A., Edmonds, J.S. and Stick, R.V. (1994) Synthesis, NMR spectra and chromatographic properties of five trimethylarsonioribosides. *Appl. Organometal. Chem.* **8**, 517–523.

Gailer, J., Francesconi, K.A., Edmonds, J.S. and Irgolic, K.J. (1995) Metabolism of arsenic compounds by the blue mussel *Mytilus edulis* after accumulation from seawater spiked with arsenic compounds. *Appl. Organometal. Chem.* **9**, 341–355.

Glover, J.W. (1979) Concentrations of arsenic, selenium and ten heavy metals in school shark, *Galeorhinus australis* (Macleay) and gummy shark, *Mustelus antarcticus* (Günther), from south-eastern Australian waters. *Aust. J. Mar. Freshwater Res.* **30**, 505–510.

Gohda, S. (1975) Valence states of arsenic and antimony in sea water. *Bull. Chem. Soc. Jap.* **48**, 1213–1216.

Gschwend P.M., MacFarlane, J.K and Newman, K.A. (1985) Volatile halogenated organic compounds released into seawater from temperate marine macroalgae. *Science* **227**, 1033–1035.

Hall, R.A., Zook, E.G. and Meaburn, G.M. (1978) *National Marine Fisheries Service Survey of Trace Elements in the Fishery Resource*, NOAA Technical Report NMFS SSRF–721, 313pp.

Hanaoka, K., Tagawa, S. and Kaise, T. (1992) The fate of organoarsenic compounds in marine ecosystems. *Appl. Organometal. Chem.* **6**, 139–146.

Howard, A.G. and Comber, S.D.W. (1989) The discovery of hidden arsenic species in coastal waters. *Appl. Organometal. Chem.* **3**, 509–514.

Ishida, Y. and Kadota, H. (1967) Isolation and identification of dimethyl-β-propiothetin from *Gyrodinium cohnii*. *Agric. Biol. Chem.* **31**, 756–757.

Jin, K., Hayashi, T., Shibata, Y. and Morita, M. (1988) Arsenic-containing ribofuranosides and dimethylarsinic acid in green seaweed, *Codium fragile*. *Appl. Organometal. Chem.* **2**, 365–369.

Johnson, D.L. (1972) Bacterial reduction of arsenate in sea water. *Nature* **240**, 44–45.

Johnson, D.L. and Burke, R.M. (1978) Biological mediation of chemical speciation. II. Arsenate reduction during marine phytoplankton blooms. *Chemosphere* **7**, 645–648.

Johnson, D.L. and Pilson, M.E.Q. (1972) Arsenate in the western North Atlantic and adjacent regions. *J. Mar. Res.* **30**, 140–149.

Kennedy, V.S. (1976) Arsenic concentrations in some coexisting marine organisms from Newfoundland and Labrador. *J. Fish. Res. Board Can.* **33**, 1388–1393.

Klumpp, D.W. (1980) Characteristics of arsenic accumulation by the seaweeds *Fucus spiralis* and *Ascophyllum nodosum*. *Mar. Biol.* **58**, 257–264.

Kurosawa, S., Yasuda, K., Taguchi, N. *et al.* (1980) Identification of arsenobetaine, a water soluble organo-arsenic compound in muscle and liver of a shark, *Prionace glaucus*. *Agric. Biol. Chem.* **44**, 1993–1994.

Lancaster, R.J., Coup, M.R. and Hughes, J.W. (1971) Toxicity of arsenic present in lakeweed. *NZ Vet. J.* **19**, 141–145.

Larsen, E.H., Pritzel, G. and Hansen, S.H. (1993) Speciation of eight arsenic compounds in human urine by high-performance liquid chromatography with inductively coupled plasma mass spectrometric detection using antimonate for internal chromatographic standardization. *J. Anal. Atom. Spectr.* **8**, 557–563.

Lawrence, J.F., Michalik, P., Tam, G. and Conacher, H.B.S. (1986) Identification of arsenobetaine and arsenocholine in Canadian fish and shellfish by high-performance liquid chromatography with atomic absorption detection and confirmation by fast atom bombardment mass spectrometry. *J. Agric. Food Chem.* **34**, 315–319.

Leatherland, T.M. and Burton, J.D. (1974). The occurrence of some trace metals in coastal organisms with particular reference to the Solent region. *J. Mar. Biol. Assoc. UK* **54**, 457–468.

Lovelock, J.E. (1975) Natural halocarbons in the air and in the sea. *Nature* **256**, 193–194.

Lunde, G. (1970) Analysis of arsenic and selenium in marine raw materials. *J Sci. Food Agric.* **21**, 242–247.

Luten, J.B., Riekwel-Booy, G. and Rauchbaar, A. (1982) Occurrence of arsenic in plaice (*Pleuronectes platessa*), nature of organo-arsenic compound present and its excretion by man. *Environ. Health Persp.* **45**, 165–170.

Maher, W.A. and Clarke, S.M. (1984) The occurrence of arsenic in selected marine macroalgae from two coastal areas of South Australia. *Mar. Polln Bull.* **15**, 111–112.

McBride, B.C. and Wolfe, R.S. (1971) Biosynthesis of dimethylarsine by methanobacterium. *Biochemistry* **10**, 4312–4317.

Maeda, S. (1994) Biotransformation of arsenic in the freshwater environment, in *Arsenic in the Environment. Part 1: Cycling and Characterization*, (ed. J.O. Nriagu), John Wiley & Sons, New York, pp. 155–187.

Maugh, T.H. II (1979) It isn't easy being king. *Science* **203**, 637.

Morita, M. and Edmonds, J.S. (1992) Determination of arsenic species in environmental and biological samples. *Pure Appl. Chem.* **64**, 575–590.

Morita, M. and Shibata, Y. (1987) Speciation of arsenic compounds in marine life by high performance liquid chromatography combined with inductively coupled argon plasma atomic emission spectrometry. *Anal. Sci.* **3**, 575–577.

Morita, M. and Shibata, Y. (1988) Isolation and identification of arseno-lipid from a brown alga, *Undaria pinnatifida* (Wakame). *Chemosphere* **17**, 1147–1152.

Morita, M. and Shibata, Y. (1990) Chemical form of arsenic in marine macroalgae. *Appl. Organometal. Chem.* **4**, 181–190.

Noguchi, K. and Nakagawa, R. (1970) Arsenic in the waters and deposits of Osoreyama hot springs, Aomori Prefecture. *Nippon Kagaku Zasshi* **91**, 127–131.

Norin, H., Ryhage, R., Christakopoulos, A. and Sandström, M. (1983) New evidence for the presence of arsenocholine in shrimps (*Pandalus borealis*) by use of pyrolysis gas chromatography–atomic absorption/mass spectrometry. *Chemosphere* **12**, 299–315.

Peden, J.D., Crothers, J.H., Waterfall, C.E. and Beasley, J. (1973) Heavy metals in Somerset marine organisms. *Mar. Polln Bull.* **4**, 7–9.

Phillips, D.J.H. and Depledge, M.H. (1985) Metabolic pathways involving arsenic in marine organisms: a unifying hypothesis. *Mar. Environ. Res.* **17**, 1–12.

Portmann, J.E. and Riley, J.P. (1964) Determination of arsenic in sea water, marine plants and silicate and carbonate sediments. *Anal. Chim. Acta* **31**, 509–519.

Powell, J.H., Powell, R.E. and Fielder, D.R. (1981) Trace element concentrations in tropical marine fish at Bougainville Island, Papua New Guinea. *Water Air Soil Polln* **16**, 143–158.

Rao, Ch.K., Chinnaraj, S., Inamdar, S.N. and Untawale, A.G. (1991) Arsenic content in certain marine brown algae and mangroves from Goa coast. *Indian J. Mar. Sci.* **20**, 283–285.

Robinson, B.H., Brooks, R.R., Outred, H.A. and Kirkman, J.H. (1995) Mercury and arsenic in trout from the Taupo Volcanic Zone and Waikato River, North Island, New Zealand. *Chem. Speciation Bioavail.* **7**, 27–32.

Sanders, J.G. (1979) Microbial role in the demethylation and oxidation of methylated arsenicals in seawater. *Chemosphere* **3**, 135–137.

Sanders, J.G. and Cibik, S.J. (1985) Adaptive behaviour of euryhaline phytoplankton communities to arsenic stress. *Mar. Ecol. Prog. Ser.* **22**, 199–205.

Sanders, J.G. and Windom, H.L. (1980) The uptake and reduction of arsenic species by marine algae. *Estuar. Coastal Mar. Sci.* **10**, 555–557.

Shibata, Y. and Morita M. (1988) A novel, trimethylated arseno-sugar isolated from the brown alga *Sargassum thunbergii*. *Agric. Biol. Chem.* **52**, 1087–1089.

Shibata, Y., Sekiguchi, M., Ohtsuki, A. and Morita, M. (in press) Arsenic compounds in zoo- and phytoplanktons of marine origin. *Appl. Organometal. Chem.*

Shiomi, K., Shinagawa, A., Igarashi, T. *et al.* (1984) Contents and chemical forms of arsenic in shellfishes in connection with their feeding habits, *Bull. Jap. Soc. Sci. Fish.* **50**, 293–297.

Shiomi, K., Kakehashi, Y., Yamanaka, H. and Kikuchi, T. (1987) Identification of arsenobetaine and a tetramethylarsonium salt in the clam *Meretrix lusoria. Appl. Organometal. Chem.* **1**, 177–183.

Shiomi, K., Sugiyama, Y., Shimakura, K. and Nagashima, Y. (1995) Arsenobetaine as the major arsenic compound in the muscle of two species of freshwater fish. *Appl. Organometal. Chem.* **9**, 105–109.

Smales, A.A. and Pate, B.D. (1952) The determination of sub-microgram quantities of arsenic by radioactivation. II: The determination of arsenic in sea water. *Analyst* **77**, 188–195.

Stoeppler, M. and Brandt, K. (1979) Comparative studies on trace metal levels in marine biota II. Trace metals in krill, krill products and fish from the Antarctic Scotia Sea. *Zeitschrift für Lebensmitteluntersuchung und Forschung* **169**, 95–98.

Sugawara, K. and Kanamori, S. (1964) The spectrophotometric determination of trace amounts of arsenate and arsenite in natural waters with special reference to phosphate determination. *Bull. Chem. Soc. Jap.* **37**, 1358–1363.

Tagawa, S. and Kojima, Y. (1976) Arsenic content and its seasonal variation in seaweed. *J. Shimonoseki Univ. Fish.* **25**, 67–74.

Vahter, M. (1994) Species differences in the metabolism of arsenic compounds. *Appl. Organometal. Chem.* **8**, 175–182.

Vahter, M., Marafante, E., Lindgren, A. and Dencker, L. (1982) Tissue distribution and subcellular binding of arsenic in marmoset monkeys after injection of ^{74}As-arsenite. *Arch. Toxicol.* **51**, 65–77.

Vahter, M., Couch, R., Nermell, B. and Nilsson, R. (1995) Lack of methylation of inorganic arsenic in the chimpanzee. *Toxicol. Appl. Pharmacol.* **133**, 262–268.

Vogel, G., Woznicka, M., Gfeller, H. *et al.* (1990) 1(3),2-Diacylglyceryl-3(1)-*O*-2′-(hydroxymethyl)(*N,N,N*-trimethyl)-β-alanine (DGTA): a novel betaine lipid from *Ochromonas danica* (Chrysophyceae). *Chem. Phys. Lipids* **52**, 99-109.

Welch, A.D. and Landau, R.L. (1942) The arsenic analogue of choline as a component of lecithin in rats fed arsenocholine chloride. *J. Biol. Chem.* 581–588.

Whyte, J.N.C. and Englar, J.R. (1983) Analysis of inorganic and organic-bound arsenic in marine brown algae. *Bot. Mar.* **26**, 159–164.

Zingde, M.D., Singbal, S.Y.S. *et al.* (1976) Arsenic, copper, zinc and manganese in the marine flora and fauna of coastal and estaurine waters around Goa. *Indian J. Mar. Sci.* **5**, 212–217.

7 *Determinants of trace metal concentrations in marine organisms*

MURRAY T. BROWN AND MICHAEL H. DEPLEDGE

7.1 INTRODUCTION

The purpose of this chapter is to provide a synthesis of the very extensive literature pertaining to the factors that influence the uptake, accumulation and handling of trace metals by marine biota. For more extensive reviews refer to the key references: Eisler (1981), Fowler (1990), Samiullah (1990), Vernet (1991), Bryan and Langston (1992) and Tessier and Turner (1995). The literature is replete with examples of the quantification of metal concentrations in an array of marine organisms collected from numerous locations. However, it is evident that information concerning the factors that determine the metal burdens of these organisms is patchy, with certain groups more thoroughly investigated than others; for example, contrast the molluscs with macroalgae.

7.2 TRACE METAL CONTAMINATION OF MARINE ECOSYSTEMS

Before focusing on specific determinants of trace metal concentrations in marine organisms, it is instructive to examine the extent of metal contamination in marine ecosystems and their components.

Reports outlining trace metal contamination of marine ecosystems, through natural and anthropogenic inputs, have been published by numerous international bodies (e.g. Gray, 1979; UNEP, 1983; Kullenberg, 1986; GESAMP, 1990). Natural inputs into the sea have been categorized by Bryan (1976a) as follows.

- Coastal supply, including inputs from rivers and from erosion produced by wave action and glaciers.

Metal Metabolism in Aquatic Environments. Edited by William J. Langston and Maria João Bebianno. Published in 1998 by Chapman & Hall, London. ISBN 0 412 80370 4

- Deep-sea supply, including metals released from deep-sea volcanism and those removed from particles or sediments by chemical processes.
- Supply which bypasses the nearshore environment; in particular, metals transported through the atmosphere as dust particles or as aerosols, and also material produced by glacial erosion in polar regions which is then transported elsewhere by floating ice.

The oceans, which provide a vital sink for many trace metals and their compounds, are intimately involved in geochemical cycling. There is growing concern that the natural cycling rates of many metals are being disrupted as a consequence of human activities (Ramade, 1987). Recent estimates indicate that up to 150 000 tonnes of Hg is released naturally per year as a result of degassing from the Earth's crust, compared with the 8000 to 10 000 tonnes released through human activities (Goyer, 1991). Anthropogenic inputs of Pb, Zn, Cd and Cu are considered to be between one and three orders of magnitude higher than natural fluxes (Schindler, 1991).

Routes of anthropogenic inputs of metals into the sea are the atmosphere and rivers (GESAMP, 1990). Metal particles released into the air at ground level are mixed vertically and consequently contaminants may be transported many thousands of kilometres from where they were first released (for further details see Pacyna *et al.*, 1991). This obviously creates problems when trying to relate effects of trace metals at one locality to a particular metal source elsewhere. The importance of human activity in redistributing trace metals is evident in the 200-fold increase in the Pb content of Greenland ice. From a low natural level about 2700 years ago, the concentration of Pb increased during the industrial age and has continued to rise rapidly since the addition of Pb to petrol during the 1920s (Goyer, 1991). Differences in global climate also result in uneven deposition of trace metals (Bruenig, 1989; McKay and Thomas, 1989). For example, the net accumulation of Hg, Cd, V and Mn in Arctic biota and ice has been attributed to the relative absence of precipitation scavenging and strong atmospheric inversions. The emission source of these metals is thought to have been industrialized temperate zones and thus considerable directional transport has occurred (Rahn and McCaffrey, 1979). In reviewing the aquatic transport of chemicals, Goldberg (1989) concluded that organic compounds in the sea play a key role in determining the extent of trace metal transport. Through the action of particulate organics, which take up metals and artificially produced radionuclides, the descent of metals through the water column is enhanced.

Comparisons of atmospheric and riverine inputs of trace metals into the sea indicate that only 2% of the Pb which eventually dissolves in seawater enters the global ocean via rivers. The primary source of most of the dissolved Cd, Cu, Fe and Zn is also the atmosphere. At a regional level, it is thought that for the trace metals Cd, Hg, Cu, Pb and Zn, 40–60% of the input into the North

Sea is via atmospheric deposition (QSRNS, 1987). Similar findings have been reported for other temperate water bodies but it would be unwise to conclude that the relative importance of these routes of entry is the same for all marine and brackish water ecosystems. Detailed information is lacking for subtropical and tropical regions, and for southern hemisphere ecosystems where different patterns of rainfall, ocean currents, prevailing winds and annual temperature cycles affect trace metal input and persistence.

Domestic effluents and urban stormwater runoff have also been identified as significant sources of trace metal input into coastal waters. Concentrations in the milligrams per litre range (Connell and Miller, 1984) can be found in domestic effluents; metabolic waste, corrosion of water pipes (Cu, Pb, Zn and Cd) and consumer products (e.g. detergent formulations containing Fe, Mn, Cr, Ni, Co, Zn, B and As) all contribute appreciable amounts.

7.3 TRACE METAL CONCENTRATIONS IN MARINE BIOTA

The resultant concentrations of trace metals in biota arise from a series of complex interactions between several processes. Thus, in any particular organism, tissue metal concentrations reflect the amount of metal taken up into the organism, the proportion of that metal which is distributed to each tissue, and the extent to which the metal enters and is retained within each tissue. With regard to the latter, the metabolic requirement (if any) for the metal, the metal concentration that can be tolerated in the tissue without initiating detoxification mechanisms, via excretion or sequestration, and whether the tissue has a role in storage of detoxified metal forms are of primary importance (Depledge, 1989a).

Organisms exhibit selectivity with regard to their body loads of metals (Simkiss and Mason, 1983). Thus, so-called bulk metals (e.g. Na, K, Mg and Ca) are found in large amounts, while trace metals are present at much lower concentrations (Simkiss and Taylor, 1989). A further distinction needs to be made between those considered essential for life and those that are non-essential. Examples of essential trace metals are Cu, Zn and Fe. These elements are vital components of enzymes, respiratory proteins and certain structural elements of organisms (Depledge and Rainbow, 1990). Thus, carbonic anhydrase, carboxypeptidase A and B and several hydrogenases contain Zn; pyruvate carboxylase contains Mn; the metalloid Se is a component of glutathione peroxidase; vitamin B12 contains Co; Cu is present in cytochrome oxidase, plastocyanin and haemocyanin; and haemoglobin and ferrodoxin contain Fe (Bryan, 1976a; DeBoer, 1981). A range of trace metals must be delivered to the tissues of an organism in order to meet the diverse metabolic and respiratory requirements whilst at the same time excessive accumulation of potentially toxic metal species must somehow be prevented. Dose–response curves highlight the significance of an appropriate metal supply to the well-being of organisms; deficiency or excess of essential metals, beyond certain threshold concentrations, gives rise to detrimental effects (Rainbow, 1985). Other trace metals, such as Cd, Hg and Pb, are considered as non-essential because they have no known biological role; these become highly toxic when found at metabolically active sites, even at relatively low concentrations (Rainbow, 1985).

The most comprehensive surveys of trace metal concentrations in marine organisms conducted to date are those by Eisler (1981), Bryan (1984) and Furness and Rainbow (1990). The vast majority of information on metal concentrations in whole organisms and tissues concerns molluscs and crustaceans, although there are now substantial databases for annelids (polychaetes), coelenterates and echinoderms and, to a lesser extent, marine algae. On closer inspection of this information it becomes apparent that certain species have received considerably more attention than others – for some species investigations have occurred only at a single locality on one occasion. This is clearly a major deficit in the current database and a cause for concern, since it may provide us with a distorted view of metal handling and toxicity in biota. A common motive for selecting particular species appears to be the need to investigate the potential for heavy metal transfer to humans through the ingestion of seafood. Not surprisingly, therefore, edible bivalves (including mussels, oysters and scallops) are particularly well represented in studies of molluscs (Chapter 8), while edible crabs, shrimps and lobsters have received most attention amongst crustaceans (Chapter 9). As a consequence many of the key species in ecosystems, which may have relevance from an ecotoxicological viewpoint as opposed to a human toxicological standpoint, have been neglected.

The number of species of phytoplankton and macrophytes investigated is quite restricted (Phillips, 1980, 1994), In general, their metal content closely reflects soluble metal bioavailability in seawater with a high degree of time integration (Bryan, 1969; Young, 1975). Field transplantation studies have provided further information on the degree of time integration of different seaweeds in coastal and estuarine sites (e.g. Myklestad *et al.*, 1978; Eide *et al.*, 1980; Ho, 1984). The apparent straightforward relationship between internal concentrations and seawater concentrations, together with similarities in the responses of a range of species to increasing metal concentrations, has led to the use of macroalgae as monitors of metal pollution (e.g. Burrows, 1971; Bryan and Hummerstone, 1973; Say *et al.*, 1990; Karez *et al.*, 1994). However, it is now known that various extrinsic and intrinsic factors (e.g. interactions between trace metals, environmental variables which influence growth rates) can significantly affect the accumulation of metals by these algae and so seaweeds may not accurately reflect metal concentrations in the surrounding water, thereby reducing their appropriateness as biomonitors. Tables 7.1 and 7.2 provide a general guide to the metal concentrations in marine organisms.

Table 7.1 Examples of trace metal concentrations (μg g^{-1} dry weight) in a range of invertebrates from estuaries and coastlines in different locations of the world (data adapted from Eisler, 1981 and Bryan, 1984)

Taxon	Location	Cd	Cr	Cu	Fe	Mn	Ni	Pb	Zn
Polychaetes									
Nereis diversicolor	SW England (estuaries)	0.03–10	0.1–10.5	10–1430	181–871	4–52	0.6–15.6	0.2–1190	91–510
Notomastus tenuis	Los Angeles (harbour)	0.6	33	46	181	13.5	7.8	1.2	59
Molluscs									
Crassostrea virginica	Connecticut	4.6–107	–	214–4304	–	4.1–29.6	–	–	59
C. virginica	California	12–40	–	273–1510	148–260	13–34	–	–	3570–19300
C. gigas	Knysna estuary (S. Africa)	3.7–9	–	32	66–128	12–16	1–5.2	1–1.8	97–396
Ostrea edulis	SW England (estuaries)	5.9–7.9	0.34	2610	219–394	7–16.7	0.2–3	2.5–8	1966–17100
O. angase	Victoria, Australia	0.9–2.6	–	39–104	612–2585	7.8–17	–	0.5–2	878–3051
Mytilus galloprovincialis	NW Mediterranean Sea	0.4–6	0.5–29	2.4–15.5	149–2220	3–70	1–14	2.7–117	97–644
M. edulis	SW England	0.8–2.6	1–2.7	4–13.6	152–401	5.2–35.5	1–3.5	30–105	57–199
M. edulis	W. USA	2.3–10.5	1–7.5	3.5–8.6	430	8.2–25	0.4–6.3	2–8.8	90–260
M. edulis	E. USA	0.6–6.2	–	6.7–13.2	–	8.9–18	–	0.4–9.5	67–189
Littorina littorea	SW England (estuaries)	0.5–2.6	0.1–1	62–194	272–784	18–133	2.2–4.1	3.7–70	45–284
Haliotis tuberculata	English Channel	5.6	0.8	28–39	303–474	8.3	13.6	2.1	98–103
Patella vulguta	SW England	3.3–27.5	0.5–2.6	10–27	897–2330	5.4–36	1–3.7	5–38	83–224
P. vulgata	Trondheim Fjord, Norway	7–22	7–17	12–30	1289–2505	–	4–11	–	127–238
Octopus vulgaris	Mediterranean Sea	1.2	–	260	140	5	–	–	150
Nucella lapillus	SW England	5.5–16	0.4–5.6	51–141	193–270	11.4–17	1.4–4	2–7	235–520
Crustacea									
Pandalus borealis	Oslo Fjord, Norway	0.9	–	95	–	–	–	2.7	103
Crangon crangon	England/Wales coasts	1.2	2.8	92	–	–	–	12.8	88
Semibalanus balanoides	Cardigan Bay, Wales	10–28	–	0.1–0.7	–	–	–	–	4500–23100

Balanus amphitrite	N. Adriatic	–	2–4	44–109	–	–	–	7–12	
Acartia clausi	Saronikis Gulf, Greece	0.6	3.3	34–107	738	9.3	–	–	1270
copepods	N. Baffin Is.	5	–	3.7	78	–	–	–	60
amphipods	N. Baffin Is.	7	–	26	87	–	–	–	43
Carcinus maenas	Scotland	2.4	–	33	1345	673	15.6	38	76
Echinoderms									
Asterias rubens	Isle of Man, Irish Sea	3.7	–	4.3	37	6.5	1.5	2.3	190
Solaster papposus	Isle of Man, Irish Sea	4.5–5.3	–	6–11	170–200	31–43	2–4	5–7	120–130
Tunicates									
Halocynthia sp.	Adriatic Sea	0.3	–	6	–	148	–	–	88
Coelenterates									
Tealia felina	Isle of Man, Irish Sea	0.7	0.4	57	730	9	3	2.5	280
Alcyonium digitatum	Isle of Man, Irish Sea	4	0.4	9.7	250	3.7	17	24	46
*Pleuyrobranchia pileus**	Black Sea	–	10	300	5000	20	50	40	9000

*μg g^{-1} ash-free dry weight

Table 7.2 Examples of trace metal concentrations (μg g^{-1} dry weight) in a range of marine macrophytes from estuaries and coastlines in different locations of the world

Taxon	Locations	Cd	Cr	Fe	Mn	Ni	Pb	Zn
Phaeophyceae								
Ascophyllum nodosum	Norwegian Fjords[1,2]	4–240	2–13	84–467	–	1–22	2–12	65–720
Fucus ceranoides	North West Spain[3]	30–900	0.5–12	20–3500	15–4500	3–60	–	100–900
Fucus vesiculosus	South West UK[4]	9–301	1–4	41–1168	128–392	–	6–31	149–1240
Fucus vesiculosus	Northern Baltic Sea[5]	2.1–8	0.1–1.4	48–522	79–306	4.2–46.4	2–11.7	159–877
Cystoseira spp.	Northern Adriatic Sea[6]	4–10	–	–	40–139	–	4–6	29–50
Padina commersonii	Singapore[7]	3.8–7.3	–	–	–	4–6.5	4.3–7.9	20.7–50.1
Sargassum spp.	Korea[8]	7–25	–	–	–	–	4.2–8.9	11–61
Chlorophyceae								
Enteromorpha spp.	East UK[9]	0.2–1260	–	–	–	–	6–322	15–437
Ulva spp.	Hong Kong[10]	8–132	–	40–12000	6–750	4–48	3–175	10–110
Ulva spp.	Goa. India[11]	6.7–13.9	–	354–1169	111–1721	7–39	15.6–17.8	2.8–23.8
Ulva spp.	Brazil[12]	2–10.2	1–8.7	–	–	–	3.6–14.5	4.7–80.2
Caulerpa spp.	Gt Barrier Reef, Australia[13]	1–3.2	–	–	–	0.8–2.6	0.7–2.9	0.9–17
Codium tomentosum	Israel[14]	2.9–5.5	–	–	–	5.2–5.5	1.9–5.6	117–218
Rhodophyceae								
Porphyra suborbiculata	Hong Kong[15]	9.2–13.1	–	68–485	14–26	1.7–5.8	2.6–5.4	11–54
Porphyra spp.	Irish Sea[16]	6.6–19.5	–	104–3800	14–93	0.2–9.6	0.8–10.5	35–177
Acanthophora spicifera	Brazil[17]	4.1–9.3	1.3–5	–	–	–	4.6–10	31–110.2
A. spicifera	Gt Barrier Reef, Australia[13]	2.6–3.5	–	–	–	3–5.1	0.6–2.9	8–13
A spicifera	Goa, India[11]	7.4–80.4	–	239–1012	63–344	9.5–14.7	6.1–17.2	11.8–28.7

Table 7.2 *Continued*

Taxon	Locations	Cd	Cr	Fe	Mn	Ni	Pb	Zn
Hypnea musciformis	Goa, India[11]	6.6–22.5	–	–	–	9.6–12.7	11.8–16	10.4–50.8
Laurencia sp.	Penang, Malaysia[18]	4.7–13.4	–	–	–	–	1.7–8.3	14.2–60.5
Angiospermae								
Seagrass spp.	Indonesia[19]	0.3–32	–	–	–	–	0.3–14.5	0.5–88.1

Sources: [1]Stenner & Nickless (1974); [2]Lande (1977); [3]Barreiro *et al.* (1993); [4]Bryan & Hummerstone (1973); [5]Soderland *et al.* (1988); [6]Munda & Hudnik (1991); [7]Bok & Keong (1976); [8]Pak *et al.* (1977); [9]Say *et al.* (1990); [10]Ho (1990); [11]Agida *et al.* (1978); [12]Lacerda *et al.* (1985); [13]Denton & Burdon-Smith (1986); [14]Roth & Hornung (1977); [15]Ho (1987); [16]Preston *et al.* (1972); [17]Karez *et al.* (1994); [18]Sivalingham (1978); [19]Nienhuis (1986).

7.4 SPECIATION AND BIOAVAILABILITY OF TRACE METALS

Quantifying the biological responses to trace metals depends upon accurately determining the exposure to the pollutant, but this is problematical since the total metal concentrations in seawater and sediment do not constitute the concentrations available to the marine biota (Luoma, 1983; Brezonik *et al.*, 1991). The term 'bioavailable' is used to refer to the proportion of a chemical in the environment that might be taken up into an organism. It has also been used to refer exclusively to the amount of chemical that actually is taken up into an organism. This chapter will use the former definition. There is also some confusion over the use of the terms 'uptake' and 'accumulation'. The former refers to the entrance of the chemical into an organism, the latter to the amount of chemical that remains in an organism following exposure over a particular period of time. To give an extreme example, accumulation of a trace metal may be negligible if excretory mechanisms are *c.* 100% effective; however, there may have been extensive uptake.

Within the marine environment trace metals exist in equilibrium among free hydrated metal ions, metal bound in organic (e.g. amines, humic acid) and inorganic (e.g. OH^-, CO_3^{2-}) complexes and metal bound to organic and inorganic particulate matter. The chemical composition of the seawater has a strong influence on the speciation of metals. Thus, in turbid estuarine waters, a large proportion of the total metal load is bound in or to organic or inorganic particulate matter (Salomons and Forstner, 1984); mixing in estuaries alters metal speciation as the ionic strength increases. Dissolved organic complexes and particulate matter may undergo flocculation and, for some metals, the result is that a large proportion of the load transported in the river water sinks to the sediments of the estuary. However, other metals (e.g. Cd) are displaced from particulate matter by chloride ions, due to the formation of chloride complexes (ElbayPoulichet *et al.*, 1987). In estuaries, the speciation of metals that remain in solution is affected by the increasing concentration of anions, particularly chloride, and for most metals free ions constitute a relatively minor proportion of the total dissolved metal concentration (Zirino and Yamamoto, 1972; Long and Angino, 1977; Mantoura *et al.*, 1978).

Certain trace metals are available for uptake into organisms from solution only as free ions, whereas others are transported across biological membranes as inorganic complexes. In experiments in which the free species of Cu and Cd were either carefully controlled by organic chelators or determined by means of ion-selective electrodes, the toxicity and bioavailability were correlated with the concentration of free metal ions rather than total dissolved metal concentration (Sunda and Guillard, 1976; Sunda *et al.*, 1978; Zamuda and Sunda, 1982; Sanders *et al.*, 1983). Such findings are consistent with the uptake and toxicity of these metals increasing with decreasing salinity (and thereby free ion concentration) in most estuarine organisms studied (McLusky *et al.*, 1986). In contrast to these two metals, inorganic Hg is thought to be transported across lipid membranes predominantly as uncharged chloride complexes (Gutknecht, 1981; Bienvenue *et al.*, 1984).

Since the chemistry of sediment is considerably more complex than that of seawater, the mechanisms by which geochemistry affects metal bioavailability from sediments are not as well understood as those influencing metal speciation in seawater. Hence, despite some attempts (Chapter 8) there are no generally applicable, reliable techniques for assessing bioavailability of metals in such media (Luoma, 1989). With regard to uptake of metals from ingested sediment, very few studies have considered the actual species of metal that are available for transport across the lining of the digestive system. Luoma (1989) speculated that the processes that determine the availability of metals from ingested sediment in the gut of deposit-feeding invertebrates would be identical to the processes that control sorption/desorption processes in sediments.

Thus, trace metals that accumulate in seawater and sediments are not necessarily freely available for uptake into biota. A proportion of the metal may be strongly bound in dissolved complexes or on sediment surfaces, or in organic films surrounding particles. This speciation of metals is extremely difficult to follow or predict either qualitatively or quantitatively (Cantillo and Segar, 1975; Turner, 1984). Alterations in physicochemical parameters of the environment can strongly influence the relative proportions of the metal species that can be taken up; alterations in pH, redox potential, salinity, temperature, etc. can all greatly influence the bioavailability of metals for uptake into marine organisms (Mantoura *et al.*, 1978). This is an important consideration for biomonitoring studies. Just because organisms do not contain especially high trace metal concentrations in their tissues, this does not preclude the possibility that metals may be present in the environment at elevated levels: they may be firmly bound in highly stable complexes. As physicochemical conditions alter – after, for example, resuspension of sediments due to turbulence – rapid conversion of metals to bioavailable ionic forms may result in higher concentrations in biota (Salomans and Forstner, 1984; Samiullah, 1990).

7.5 SOURCES OF TRACE METALS FOR UPTAKE

In general, organisms obtain metals by direct uptake from the surroundings across the entire body surface of the organism, across specialized respiratory structures (gills or lungs), across the digestive epithelium if water is imbibed, via ingested food, or by a combination of routes. For marine algae, metals are derived almost exclusively from the aqueous phase, although it has been proposed by some (e.g. Luoma *et al.*, 1982) that metals can be taken up directly from sediments or suspended inorganic particulates. While these routes cannot be totally ruled out, it is difficult to confirm since algal surfaces are heavily contaminated by organic material (Patrick and Loutit, 1977; Holmes *et al.*, 1991) and inorganic particulate material (Barnett and Ashcroft, 1985; Bryan *et al.*, 1985) that contains significant quantities of metals. Submerged angiosperms may absorb metals from sediments via root–rhizome systems and from water via leaves but the relative contributions of these routes to total metal burden are not fully understood (Guilizzoni, 1991; Ward, 1989). Problems associated with surface contamination in salt marsh plants have been highlighted by Beeftink and Niewenhuize (1986); the observed differences between species may relate to surface slime and mucus production, available surface area for deposition, or to the complex microstructure and chemistry of surfaces. Whatever the cause, surface contamination is a potential source of interference in the use of marine plants and algae as monitors of metal pollution. To date there is no standardization of washing techniques or use of correction factors to quantify this problem. Charged polysaccharides (e.g. polygalacturonic acids, alginate and carrageenan) associated with plant and algal cell walls which are in direct contact with metals in solution have a high affinity for some trace metals (e.g. Pb > Cu > Zn) and can adsorb them in a process akin to ion exchange (Haug, 1961; Veroy *et al.*, 1980; Ernst *et al.*, 1992). While of importance, the amount of metals bound to cell walls is often only a small proportion of the total cellular content (Rai *et al.*, 1981).

For many animals it is often not clear which route is the more important, although this may have great significance for subsequent inter-organ distribution of metals and, indeed, toxicity. Since marine animals are constantly bathed in metal-containing water, and often pass large volumes of water over respiratory surfaces for the purpose of gas exchange, it is often assumed that uptake across the body surface is the predominant route of entry (Rainbow, 1988). However, for invertebrates and vertebrates (fish and mammals) direct uptake from water may be only of minor importance. On reviewing the available evidence, Bryan (1984) concluded that for many molluscs (see also Chapter 8), crustaceans and annelids, metal uptake via the food may still be of greater significance. There are numerous examples to illustrate that the route of metal uptake influences both distribution of metals in tissues of an organism and toxicity of the metal. For example, the accumulation patterns of Cu in the starfish *Asterias rubens* differ markedly depending on whether animals are exposed to Cu in seawater or in their food (Depledge and Payet, unpublished). In the talitrid amphipod *Orchestia gammarellus*, Weeks and Rainbow (1990) reported that Zn accumulated from a dietary source can be re-excreted after 24 hours, whereas Zn accumulated from seawater is retained with 100% efficiency (Weeks and Rainbow, 1991).

7.6 UPTAKE MECHANISMS

Uptake of many dissolved class B and transition metals across cell membranes is considered to be by facilitated diffusion rather than active transport, the diffusion gradient being maintained by binding metals to intracellular ligands (see also Chapter 1). Simkiss and Taylor (1989) proposed that non-ionic, inorganic species and organic derivatives may also diffuse into organisms due to their high lipid solubility. Uptake kinetics by passive diffusion are described by Fick's Law. (For a detailed description of the factors involved, see Depledge and Rainbow, 1990.) Fick's equation may require some modification to take account of the electrical potential differences that exists across most cellular barriers. Such potentials might influence to different extents the trans-barrier distribution of metal ions of different size and valency (Depledge and Rainbow, 1990).

Some metals enter by active transport. For example, Cd may enter a variety of crustaceans, molluscs and fish via active transport through Ca ion pumps, while pinocytosis has been shown to be involved in the uptake of metal-rich particles in the gills and pharynx of some molluscs and ascidians respectively (Kalk, 1963; Hobden, 1967). It is important to note that metal ions entering organisms along these routes are bioavailable, but would not have entered the organism were it not for these energy-consuming biological processes. This serves to emphasize the importance of interactions between the bioavailable fraction of metals in the environment and the biological characteristics of exposed organisms in jointly determining the amount of metal that accumulates in tissues.

Thus, even under similar ambient conditions, variations in surface area available for absorption, permeability of cells/tissues, number and nature of binding sites (intra- and extracellular) and metabolic rate can result in differences in metal uptake between species and even between individuals of the same species.

7.7 ACCUMULATION OF TRACE METALS

Once in the organism, metals may become associated with ligands having a strong binding capacity, resulting in their accumulation. In animals, metallothioneins are one of the key determinants of the ability to withstand exposure to trace metals such as Cu and Cd. The properties of these small (*c.* 10 kDa) cysteine-rich proteins were reviewed by Engel and Brouwer (1989), Benson *et al.* (1990) and Petering *et al.* (1990). They are normal constituents of most cells investigated so far and probably serve as intracellular storage sites for essential metals to fulfil metabolic requirements. They have been detected in Echinodermata, Annelida, Mollusca, Arthropoda and a wide range of vertebrates (for example, Chapters 8, 9 and 10). Metallothioneins are inducible by exposure to raised environmental concentrations of some heavy metals (e.g. Cd, Cu and Zn). Consequently, the detection of elevated metallothionein concentrations in selected tissues of organisms can be used to map metal exposure (Engel and Roesijadi, 1987; Benson *et al.*, 1990). However, before using these biomarkers of metal exposure, natural fluctuations in tissue metallothionein concentrations must first be well characterized (Chapter 8).

Although the genes encoding metallothioneins are also present in plants and algae, a family of thiol-rich polypeptides, known as phytochelatins, appears to play an important role in the homeostatic control and sequestration of metal ions in higher plant, fungal and algal cells, although they do not necessarily convey differential tolerance (for review, see Rauser, 1995). Unlike metallothioneins, phytochelatins are not direct gene products arising from translation of mRNA but are instead derived from enzymatic modifications of glutathione or its precursor (Grill *et al.*, 1987). Phytochelatins are inducible by various heavy metals (e.g. Ag, Au, Bi, Cd, Cu, Hg, Pb, Zn) and also by the metalloids Se and As in all the species of plants investigated to date (Grill *et al.*, 1987; Steffens, 1990), though the induction ability of metals varies (Ernst *et al.*, 1992). Phytochelatins, too, have potential as biomarkers of trace metal exposure in plants but, once again, natural variations in concentration must first be characterized. A report of measurable levels of phytochelatins in natural populations of marine phytoplankton at environmentally relevant free metal ion concentrations (Ahner *et al.*, 1994) lends support to the idea that phytochelatins may be useful quantitative markers of metal exposure in the marine environment.

As well acting as an antioxidant, glutathione can also bind and therefore potentially detoxify a number of metals including Cu, Cd, Cr, Fe, Hg, Ni, Pb, Zn (Christie and Costa, 1984). Induction of metallothioneins is relatively slow and it has been suggested that glutathione is important in preventing toxicity during the initial stages of acute exposure (e.g. Freedman *et al.*, 1989). Furthermore, glutathione can reduce the toxicity of metals such as Cu and Fe by assisting in the removal of toxic oxygen species (e.g. superoxide ions and peroxides) which react with these metals to form hydroxyl radicals (Halliwell and Gutteridge, 1984; Hanna and Mason, 1992). In addition to glutathione, a number of other low molecular weight metal-binding complexes have been identified. It has been suggested that the appearance of these indicates the onset of metal-induced stress (Sanders and Jenkins, 1984; Jenkins and Mason, 1988). Their precise identities have not been fully resolved, but several candidate molecules have been suggested, including thiol esters (Coombs, 1974), amino acids (Rijstenbil and Wijnholds, 1991) and N heterocycles (Fayi and George, 1985).

Metal-containing 'granules' have been observed in a variety of different marine organisms, e.g. annelids, arthropods, cnidaria, molluscs and nematodes. They can be intracellular or extracellular, are highly variable in form and can be associated with a range of cellular locations which are apparently species specific but are most commonly in the cells of the digestive and excretory epithelia. The deposition of metals in these structures has been used as evidence for their role in metal detoxification and metabolism, though direct proof for this is lacking (for reviews, see Mason and Nott, 1981; Brown, 1982; Nott, 1991; and Chapters 8 and 12). Locations for intracellular bound metals in marine plants and algae are less clear. While vacuoles are considered to be important storage sites in terrestrial angiosperms (Ernst *et al.*, 1992), deposition within nuclei and cytosol-located polyphosphate granules and physodes (rich in polyphenols) have been observed in certain algal species (Silverberg *et al.*, 1976; McLean and Williamson, 1977; Lignell *et al.*, 1982; Walsh and Hunter, 1992).

7.8 ELIMINATION OF TRACE METALS

An alternative approach for detoxifying metals once inside a cell is to eliminate them. Marine organisms have developed a number of routes and modes for excreting metals. Active efflux transport systems would provide an effective detoxification mechanism. Class A metals can be selectively and actively excreted against concentration gradients by permeases and the Na/K and Ca ATPases (Mason and Jenkins, 1995). Comparable data for class B metals are less common, though there is evidence from studies on prokaryotes (Silver and Misra, 1988). A more common method in eukaryotes is either active or passive excretion of metals in a complexed form. Common routes in animals include the urine, faeces and bile. In fish, for example, metals are excreted via the kidney or chloride cells of the gills (Crespo *et al.*, 1981), whereas in crustaceans the hepatopancreas or the green gland is the primary route (Mason and Jenkins, 1995). The kidneys of some bivalves contain large quantities of metal-rich granules which are assumed to be passed into the urine and eliminated; the high metal concentrations associated with molluscan kidneys are due to these granules (Chapter 8). Other more peculiar examples of elimination include, the diapedic secretion of metal-laden amoebocytes by *Littorina littorea* (e.g. Marigomez *et al.*, 1990) and the expulsion of metal-rich endosymbiotic zooxanthellae by the coelenterate *Anemonia viridis* (Harland and Nganro, 1990). In some cases, rather than being excreted to the surrounding environment, metal complexes are secreted into mineralized and organic extracellular structures such as shells, exocuticle or byssal threads (e.g. Martin, 1973; Bourget, 1974; Coombs and Keller, 1981).

7.9 REGULATION OF TRACE METAL ACCUMULATION

Changes in bioavailability may not always be reflected in changes in tissue metal concentrations. For example, as seawater Zn concentrations increase within a certain range, whole body Zn concentrations in the shrimp *Palaemon elegans* remain unchanged (White and Rainbow, 1984; Nugegoda and Rainbow, 1987, 1989); Zn uptake increases with increasing Zn bioavailability, but increased Zn excretion results in maintenance of a relatively constant body load. This has been termed 'metal regulation' and the organism involved is said to be a 'regulator' of Zn (Rainbow, 1988; see also Chapter 9). This 'regulatory ability' apparently breaks down at high environmental Zn concentrations, and body Zn concentrations rise with increasing Zn bioavailability. The significance of regulation breakdown is not clear since animals appear to survive without harm long after regulatory ability is exceeded. An example of an organism which is apparently unable to 'regulate' body concentrations of essential metals independent of environmental concentrations is the barnacle *Elminius modestus*. This species simply accumulates Zn at all exposure concentrations above background concentrations and excess Zn is stored in granules bound to pyrophosphate, rather than being excreted (Pullen and Rainbow, 1991). It has been proposed that species which respond in a similar way should be designated as accumulators, and such species are considered to be particularly suitable for inclusion in biomonitoring studies (Rainbow, 1992). Whilst this approach may be helpful in providing a pragmatic solution to the problem of selecting suitable biomonitoring organisms from the array of species present in ecosystems, it should be recognized that it is of limited scientific value and may even obscure understanding of the mechanisms involved in heavy metal handling (Depledge and Rainbow, 1990). The designation of organisms as either regulators or accumulators disguises the fact that there is a range of intermediate possibilities. Furthermore, while some organisms 'regulate' one metal, others are not regulated. For example, *Nereis diversicolor* accumulates Cd, Cu, Co, As, Ag, Hg and Pb in proportion to environmental concentrations, whereas concentrations of Fe and Zn are unaffected by increasing levels of pollution (Bryan *et al.*, 1985; Bryan and Langston, 1992). It is therefore important to specify for which metal the 'regulatory' ability applies. Moreover, for a particular species, the 'regulatory' ability for a given metal may vary with route of uptake.

The regulator/accumulator classification is also misleading from a mechanistic viewpoint. In a number of scientific articles, organisms have been referred to as handling heavy metals by 'regulator strategies' or 'accumulator strategies'. This terminology should be avoided in future as 'strategy' implies an active component in which an organism chooses a particular course of action (cf. *Oxford English Dictionary*). Clearly, this is not the case. Furthermore, 'accumulators' of heavy metals avoid excessively high intracellular concentrations at metabolically active sites either by sequestering metals in, for example, inert granules or by eliminating them from metabolically active tissues and storing them in inert tissues, such as shells or carapaces (Simkiss, 1976; Coombs and George, 1978). The ability to store metals in this way appears to be common to all the major invertebrate phyla, with a prime site of granule storage being the digestive gland. To prevent the transfer of potentially toxic metals into the blood of an animal, the cells of the gut lining appear either to excrete assimilated metals back into the gut lumen or to bind them for storage in a metabolically unavailable form. Given the high availability of proteinaceous binding sites intracellularly, excretion of metals back into the lumen might involve energy-requiring mechanisms working against a concentration gradient. Thus, the intracellular storage of metals in either a soluble or insoluble form may be energetically less expensive.

One final piece of evidence in the case against using the regulator/accumulator terminology is that it can obscure biologically significant events.

Many studies consider only the ways in which whole organism metal concentrations alter with seawater metal concentrations and neglect to take account of the internal tissue distribution of metals and metal species, which are likely to be key determinants of metal toxicity and adverse effects. For example, a situation can be envisaged in which increased metal exposure results in excess uptake of metal with the need for enhanced excretion to maintain an approximately constant whole body concentration. Such 'regulation' is unlikely to be perfect, and so metal ions might accumulate in some target tissue in the body. If the site of accumulation happens to be in nervous tissue, heart or even respiratory structures (gills and lungs), which constitute a relatively small proportion of the whole body weight in many animals, metal concentrations might increase markedly without this being reflected in a statistically significant change in whole body load. This situation could easily occur because usually the 'normal' range for whole body trace metal concentrations is rather large. To give an example, the midgut gland of a typical brachyuran crab constitutes approximately 1–2% of the dry weight of the animal and contains typical Cu concentrations of *c.* 80 μg Cu g^{-1} dry weight. This value could increase by an order of magnitude without a statistically significant departure of the whole body Cu concentration from the normal range for decapods. Thus, in a biomonitoring study, such a change would go undetected even though it may be of biological importance.

The key weakness of the regulator/accumulator concept is that its mechanistic basis has yet to be fully established. If an organism can 'regulate' its whole body metal load then this implies that it is capable of detecting metals, ascertaining what its whole metal body load is at any moment in time, and then acting to ensure that metal is either taken up or excreted, to maintain the *status quo*. Furthermore, since several metals and metalloids are considered as essential, it also implies that this system is either very versatile or is replicated many times. There is no evidence to support such a hypothesis. Metal receptors and associated feedback mechanisms that could fulfil such roles have not been found.

The regulator/accumulator terminology was developed primarily on the basis of laboratory studies in which trace metal uptake from seawater alone had been considered. There is no evidence to suggest that similar biological responses pertain in organisms *in situ*. Indeed, Alliot and Frenet-Piron (1990) found that Cu, Zn, Cd and Pb concentrations in the shrimp *Palaemon serratus* from the Brittany coast fluctuated in accordance with changes in heavy metal concentrations in the sea. This indicates that although shrimps of this genus were designated Zn regulators by Rainbow (1992) and yet 'accumulate' the non-essential metal Cd, whole body metal concentrations can still be used for monitoring fluctuating environmental bioavailability of metals. It should be noted that further verification of the findings of Alliot and Frenet-Piron (1990) is required, as the rates of metal depuration from the shrimps that are implied by their study are very high compared with depuration rates actually measured in laboratory studies.

7.10 RESISTANCE MECHANISMS

In many marine organisms, exposure to elevated metal concentrations can induce a degree of resistance that will have a bearing on the metal concentration in an organism. If exposure is to long-term chronic contamination, this resistance may have a genetic component. The mechanisms by which resistance are conferred will involve either an enhanced ability to detoxify the metal internally, or the reduction of tissue metal burdens involving the release of compounds that chelate metals and hence reduce their bioavailability, exclusion through reduced permeability or increased excretion. Internal detoxification may result in the enhanced uptake of metals due to the maintenance of diffusion gradients. For example, the accumulation rates and body burdens of Cd are greater in molluscs that produce metallothionein readily compared with those where induction is weak (Langston and Spence, 1995).

Many marine micro- and macroalgae release metal-complexing organic compounds (e.g. sulphated polysaccharides, uronates, polyphenols, amino acids, polypeptides and proteins) which can reduce the bioavailability and toxicity of metals (Romeo and Gnassia-Barelli, 1993). However, there are cases where the extracellular material does not always have a protective role. For example, Hall *et al.* (1979) found that the organic material produced by a Cu-tolerant strain of *Ectocarpus siliculosus* did not protect the non-tolerant strain from the toxic effects of Cu. Mucus production as a means of reducing metal uptake is common amongst animal species subjected to metal exposure; increased production arises by enhanced rates of secretion and increased numbers of mucus-producing epithelial cells of the skin, gills and intestine.

There are several examples where resistance is related to a reduction in membrane permeability. Tolerant strains of the ship-fouling macroalga *Ectocarpus siliculosus* accumulated significantly less Cu than non-tolerant strains as a result of altered membrane permeability (Hall *et al.*, 1979). Differential Hg tolerance of *Fundulus heteroclitus* embryos is due to lowered permeability of the eggs (Toppin *et al.*, 1987).

While the results from various studies provide support for the evolution of resistance in populations inhabiting metal-polluted sites, others indicate the lack of a genetic adaptation (for review, see Mulvey and Diamond, 1991). For example, populations of *Nereis diversicolor* from Cd and Zn contaminated sites were no more resistant than their counterparts from control sites (Bryan, 1976b). Costs may be incurred in association with the development of resistance. Reduced growth and/or reproduction may result from the increased energy expenditure on synthesis of detoxifying ligands or maintenance of metal exclusion. If there is an inherited component to the resistance, a reduction in the overall fitness of the population to respond to any additional perturbations may result.

Even after due consideration of the various factors outlined above, there may still be inherent variability in metal burdens due to genetic differences which are not necessarily related to metal resistance *per se*.

7.11 ALLOMETRY

Measured body or tissue metal burdens can be influenced by alterations in body size or condition since metal concentrations in organisms are a function of net accumulation, determined on a weight basis (for reviews, see Newman and Heagler, 1991; Langston and Spence, 1995). From laboratory studies on bivalves, crustaceans and fish, there are numerous examples illustrating that accumulation occurs more rapidly in smaller individuals, due to their relatively large ratio of surface area to volume, but from field studies there is a lack of consistency in the effects of body size on metal burdens. Growth rate can be particularly influential, but the effects are specific to species, metal and often site. If new tissue is incorporated faster than metal is accumulated, then growth will, in effect, dilute metal concentrations. This will be particularly marked in younger actively growing organisms or tissues. Such effects of growth and age are thought to account for differential metal loading within many marine macroalgae. For example, metal concentrations are often higher in older parts of fucoid algae, whereas rapid growth, and relatively slow incorporation of metals, results in lowest concentrations in the actively growing tips. Results are not consistent for all metals and sites. Thus, any factor that influences growth rate will result in additional variability in the metal content of the organism. At present, a unifying theory to explain the diversity in metal-size relationships is lacking.

7.12 INTERACTIONS AMONG TRACE METALS

Most studies of trace metal concentrations in biota ignore the influence that one metal may have on the uptake, accumulation and toxicity of other metals or, indeed, other pollutants. Despite this, a number of interactions have been identified (see also Chapter 11). One of the most quoted examples is that of the antagonistic interaction between Hg and Se, first demonstrated in the rat kidney (Parizek and Ostadalova, 1967). In many marine vertebrates, Hg and Se concentrations are positively correlated in specific tissues (Koeman *et al.*, 1973, 1975; Mackay *et al.*, 1975; Norheim, 1987). The mechanism by which Se ameliorates Hg toxicity in vertebrates remains obscure (Pelletier, 1985), though some possibilities are discussed in Chapter 11. Cd–Se interactions have also been reported (Magos and Webb, 1980) and Fe–Hg interactions have been described in the bivalve *Mercenaria mercenaria* where Hg concentrations in the range 0.1–1.0 mg l^{-1} resulted in reduced Fe concentrations in the mantle fringe tissues of the clams (Fowler *et al.*, 1975). Also, low Se concentrations reduce Hg uptake in *Mytilus edulis* (Davies and Russell, 1988). Cu–Mn interactions may have special significance with regard to phytoplankton growth. Thus, with increasing Mn ion availability, the growth rate of *Thalassiosira pseudonana* increases but, if Cu is added simultaneously, competitive inhibition of Mn uptake results and as a consequence growth rates decline (Sunda and Huntsman, 1983). Reduced Mn uptake in the presence of Cu, Zn and Cd has also been observed in the brown seaweeds *Fucus vesiculosus* and *Ascophyllum nodosum* (Morris and Bale, 1975; Foster, 1976). Competition between metals for binding sites, or possibly uptake sites, has been put forward as an explanation (Phillips, 1977).

One metal may also influence the tissue concentration of another by exchange. For example, it has been proposed that during the moult cycle of crabs, as Cu availability increases at ecdysis, possibly as a result of haemocyanin catabolism and Cu release, the metal is bound to metallothionein in the midgut gland and displaces Zn as it does so (Engel and Brouwer, 1989). Consequently, as Cu concentration increases, Zn concentration falls.

Bulk metals, such as Ca, have been shown to influence heavy metal concentrations in tissues. For example, Bjerregaard and Depledge (1989) demonstrated for the gastropod *Littorina littorea* that Ca ion concentration exerts a greater effect on Cd uptake than does altering salinity. As complexation of Cd ions by chloride falls with reduction in salinity, one might expect Cd uptake to increase, but if the concentration of Ca in seawater is increased as salinity is reduced, then Cd uptake also falls. This is thought to be the result of competition between Cd and Ca ions for active uptake by Ca pumps. In other species – for example, *Carcinus maenas* and *Mytilus edulis* – uptake of Cd by active transport appears to be of less significance and so an increase in Cd uptake occurs as salinity is reduced, whether or not Ca concentration in the seawater is increased.

One final consideration when studying interactions among trace metals in the environment is to establish whether observed correlations between, for example, concentrations of two metals in an organism are causative or merely reflect a common dependency on a third factor, such as the weight or age of the organism. Statistical techniques are available for dealing with this problem (Packard and Boardman, 1987).

7.13 SEASONAL VARIATION

The accumulation of metals by marine organisms can vary through the year. Such seasonal changes can be caused by a combination of parameters including growth, reproduction and moulting cycles, food supply and environmental conditions (e.g. temperature, irradiance, nitrogen and phosphorus concentrations) acting directly on uptake or indirectly on growth rates. For example,

Alliot and Frenet-Piron (1990) found marked differences in Zn, Pb, Cu and Cd concentrations in the shrimp *Palaemon serratus* off the coast of Brittany at different times of the year. While these changes were attributed solely to variations in boating activities and associated tourism, it seems likely that other factors such as seasonal differences in temperature, wave action, etc. were also responsible. Interestingly, concentrations measured in the shrimp were in accordance with changes in trace metal concentrations in the sea, suggesting that following uptake, these shrimps are able to depurate excess metal when seawater concentrations decline. Furthermore, the concentrations of essential metals to which the shrimps were exposed were well within the range at which Rainbow (1988) claims shrimps regulate body loads independent of environmental concentrations, while non-essential metals (which Rainbow found to accumulate) were in fact depurated when ambient seawater concentrations fell. This demonstrates an important principle, namely, that simplified laboratory experiments often do not mimic the responses of organisms in the natural environment and consequently great care should be taken when predicting what will happen in the field from laboratory data alone.

Temporal fluctuations in metal burdens are often superimposed upon variations due to alterations in body or tissue weight. For example, seasonal variations can occur due to weight-related effects of spawning (Zaroogian, 1980) and utilization of glycogen (Boyden and Phillips, 1981). Seasonal fluctuations in metal content of *Semibalanus balanoides* are due to the combined effects of reproductive state and the availability of phytoplankton (Ireland, 1974; Powell and White, 1990). Seasonal changes in metal burdens in scallops (*Pecten maximus* and *Chlamys opercularis*) are due to metal concentrations in, and availability of, phytoplankton (Bryan, 1973). Metal-specific seasonal changes are observed in *Euphausia superba*; for example, elevated concentrations of Fe, Mn, Zn, Pb and Hg are observed during summer (January and February). Changes may be linked to a reduced tissue water content and an increased intake of metal via ingested food. Conversely, concentrations of Cd, Ni and Co increase in winter and decrease during spring (Yamamoto *et al.*, 1987).

Seasonal fluctuations in metal concentrations of marine algae have been reported and are considered to be a result of variations in growth during the year and the requirement for metabolic energy for the uptake of those metals. Periods of high growth serve to 'dilute' the accumulated metals and hence reduce their concentrations, whereas the reverse occurs when growth is slow (Lobban and Harrison, 1994; Phillips, 1994).

7.14 SEX, REPRODUCTIVE STATUS AND MOULTING

Few studies have ascertained differences in metal concentrations between sexes. In some cases differences can be explained by different concentrations of metals in the gonads of males and females (e.g. *Mytilus edulis*; Lobel *et al.*, 1991). In contrast, in *Donax trunculus,* high concentrations of Mn found in females are associated with the formation of renal stones (Mauri and Orlando, 1982). Studies on fish also reveal differences between sexes. For example, reduced Cu and Cd concentrations occur in females of the teleost *Blennius pholis* during spawning when these metals are incorportated into oocytes, which are ultimately shed (Shackley *et al.*, 1981); this contrasts with most invertebrates which shed metal-poor eggs.

Moulting cycles, particularly in decapod crustaceans, contribute significantly to variations in tissue metals concentrations, even in unpolluted sites. Nugegoda and Rainbow (1988) have highlighted the close association between metal content and the physiological changes that occur during moulting.

7.15 INTER-INDIVIDUAL VARIABILITY

Depledge (1990a) has emphasized the importance of considering inter-individual differences among the representatives of a population in ecotoxicological studies. With regard to trace metal accumulation, it is interesting to note that even when similar-sized individuals of a particular species are exposed to a trace metal in controlled, uniform conditions, inter-individual differences in tissue metal concentrations are still evident. This points to the importance of biological determinants of trace metal uptake and accumulation. There have been a number of studies in which the physiological/nutritional state of the organism has been shown to have a very marked influence on the uptake, distribution and effects of heavy metals (McLusky *et al.*, 1986; Depledge, 1989b, 1990b; Donker, 1992).

Depledge and Bjerregaard (1989) and Lobel *et al.* (1989) pointed out that, within a given population, the frequency distributions of particular trace metal concentrations are often not 'normal'. This fact is almost invariably ignored in biomonitoring surveys. Depledge and Bjerregaard (1989) found that in midgut gland samples from individuals of a *Carcinus maenas* population, the concentrations of Cu, Zn and Cd showed a skewed distribution. A similar picture emerged for Cd in muscle tissue whereas Cu and Zn concentrations were normally distributed. The problem that arises by assuming a 'normal' distribution when the data are skewed has been elegantly demonstrated by Lobel *et al.* (1982). For Zn in *Mytilus edulis* collected from the Tyne estuary, UK, the mean Zn concentration was only 75% of the mid-range value. Furthermore, in a comparison of three sites that were contaminated to different extents by Zn, lowest tissue concentrations recorded at each site were very similar (0.83, 1.5 and 1.11 μmol Zn g^{-1}) whereas highest concentrations were markedly different (3.32, 10.0 and 20.5 μmol Zn g^{-1}). The distributions of tissue concentrations from animals at each site were positively skewed. Statistical techniques are available for quantifying and comparing the residual variability of trace metal concentrations in biological tissues (Lobel *et al.*, 1989).

7.16 INFLUENCE OF CLIMATE AND GEOGRAPHY

Temperature is a key factor influencing the kinetics of metal uptake and excretion in marine biota. Other factors such as salinity, oxygen tension and particulate content of air and water are also known to affect metal bioavailability. Since these factors vary with climate and among different geographic regions, they may give rise to considerable variability in tissue metal concentrations. This is of importance when comparing metal pollution levels at different localities. For example, there is evidence to suggest that essential metal concentrations are lower in the tissues of some oceanic marine invertebrates and selected crustaceans from subtropical and tropical inshore waters than those in temperate species from relatively clean areas (Hungspreugs, 1988). When compared with temperate scallop species (Eisler, 1981), the concentrations of Fe, Zn, Cd, Cu, Mn, Pb and Ni found in *Adamussium colbecki* collected from the Antarctic are much lower (Berkman and Nigro, 1992). Under such circumstances it is vitally important to know what the 'normal' metal concentrations in organisms are before attempting to recognize abnormal concentrations that might result from pollution (Depledge *et al.*, 1992). Rarely would it be appropriate to use normal values obtained for one species to compare with values obtained for another species (or even the same species) from a different climatic zone.

7.17 UNUSUAL CASES OF METAL ACCUMULATION

Most marine organisms from clean and polluted sites contain metals in the concentration ranges mentioned so far, differences between contaminated and non-contaminated sites typically varying over two to four orders of magnitude. However, there are some instances where extraordinarily high heavy metal concentrations have been measured, either in whole organisms or in specific tissues. In certain cases these extremes cannot be attributed to pollutant exposure, but appear to occur naturally. Obviously, it is important to distinguish these exceptions from the general trends to avoid misinterpretation of the seriousness of pollution threats. Examples of the highest reported metal concentrations are: $18500\ \mu g$ Al g^{-1} dry weight reported for the sponge *Dysidea crawshayi* (Bowen and Sutton, 1951); $500\ \mu g$ Cd g^{-1} dry weight and $40\ 800\ \mu g$ Zn g^{-1} dry weight measured in the excretory organs of *Pecten maximus* and *Chlamys opercularis* (Bryan, 1976a); extraordinarily high concentrations of As ($> 2000\ \mu g\ g^{-1}$) in the gastropod *Hemifusus ternatanus* and in the feeding palps of the polychaete *Tharyx marioni* (Gibbs *et al.*, 1983; Phillips and Depledge, 1986). Barnacles from the Thames estuary contain Zn concentrations of $153\ 000\ \mu g$ Zn g^{-1} dry weight, which is equivalent to 15% of the dry weight of the animal (Rainbow, 1987). The highest concentration so far reported for Cd in a barnacle is $156\ \mu g$ Cd g^{-1} dry weight in *Chthamalus stellatus* from the Azores (Weeks *et al.*, in press).

7.18 CONCLUSIONS

While chemical speciation can be considered to have an overriding influence on the accumulation and toxicity of metals, the significance of the various biological parameters highlighted in this review should not be underestimated. The considerable degree of variability in metal concentrations between species and between individuals of the same species collected at single locations results from these latter factors. Assuming bioavailability remains unaltered, the degree to which metals accumulate will depend upon the kinetics of uptake and loss in relation to the weight of the organism. These processes, too, are influenced by biological variables: uptake and loss are determined by, for example, the permeability of external surfaces, the efficiency of excretory systems and the types and number of ligands in the cytosol, whereas body weight may be modified by growth, reproductive development, nutrition and season. Hence, in order to gain any meaningful information from measurements of metal body burdens, an understanding of the physiological state of the organism and the underlying processes involved in metal accumulation is essential. Therefore, in biomonitoring programmes, the appropriate organism(s) for the type of contamination under study (dissolved/particulate/diet) must be selected, and optimal sampling procedures should be employed to effectively reduce the influence of biological variables (Langston and Spence, 1995).

REFERENCES

Agida, V.V., Bhosle, N.B. and Untawale, A.G. (1978) Metal concentrations in some seaweeds of Goa (India). *Botanica Marina* **21**, 247–250.

Ahner, B.A., Price N.M. and Morel, F.M.M. (1994) Phytochelatin production by marine phytoplankton at low free metal ion concentrations: laboratory studies and field data from Massachusetts Bay. *Proceeedings of the National Academy of Sciences of the USA* **91**, 8433–8436.

Alliot, A. and Frenet-Piron, M. (1990) Relationship between metals in sea-water and metal accumulation in shrimps. *Marine Pollution Bulletin* **21**, 30–33.

Barnett, B.E. and Ashcroft, C.R. (1985) Heavy metals in *Fucus vesiculosus* in the Humber Estuary. *Environmental Pollution (B)* **9**, 193–213.

Barreiro, R., Real, C. and Carballeira, A. (1993) Heavy metal accumulation by *Fucus ceranoides* in a small estuary in north-west Spain. *Marine Environmental Research* **36**, 39–61,

Beeftink, W.G. and Niewenhuize, J.(1986) Monitoring trace metal contamination in salt marshes of the Westerschelde Estuary. *Environmental Monitoring Assessment* **7**, 233–248.

Benson, W.H., Baer, K.N. and Wilson, C.F. (1990) Metallothionein as a biomarker of environmental metal contamination, in *Biomarkers of Environmental Contamination*, (eds J.F. McCarthy and L.R. Shugart), Lewis, Boca Raton, pp. 255–266.

Berkman, P.A. and Nigro, M. (1992) Trace metal concentrations in scallops around Antarctica. *Marine Pollution Bulletin* **24**, 322–323.

Bienvenue, E., Boudou, A., Desmazs, J.P. *et al.* (1984) Transport of mercury compounds across bimolecular lipid membranes: effect of lipid composition, pH and chloride concentration. *Chemical Biological Interactions* **48**, 91–101.

Bjerregaard, P. and Depledge, M.H. (1989) Effect of salinity and calcium concentration on cadmium uptake in *Littorina littorea* (L.), in *Collected Abstracts of the 1st European Conference on Ecotoxicology*, Copenhagen, Denmark, p. 68.

Black, W.A.P. and Mitchell, R.L. (1952) Trace elements in the common brown algae and seawater. *Journal of the Marine Biological Association of the UK* **30**, 575–584.

Bok, C.S. and Keong, W.M. (1976) Heavy metals in marine biota from coastal waters around Singapore. *Journal Singapore National Academy Science* **5**, 47–53.

Bourget, E. (1974) Environmental and structural control of trace metals in barnacle shells. *Marine Biology* **28**, 27–36.

Bowen, V.T. and Sutton, D. (1951) Comparative studies of mineral constituents of marine sponges. *Journal of Marine Research* **10**, 153–167.

Boyden, C.R. and Phillips, D.J.H. (1981) Seasonal variation and inherent variability of trace elements in oysters and their implications for indicator studies. *Marine Ecology Progress Series* **5**, 29–40.

Brezonik, P.L., King, S.O. and Mach, C.E. (1991) The influence of water chemistry on trace metal bioavailability and toxicity to aquatic organisms, in *Metal Ecotoxicology Concepts and Applications*, (eds M.C. Newman and A.W. McIntosh), Lewis, Boca Raton, pp. 1–31.

Brown, B.E. (1982) The form and function of metal containing 'granules' in invertebrate tissues. *Biological Review* **57**, 621–667.

Bruenig, E.F. (1989) Ecosystem of the World, in *Ecotoxicology and Climate*, (eds P. Bourdeau, J.A. Haines, W. Klein and C.R.T. Murti), SCOPE 38, John Wiley, Chichester, pp. 29–40.

Bryan, G.W. (1969) The absorption of zinc and other metals by the brown seaweed *Laminaria digitata*. *Journal of the Marine Biological Association of the UK* **49**, 225–243.

Bryan, G.W. (1973) The occurrence and seasonal variation of trace metals in the scallops *Pecten maximus* (L.) and *Chlamys opercularis* (L.). *Journal of the Marine Biological Association of the UK* **53**, 145–166.

Bryan, G.W. (1976a) Heavy metal contamination in the sea, in *Marine Pollution*, (ed. R. Johnson), Academic Press, London, pp. 185–302.

Bryan, G.W. (1976b) Some aspects of heavy metal tolerance in aquatic organisms, in *Effects of Pollutants on Aquatic Organisms*, (ed. A.P.M. Lockwood), Cambridge University Press, Cambridge, pp. 7–34.

Bryan, G.W. (1984) Pollution due to heavy metals and their compounds, in *Marine Ecology*, (ed. O. Kinne), John Wiley, Chichester, pp. 1289–1431.

Bryan, G.W. and Hummerstone, L.G. (1971) Adaptation of the polychaete *Nereis diversicolor* to estuarine sediments containing high concentrations of heavy metals. I. General observations and adaption to copper. *Journal of the Marine Biological Association of the UK* **51**, 845–863.

Bryan, G.W. and Hummerstone, L.G. (1973) Brown seaweed as an indicator of heavy metals in estuaries in south-west England. *Journal of the Marine Biological Association of the UK* **53**, 705–720.

Bryan, G.W. and Hummerstone, L.G. (1973) Adaptation of the estuarine polychaete *Nereis diversicolor* to estuarine sediments containing high concentrations of zinc and cadmium. *Journal of the Marine Biological Association of the UK*, **53**, 839–857.

Bryan, G.W. and Langston, W.J. (1992) Bioavailability, accumulation and effects of heavy metals in sediments with special reference to United Kingdom estuaries: a review. *Environmental Pollution* **76**, 89–131.

Bryan, G.W., Langston, W.J., Hummerstone, L.G. and Burt, G.R. (1985) A guide to the assessment of heavy metal containation in estuaries using biological indicators. *Occasional Publications of the Marine Biological Association of the UK* **4**, 1–92.

Burrows, E.M. (1971) Assessment of pollution effects by the use of algae. *Proceedings of the Royal Society of London B*, **177**, 295–306.

Cantillo, A.Y. and Segar, D.A. (1975) Metal species identification in the environment: a major challenge for the analyst, in *International Conference. Heavy Metals in the Environment*, Toronto, CEP Consultants, Edinburgh, pp. 183–204.

Christie, N.T. and Costa, M. (1984) *In vitro* assessment of the toxicity of metal compounds. IV. Disposition of metals in cells: interactions with membranes, glutathione, metallothionein and DNA. *Biological Trace Metal Research* **6**, 139–158.

Clark, D.R. Jr (1979) Lead concentrations: bats vs terrestrial small mammals collected near a major highway. *Environmental Science and Technology* **13**, 338–340.

Connell, D.W. and Miller, G.J. (1984) *The Chemistry and Ecotoxicology of Pollution*, John Wiley, New York, 444 pp.

Coombs, T.L. (1974) The nature of zinc and copper complexes in the oyster *Ostrea edulis*. *Marine Biology* **28**, 1–10.

Coombs, T.L. and George, S.G. (1978) Mechanisms of immobilization and detoxification of metals in marine organisms, in *Physiology and Behaviour of Marine Organisms*, (eds. D.S. McLusky and A.J. Berry), Pergamon Press, Oxford, pp. 179–187.

Coombs, T.L. and Keller, P.J. (1981) *Mytilius* byssal threads as an environmental marker for metals. *Aquatic Toxicology* **1**, 291–300.

Crespo, S., Soriano, E., Sampera, C. and Balasch, J. (1981) Zinc and copper distribution in excretory organs of the dogfish *Scyliorhinus canicula* and chloride cell response following treatment with zinc sulphate. *Marine Biology* **65**, 117–123.

Davies, I.M. and Russel, R. (1988) The influence of dissolved selenium compounds on the accumulation of inorganic and methylated mercury compounds from solution by the mussel *Mytilus edulis* and the plaice *Pleuronectes platessa*. *Science of the Total Environment* **68**, 197–205.

DeBoer, J.A. (1981) Nutrients, in *The Biology of Seaweeds*, (eds C.S. Lobban and M.J. Wynne, M.J.), Blackwell Scientific, pp. 356–391.

Denton, G.R. and Burdon-Smith, C. (1986) Trace metals in algae from the Great Barrier Reef. *Marine Pollution Bulletin* **17**, 98–107.

Depledge, M.H. (1989a) Re-evaluation of copper and zinc requirements in decapod crustaceans. *Marine Environmental Research* **27**, 115–126.

Depledge, M.H. (1989b) Studies on copper and iron concentrations, distributions and uptake in the brachyuran, *Carcinus maenas*, following starvation. *Ophelia* **30**, 187–189.

Depledge, M.H. (1990a) Interactions between heavy metals and physiological processes in estuarine invertebrates, in *Estuarine Ecotoxicology*, (eds P.L. Chambers and C.M. Chambers), Japaga, Wicklow, Ireland, pp. 89–100.

Depledge, M.H. (1990b) New approaches in ecotoxicology: can inter-individual physiological variability be used as a tool to investigate pollution effects? *Ambio* **19**, 251–252.

Depledge, M.H. and Bjerregaard, P. (1989) Explaining variation in trace metal concentrations in selected marine invertebrates: the importance of interactions between physiological state and environmental factors, in *Phenotypic Response and Individuality in Aquatic Ectotherms*, (ed. J.C. Aldrich), Japaga, Wicklow, Ireland, pp. 121–126.

Depledge, M.H. and Rainbow, P.S. (1990) Models of regulation and accumulation of trace metals in marine invertebrates: a mini-review. *Comparative Biochemistry and Physiology* **97C**, 1–7.

Depledge, M.H., Forbes, T.L. and Forbes, V.E. (1992) Evaluation of cadmium, copper, zinc and iron concentrations and tissue distributions in the benthic crab, *Dorippe granulata* (De Haan, 1841) from Tolo Harbour, Hong Kong. *Environmental Pollution* **81**, 15–19.

Donker, M. (1992) Physiology of metal adaptation in the isopod *Porcellio scaber*. PhD thesis, Free University of Amsterdam, Netherlands, 117 pp.

Eide, I., Myklestad, S. and Mesom, S. (1980) Long-term uptake and release of heavy metals by *Ascophyllum nodosum* (L.) Le Jol. (Phaeophyceae) *in situ*. *Environmental Pollution (A)* **23**, 19–28.

Eisler, R. (1981) *Trace Metal Concentrations in Marine Organisms*, Pergamon Press, Oxford, 685 pp.

Elbay-Poulichet, F., Martin, J.M., Huang, W.W. and Zhu, J.X. (1987) Dissolved Cd behaviour in some selected French and Chinese estuaries. Consequences on Cd supply to the ocean. *Marine Chemistry* **322**, 125–136.

Engel, D.W. (1988) The effect of biological variability on monitoring strategies: metallothioneins as an example. *Water Resources Bulletin* **24**, 981–987.

Engel, D.W. and Brouwer, M. (1989) Metallothionein and metallothionein-like proteins: physiological importance, in *Advances in Comparative and Environmental Physiology*, Vol. 5, Springer-Verlag, Berlin, pp. 53–75.

Engel, D.W. and Roesijadi, G. (1987) Metallothioneins: a monitoring Tool, in *Pollution and Physiology of Estuarine Organisms*. (eds F.J. Vernberg, F.P. Thurberg, A. Calabrese and W.B. Vernberg), University of South Carolina Press, Columbia, pp. 421–437.

Ernst, W.H.O., Verkleij, J.A.C. and Schat, H. (1992) Metal tolerance in plants. *Acta Botanica Neerlandica* **41**, 229–248.

Fayi, L. and George, S.G. (1985) Purification of very low molecular weight Cu-complexes from European oyster, in *Marine Pollution and Physiology: Recent Advances*, (eds F.J. Vernberg, F.B. Thurberg, A. Calabrese and W. Vernberg), Belle W. Baruch Library in Marine Science, University of South Californai Press, Columbia, pp. 145–155.

Foster, P. (1976) Concentrations and concentration factors of heavy metals in brown algae. *Environmental Pollution* **10**, 45–53.

Fowler, S.W. (1990) Critical review of selected heavy metal and chlorinated hydrocarbon concentrations in the marine environment. *Marine Environmental Research* **29**, 1–64.

Fowler, B.A., Wolfe, D.A. and Hettler, W.F. (1975) Mercury and iron uptake by cytochromes in mantle epithelial cells of quahog clams (*Mercenaria mercenaria*) exposed to mercury. *Journal of the Fisheries Research Board of Canada* **32**, 1767–1775.

Freedman, J.H., Ciriolo, M.R. and Peisach J. (1989) The role of glutathione in copper metabolism and toxicity. *Journal of Biological Chemistry* **264**, 5598–5605.

Furness R.W. and Rainbow, P.S. (1990) *Heavy Metals in the Marine Environment*, CRC Press, Boca Raton, 256 pp.

GESAMP (1990) *The State of the Marine Environment*, Blackwell Scientific Publications, Oxford, 146 pp.

Gibbs, P.E., Langston, W.J., Burt, G.R. and Pascoe, P.L. (1983) *Tharyx marioni* (Polychaeta): a remarkable accumulator of arsenic. *Journal of the Marine Biological Association of the UK* **63**, 313–325.

Goldberg, E.D. (1989) Aquatic transport of chemicals, in *Ecotoxicology and Climate*, (eds P. Bourdeau, J.A. Haines, W. Klein and C.R.T. Murti), SCOPE 38, John Wiley, Chichester, pp. 51–64.

Goyer, R.A. (1991) Toxic effects of metals, In *Casarett and Doull's Toxicology*, 4th edn, (eds M.O. Amdur, J. Doull and C.D. Klaasen), Pergamon Press, New York, pp. 623–680.

Gray, J.S. (1979) Pollution-induced changes in populations. *Philosophical Transactions of the Royal Society, Series B* **286**, 545–561.

Grill, E., Winnacker, E.L. and Zenk, M.H. (1987) Phytochelatins, a class of heavy-metal binding proteins, are functionally analagous to metallothioneins. *Proceedings of the National Academy of Sciences of the USA* **84**, 439–443.

Guilizzoni, P. (1991) The role of heavy metals and toxic materials in the physiological ecology of submersed macrophytes. *Aquatic Botany* **41**, 87–109.

Gutknecht, J. (1981) Inorganic mercury (Hg^{++}) transport through lipid bilayer membranes. *Journal of Membrane Biology* **61**, 61–66.

Hall, A., Fielding, A.H. and Butler, M. (1979) Mechanisms of copper tolerance in the marine fouling alga *Ectocarpus siliculosus* – evidence for an exclusion mechanism. *Marine Biology* **54**, 195–199.

Halliwell, B. and Gutteridge, J.M.C. (1984) Oxygen toxicity, oxygen radicals, transition metals and disease. *Biochemical Journal* **219**, 1–14.

Hanna, P.M. and Mason, P.R. (1992) Direct evidence for inhibition of free radical formation from Cu(1) and hydrogen peroxide by glutathione and other potential ligands using the EPR spin-trapping technique. *Archives of Biochemistry and Biophysics* **295**, 205–213.

Harland, A.D. and Nganro, N.R. (1990) Copper uptake by the sea anemone *Anemonia viridis* and the role of zooxanthellae in metal regulation. *Marine Biology* **104**, 297–301.

Haug, A. (1961) The affinity of some divalent metals to different types of alginates. *Acta Chemica Scandanavia* **19**, 1221–1226.

Ho, Y.B. (1984) Zn and Cu concentrations in *Ascophyllum nodosum* and *Fucus vesiculosus* (Phaeophyta, Fucales) after transplantation to an estuary contaminated with mine wastes. *Conservation and Recycling* **7**, 329–337.

Ho, Y.B. (1987) Metals in 19 intertidal macroalgae in Hong Kong waters. *Marine Pollution Bulletin* **18**, 564–566.

Ho, Y.B. (1990) Metals in *Ulva lactuca* in Hong Kong intertidal waters. *Bulletin of Marine Science* **47**, 79–85.

Hobden, D.J. (1967) Iron metabolism in *Mytilus edulis*. I. Variation in total content and distribution. *Journal of the Marine Biological Association of the UK* **47**, 597–606.

Holmes M.A., Brown, M.T., Loutit, M.W. and Ryan K. (1991) The involvement of epiphytic bacteria in zinc concentration by the red alga *Gracilaria sordida*. *Marine Environmental Research* **31**, 55–67.

Hungspreugs, M. (1988) Heavy metals and other non-oil pollutants in south-east Asia. *Ambio* **17**, 178–182.

Ireland, M.P. (1974) Variations in the zinc, copper, manganese and lead content of *Balanus balanoides* in Cardigan Bay, Wales. *Environmental Pollution* **7**, 65–75.

Jenkins, K.D. and Mason, A.Z. (1988) Relationships between subcellular distributions of cadmium and pertubations in reproduction in the polychaete *Neanthes arenaceodentata*. *Aquatic Toxicology* **12**, 229–244.

Kalk, M. (1963) Absorption of vanadium by tunicates. *Nature London* **198**, 1010–1011.

Karez, C.S., Magalhaes, V.F., Pfeiffer W.C. and Amado Fiho, G.M. (1994) Trace metal accumulation by algae in Sepetiba Bay, Brazil. *Environmental Pollution* **83**, 351–356.

Koeman, J.H., Peeters, W.H.M., Koudstaal-Hol, C.H.M. *et al.* (1973) Mercury–selenium corelations in marine mammals. *Nature, London* **245**, 385–386.

Koeman, J.H., van de Ven, W.S.M., de Goeij, J.J.M. *et al.* (1975) Mercury and selenium in marine mammals and birds. *Science of the Total Environment* **3**, 279–287.

Kullenberg, G. (1986) The IOC programme on marine pollution. *Marine Pollution Bulletin* **17**, 341–352.

Lacerda, L.D. de, Teixeira V.L. and Guimaraes J.R.D. (1985) Seasonal variation of heavy metals in seaweeds from Conceicao de Jaceri (R.J.), Brazil. *Botanica Marina* **28**, 339–343.

Lande, E. (1977) Heavy metal pollution in Trondheimsfjorden, Norway and the recorded effects on the fauna and flora. *Environmental Pollution* **12**, 187–198.

Langston, W.J. and Spence, S.K. (1995) Biological factors involved in metal concentrations observed in aquatic organisms, in *Metal Speciation in Aquatic Systems*, (eds A. Tessier and D.R. Turner), John Wiley, pp. 407–478.

Lignell, A., Roomans, G.M. and Pedersen, M. (1982) Localization of cadmium in *Fucus vesiculosus* L. by X-ray microanalysis. *Zeichschrift fur Pflanzenphysiologie* **105**, 103–109.

Lobban C.S. and Harrison P.J. (1994) *Seaweed Ecology and Physiology*, Cambridge University Press, Cambridge, 366 pp.

Lobel, P.B., Mogie, P., Wright, D.A. and Wu, B.L. (1982) Metal accumulation in four molluscs. *Marine Pollution Bulletin* **13**, 170–174.

Lobel, P.B., Belkhode, S.P., Jackson, S.E. and Longerich, H.P. (1989) A universal method for quantifying the residual variability of element concentrations in biological tissues using 25 elements in the mussel *Mytilus edulis* as a model. *Marine Biology* **102**, 513–518.

Lobel, P.B., Bajdik, C.D., Belkhode, S.P. *et al.* (1991) Improved protocol for collecting mussel watch specimens taking into account sex, size, condition, shell shape and chronological age. *Archives Environmental Contamination and Toxicology* **21**, 409–414.

Long, D.T. and Angino, E.E. (1977) Chemical speciation of Cd, Cu, Pb, and Zn in mixed freshwater, seawater, and brine solutions. *Geochimica Cosmochimica Acta* **41**, 1183–1191.

Luoma, S.M. (1983) Bioavailability of trace metals to aquatic organisms – a review. *Science of the Total Environment* **28**, 1–22.

Luoma, S.M. (1989) Can we determine the biological availability of sediment-bound trace metals? *Hydrobiologia* **176/177**, 379–396.

Luoma, S.M., Bryan, G.W. and Langston, W.J. (1982) Scavenging of heavy metals from particulates by brown seaweed. *Marine Pollution Bulletin* **13**, 394–396.

Mackay, N.J., Kazacos, M.N., Williams, R.J. and Leedow, M.I. (1975) Selenium and heavy metals in black marlin. *Marine Pollution Bulletin* **6**, 57–61.

Magos, L. and Webb, M. (1980) The interactions of selenium with cadmium and mercury. *CRC Critical Reviews in Toxicology* **8**, 1–42.

Mantoura, R.F.C., Dickson, A. and Riley, J.P. (1978) The complexation of metals with humic materials in natural waters. *Estuarine and Coastal Marine Science* **6**, 387–408.

Marigomez, J.A., Cajaraville, M.P. and Angulo, E. (1990) Histopathology of the digestive gland–gonad complex of the marine prosobranch *Littorina littorea*. *Diseases of Aquatic Organisms* **9**, 229–238.

Marina, M and Enza, O. (1983) Variability of zinc and manganese concentrations in relation to sex and season in the bivalve *Donax trunculus*. *Marine Pollution Bulletin* **14**, 342–346.

Martin, J.L.M. (1973) Iron metabolism in *Cancer irroratus* (Crustacea, Decapoda) during the intermoult cycle with special reference to iron in the gills. *Comparative Biochemistry and Physiology* **46A**, 123–129.

Mason, A.Z. and Jenkins, K.D. (1995) Metal detoxification in aquatic organisms, in *Metal Speciation and Bioavailability in Aquatic Systems*, (eds A. Tessier and D.R. Turner), John Wiley, pp. 479–608.

Mason, A.Z. and Nott, J.A. (1981) The role of intracellular biominerilized granules in the regulation and detoxification of metals in gastropods with special reference to the marine prosobranch *Littorina littorea*. *Aquatic Toxicology* **1**, 239–256.

Mauri, M. and Orlando, E. (1982) Experimental study on renal concretions in the wedge shell *Donax trunculus* (L). *Journal of Marine Biology and Ecology* **63**, 47–57.

McKay, G.A. and Thomas, M.K. (1989) Climates of the World seen from an ecotoxicological perspective, in *Ecotoxicology and Climate*, (eds P. Bourdeau, J.A. Haines, W. Klein and C.R. Krishna Murti), SCOPE 38, John Wiley, Chichester, pp. 15–28.

McLean, M.W. and Williamson, F.B. (1977) Cadmium accumulation by the marine red alga *Porphyra umbilicalis*. *Physiologia Plantarum* **41**, 268–272.

McLusky, D.S., Bryant, V. and Campbell, R. (1986) The effects of temperature and salinity on the toxicity of heavy metals to marine and estuarine invertebrates. *Oceanography and Marine Biology Annual Review* **24**, 481–520.

Morris, A.W. and Bale, A.J. (1975) The accumulation of cadmium, copper, manganese and zinc by *Fucus vesiculosus* in the Bristol Channel. *Estuarine and Coastal Marine Science* **3**, 153–163.

Mulvey, M. and Diamond, S.A. (1991) Genetic factors and tolerance acquisition in populations exposed to metals and metalloids, in *Metal Ecotoxicology Concepts and Applications* (eds M.C. Newman and A.W. McIntosh), Lewis, Boca Raton, Michigan, pp. 301–321.

Munda I.M. and Hudnik, V. (1991) Trace metal content in some seaweeds from the northern Adriatic. *Botanica Marina* **34**, 2241–2249.

Myklestad, S., Eide, I. and Melsom, S. (1978) Exchange of heavy metals in *Ascophyllum nodosum* (L.) Le Jol. *in situ* by means of transplantation experiments. *Environmental Pollution* **16**, 277–284.

Newman, M.C. and Heagler, M.G. (1991) Allometry of metal accumulation and toxicity, in *Metal Ecotoxicology Concepts and Applications* (eds M.C. Newman and A.W. McIntosh), Lewis, Boca Raton, Michigan, pp. 91–130.

Nienhuis, P.H. (1986) Background levels of heavy metals in nine tropical seagrass species in Indonesia. *Marine Pollution Bulletin* **17**, 508–511.

Norheim, G. (1987) Levels and interactions of heavy metals in seabirds from Svalbard and the Antarctic. *Environmental Pollution* **47**, 83–94.

Nott, J.A. (1991) Cytology of pollutant metals in marine invertebrates: a review of microanalytical applications. *Scanning Microscopy* **5**, 191–205.

Nugegoda, D. and Rainbow, P.S. (1987) The effect of temperature on zinc regulation by the decapod crustacean, *Palaemon elegans*, Rathke. *Ophelia* **27**, 17–30.

Nugegoda, D. and Rainbow, P.S. (1988) Zinc uptake and regulation by the sublittoral prawn *Pandalus montagui* (Crustacea: Decapoda). *Estuarine and Coastal Shelf Science* **26**, 619–632.

Nugegoda, D. and Rainbow, P.S. (1989) Effects of salinity changes on zinc uptake and regulation by the decapod crustaceans, *Palaemon elegans* and *Palaemonetes varians*. *Marine Ecology Progress Series* **51**, 57–75.

Packard, G.C. and Boardman, T.J. (1987) The misuse of ratios to scale physiological data that vary allometrically with body size, in *New Directions in Ecological Physiology* (eds M.E. Feder, A.F. Bennett, W.W. Burggren and R.B. Huey), Cambridge University Press, pp. 216–239.

Pacyna, J.M., Munch, J. and Axanfeld, F. (1991) European inventory of trace metal emissions to the atmosphere, in *Heavy Metals in the Environment*, (ed. J.P. Vernet), Elsevier, Amsterdam, pp. 1–20.

Pak, C.K., Yang, K.R. and Lee I.K. (1977) Trace metals in several edible marine algae of Korea. *Journal Oceanographical Society, Korea* **12**, 41–47.

Parizek, J. and Ostadalova, I. (1967) The protective effect of small amounts of selenite on sublimate intoxication. *Experimentia*,**23**, 142–143.

Patrick, F.M. and Loutit, M.W. (1977) The uptake of heavy metals by epiphytic bacteria of *Alisma plantago-aquatica*. *Water Research* **11**, 333–335.

Pelletier, E. (1985) Mercury–selenium interactions in aquatic organisms: a review. *Marine Environmental Research* **18**, 111–132.

Petering, D.H., Goodich, M., Hodgman, W. *et al.* (1990) Metal-binding proteins and peptides for the detection of heavy metals in aquatic organisms, in *Biomarkers of Environmental Contamination* (eds J.F. McCarthy and L.R. Shugart), Lewis, Boca Raton, pp. 239–254.

Phillips, D.J.H. (1977) The use of biological indicator organisms to monitor trace metal pollution in marine and estuarine environments – a review. *Environmental Pollution* **13**, 281–317.

Phillips, D.J.H. (1980) *Quantitative Aquatic Biological Indicators*, Applied Science Publishers, London, 488 pp.

Phillips, D.J.H. (1994) Macrophytes as biomonitors of trace metals, in *Biomonitoring of Coastal Waters and Estuaries*, (ed. K.J.M. Kramer), CRC Press Inc., Boca Raton, pp. 85–103.

Phillips, D.J.H. and Depledge, M.H. (1986) Distibution of inorganic and total arsenic in tissues of the marine gastropod, *Hemifusus ternatanus*. *Marine Ecology Progress Series* **34**, 261–266.

Powell, M.I. and White, K.N. (1990) Heavy metal accumulation by barnacles and its implications for their use as biological monitors. *Marine Environmental Research* **30**, 91–118.

Preston, A., Jefferies, D.F., Dutton, J.W.R. *et al.* (1972) British Isles coastal waters, the concentrations of selected heavy metals in seawater, suspended matter and biological indicators – a pilot survey. *Environmental Pollution* **3**, 69–82.

Pullen, J.H.S. and Rainbow, P.S. (1991) The composition of pyrophosphate heavy metal granules in barnacles. *Journal of Experimental Marine Biology and Ecology* **150**, 249–266.

QSRNS (1987) *Report of the Scientific and Technical Working Group*, Quality Status Report on the North Sea, HMSO, Department of the Environment, UK, 88 pp.

Rahn, K.A. and McCaffrey, R.J. (1979) Long-range transport of pollution aerosols to the Arctic: a problem without borders, in *Proceedings of WMO Symposium on*

Long-Range Transport of Pollutants and its Relation to General Circulation including Stratospheric/Tropospheric Exchange Processes, World Meteorological Office, Geneva, pp. 25–36.

Rai, L.C., Gaur J.P. and Kumar H.D. (1981) Phycology and heavy metal pollution. *Biological Review of the Cambridge Philosophical Society* **50**, 99–151.

Rainbow, P.S. (1985) The biology of heavy metals in the sea. *International Journal of Environmental Studies* **25**, 195–211.

Rainbow, P.S. (1987) Heavy metals in barnacles, in *Barnacle Biology*, (ed. A.J. Southward), A.A. Balkema, Rotterdam, pp. 405–417.

Rainbow, P.S. (1988) The significance of trace metal concentrations in decapods. *Symposia of the Zoological Society of London* **59**, 291–313.

Rainbow, P.S. (1992) The significance of trace metal concentrations in marine invertebrates, in *Ecotoxicology of Metals in Invertebrates*, (eds. R. Dallinger and R. Rainbow), Lewis, Boca Raton. pp. 3–23.

Ramade, F. (1987) *Ecotoxicology*, John Wiley, New York, 262 pp.

Rauser, W.E. (1995) Phytochelatins and related peptides: structure, biosynthesis, and function. *Plant Physiology* **109**, 1141–1149.

Rijstenbil, J.W. and Wijnholds, J.A. (1991) Copper toxicity and adaptation in the marine diatom *Ditylum brightwellii. Comparative Biochemistry and Physiology* **100C**, 147–150.

Romeo, M. and Gnassia-Barelli, M. (1993) Organic ligands and their role in complexation and transfer of trace metals (micronutrients) in marine algae, in *Macroalgae, Eutrophication and Trace Metal Cycling in Estuaries and Lagoons* (eds J.W. Rijstenbil and S. Haritonidis), Symp. Proc. Sub Group III, Action COST-48, NIOO-CEMA, Yerseke, The Netherlands, pp. 121–135.

Salomons, W. and Forstner, U. (1984) *Metals in the Hydrocycle*, Springer Verlag, Berlin, 349 pp.

Samiullah, Y. (1990) *Biological Monitoring of Environmental Contaminants: Animals*, MARC Report Number 37, Global Environmental Monitoring Programme, pp. 767.

Sanders, B.M. and Jenkins, K.D. (1984) Relationships between free cupric ion concentrations in seawater and copper metabolism and growth in crab larvae. *The Biological Bulletin* **167**, 704–712.

Sanders, B.M., Jenkins, W.G., Sunda, W.G. and Costlow, J.D. (1983) Free cupric ion activity in seawater: effects on metallothionein and growth in crab larvae. *Science* **222**, 53–55.

Say, P.J. Burrows, I.G. and Whitton, B.A. (1990) *Enteromorpha* as monitor of heavy metals in estuaries. *Hydrobiologia* **195**, 119–126.

Schindler, P.W. (1991) The regulation of heavy metal concentrations in natural aquatic systems, in *Heavy Metals in the Environment* (ed. J.P. Vernet), Elsevier, Amsterdam, pp.95–123.

Shackley, S.E., King, P.E. and Gordon, S.M. (1981) Vitellogenesis and trace metals in a marine teleost. *Journal of Fish Biology* **18**, 349–352.

Silver, S. and Misra, T.K. (1988) Plasmid-mediated heavy metal resistances. *Annual Review of Microbiology* **42**, 717–743.

Silverberg, B.A., Stokes, P.M. and Ferstenberg L.B. (1976) Intranuclear complexes in a copper-tolerant green alga. *The Journal of Cell Biology* **69**, 210–214.

Simkiss, K. (1976) Intracellular and extracellular routes in biomineralization. *Symposium of the Society of Experimental Biology* **30**, 423–444.

Simkiss, K. and Mason, A.Z. (1983) Metal ions: metabolic and toxic effects, in *The Mollusca*, Vol. 2, (ed. P.W. Hochachka), Academic Press, New York, pp. 101–164.

Simkiss, K. and Taylor, M.G. (1989) Metal fluxes across membranes of aquatic organisms. *Reviews in Aquatic Science* **1**, 173–188.

Sivalingum, P.M. (1978) Biodeposited trace metals and mineral content studies of some tropical marine algae. *Botanica Marina* **21**, 327–330.

Soderland, S., Forsberg, A. and Pederson, M. (1988) Concentrations of cadmium and other metals in *Fucus vesiculosus* L. and *Fontinalis dalecarlica* Br. Eur. from the northern Baltic Sea and the southern Bothnian Sea. *Environmental Pollution* **51**, 197–212.

Steffens, J.C. (1990) The heavy metal-binding peptides of plants. *Annual Review of Plant Physiology and Plant Molecular Biology* **41**, 553–575.

Stenner, R.D. and Nickless G (1974) Distribution of some heavy metals in organisms in Hardangerfjord and Skjerstadfjord, Norway. *Water, Air, Soil Pollution* **3**, 279–291.

Sunda, W.G. and Guillard, R.R. (1976) The relationship between cupric ion activity and the toxicity of copper to phytoplankton. *Journal of Marine Research* **34**, 511–529.

Sunda, W.G. and Huntsman, S.A. (1983) Effect of competitive interactions between manganese and copper on cellular manganese and growth in estaurine and oceanic species of the diatom *Thalassiosira*. *Limnology and Oceanography* **28**, 924–934.

Sunda, W.G., Engel, D.W. and Thuotte, R.M. (1978) Effect of chemical speciation on toxicity of cadmium to grass shrimp *Palaemonetes pugio*: importance of free cadmium ion. *Environmental Science and Technology* **12**, 409–413.

Tessier, A. and Turner, D.R. (1995) *Metal Speciation and Bioavailability in Aquatic Systems*, John Wiley, 679 pp.

Toppin, S.V., Heber, M., Weis, J.S. and Weis, P. (1987) Changes in reproductive biology and life history in *Fundulus heteroclitus* in a polluted environment, in *Pollution Physiology of Estuarine Organisms* (eds W. Vernberg, A. Calbrese, F. Thurberg and F.J. Vernberg), University of South Carolina Press, Columbia, pp. 171–184.

Turner, D.R. (1984) Relationships between biological availability and chemical measurements, in *Metal Ions in Biological Systems, Vol 18, Circulation of Metals in the Environment*, (ed. H. Sigel), Marcel Dekker, New York, pp. 137–164.

UNEP (1983) *A Review and the Prospects for Open Ocean Pollution Monitoring*, internal report prepared for the Regional Seas Programme Activity Centre, UNEP by the Monitoring and Assessment research centre, University of London, 11 pp.

Vernet, J.P. (1991) *Heavy metals in the Environment*, Elsevier, Amsterdam, 405 pp.

Veroy, R.L., Montano, N., de Guzman, M.L.B. *et al.* (1980) Studies on the binding of heavy metals to algal polysaccharides from Philippine seaweeds. I. Carrageenan and the binding of lead and cadmium. *Botanica Marina* **23**, 59–62.

Walsh, R.S. and Hunter, K.A. (1992) Influence of phosphorus storage on the uptake of cadmium by the marine alga *Macrocystis pyrifera*. *Limnology and Oceanography* **37**, 1361–1369.

Ward, T.J. (1989) The accumulation and effects of metals in seagrass habitats, in *Biology of Seagrasses: a Treatise on the Biology of Seagrasses with Special Reference to the Australian Region*, (eds A.W.D. Larkum, A.J. McComb and S.A. Shepherd), pp. 797–820.

Weeks, J.M. and Rainbow, P.S. (1990) A dual-labelling technique to measure the relative assimilation efficiencies of invertebrates taking up trace metals from food. *Functional Ecology* **4**, 711–717.

Weeks, J.M. and Rainbow, P.S. (1991) The uptake and accumulation of zinc and copper from solution by two species of talitrid amphipods (Crustacea). *Journal of the Marine Biological Association of the UK* **71**, 811–826.

Weeks, J.M., Rainbow, P.S. and Depledge, M.H. (in press) Barnacles (*Chthamalus stellatus*) as biomonitors of trace metal bioavailability in the waters of Sao Miguel (Azores). *Proceedings of the Second International Workshop of Malacology and Marine Biology*, Sao Miguel, Azores.

White, S.L. and Rainbow, P.S. (1984) Regulation of zinc concentration in *Palaemon elegans* (Crustacea: Decapoda): zinc flux and effects of temperature, zinc concentration and moulting. *Marine Ecology Progress Series* **16**, 135–147.

Yamamoto, Y., Honda, K. and Tatsukawa, R. (1987) Heavy metal accumulation in Antarctic krill *Euphausia superba*. *Proceedings NIPR Symposium Polar Biology* **1**, 198–204.

Young, M.L. (1975) The transfer of ^{65}Zn and ^{59}Fe along a *Fucus serratus* (L.) *Littorina obtusata* (L.) food chain. *Journal Marine Biological Association of the United Kingdom*, **55**, 583–610.

Zamuda, C.D. and Sunda, W.G. (1982). Bioavailability of dissolved copper to the american oyster *Crassostrea virginica*. I. Importance of chemical speciation. *Marine Biology*, **66**, 77–82.

Zaroogian, G.E. (1980). *Crassostrea virginica* as an indicator of cadmium pollution. *Marine Biology*, **58**, 275–284.

Zirino, A. and Yamamoto, S. (1972). A pH-dependent model for the chemical speciation of copper, zinc, cadmium and lead in seawater. *Limnology and Oceanography* **17**, 661–671.

8 *Metal handling strategies in molluscs*

WILLIAM J. LANGSTON, MARIA JOÃO BEBIANNO AND GARY R. BURT

8.1 INTRODUCTION

The concentration and distribution of metals in molluscan tissues, as in any other organism, is highly dependent on the biochemical processes of metal metabolism occurring within cells. Many of these processes and associated metal-binding systems are common to a broad range of phyla, reflecting their evolutionary success in regulating the essential elements (e.g. Cu, Zn, Fe, Mn, Co) for physiological, biochemical and morphological purposes. The involvement of metallothionein (MT) in buffering intracellular metal ions is one of the best examples of such a ubiquitous metal sequestration system. Proteins of a metallothionein-like nature are represented in many branches of the phylogenetic tree, from microorganisms to humans, and consequently are regarded as a central constituent of metal metabolism. Other sequestration mechanisms are relatively unique to individual taxonomic groups so that even within a single phylum, exemplified here by molluscs, there may be a large diversity in the nature and expression of metal metabolism, according to metal requirements. The result is manifested by a wide range of behaviour in terms of uptake, detoxification and storage – the major components of metal bioaccumulation.

Basic chemical rules dictate that the same regulatory mechanisms are usually prominent in dealing with those, usually rare, metals regarded as pollutants (e.g. Hg, Cd, Ag) but which have become elevated, locally, as a result of industrial and domestic activities in recent times. Understanding the strategies adopted by molluscs in their involvement with metals, whether they be essential, pollutant or excess essential metals, is the key to assessing the ecotoxicological significance of accumulated metal burdens, data on which are commonly collected in monitoring schemes.

Metal Metabolism in Aquatic Environments. Edited by William J. Langston and Maria João Bebianno. Published in 1998 by Chapman & Hall, London. ISBN 0 412 80370 4

8.2 BIOAVAILABILITY AND ASSIMILATION PATHWAYS

The metal concentration in any organism reflects the net product of a combination of processes including uptake, elimination, storage and transformation and will inevitably vary between metals, species and even individuals, because of differences in permeability, metabolic rate and the quantities and types of metal-binding ligands present either at the surface of the organism or intracellularly. Although it may not be essential to quantify all the processes precisely, some appreciation of their relative contribution to the overall pattern of metal turnover is often the only way of interpreting tissue residue data. First, though, a consideration of assimilation pathways in molluscs is needed.

8.2.1 DISSOLVED METALS AND CHEMICAL SPECIATION

The incorporation of metals into cells from solution is common to all aquatic molluscs, whether it be at the surface (from surrounding water) or in the alimentary tract following intestinal digestion of food. Consequently, dissolved metals have so far received most attention in kinetic studies on uptake and loss. Net accumulation usually follows either exponential or linear patterns. Exponential accumulation arises when uptake is initially rapid in relation to loss, often signifying a preliminary adsorption phase which is dependent on the surface characteristics of the organism (Chapter 1). Once external adsorption sites become filled, net accumulation slows (limited by the rate of inward diffusion and provision of internal ligand systems) and loss of metal becomes increasingly significant. Eventually uptake and loss rates may cancel each other out and the organism's body burden approaches steady state with respect to ambient concentrations (often proportionality is observed). Steady state concentrations in tissues are usually characterized in this scenario by the simple first-order equation:

$$C_t = C_{ss} (1 - e^{-\gamma t}) \qquad (8.1)$$

where C_t and C_{ss} are concentrations in the organism at time t and at steady state, respectively; elimination coefficient $\gamma = 0.693/t_{1/2}$ ($t_{1/2}$ being the biological half-life).

Linear accumulation over time arises when excretion of metal is negligible or slow in relation to uptake, and therefore body burdens continue to increase during exposure, provided that metal-binding ligands do not become saturated, or the rate of organism growth does not outstrip that of metal assimilation. Steady state may take a considerable time, if it occurs at all. This type of model is common in molluscs exposed to Cd, and in a number of cases is caused by the metal becoming irreversibly bound to inducible ligands such as metallothionein (MT). However, there are few hard and fast rules in molluscs: where MT involvement in Cd metabolism is limited, as in *Macoma balthica*, for example, some excretion of the metal is achievable and the turnover of Cd is reflected by an exponential-type model (Langston and Zhou, 1987b). Cd burdens and uptake rates can vary by more than an order of magnitude between mollusc species maintained in similar conditions, due primarily to differences in the level of MT induction (Langston and Spence, 1995).

Occasionally, accumulation may appear biphasic as described for Hg in oysters (Mason *et al.*, 1976) and for Hg and Cd in freshwater clams (Balogh and Salanki, 1984). This signifies a combination of initial rapid (reversible) adsorption, followed by a largely irreversible linear phase of much longer duration. The presence of different metal pools that vary in exchange rates, leading to such biphasic patterns, may necessitate somewhat more complex models than those described above.

Thus, for many dissolved Class B and transition metals, entry into molluscs is thought to involve passive diffusion, perhaps facilitated by carrier molecules (Chapter 1), and passage along a gradient that is maintained by intracellular ligands of increasing binding strength, some of which (such as MT) may be inducible. Metal forms with the highest free energy should be most readily bound to carriers and therefore most bioavailable. Hence, uptake is usually envisaged as being proportional to the free ion concentration or activity, at least for divalent cations such as Cd^{2+}, Cu^{2+} and Zn^{2+}. The negative effect of increasing salinity on metal (Cd, Cu, Hg, Zn) uptake is well established in, for example, oysters (Engel *et al.*, 1981; Wright and Zamuda, 1987) and the assumption made generally is that increasing complexation of the free ion (principally with chloride) at high salinities reduces availability. Competition from Ca, or osmotic and other physiological changes, cannot be overruled in many of these examples, however.

Strong complexing agents such as NTA and EDTA reduce the availability of Cu and Zn in the oyster *Crassostrea virginica,* apparently supporting the free ion hypothesis of uptake (Zamuda and Sunda, 1982; Harrison, 1979). However, complexation sometimes enhances uptake: Cd bioaccumulation rates in the mussel *Mytilus edulis* are doubled by prior sequestration with EDTA, pectin or humic and alginic acids, when compared with the ionic form, suggesting that the uncharged forms are transported across membranes preferentially (George and Coombs, 1977). In this example the nature of the organic ligand is of little significance, though incorporation of other metals is sometimes highly dependent on the nature of the association. Thus, enhanced bioavailability of Cr(III) bound to proteins in tannery waste accounts for the exceptional body burdens in mussels collected from sites close to these outfalls, despite the fact that $Cr(OH)_3$ is by far the most dominant Cr species in the effluent (Walsh and O'Halloran, 1997). Most importantly, some of the most ecotoxicologically relevant metal-related events have been caused by metal–organic moieties: increased lipid solubility of organometals – including methyl mercury, alkyl lead and tributyl tin – facilitates entry across membranes. The presence of chemical species of reduced polarity ($AgCl^0$; $HgCl_2^0$) may result in enhanced membrane transport for similar reasons (reviewed in Bryan and Langston, 1992).

In complex field situations it is perhaps not surprising that simple models, based on diffusion of the free ion, have yet to be demonstrated extensively for molluscs. Interpretation could be further complicated where pinocytosis of macromolecular metal species at the gill surface takes place, as seems likely in bivalves such as mussels. Nevertheless, some attempts have proved fruitful. By applying surface complexation concepts, Tessier *et al.* (1993) were able to predict Cd body burdens in the freshwater mussel *Anodonta grandis* from estimations of Cd^{2+} in overlying lake waters. Deterministic models applicable in nature are therefore achievable under some conditions.

Another mechanism for inward transport of metals (particularly Sr^{2+}, Ba^{2+} and Cd^{2+} but also Mn^{2+}, Ni^{2+}, Co^{2+}, Zn^{2+} and Pb^{2+}) in cells of higher animals involves adventitious use of voltage-sensitive Ca^{2+} channels, leading ultimately to disruption of Ca or Ca/Mg ATPases. Though Cd uptake in oyster gills has been shown to be inhibited by Ca channel blockers (Roesijadi and Unger, 1993), evidence for Cd and Ni indicates that this route may not be quantitatively important in bivalves (Zaroogian and Anderson, 1995). Nevertheless it at least questions the widely held view that passive diffusion is the sole mechanism for uptake of dissolved metals; it would be interesting to test whether there is any sharing of regulatory mechanisms between Ca and other metals in molluscs, as suspected in higher aquatic phyla such as crustaceans and fish.

Uptake of metal from water can take place across the whole body surface, in addition to the gills, in types lacking external shells (e.g. cephalopods, nudibranchs) and in all molluscs some absorption of dissolved metal across the digestive epithelium undoubtedly takes place following intestinal digestion or if water is imbibed; nevertheless gills are obvious sites of high permeability to metals as a result of their respiratory/nutritional functions, and increased metal levels are often encountered here, relative to other parts of the body, especially where water is the principal source of contamination. Enhanced inducibility of MT and other ligands may contribute to elevated metal assimilation in gills, relative to other tissues (see, for example, *Ostrea edulis* in Figure 8.1).

The source of ventilated water is likely to be influential in determining accumulated metal burdens in sediment-dwelling molluscs. Only 4% of the water processed by the clam *Macoma nasuta* is interstitial (Winsor *et al.*, 1990) suggesting that this vector may be small in bivalve species that are able to isolate themselves from the surrounding sediment chamber. However, pore waters are often much enriched in metals compared with overlying water and their relative contribution needs to be quantified in a broader range of species. This leads us to the broader issue of separating sources of metals since, in nature, aquatic molluscs accumulate metals from a combination of water and diet (including sediments) in varying proportions. Resolving assimilation pathways is central to improved prediction of bioaccumulation.

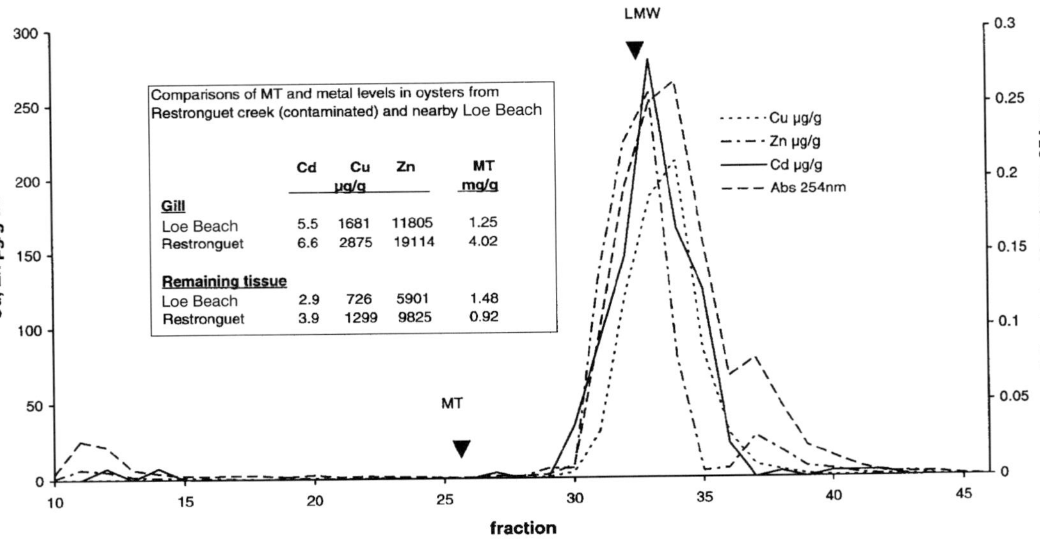

	Cd	Cu	Zn	MT
		µg/g		mg/g
Gill				
Loe Beach	5.5	1681	11805	1.25
Restronguet	6.6	2875	19114	4.02
Remaining tissue				
Loe Beach	2.9	726	5901	1.48
Restronguet	3.9	1299	9825	0.92

Figure 8.1 *Ostrea edulis*, Restronguet Creek. Cytosolic metal profiles indicate Cd, Cu and Zn bound primarily to low molecular weight (LMW) ligands with limited involvement of metallothionein (MT).

8.2.2 PARTICULATE METALS AND THE SIGNIFICANCE OF GEOCHEMICAL ASSOCIATIONS: EQUILIBRIUM MODELS

Direct assimilation of particulate metals – by intracellular digestion in the gut or pinocytosis at the gill surface – may be a significant feature in filter-feeding molluscs. The review of interactions between sediments and biota (Chapter 2) highlights the complexities involved in defining metal bioavailability from solid phases. Some of the earliest indications of the assimilation of particulate metals by molluscs, and its dependency on the physicochemical nature of the solid-phase/metal association, are provided by Luoma and Jenne (1976, 1977). These studies demonstrate the importance of organic matter in suppressing sediment Cd bioavailability in the clam *Macoma balthica*. In contrast, Ag, Co and Zn are assimilated from detrital organics, though coprecipitation with Fe and Mn oxyhydroxides inhibits the uptake of the latter two metals: thus, the general hypothesis is established that bioavailability declines as the strength of binding to various sediment components increases, in line with principles derived from complexation studies with dissolved metals.

Unfortunately, speciation techniques for oxidized sediments are much less advanced than for metals in water, and are primarily based on operational methods of uncertain selectivity to extract specific (labile) metal forms. Nevertheless, such partial extractions are often considered to be of more predictive value than total metal analysis, because they do not include residual fractions that may not be accessible to biota: consequently, they provide some of the most meaningful and practical assessments of sediment metal availability to molluscs (Table 8.1). For example, 0.05 M HCl or EDTA is much more effective than total sediment digestion in predicting bioaccumulation of Zn, Pb and Cd in deposit-feeding gastropods *Velacumantis australis* and *Pyrazus ebeninus*, from sediments of similar physical and chemical characteristics. However, in *P. ebeninus*, only Pb bioavailability can be predicted directly by these extracts from a range of field sediments of different types: for Zn (and Cu), normalization of the EDTA extract with respect to organic carbon greatly improves correlations, suggesting that organics modify bioavailability (Ying *et al.*, 1992; Table 8.1). Thus, by examining the goodness of fit between extracted metals and body burdens in ubiquitous benthic species, over a wide range of sites and conditions, it is often possible to quantify the influence of sediment contamination and to evaluate the strength of anthropogenic contributions from different sources. Scrutiny of outliers in the data can also help to highlight the more important geochemical parameters that modify bioavailability of the metal in question.

Particularly useful relationships for predicting metal bioavailability in surface sediments have arisen from extensive field-based studies with marine and freshwater infaunal clams. These confirm the role of the major metal-binding components (x) (Fe and Mn oxyhydroxides or organic matter) in

Table 8.1 Examples of relationships between metals in sediments and molluscs (based mainly on field studies)

Metal	Sediment extract best predicting availability	R value
Scrobicularia plana[a–g]		
Ag	Ag in 1 M HCl	0.71
As	As/Fe in 1 M HCl	0.91
Cr	Cr in HNO_3 extract	0.91
Hg	Hg in HNO_3 extract/% organics	0.80
Pb	Pb/Fe in 1 M HCl	0.80
Sn	Sn in 1 M HCl	0.87
TBT	Sn in hexane/HCl extracts	0.81
Zn	Zn in 1 M NH_4OAc	0.62
Macoma balthica[a–g]		
Ag	Ag in 1 M HCl	0.83
As	As/Fe in 1 M HCl	0.94
Cd	Cd in 1 M HCl	0.46
Hg	Hg in HNO_3 extract/% organics	0.83
Pb	Pb/Fe in 1 M HCl	0.76
Zn	Zn in 1 M HCl	0.56
Elliptio complanata[h,i]		
Cu	Cu/Fe in 0.04 M $NH_2OH.HCl$	0.97
Zn	Zn/Fe in 30% H_2O_2 and 0.04 M $NH_2OH.HCl$ respectively	0.88
Pb[l]	Pb/Fe in 0.04 M $NH_2OH.HCl$	0.96
Anodonta grandis[h,i]		
Cu	Cu/Fe in 0.04 M NH_2 OH.HCl	0.92
Zn	Zn/Fe in 30% H_2O_2 and 0.04 M NH_2 OH.HCl, respectively	0.90
Pb[l]	Pb/Fe in 0.04 M NH_2 OH.HCl	0.87
Mytilus edulis[j]		
Pb	Pb in 30% H_2O_2 (pH 2)/Total S	0.916
Pb	Pb in 1 M NaOAc/Fe in 0.04 M $NH_2OH.HCl$	0.896
Velacumantis australis[k]		
Pb	Pb in 0.05 M EDTA	0.98
Zn	Zn in 0.05 M HCl	0.99
Cd	Cd in 0.05 M EDTA	0.92
Pyrazus ebeninus[k]		
Cu	Cu in 0.05 M EDTA/TOC	0.81
Pb	Pb in 0.05 M HCl	0.87
Zn	Zn in 0.05 M EDTA/TOC	0.72

Sources: [a]Bryan and Langston, 1992; [b,c]Luoma and Bryan, 1978, 1982; [d–g]Langston 1980, 1982, 1986, 1990; [h,i]Tessier *et al.*, 1984, 1993; [j]Bourgoin *et al.*, 1991; [k]Ying *et al*, 1992. Relationships refer to whole organism except [l]gills.

mediating uptake. Thus, for many of the examples in Table 8.1, metal burdens in molluscs [M$_b$] are best described by the linear equation:

$$[M_b] = (m[Ma]x/[x]) + c \tag{8.2}$$

where [*Ma*]*x* is the concentration of metal associated with sediment component *x*; [*x*] is the concentration of the metal-binding component *x*; *m* and *c* are slope and intercept.

Normalizing routines quantitatively account for the influence of these major geochemical parameters and indicate that metal impact in benthic molluscs is unlikely to be the same in all sediments. The success of extractable Fe and organic content as normalizers (Table 8.1) reflects their importance in the partitioning of adsorbed (non-detrital) fractions in oxidized surface sediments, which commonly form a major part of the diet of deposit and suspension feeders. Intuitively, one would expect sulphides to exert more control under anoxic conditions and they might conceivably determine bioavailability of metals from reducing sediments (see below); but the validity of traditional extraction schemes is questionable in these circumstances and the relevance of buried sediments as a metal source for deposit feeders is largely unknown.

Despite the excellent correlations between tissue burdens and sediment fractions that are sometimes achieved (Table 8.1), there is still much speculation as to the mechanisms of uptake and accompanying geochemical controls. The simplest explanations assume that assimilation of particulate metal takes place in the gut and that metal-binding sediment phases either compete with uptake sites in the digestive epithelium or render the metal less labile. Nevertheless, even in deposit-feeding clams such as *Scrobicularia plana* and *M. balthica*, which are thought to derive most metals from ingested surface sediments, accumulation of some metal from interstitial and overlying water cannot be ruled out (Langston and Spence, 1995). It can be argued from a regulatory standpoint that it may not be important to separate uptake sources since, if the system is at adsorptive equilibrium, the amount of soluble metal [M] will be a function of that which is adsorbed [M*a*] to a complexing sediment phase *x* (i.e. [M*a*]*x*), together with the concentration of the solid phase(s) responsible for binding that metal [*x*]. The ratio [*Ma*]*x*/[*x*] (also used in equation 8.2 as a predictor of bioavailable sediment metal) would therefore represent a surrogate of dissolved metal according to the equilibrium model (Tessier *et al.*, 1993). (Note: if more than one phase is important in complexation then additivity of this term, for each phase, is assumed.)

Consequently, the normalized sediment parameters displayed in Table 8.1, such as *As/Fe* and *Hg/% organics*, are plausibly a function of dissolved As and Hg, respectively, as well as representing bioavailable sediment fractions – provided the system is at equilibrium. This hypothesis receives some support from experimental studies with artificially produced sediment (Luoma and Jenne, 1977) which indicate that sedimentary sinks from which metal bioaccumulation (in *M. balthica*) is greatest also exhibit the greatest rates of sediment-to-water desorption (lowest K_d values). Where such sinks occur in the field, bioavailability of particulate metals could be enhanced both through increased assimilation from ingested material and from the higher concentrations of dissolved metal that arise from the increased desorption from sediment.

Interestingly, in an extensive field study of Cd accumulation in *A. grandis* from lakes of varying pH and metal contamination, tissue burdens are indicated to be closely related to dissolved Cd – notably, as mentioned previously, the computed activity of the free ion (K Cd^{2+}; $r = 0.9$, $P < 0.001$; Tessier *et al.*, 1993). Using data from the same study, comparable predictions of Cd bioavailability can be made based on Cd partitioning into particulate Fe (oxyhydroxides) and organics, though the required models in this example are somewhat more complex than the simple normalizations described above: Cd uptake in *A. grandis* is pH dependent and the latter variable needs to be incorporated into the equation (Tessier *et al.*, 1993). In this example, measurements of dissolved Cd might be expected to be the best predictor of bioavailability, since the particle reactivity of this metal is relatively low. Assumption of steady state conditions would be less realistic for metals whose behaviour is dominated by sediment geochemistry, and kinetic models (see below) may be suitable alternatives, or companions, to equilibrium partitioning in the indirect assessment of assimilation pathways.

Not surprisingly, attempts to predict metal bioavailability, based on equilibrium partitioning, are sometimes unsuccessful. Cu levels in clams (*S. plana* and *M. balthica*) and various sediment extracts are not always significantly related, even when normalized with respect to Fe or organics, and other geochemical parameters are presumed to have a dominant influence on Cu accumulation. These are probably a function of redox, together with the presence of sulphides in the sediment, since anomalously high levels of Cu are observed in clams from relatively anoxic sites even though sediment contamination is not evident (Bryan and Langston, 1992). The presence of high levels of (dilute HCl-extractable) acid volatile sulphide (AVS) in sediments is known to modify the impact of Cu and Ni in amphipod bioassays (Di Toro *et al.*, 1992) though in this case the higher levels of sulphide reduce bioavailability and toxicity (where S : metal ratio > 1, metal is assumed to precipitate as insoluble, unavailable, metal sulphide). High total S content of sediments (close to a Pb/Zn smelting complex) also reduces Pb availability to mussels *M. edulis* in a predictable way (Bourgoin *et al.*, 1991) though whether this is a competitive phenomenon or the result of the anomalously sulphide-rich nature of ore-impacted sediment is unknown. The enhanced Cu burdens observed in *S. plana* and *M. balthica* from anoxic field sites await further explanation. An increase in bioavailable cupric ions during bouts of anoxia is possible; alternatively, enhanced Cu burdens could result from immobilization in tissues, as metal sulphide, following an influx of hydrogen sulphide; complexation with metal ions could even be a means of detoxifying this excess H_2S.

In summary, metal levels in benthic molluscs from different sites are seldom accurately predicted from standard measurements of total sediment metal, but are subjected to a variety of geochemical influences. Simplified, the bioavailability of metals in anoxic (usually subsurface) sediments appears to fall under the control of sulphide reactivity: metal ions whose solubility is less than FeS (e.g. Cd, Ni, Cu, Zn, Hg and Pb) may precipitate as insoluble metal sulphides which are relatively unavailable, at least in laboratory assays (see Di Toro *et al.*, 1992). In nature, the picture may be more complex and we need a better insight as to whether or not subsurface, reducing sediments represent an important source of uptake, for a range of species. This is more certain for aerobic surface sediments where metal availability is often determined, predictably, by adsorption/desorption characteristics on Fe/Mn oxyhydroxide and organic coatings. However, equilibrium partitioning models are not universally applicable and do not necessarily help to distinguish between uptake from solid or solute phases, which may vary according to feeding strategy and habitat. Here, kinetic models, where individual pathways are treated additively, assume increasing importance.

8.2.3 SEPARATING DIETARY AND SOLUTE PATHWAYS OF METAL ASSIMILATION : KINETIC MODELS

The relative importance of food or water as metal sources depends on the feeding habit of the mollusc, combined with the physical and chemical form of the metal in question. (An example of this, discussed later with respect to Figure 8.4, concerns particulate Ag in the East Looe estuary that is readily available to the deposit-feeding clam *S. plana* but not to the suspension-feeding mussel *M. edulis*). Thus, although metals in solution are usually considered to be more bioavailable than solid-phase metal, the higher concentrations in the latter often render dietary vectors more important.

The distinction between uptake routes is poorly categorized for the majority of molluscs and is often blurred for a number of reasons. For example, dissolved metals can become incorporated into the mucus layers on the gills of filter feeders and ultimately become available for assimilation in the gut, whilst metals in imbibed water can also be absorbed by the digestive epithelium. It is difficult to conceive that any aquatic type (including a deposit-feeder) avoids at least some uptake from ambient water; equally, it is hard to imagine that heterotrophic molluscs do not assimilate at least a small fraction of the metal content of their food. To date, most of our knowledge of pathways reflects this uncertainty. Phillips (1979) has suggested that poor correlations between metal levels in mussels and surrounding water are an indication that food (phytoplankton) is an important source for most metals, other than Cd. Dietary uptake has been proposed in a number of other studies with mussels (Fe, Hg), clams (Ag), winkles (As, Zn) and oysters (Cr), though often these comprise indirect estimates at best (reviewed by Luoma, 1983). Thus, metal uptake in filter-feeding bivalves is frequently correlated with feeding activity (indicative of dietary uptake) but it is not always certain whether metals are directly assimilated from food or, adventitiously, from solution as a result of enhanced filtration (reviewed in Langston and Spence, 1995).

Further indirect evidence of assimilation pathways is invoked from tissue distribution, which may vary with the source of metal. Field data suggest that most elements partition preferentially towards internal tissues such as digestive gland and muscle when accumulated from food, and to gills and mantle when the major vector is water. This is supported by a limited number of experimental observations: for example, bacterially associated ^{57}Co, in contrast to dissolved ^{57}Co, is incorporated preferentially into digestive gland of *M. balthica*, confirming that the chemical form of assimilated metal influences tissue distributions (Harvey and Luoma, 1985).

Observations of metal uptake routes in carnivorous molluscs are rare and available evidence is based largely on tissue distributions. On this basis, species that are high in the food chain, including cephalopods *Eledone cirrhosa* (octopus) and *Sepia officinalis* (cuttlefish), appear to derive much of their metal from food: > 80% of the body burden of Ag, Cd and Co, and 40–80% Cr, Mn, Ni, Pb, V, Zn, Cu and Fe is situated in the digestive gland (Miramand and Bentley, 1992). Similar tissue distributions occur in the carnivorous gastropod *Nucella lapillus* and there is some experimental evidence confirming the importance of dietary metals. In the field, body burdens in *N. lapillus* appear to be influenced (by up to two-fold) by the nature and metal-accumulating ability of prey species (barnacles, limpets or mussels) which, in turn, reflect levels of environmental contamination (reviewed by Bryan *et al.*, 1985).

There have been very few comprehensive assessments of metal assimilation pathways in molluscs. Nevertheless, it is important to attempt quantification of uptake and loss from sediment/food and water vectors for at least a few key species, in order to supply accurate data for kinetic models that predict bioaccumulation and transfer through aquatic ecosystems. Direct methods employed previously include the use of dialysis bags and filter chambers. Animals housed within these devices, which exclude particulates, accumulate metals from solution only, whilst those outside can derive metals from both soluble and particulate vectors. Using such an approach for the clam *M. balthica*, solute uptake is shown to account for the majority of Zn (77%) and Cd (61%) accumulation, whilst Co uptake is largely (> 80%) from particulates (suspended labelled bacteria; Harvey and Luoma, 1985). This variation is, perhaps, predictable in environmentally controlled experiments, based on the partitioning characteristics of the metals involved, but it is not necessarily indicative of behaviour under complex natural conditions. More research on this topic is needed in order to provide reliable budgets of metal uptake from different sources.

An indirect guide to the relative importance of uptake routes involves the comparison of organism concentration factors ($CF_{organism/water}$) derived from laboratory exposures to dissolved metals or radioisotopes with field values ($\Sigma \, CF_{organism}$). The difference, as indicated by equation 8.3, may represent the contribution from food which, in studies with mussels at least, appears important for Cd, Zn, Co, Fe and Se (reviewed by Fisher and Reinfelder, 1995). The technique is, inevitably, an approximation, due to assumptions of steady state and linearity across all concentration ranges. This is particularly problematic when considering elements that may be regulated. A more valid if still indirect approach to quantifying pathways involves the inclusion of kinetic parameters of uptake and loss, incorporating a term for assimilation efficiency of ingested food (Fisher and Reinfelder, 1995). Starting with the premise that:

$$\Sigma \, CF_{organism} = CF_{organism/water} + CF_{organism/food} \qquad (8.3)$$

the steady-state concentration factor for metal accumulated from the dissolved phase ($CF_{organism/water}$) can be measured experimentally as uptake rate of dissolved metal ($\mu g \ g^{-1}$ per day) $\div$ elimination rate (d^{-1}). The steady-state concentration factor from food ($CF_{organism/food}$) is equivalent to fCF_{food}, where CF_{food} is the equilibrium concentration factor for the metal in food (with respect to that in surrounding water) and function f is expressed as:

$$f = (AE \times I)/(k + G) \qquad (8.4)$$

where metal assimilation efficiency (AE) = % ingested metal retained, I = ingestion ratio of food $\div$ weight of animal, k = turnover rate (d^{-1}) and G = growth constant (d^{-1}).

Thus, by assuming additive behaviour for metal kinetics in the different vectors, model predictions can be applied to the field to assess, indirectly, the relative impact of different pathways. Accuracy has been improved by recently developed techniques to quantify assimilation efficiencies (AE) for solid-phase metals – a major variable affecting metal accumulation from the diets of sediment-dwelling bivalves (Decho and Luoma 1991; Luoma *et al.*, 1992). Traditionally this has involved difficult measurements of ingested and egested material in order to derive mass balances. Using more accurate methods based on dual-label, pulse-chase experiments, assimilation of an isotope X can be compared with that of a relatively inert isotopic form, Y (e.g. ^{241}Am), eliminating the need for mass-balance recoveries:

$$AE_X = [(X/Y \text{ in food}) - (X/Y \text{ in faeces})]/(X/Y \text{ in food}) \qquad (8.5)$$

Highly efficient uptake of Se ($AE = 93\%$) from food (particularly diatoms) by *M. balthica* has been demonstrated by this technique: extrapolation to contaminated sites in San Francisco Bay enabled Luoma *et al.* (1992) to highlight the resultant hazards of accumulated Se burdens in *M. balthica* to consumers such as diving ducks. Because uptake of Se by clams is predominantly from particulates (99%), predictions of body burdens based solely on concentration factors for soluble Se would not have revealed such a threat, clearly illustrating the need for accurate identification of exposure routes.

A key factor in modifying *AE* is the partitioning of the element in the food of molluscs. Thus, the ease of transfer of metals from phytoplankton (*Isochrysis galbana*) to larval clams (*Mercenaria mercenaria*) and oysters (*C. virginica*) is directly related to the proportion present in the easily digested cytoplasmic fraction of the algae: elements bound to cell walls are more likely to pass through the digestive tract unassimilated (Fisher and Reinfelder, 1995). In oyster larvae, efficiency of assimilation (%) increases in the sequence Am (8) < Co (19) < Ag (33) < Cd (61) < Zn (79) < Se (97), and for clam larvae Am (5) < Ag (22) < Co (26) < Cd (33) < Zn (41) < Se (100). This pattern is not dissimilar to that observed in adults (and other bivalves, including mussels and deposit-feeding clams) and suggests that essential elements (e.g. Se and Zn) have the highest *AE*. In some species Cd, Ag and, to some extent, Zn are assimilated more readily in adults than in larvae, possibly due to a longer gut retention time for food, coupled with a more rigorous digestive regime which is capable of attacking some of the more refractory forms (Fisher and Reinfelder, 1995).

Unassimilated elements, whether bound to phytoplankton cell walls, granules (see Chapter 12) or other non-cytoplasmic debris, will become enriched in the faeces of molluscs (relative to their diet), as demonstrated by Brown (1986) for benthic sediment-feeding gastropods (*Hydrobia ulvae*) and bivalves (*Mya arenaria, Cerastoderma edule, S. plana, M. balthica*) . This enhancement may be magnified by the selection, by filter feeders, of certain particle types, relative to bulk sediments, and illustrates the importance of food quality in determining metal accumulation. For example, the presence of adherent bacteria and extra-cellular polymers on the sediment particles selected by *M. balthica* enhances the digestibility and bioavailability of Ag, Cd and Zn (Harvey and Luoma, 1985). Mercury-resistant bacteria also have a stimulating influence on Hg uptake by oysters (*Crassostrea virginica*). The presence and quality of food can sometimes directly influence accumulation of metals in filter-feeding bivalves even though the probable vector for uptake is water: a twofold increase in Cd concentrations has been observed in the gills of *Mytilus edulis* as a result of increased ventilatory activity in the presence of food (Riisgard *et al.*, 1987). A similar explanation may account for faster rates of Cd and Zn loss from fed vs. starved clams (Luoma and Jenne, 1976).

Clearly, *AE* is element specific and dependent on food/sediment type, together with the digestive strategy of the mollusc under investigation. Comparisons of ingestion, retention and egestion rates of metal and food material (^{51}Cr- and ^{14}C-labelled bacteria) in suspension-feeding and deposit-feeding bivalves (*Potamocorbula amurensis* and *M. balthica*, respectively) reveal major differences in the relative contributions of extra- and intracellular digestion between the two feeding types. The first phase (extracellular digestion followed by intestinal uptake) is relatively short and is more efficient at digesting bacterial cells in the deposit feeder *M. balthica* than in *P. amurensis*, though little metal (^{51}Cr) assimilation takes place during this phase. The second, slower phase (intracellular digestion) is quantitatively more significant in absorbing nutrition (^{14}C) and metal (^{51}Cr): during this stage ^{51}Cr absorption in *P. amurensis* is double that of *M. balthica*, primarily because more food is processed. Offsetting this, ^{51}Cr loss rates, following extra- and intracellular phases, are lower in *M. balthica* than in *P. amurensis* (Decho and Luoma, 1991).

Digestive processing and release strategies of ingested particles therefore determine accumulation and response to metals. Because of the more efficient absorption of Cr during intracellular digestion, *P. amurensis* might be suspected of being more vulnerable to contamination (Decho and Luoma, 1991); put another way, the slow assimilation of metal by the detrital feeding infaunal clam *M. balthica* could be an adaptation to avoid toxicity in its naturally metal-enriched sedimentary environment. Experiments that indicate modification of digestive strategies in molluscs exposed to metal (Cr), notably a reduction in the proportion of particles subjected to intracellular digestion (together with a reduction in ingestion rates), appear to support the idea of such an adaptive mechanism (Decho and Luoma, 1996). The influence of food quality on assimilation is also well illustrated in this example, ^{51}Cr^{3+} assimilation being much more efficient from bacteria and polymers than from diatoms or sediments.

Comparable studies with *P. amurensis* and *M. balthica* using ^{109}Cd show that *AE* values vary between metals and particle types (Fe oxyhydroxides or silica particles). Values for Cd range between 9 and 56%, whilst those of Cr^{3+} are consistently < 11%, irrespective of particle type (Decho and Luoma, 1994). In the mussel *M. edulis* fed on labelled diatoms, *AE* is highest for Se (70–80%), intermediate for Co, Cd and Zn (15–50%) and low for Ag and Am, though in all cases *AE* increases linearly with ingestion rate and food concentration (Wang *et al.*, 1995). Interestingly, the same paper gives a possible insight into the mechanism of Zn regulation in mussels, by demonstrating that *AE* for Zn decreases when the metal is abundant. From modelling, the primary uptake route for this metal is indicated to be the diet (80%).

Recent techniques have thus improved our ability to predict metal bioavailability in molluscs. Both equilibrium partitioning and kinetic models are achievable and realistic, though they require painstaking efforts in data collection for validation: for example, each kinetic model should contain reasonable estimates of organism–metal fluxes and assimilation efficiencies for each vector – including interstitial and surface water, together with particulate food source(s).

8.2.4 BIOACCUMULATION VS. REGULATION

The underlying control on tissue burdens is determined by the overall balance between inward and outward fluxes of metal. By sequestering metals within cells, the free metal ion concentration can be kept below that outside, so maintaining the gradient for continuing diffusion inwards. This is an important requirement for essential elements but can become problematic for 'pollutant' metals (e.g. Ag, Cd, Hg) or when essential metals reach contaminant proportions in the environment: normal sequestration systems may then need to assume detoxifying roles.

Elimination of most metals can be demonstrated by moving molluscs from contaminated to low-metal environments, either in the laboratory or the field. Observed rates are species, metal and, sometimes, site specific. Loss may occur as a result of desorption from the entire body surface, excretion through permeable membranes such as those of the gill and kidney (both usually rapid), and by egestion of unassimilated material and exocytosis of metal-rich concretions into the digestive tract or urine (often slow): thus, even metals that are stored in immobilized form – as granules or membrane-lined vesicles – may be lost from the organism, eventually. Relative combinations of these pathways, some of which are described in later sections, will dictate the overall kinetic pattern of elimination.

Generally, metals that are exchanged rapidly are accumulated far less effectively than those that are exchanged slowly. For most molluscs, metal uptake is often faster than elimination and so accumulation usually proceeds in proportion to external concentrations. Metal regulation is achieved when elimination and uptake rates are balanced across a range of ambient concentrations, such that net bioaccumulation does not ensue (see also Chapter 9). It is one means by which organisms are able to tolerate a potentially deleterious influx of metal, though regulation is the exception rather than the rule for molluscs: where present, regulatory powers tend to be limited compared with those displayed by polychaetes, crustaceans and fish. Indeed, active, inducible efflux systems for group B metals have yet to be demonstrated widely in molluscs, perhaps explaining why they are a relatively metal-sensitive taxon – one of the first to be eliminated from heavily polluted environments. This limited regulatory ability is one of the main criteria for including molluscs in biomonitoring schemes, which require a linear response between body burdens and environmental contamination levels. Uptake of non-essential elements in molluscs is almost universally determined by the degree of exposure. In contrast, body burdens of some essential metals may be less influenced by external concentrations, suggesting varying degrees of homeostasis. This is most evident in gastropods – for example, winkles (*Littorina* spp.), limpets (*Patella* spp.) and the dogwhelk *N. lapillus* (Bryan *et al.*, 1985) – whose Zn burdens are often largely independent of environmental levels and whose Cu loading may be buffered according to requirements for the Cu-containing respiratory pigment, haemocyanin (HCY). Zn and Cu homeostasis is indicated in some bivalve species, including *Mytilus edulis* and *Perna viridis* (*Mytilacea*), though Cu regulation, unlike Zn, is partial and uptake is not totally unrelated to exposure, particularly at higher concentrations (Bryan *et al.*, 1985; Chan, 1988).

Most of the evidence for metal regulation in molluscs is indirect and stems from observations of non-linearity between body burdens and environmental contamination. Detailed studies are now required to demonstrate the mechanisms involved. There are two basic possibilities by which molluscs might tolerate an excessive influx of metal. The first involves immediate elimination (or exclusion) of metal in a dose-dependent manner, i.e. the classical notion of regulation. A second strategy is to store surplus metal in a metabolically inert form. To some extent these processes overlap, as both suppress biochemically reactive forms; furthermore, storage is often reversible in the case of essential metals – mobilization occurring in response to metabolic requirements – though in other circumstances, storage can lead to elevated burdens and is thus not considered as true regulation.

Uptake and accumulation of organometals do not appear to be regulated, but some species are capable of reducing the toxicity of, for example, methylmercury, by conversion and storage in inert – usually inorganic – form (Chapter 5). Tributyltin (TBT) is detoxified by stepwise debutylation (eventually to inorganic Sn), involving the formation of water-soluble hydroxylated metabolites – mediated by the microsomal cytochrome P450 monooxygenase system at rates which are species specific (reviewed in Langston, 1996). This can give rise to characteristic differences in body burdens and ratios of TBT : metabolites, as illustrated by a comparison of estuarine bivalves. Thus, organotin in *Mercenaria mercenaria* and *Mya arenaria* is principally as TBT, the parent compound, with limited signs of breakdown to dibutyltin (DBT); consequently, body burdens are high in these organisms. In contrast, some 70% of organotin in another sediment-dwelling clam (*Petricola*) is present as the metabolite DBT – indicative, along with the correspondingly low organotin burdens, of rapid degradation. Differences in metabolic activity may, therefore, contribute to the large variations in CF_{TBT} observed between species such as *Mya* (500 000) and *Petricola* (12 000), even though these species occupy almost identical ecological niches.

8.2.5 ALLOMETRIC CONSIDERATIONS

Metal concentrations are a function of net metal content, expressed on a weight basis; therefore any change in growth or condition, relative to net accumulation, will influence measured body concentrations. The importance of allometric parameters on metal concentrations in molluscs is exemplified by field studies (e.g. Boyden, 1977) relating body weight (W) and metal content (Y). Relationships may often be described using the transformed growth equation:

$$\log Y = \log a + b \log W \qquad (8.6)$$

If the slope (b) = 1, accumulation and turnover of the metal is taking place at a rate that is proportional to growth, implying a connection with metabolism. If the slopes are < 1, accumulation is more rapid in smaller animals, relative to growth, possibly because of the larger ratio of surface area to volume. A slope > 1 suggests net accumulation throughout the life of the organism, which does not reach steady state: growth may be rapid in young individuals, effectively diluting metal concentrations, but slows down with age and metal is incorporated at a faster rate than new tissue.

The majority of slopes for Cd, Co, Cu, Fe, Mn, Ni, Pb and Zn were found to be near 1 (metal content independent of size) or < 1 (contents higher in smaller individuals) in Boyden's studies of oysters (*Ostrea edulis, Crassostrea gigas*), clams (*Mercenaria mercenaria, Venerupis decussata*), scallops (*Chlamys opercularis, Pecten maximus*), mussels (*Mytilus edulis*) and gastropods (*Crepidula fornicata, Patella* spp., *Littorina littorea, Buccinum undatum* and *Scaphander lignarius*). Subsequent work with the clam *Scrobicularia plana* has shown that size/concentration relationships may vary markedly between locations, reflecting site-specific differences in growth rates. Relationships can even change over the organism's life span, indicating that the rate processes governing metal content are not necessarily uniform over time. Furthermore, since size relationships for different metals can vary so greatly, there would appear to be a variety of subtle factors in operation, rather than just a straightforward metabolic explanation for size effects. The scale and form of contamination contributes to the variability imparted by spatial variations in growth and, in general, relationships between size and metal content are more positively sloped (i.e. fail to achieve steady state) in metal-rich environments (reviewed in Langston and Spence, 1995). Compensation for growth-related variability in metal concentrations is now a major consideration in improving protocols for monitoring (Chapter 7).

8.3 SUBCELLULAR COMPONENTS OF METAL METABOLISM

The chemical properties of the elements have been harnessed to fulfil a variety of catalytic, enzymatic and structural roles and their beneficial (or adverse) effects are a reflection of their interaction with cellular components. Zn is the most universally important and is involved with virtually all aspects of metabolism. In addition to its role in catalysis and the configuration of regulatory DNA-binding proteins, Zn stabilizes the structure of a variety of proteins and organelles, participates in transport processes and is involved in immuno-response (Vallee, 1991). In general, Class A metals predominantly form ionic compounds (associated with O- and N-based ligands); class B metals tend to form metal–ligand complexes (notably with S); borderline metals display intermediate properties of both these classes. In most cases, deleterious biological effects are the result of antagonistic behaviour, where the exceptional presence of one metal alters the requirements for another. While we are accustomed to the concept of non-essential metals such as Hg, Cd and Pb having no known biological role, *in vitro* studies with mammalian preparations indicate activation of some metal-dependent enzymes by these pollutant metals at low doses: conversely, intracellular Cu and Zn , despite being needed for upwards of 60 enzymes (Dixon and Webb, 1979), can prove highly toxic if unregulated. The distinction between essential and non-essential metal is therefore not always clear-cut and is perhaps better described as a continuum, dependent on the concentration and form of metal association.

Molluscs, like other organisms, have evolved a number of subcellular systems for accumulation, regulation and immobilizion of essential metals during phases of excess (George *et al.*, 1979; Viarengo and Nott, 1993); indeed, of all aquatic invertebrates, molluscs probably exhibit the most extensive range of ligands. These commonly involve soluble (cytosolic) molecules such as metallothionein (MT) and other metalloproteins, or compartmentalization into lysosomes, granules and membrane-bound vesicles. The same ligands may be involved in immobilizing non-essential elements. The properties and ecotoxicological implications of sequestration systems, and the variation displayed among different molluscs, are reviewed here; the extent to which metal-binding characteristics affect trophic transfer is discussed in Chapter 12.

8.3.1 METALLOTHIONEIN (MT)

MT was first reported almost 40 years ago as the Cd-, Cu- and Zn-binding protein responsible for metal accumulation in the horse kidney (Margoshes and Vallee, 1957). Mammalian MT has a molecular weight of 6800 Da, based on amino acid analysis (60–62 amino acid residues), and an apparent molecular weight of about 12–15 kDa based on size-exclusion chromatography, due to its prolate ellipsoid shape; metals are co-ordinated with cysteine in two metal thiolate clusters (a total of 6–7 gram-atoms Cd or Zn mol^{-1} MT; George and Langston, 1994). The high cysteine content (30%), low molecular weight, heat stability and non-enzymatic nature, together with a strong affinity to bind Class B metal cations such as Ag, Cd, Cu, Hg and Zn (Table 8.2) enable MT to be differentiated from most other proteins. Though occurring mainly in the cytoplasm, MT has also been detected in the nucleus and lysosomes. Several overviews on the chemistry, molecular biology, cellular physiology and toxicology of MT in aquatic organisms have been published (Roesijadi, 1992; Viarengo and Nott, 1993; Roesijadi and Robinson, 1994; George and Langston, 1994).

MT is now thought to be almost ubiquitous among aquatic organisms – based, it should be noted, on rather limited evidence: the occurrence of MT (or MT-like proteins) has been reported for some 50 different aquatic invertebrates (from five phyla), three-quarters of which are molluscs or crustaceans (Table 8.2). Although the function of MT in many organisms is still unresolved, metal-binding sequences are surprisingly well conserved, in evolutionary terms, and a fundamental involvement in metal pathways seems certain. One proposed role of mammalian MT is as an intracellular metal reservoir which is

Table 8.2 Occurences of metallothionein-like proteins in molluscs

Species	Tissue	Exposure route	Metals bound by MT	Reference
Polyplacophora				
Cryptochiton stelleri	Digestive gland	Field	Cd	Olafson *et al.*, 1979
Gastropoda				
Crepidula fornicata	Whole animal	Lab.	Hg	Harrison *et al.*, 1987
*Patella granularis**	Whole animal	Lab.	Zn	Hennig, 1986
Patella intermedia	Whole animal	Field	Cd, Cu	Howard and Nickless, 1977
*Patella vulgata**	Whole animal	Field	Cd, Cu, Zn	Howard and Nickless, 1977
	Whole animal*	Field	Cd, Cu, Zn	Noel–Lambot *et al.*, 1980
	Viscera	Field	Cd, Cu, Zn	Noel–Lambot *et al.*, 1980
	Foot	Field	Cd, Cu, Zn	Noel–Lambot *et al.*, 1980
Littorina littorea[1]	Whole animal	Field	Cd, Cu, Zn	Howard and Nickless, 1978
	Whole animal	Field	Ag, Cd, Cu, Hg	Langston and Zhou, 1986
	Gills	Lab.	Cd	Langston and Zhou, 1987a
	Gills	Lab.	Cd, Cu	Gully and Mason, 1993
	Digestive gland	Lab.	Cd	Langston and Zhou, 1987a
	Kidney	Lab.	Cd	Langston and Zhou, 1987a
	Headfoot	Lab.	Cd	Langston and Zhou, 1987a
Nassarius reticulatus	Gills	Lab.	Cd, Cu	Hylland *et al.*, 1994
	Intestine	Lab.	Cd, Cu, Zn	Hylland *et al.*, 1994
	Hepatopancreas	Lab.	Cd, Cu, Zn	Hylland *et al.*, 1994
	Headfoot	Lab.	Cd, Cu	Hylland *et al.*, 1994
Batillus cornutus	Hepatopancreas	Field	Cd	Dohi *et al.*, 1986
*Bullia digitalis**	Whole animal	Field	Cd, Cu, Zn	Hennig, 1986
Bivalvia				
Crassotrea virginica	Whole soft tissues	Lab.	Cd	Casterline and Yip, 1975
	Whole soft tissues	Lab.	Cd, Cu, Zn	Ridlington and Fowler, 1979
	Whole soft tissues*	Lab.	Cd	Roesijadi *et al.*, 1989
	Gills	Lab.	Cd	Roesijadi and Klerks, 1989
	Gills**	Lab.	Cd	Unger *et al.*, 1991
	Embryos	Lab.	Cd, Cu, Zn	Ringwood and Brouwer, 1993
Crassostrea gigas	Digestive gland	Field	Cd, Cu	Imber *et al.*, 1987
	Whole animal	Lab.	Cd, Cu	Frazier and George, 1983
	Whole animal	Lab.	Ag	Berthet, 1990
Ostrea edulis[2]	Whole animal	Lab.	Cd, Cu	Frazier and George, 1983
	Digestive system	Field	Cd, Cu, Zn	Julsham and Andersen, 1983
	Liver	Field	Cd, Cu, Zn	Julsham and Andersen, 1983
Mytilus edulis	Whole soft tissues	Lab.	Cd, Cu, Zn	Noel–Lambot, 1976
	Whole soft tissues	Lab.	Cd	Bebianno and Langston, 1991
	Whole soft tissues*	Lab.	Cd	Mackay *et al.*, 1993
	Whole soft tissues*	Lab.	Cd, Cu, Zn	Frankenne *et al.*, 1980
	Gills	Field	Cd	Marshall and Talbot, 1979
	Gills*	Lab.	Hg	Roesijadi and Hall, 1981
	Kidney	Lab.	Cd, Zn	George and Pirie, 1979
	Mantle	Lab.	Cd	Carpenè *et al.*, 1979
	Midgut gland	Lab.	Cd	Scholz, 1980
	Digestive gland*	Lab.	Cd, Cu, Zn	George *et al.*, 1979

Table 8.2 *Continued*

Species	Tissue	Exposure route	Metals bound by MT	Reference
	Digestive gland	Lab.	Cu	Harrison and Lam, 1985
	Digestive gland	Lab.	Cd	Bebianno and Langston, 1991
	Hepatopancreas	Lab.	Cd	Nolan and Duke, 1983a
	Larvae*	Lab.	Hg	Roesijadi *et al.*, 1982
Mytilus galloprovincialis	Whole soft tissues	Lab.	Cd	Bebianno and Langston, 1992
	Gills	Lab.	Cu, Zn	Viarengo *et al.*, 1980, 1981
	Gills	Lab.	Cd	Pavicic *et al.*, 1985
	Digestive gland*[3]	Lab.	Cu, Zn	Viarengo *et al.*, 1981, 1985, 1989
	Digestive gland	Lab.	Cd	Pavicic *et al.*, 1985
	Digestive gland	Lab. and Field	Hg	Tusek-Znidaric *et al.*., 1986
	Mantle	Lab.	Cu, Zn	Viarengo *et al.*, 1981
	Muscle	Lab.	Cd	Carpenè *et al.*, 1983
	Foot	Lab.	Cd	Carpenè *et al.*, 1983
	Gonads	Lab.	Cd	Pavicic *et al.*, 1985
	Eggs	Lab.	Cd	Pavicic *et al.*, 1985
	Larvae	Lab.	Cd	Pavicic *et al.*, 1985
	Embryos	Lab.	Cd, Cu, Zn, Hg	Pavicic *et al.*, 1984
*Chromytilus meridionalis**	Whole animal	Lab.	Zn	Hennig, 1986
Dreissena polymorpha	Whole soft tissues	Lab.	Cd, Cu, Zn	Tessier and Blais, 1996
Ruditapes decussatus	Whole soft tissues	Lab.	Cd, Cu, Zn	Bebianno *et al.*, 1993
	Gills	Lab.	Cd, Cu	Bebianno *et al.*, 1993; Romeo and Gnassia-Barelli, 1995
	Digestive gland	Lab.	Cd, Cu	Bebianno *et al.*, 1993; Romeo and Gnassia-Barelli, 1995
Protothaca staminea	Gills	Lab.	Cu, Zn	Roesijadi, 1980
	Kidney	Lab.	Cd, Zn, Cu	Roesijadi, 1980
	Muscle	Lab.	Cu, Zn	Roesijadi, 1980
	Viscera	Lab.	Cu, Zn	Roesijadi, 1980
Mercenaria mercenaria	Kidney	Lab.	Cd, Cu, Zn	Robinson *et al.*, 1985
Calyptogena magnifica		Lab.	Cd, Cu	Roesijadi *et al.*, 1985
Placopecten magellanicus	Kidney	Lab.	Cd, Cu, Zn	Fowler and Gould, 1988
Anodonta grandis	Whole organism	Field	Cd	Couillard *et al.*, 1993
	Gills	Field	Cd	Couillard *et al.*, 1993
Adamussium colbecki	Digestive gland	Field	Cd, Cu	Vairengo *et al.*, 1993
Pecten jacobaeus	Digestive gland	Field	Cd, Cu	Viarengo *et al.*, 1993
Scapharca inaequivalvis	Gills	Lab.	Cd	Carpenè, 1993
	Viscera	Lab.	Cd, Zn	Serra *et al.*, 1995
	Kidney	Lab.	Cd	Serra *et al.*, 1995
	Hepatopancreas	Lab.	Cd	Serra *et al.*, 1995
	Mantle	Lab.	Cd	Serra *et al.*, 1995
Mizuhopecten yessiensis	Gills	Lab.	Cd, Zn	Evtushenko *et al.*, 1986
	Kidney	Field	Cd	Evtushenko *et al.*, 1990
	Hepatopancreas	Lab.	Cd, Cu, Zn	Evtushenko *et al.*, 1986

Table 8.2 *Continued*

Species	Tissue	Exposure route	Metals bound by MT	Reference
Tridacna crocera	Gills	Lab.	Cd, Zn	Dusquesne *et al.*, 1995
	Kidney*	Lab.	Cd, Zn	Dusquesne *et al.*, 1995
	Mantle	Lab.	Cd	Dusquesne *et al.*, 1995
Cephalopods				
Ommastrephes bartrami	Liver	Field	Cd[4]	Castillo and Maita, 1991
Sepia officinalis	Liver	Field	Ag, Cu, Zn	Martoja and Marcaillou, 1993
Nototodarus gouldi	Digestive gland	Field	Cd, Cu	Finger and Smith, 1987
Tadorodes pacificus	Hepatopancreas	Field	Cd	Dohi *et al.*, 1986

Other known metal sequestration systems: [1]phosphate granules in digestive gland cells
[2]Amoebocytes (Cu, Zn)
[3]Lysosomal Cu–MT
[4]Cd-binding proteins significantly different from mussel and oyster MT

*Amino acid characterization; **primary structure (cDNA sequence)

capable of modulating homeostasis and transfer of essential elements (Zn and Cu) to and from metalloenzymes (Kojima and Kagi, 1978). Zinc-MT is particularly efficient in donating its metal to the apoproteins of various zinc-dependent enzymes (e.g. carbonic anhydrase, alkaline phosphatase, alcohol dehydrogenase, aldolase, thermolysin), thereby restoring their activity (Udom and Brady, 1980). Significantly, the involvement of MT in Zn metabolism also concerns the metalloenzymes involved in nucleic acid transcription – for example, as a catalytically essential component of DNA polymerase (Slater *et al.*, 1971) – and perhaps extends to a regulatory role in gene expression (by virtue of its ability to remove Zn from Zn finger proteins), particularly during embryogenesis (Ohtake *et al.*, 1983; Mackay *et al.*, 1993).

According to most models of MT operation, the demand for essential metals is sustained by maintaining a small store as the thionein complex. If the metal balance is positive and binding sites on MT are fully occupied, uncomplexed metal stimulates the production of MT mRNA and further MT protein: homeostasis is achieved by regulating the free metal ion content of the cell, which still performs optimally even when faced with an excess of metal. Subsequent fate depends on the metal requirements of the animal but, eventually, the protein–metal complex degrades to release the metal for essential usage or, if not required, for partial excretion. If greatly in excess, the released metal ions may stimulate yet further MT production, implying an element of recycling (ideal for a regulatory system). Subsequently, some of this excess (including MT-bound metal) may be diverted to insoluble, less reactive stores, such as the various concretions described below and in Chapter 12.

Similar pathways apply to pollutant metals, though of course there is no minimum requirement for their presence in cells. Nevertheless, because of their chemical similarity to essential metals and passive assimilation, elements such as Hg, Cd and Ag will compete for prosthetic groups on enzymes and other metalloproteins, whose conformation and activity will eventually be impaired in conditions of increased bioavailability (such non-specific binding is the principal toxic mode of action of pollutant metals). At low concentrations, tolerance may be acquired by sequestration with pre-existing or newly induced MT, but the efficiency of the system eventually declines as contamination increases and deleterious effects may follow.

The functions of MT therefore contain elements of homeostasis and detoxification. The latter was considered, initially, to be the primary role of MT because of its unique status as a Cd-containing protein. However, there is now evidence that some Cd-MTs originate as Cu- and Zn-MTs in unstressed animals (displacement occurring in response to intracellular Cd), supporting the concept of dual functionality. Aquatic ecotoxicologists have subsequently continued to devote much effort to clarifying this particular dilemma. The following is a synthesis of current understanding of the occurrence and role(s) of MT in metal metabolism in different molluscs.

The first invertebrate MTs to be described were those of the oyster *C. virginica* and mussel *M. edulis* (Casterline and Yip, 1975; Noel-Lambot, 1976). Since then, MT-like proteins have been reported in a range of aquatic mollusc tissues, binding several essential and non-essential metals (Table 8.2). Few molluscan MTs have been rigorously characterized, though evidence from amino acid sequencing indicates that those of *M. edulis, C. virginica* and the gastropod *Patella vulgata* are Class I MTs, exhibiting considerable similarity to mammalian MT (Noel-Lambot *et al.*, 1980; Unger *et al.*, 1991; Mackay *et al.*, 1993). Oyster MT is perhaps closest in structure to that of vertebrates: all cysteines in the first 27 residues of the oyster MT align with those in the mammalian form (Roesijadi *et al.*, 1989; Unger *et al.*, 1991). The lack of methionine is consistent with mammalian MT, but the high glycine content is unusual and may represent a characteristic feature of molluscan MT (Roesijadi, 1986).

Molecular mass variants of 10 and 20 kDa, suggestive of monomeric and dimeric forms (MT10 and MT20, respectively), are present in several molluscan species. In mussels, dimerized MTs have two additional cysteine residues which may be involved in linking monomers or in the provision of additional chelating capacity (Roesijadi, 1992; Mackay *et al.*, 1993). MT exists predominantly in monomeric form in the embryos of some oysters (e.g. *C. virginica*), in contrast to the dimeric form found in parental stock – possibly signifying specific but as yet unknown biological functions.

Multiple isoforms have been identified, either through isolation and purification of the protein, detection of multiples genes, or separation of cDNA sequences that code different MTs (Roesijadi *et al.*, 1988; Unger *et al.*, 1991;

Unger and Roesijadi, 1993). Polymorphisms appear to be particularly frequent in some molluscs : the nine Cd-induced MT isoforms identified in *M. edulis* are the highest number recorded in any invertebrate species, whereas polymorphism is less common in mammals. This may reflect the greater incidence of heterozygosity in invertebrates, such as mussels, or the fact that a large number of copies are required to provide an adequate response to metals (Mackay *et al.*, 1993).

(a) Functions of MT: constitutive and detoxifying roles

Functions of MT, as described earlier, include: the control of intracellular availability of essential metals to satisfy the requirements of metal-dependent components (metalloenzymes, respiratory pigments, nucleic acids and membranes); and gene regulation (Roesijadi, 1994b). Thus, MTs are capable of donating Cu or Zn to appropriate receptor molecules (metalloenzymes and transcriptional factors), thereby regulating metal-dependent activities which are essential for cell growth and development, through highly specific molecular interactions (Roesijadi, 1994b). Functions also include detoxification: faced with an excess of essential or pollutant metal, binding to MT offers protection against toxicity by limiting metal availability at inappropriate sites. MT can also be involved, indirectly, with elimination: for example, Hg is eliminated as Hg-MT from exposed mussels, via gametes (Roesijadi *et al.*, 1988). Processes that result in increased capacity for MT induction, gene amplification and gene duplication will, theoretically, confer cells or individuals with increased resistance to metal toxicity.

Thus two MT 'pools' may be envisaged: one (constitutive) represented by basally synthesized protein involved in essential-metal regulation, particularly of Cu and Zn; the other, characterized by induced protein, involved in metal detoxification (Roesijadi, 1994b). The continuing debate over the relative significance of these two pools can be resolved partially if it is assumed that different isoforms are involved in binding different metals, and therefore fulfil individual roles in metal metabolism. In oyster embryos there is evidence of both constitutive and detoxifying isoforms (see below) though this has not been widely demonstrated in other species. The issue may be clouded by competition and displacement reactions between metals: the question of dual functionality of the protein will require more extensive study before generalizations can be made.

(b) MT induction in molluscs and responses to metal exposure

MT induction can be quantified as the concentration (or rate of formation) of the MT mRNA, as MT protein concentration, or as a function of the levels of MT-bound metals. Each of these processes provides different information on the inductive process and may display different dynamics (Roesijadi and

Robinson, 1994; Roesijadi, 1994b). MT induction has been demonstrated in individuals from polluted populations or following laboratory exposure to metals such as Ag, Cd, Cu, Hg and Zn (Table 8.2) and appears to be a response common to most molluscs. The degree of MT expression may vary considerably between species (and may even be undetectable in some pectinid and tellinid bivalves) and the same variation is apparent between tissues: MT involvement in metal sequestration is most evident in gills, digestive gland and kidney, reflecting the significance of these tissues in uptake, storage and excretion of metals.

Induction of MT is regarded as a specific biochemical ('biomarker') response to metal exposure, though there are several other intrinsic and extrinsic (biotic and abiotic) factors which influence MT levels and, hence, metal metabolism in tissues. For example, both Cu- and Cd-binding characteristics in *C. virginica* change according to season and reproductive status: pre-spawning/spawning phases (during spring and summer) are the times of greatest flux and are presumably best avoided from a monitoring standpoint (Fowler *et al.*, 1986; Engel, 1988). Significant changes in metal-binding behaviour also take place during development; thus, in contrast to adults (where MT involvement is often limited), amounts of Cu, Cd and, especially, Zn associated with MT in larval oysters far exceed those present in other pools (Ringwood and Brouwer, 1993). Changes due to ontogeny are not always uniform: amounts of MT in limpets are positively correlated with age/size, a reversal of the trend seen in oysters (Noel-Lambot *et al.*, 1980). With these and other sources of variation in mind, summaries of the characteristics of MT, including potential value as an indicator of metal exposure, are presented below for some of the best-studied molluscs.

Oysters (Bivalvia)

Metal-binding has been investigated in several oyster species and results indicate the formation of MT-like proteins capable of binding Ag, Cd, Cu and, occasionally, Zn (Table 8.2). Based on amino acid analysis, the mass (7.5 kDa) and structure of *C. virginica* MT resembles that of vertebrates (64–83% corresponding residues and 29% cysteine; Unger *et al.*, 1991); 10 and 20 kDa variants have been observed, suggesting some dimerization. MT is induced by Cd exposure in *Crassostrea gigas* and *O. edulis,* in a dose-related fashion but at rates that are species specific and dependent on previous metal exposure (Frazier and George, 1983). Thus, the protein has only a minor role in sequestering Cd, Cu and Zn (< 5%) at highly contaminated sites. This is illustrated in Figure 8.1 by metal profiles and MT measurements in *O. edulis* collected from Restronguet Creek, UK, which show that cytosolic Cd, Cu and Zn are predominantly associated, in high concentrations, in a readily exchangeable form, with very low molecular weight ligands (< 5 kDa). Interpretation of the involvement of MT in metal-binding under such circumstances is further complicated by an apparent competitive interaction between Cd and Cu for binding sites (Frazier and George, 1983). Cu alone does not appear to induce MT production significantly in these oysters. Rather, if present, Cu may displace Cd from Cd-induced MT. Metallothionein is induced by Ag in *C. gigas,* although the bulk of this metal is associated with insoluble ligands (Berthet, 1990).

Exposure of the American oyster *C. virginica* to Cd results in induction and increased binding to MT in preference to other ligands (Roesijadi, 1992); and under polluted conditions in the field (Patuxent River) MT binds up to 20% of the Cd in gills of *C. virginica* (Roesijadi, 1994a) – a much higher proportion than that found in *O. edulis* from Restronguet Creek. Amounts of Cu (1%) and Zn (0.3%) bound to MT are negligible in both species. The results of Ridlington and Fowler (1979) confirm that MT is not induced by Cu in adult *C. virginica*, and that Zn is associated primarily with non-thionein pools. However, MT in *C. virginica* embryos binds more Cu and Zn than any other cytosolic pool, indicating significant ontogenetic variation in protein expression. Six putative embryonic MT isoforms have been isolated – probably the product of distinct genes. Four of these isoforms (three Zn-MTs and one ZnCu-MT) represent constitutive MTs that play an important role in Zn homeostasis, crucial to successful development of the embryo. The two others, encoded by a gene that is inducible by Cd and Cu, are formed when embryos are exposed to these metals, implying a detoxifying role (Ringwood and Brouwer, 1993). A number of metal interactions have been observed in these and other experiments (Chapter 11) which may help to interpret field results; for example, exposure of embryos to Cu (20 μg l^{-1}) results in Zn loss from the constitutive MT pool, whilst exposure to Cu and Cd, in combination, reduces Cd binding to MT – suggesting either displacement, or that Cu might be a potent inhibitor of CdMT, at least in embryos (Ringwood and Brouwer, 1995). There is also evidence of competition between Cd and Zn for MT binding sites in this species (Engel, 1983). Clearly, the relationship between bioaccumulation and induction of MT is not uniform under all conditions.

Recently, the presence of two MT forms, whose sole difference revolves around the presence or absence of an *N*-acetyl group, has been established in adult *C. virginica*. The non-acetylated form is present in minor amounts, relative to the acetylated form, in oysters from contaminated (Ag, Cd, Cu and Zn) sites in the Patuxent River, and is only inducible in the laboratory following exceptionally high Cd concentrations (> 4.4 $\times$ 10^{-1} μM). The appearance of this non-acetylated form may represent a consequence of Cd toxicity. For other forms, an increase in MT and MT mRNA over basal levels can generally be determined above 1 $\times$ 10^{-1} μM Cd (Roesijadi, 1994b).

Polarographic quantification of MT in adult oysters (*C. gigas)* was used by Imber *et al.* (1987) in an attempt to assess the assay's value as a biological indicator of metal inputs in British Columbian coastal waters. Differences in MT concentrations in the digestive glands of oysters from reference (5.02, 5.17 mg g^{-1} dw) and polluted sites (6.9, 7.18 mg g^{-1}) were small but statistically significant and were positively related to cytosolic Cu concentrations (rather than Cd or Zn, which exhibited negative trends due, possibly, to displacement). Copper and a small amount of Cd were associated with the 20 kDa (dimerized) MT-like protein and it was concluded that the MT determinations reflect differences in environmental exposure to Cu. Using a similar approach, we have observed small increases in MT induction in gills (but not remaining tissues) of *O. edulis* from inside Restronguet Creek, compared with oysters outside the estuary at Looe Beach (inset, Figure 8.1). This trend reflects contamination in tissues and overlying water, though it is not known which metal is responsible for triggering the response. Where multiple contaminants occur, there remains a question mark as to which intracellular metal component is responsible for initiating MT induction – correlation with one or more metals is not necessarily proof of cause and effect.

Mussels (Bivalvia)

Metallothioneins of both molecular mass variants (MT10 and MT20) have been purified and sequenced in *M. edulis*. MT10 includes four isoforms (72 amino acids) exhibiting substantial similarity to other molluscan MTs, and the dimeric MT20 variant can be resolved into five isoforms (monomeric equivalents each containing 71 amino acids, but with two additional cysteine residues perhaps conferring greater chelating capacity). As with oyster embryos, the large number of isoforms in mussels probably reflects different metal-binding affinities and different specificities of the promoter region of the gene (Mackay *et al.*, 1993). MTs present in *Mytilus* spp. bind, and are induced by, Ag, Cd, Cu and Hg (Table 8.2). Zn and Pb do not induce MT strongly, though they may become sequestered by pre-existing protein. The use of mussel MT to assess bioavailabilty of Pb and Zn from contaminated environments is therefore questionable (Roesijadi *et al.*, 1988; Raspor and Pavicic, 1991).

Synthesis of MT induced by Cu, Cd and Hg has been confirmed by a variety of methods including ^{35}S-cysteine incorporation, immunoassay and polarography (Viarengo *et al.*, 1980; Roesijadi *et al.*, 1988; Bebianno and Langston, 1991, 1992) and the use of mussel MT as an indicator for Ag, Cd, Cu and Hg pollution seems justified, particularly if backed up by metal-binding studies. Mussels are used worldwide in contaminant assessment, and are an obvious candidate for monitoring MT induction. It is therefore interesting to note that basal levels of MT (and induction rates) in different mytilid species are similar: acute Cd exposure of *M. edulis* and *M. galloprovincialis* (basal levels 2–3 mg MT g^{-1} dw, whole soft tissues) gives rise to a 3–4-fold increase in MT after 1 month – equivalent to induction rates of 0.23 and 0.33 mg g^{-1} per day, respectively (Figure 8.2a). Direct comparison of MT biomonitoring data for *M. edulis* and *M. galloprovincialis,* over a large geographical range, would appear to be valid and feasible. This finding is important, since separation of *Mytilus* spp. may be difficult where ranges overlap and hybridization occurs.

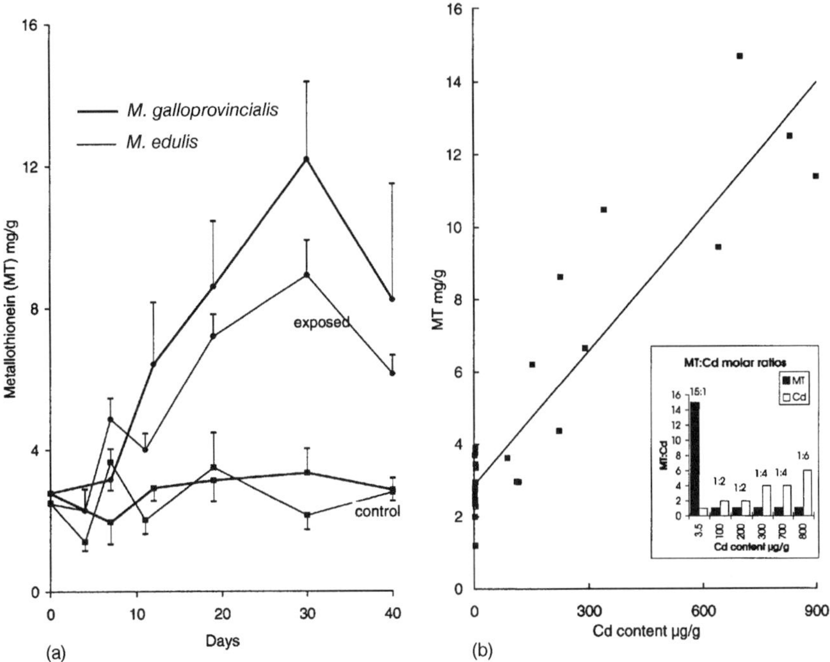

Figure 8.2 (a) Induction of metallothionein (mg g^{-1} dry weight) with time in two species of mussels (*Mytilus edulis* and *M. galloprovincialis*) exposed to cadmium at a concentration of 400 µg l^{-1}, compared with controls. (b) Relationship between MT and Cd concentrations (µg g^{-1} dw) in tissues of *M. galloprovincialis*. (Redrawn from Langston *et al.*, 1989; Bebianno and Langston, 1992).

Responses are tissue specific. Basal MT concentrations in the digestive gland of *M. edulis* and *M. galloprovincialis* are relatively high ($\sim$ 8 mg g^{-1} dw, by polarography) compared with other tissues and may reflect a role in Cu metabolism in addition to protecting against cytotoxicity. Displacement of essential metals from this substantial pre-existing MT pool may occur initially, following acute exposure to pollutant metals such as Cd (0.2–0.4 mg l^{-1}), though eventually a doubling of MT concentrations is observed, reflecting *de novo* synthesis (Bebianno and Langston, 1991; Pavicic *et al.*, 1993).

MT response is most significant in the gills of mussels, as in other filter feeders, and this tissue is probably the most appropriate for assay. Gills function as both a site for metal uptake and as an important reservoir for metal storage, and MT sequesters a significant proportion of accumulated Cd, Cu and Hg. Using a sensitive ELISA assay, basal levels of MT in the gills of *M. edulis* (0.5 µg g^{-1} ww) are shown to increase in linear fashion, to 257 µg g^{-1} and 1.78 mg g^{-1}, in response to increasing concentrations of Cd (1–50 µg l^{-1}) and Hg (0.05–5 µg l^{-1}), respectively (Roesijadi *et al.*, 1988; Roesijadi and

Felligham, 1987). Cu bound to gill MT is a direct reflection of total body burden, though at low levels of exposure this relationship is perhaps not so obvious and may be related to a certain capacity for Cu regulation. Nevertheless, a seven-fold increase in MT (3.7 μg g^{-1}) is detected following exposure of mussels to 5 μg Cu l^{-1} (Roesijadi *et al.*, 1988). Zn is effectively regulated and little bioaccumulation occurs, even at exposure levels of 250 μg l^{-1}. Limited MT induction (2.6 μg g^{-1}) is observed at this level, and even less at lower doses (1.3 and 0.9 μg MT g^{-1} at 10 and 50 μg Zn l^{-1}, respectively; Roesijadi *et al.*, 1988), reflecting the fact that Zn is primarily associated with HMW proteins, including metalloenzymes. The relative inducibility of MTs in the gills of *M. edulis* is ranked Hg > Cd > Cu > Zn, i.e. highest for non-essential metals, confirming a detoxification role. A faster rate of turnover might explain the low net increase of MT associated with essential metals.

A central theme to the debate on the role of MT as a defence system/stress response thus concerns its rate of synthesis, turnover and affinity for metals: if the protein is to be effective as a detoxifying agent, synthesis must be sufficiently rapid, and binding sufficiently strong, to prevent excessive binding to more sensitive molecules. Field data at contaminated sites confirm that Cd and other metals are effectively bound by mussel MT (see references in Table 8.2) and MT production in experiments appears to keep pace with the influx of Cd, even at extremely high exposure levels – giving rise to the linear relationship in Figure 8.2b. Inevitably some association of Cd with other intracellular ligand pools takes place ('spillover'). These disturbances in partitioning are usually a function of exposure concentration and time – a reflection of the net inward flux of metal in relation to availability of ligands – and may be helpful in the toxicological assessment of body burdens. Thus, under conditions of acute exposure, following initial induction, net MT production and Cd accumulation eventually slow down (at around 12 mg MT g^{-1} and 1 mg Cd g^{-1}, respectively, in whole mussels), indicating the approach of equilibrium and saturation of MT with Cd (inset, Figure 8.2b). Upon further exposure, 'spillover' occurs, coinciding with the onset of mortalities. At more environmentally realistic (chronic) levels, however, MT is likely to afford adequate protection against Cd.

The kinetics of MT metabolism and associated metal flux are clearly important in determining Cd body burdens and toxicity. Turnover rates for MT and Cd in *M. edulis*, expressed as half-lives, are 25 and 300 days respectively. The slow rate of Cd elimination is explained by the fact that, as MT degrades, the released Cd induces synthesis of new protein, to which the metal becomes re-sequestered (Bebianno and Langston, 1993). A similar process is indicated for the oyster *C. virginica*, though turnover rates for MT (4–20 days) and Cd (70 days) are slightly faster than in mussels (Roesijadi, 1994b). Localization of Cu-MT within lysosomes has been reported for the digestive glands of mussels (Viarengo *et al.*, 1989), suggesting that this is the principal site of degradation of the metal-binding protein. Differences in the distributions and fluxes of Cd, Cu and Zn may, therefore, be due to differential susceptibility to lysosomal degradation and varying rates of turnover of their respective MTs (George, 1990); this may explain why Cu is eliminated more rapidly from mussels than Zn or Cd (half-lives 10, 60 and 300 days, respectively).

Studies with *Mytilus* spp. were among the first, in invertebrates, to establish the link between MT and metal resistance, and to show that induction of MT by one metal can increase tolerance to another. Thus, for example, exposure of mussels to sublethal levels of Hg (and also Cu, Cd and Zn) results in increased tolerance to high doses of Hg – due partly to the involvement of MT (Roesijadi and Felligham, 1987). The precise mechanism by which MT induction confers protection to subsequent metal insult may follow several pathways: enlargement of the MT pool during pre-exposure, and eventual displacement by metals with a stronger affinity for MT; mobilization and increased sensitization of pathways for MT synthesis at the low dose; or a combination of both (reviewed by Roesijadi, 1992).

Recently, variations in mussel (*M. galloprovincialis*) embryo-larval tolerance to metals have been attributed to changes in the efficiency of protection afforded by MT, which is thought to increase during development (Pavicic *et al.*, 1994). Thus, although pre-existing basal pools of Zn-MT (available for metal-substitution reactions) confer some protection to the developing embryo, subsequent veliger larvae are more tolerant to Cd and Zn by virtue of their increased capacity for MT synthesis. Variable tolerance to different metals can be related to their rate of influx and the efficiency with which they are bound to MT: the high toxicity of Zn in early (pre-veliger) developmental stages is explained by a faster rate of accumulation and relatively smaller MT-detoxified pool (12%), compared with Cd (50%). Simultaneous exposure to Cd and Zn results in less than additive toxicity (antagonism) possibly due to the higher level of MT, induced principally by Cd (Pavicic *et al.*, 1994).

MT biosynthesis is inducible in mussels of freshwater origin, including *Anodonta grandis*, which responds to Cd (but not Cu or Zn) in a dose-dependent manner, reflecting metal bioavailability in the environment. Inter-site variation in MT levels in *A. grandis* (collected along geochemical gradients in a series of Canadian lakes) appear to be closely related to Cd^{2+} activity at the sediment–water interface (Couillard *et al.*, 1993). MT induction in the whole soft tissues of the freshwater zebra mussel *Dreissena polymorpha* is also sensitive to Cd, and occurs following exposure to concentrations in the range 0.2–20 µg Cd l^{-1} (Tessier and Blais, 1996).

Scallops and clams (Bivalvia)

Cd in the digestive gland cells of the pectinids *Adamussium colbecki* and *Pecten jacobaeus* is partially in a detoxified form, ascribed as MT (Viarengo *et al.*, 1993). Another scallop, *Pecten maximus*, has one charge form of MT, with similar properties and amino acid composition to that of *Mytilus* MT.

However, this protein sequesters only 15% of the cytosolic Cd in *P. maximus*; the major Cd-binding component – which also binds Cd through cysteine thiolate groups, but less strongly – has a molecular weight of 55 kDa (Stone *et al.*, 1986). The participation of HMW proteins in Cd-binding during acute exposure is also a characteristic of other pectinids, including *Mizuhopecten yessoensis* (Lukyanova *et al.*, 1993), though after depuration or at reduced exposure levels (and in controls) MT assumes the principle role in detoxification in this species.

A Class I MT is present in various tissues of the arcid clam *Scapharca inaequivalvis* (Serra *et al.*, 1995; Table 8.2). Two MT isoforms are identifiable in the viscera (compared with one in the kidney) and induction in response to Cd is linear, reaching 1 mg MT g^{-1} dw. The Cd bound to MT (i.e. detoxified) is relatively stable (T$_{1/2}$ for Cd-MT = 2 months; Serra *et al.*, 1995).

Basal levels of 2 mg MT g^{-1} dw have been determined by polarography in the venerupid clam *Ruditapes decussatus* (whole soft tissues). Exposure to Cd and Cu results in induction of MT (Table 8.2), though at rates that are considerably slower than in *Mytilus* spp. Rates of Cd accumulation and loss are correspondingly slow in this species. MT concentrations in *R. decussatus* gills are most responsive, and increase by twofold after acute Cd exposure (from an initial 1.97 mg MT g^{-1}), whilst in the digestive gland and remaining tissues MT concentrations remain unchanged (4.7 and 1.39 mg g^{-1}, respectively; Bebianno *et al.*, 1993). Using fluorometric techniques, Romeo and Gnassia-Barelli (1995) have also investigated the monitoring potential of *R. decussatus* gill, and observed MT induction after treatment of clams with Cu; however, little response could be detected with Cd, possibly due to the limited exposure period (7 days) – MT levels in the digestive gland were unaffected by Cd or Cu treatment. In another venereid clam, *Protothaca staminea,* Cu-, Cd- and Zn-binding MT-like proteins (10.5 kDa), together with a larger protein (14 kDa) described as copper chelatin, are thought to fulfil a protective role, though the presence of these proteins in controls indicates additional involvement in routine Cu metabolism (Roesijadi, 1980).

A detectable proportion of cytosolic Ag, Cu and Zn is associated with MT-like proteins in the tellinid clam *Macoma balthica* collected from San Francisco Bay, and is most significant at impacted sites, consistent with a detoxifying function for the protein (Johansson *et al.*, 1986). In contrast, a study of metal-binding in *M. balthica* populations from several sites in the UK, encompassing laboratory exposures to Cd and Hg, failed to reveal any major comparable role for MT, suggesting possible intraspecific divergence in the capacity for induction (Langston and Zhou, 1987b). Limited or negligible involvement of MT in metal-binding has been reported for several other bivalves, including *S. plana* (own unpublished results), *A. cygnea* and *Unio elongatulus* (Cassini *et al.*, 1986; Tallandini *et al.*, 1986), confirming that metal exposure does not always result in detectable induction, and emphasizing the point that the involvement of MT in regulation and detoxification is not always consistent.

Periwinkles (Gastropoda)

The structure and function of gastropod metallothioneins has received less attention than bivalve counterparts (Table 8.2), though characteristics of molecular weight (10 and 20 kDa forms), absorbance, metal-binding properties and amino acid content are usually compatible with confirmed molluscan MTs.

Metal metabolism has been particularly well-studied in the periwinkle *Littorina littorea* (Mason and Nott, 1981; Mason *et al.*, 1984; Langston and Zhou, 1986, 1987b; Gully and Mason, 1993). Metals bound to MT in *L. littorea* include Ag, Cd, Cu and Hg, but not Zn (Langston and Zhou, 1986). Of these metals it is Cd that is bound to MT in the most significant proportions (> 80% in animals from clean sites). The percentage of MT-bound Cd decreases with increasing Cd contamination, coinciding with 'spillover' of Cd to high molecular weight (HMW) ligands, including haemocyanin (HCY, a Cu-containing respiratory pigment that is also characteristically involved in the transport and redistribution of several other metals, including Fe, Zn, Ca, Mg, Sr and Ba). Accumulation of Cd burdens in *L. littorea* is ostensibly a continuous and irreversible process, though the degree and rate of saturation of MT is tissue dependent, and reflects exposure history (Langston and Zhou, 1986, 1987a; Bebianno and Langston, 1995). Induction of MT in kidney and gill is proportional to accumulated Cd, resulting in a three- to fourfold increase in protein (from basal levels of $\sim$ 3 mg MT g^{-1} dw), following acute exposure (0.1 mg Cd l^{-1}). Similar treatment initiates limited *de novo* synthesis of MT in the digestive gland (a 10 kDa variant initially, eventually dimerizing; Figure 8.3); however, despite a large influx of Cd, there is little net increase in MT concentration, and most of the accumulated Cd is associated with the high levels of MT (> 10 mg g^{-1}) inherent in this tissue (involved, presumably, in regulation of Cu during HCY turnover). Initially, this constitutive MT is undersaturated with respect to Cd, but becomes increasingly saturated as a function of Cd exposure. During depuration of Cd-laden winkles in clean seawater, Cd associated with HMW proteins in kidney, gill and head-foot is offloaded, predominantly to MT in the digestive gland, and the latter once more becomes the major store for Cd; some reduction in CdMT is indicated for gills, but not other tissues (Figure 8.3). Despite subcellular and tissue redistribution, there is virtually no reduction in the total Cd body burden, even after one year in Cd-free water (Langston and Zhou, 1987a). This long Cd half-life (> 300 days) is explained by recycling of the MT–Cd complex, coupled with very slow turnover of the MT protein, especially in digestive gland.

Limpets and whelks (Gastropoda)

Two similar MT isoforms which bind Cd and Cu have been isolated from the limpet *Patella vulgata*. MT is present, as MT10 and MT20, in *P. vulgata* (and *P. intermedia*) from Cu-contaminated environments, though amino acid analysis reveals a slightly lower cysteine content (20–21%) and lower metal-binding capacity (3.4 gram-atoms mol^{-1}), than oyster or mussel MT (Howard

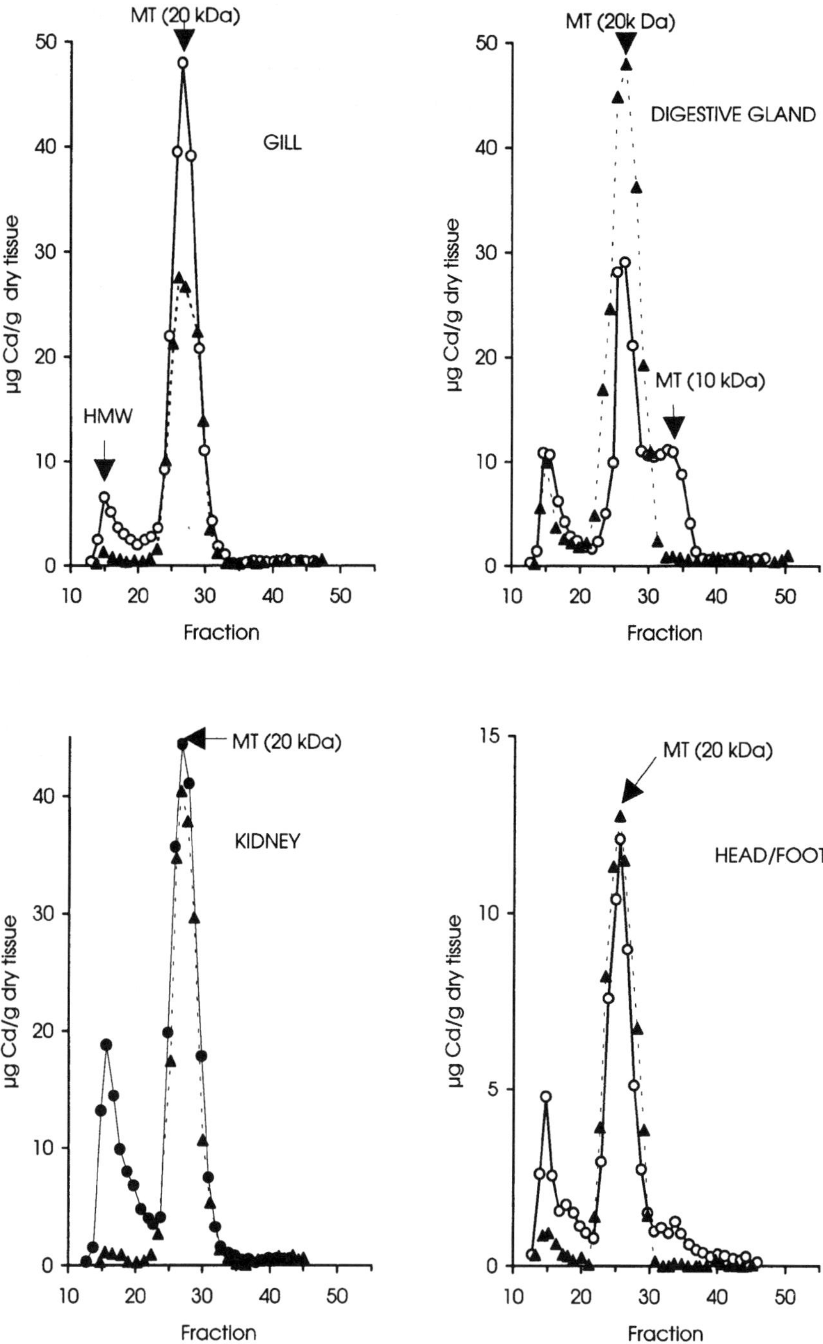

Figure 8.3 Distribution of Cd in cytosol of individual tissues following exposure of *Littorina littorea* to 0.4mg Cd l^{-1} for 39 days (solid line) and subsequent depuration for 197 days (dotted line). (Redrawn from Langston and Zhou, 1987.)

and Nickless, 1977). MT in limpets is thought to play a minor role in Zn storage, but is responsible for tolerance to Cd and Cu in contaminated environments (Howard and Nickless, 1977; Noel-Lambot *et al.*, 1980). In the slipper limpet, *Crepidula fornicata*, Hg bound to MT (8 kDa) is largely independent of contamination (Harrison *et al.*, 1987) and the Hg-MT association probably involves the displacement of other metals, rather than a net increase in MT; this is analogous to the situation for Cd-MT in the digestive gland of *L. littorea*, described above.

Cd-binding proteins are present in gills, hepatopancreas, intestine and head-foot of the whelks *Nassarius recticulatus* and *Nucella lapillus* (Table 8.2) (Noel-Lambot *et al.*, 1978). Although these proteins have yet to be fully characterized, they have Cd-thiolate bonds and a scarcity of aromatic amino acids, consistent with MT.

Cephalopods

The few studies conducted on cephalopods (mainly digestive gland) imply species-specific metal-binding mechanisms. Cu and Cd are bound to MT-like proteins of approximately 10 and 20 kDa in squid *Nototodarus gouldi* (Finger and Smith, 1987), whilst in the oceanic species *Ommastrephes bartrami* Cd-binding proteins are markedly different from most bivalve and gastropod MTs by virtue of their lower cysteine and higher glycine content (Castillo and Maita, 1991). Cu in the cuttlefish *Sepia officinalis* is associated (along with Ag and Zn) with MT-like proteins which are present in spherulae produced by the basal cells of the digestive gland. The spherulae may be a byproduct of the turnover of HCY in this tissue (Martoja and Marcaillou, 1993).

8.3.2 NON-THIONEIN PROTEINS AND OTHER CYTOSOLIC METAL COMPLEXES

The chemical properties of essential metals have been harnessed for specific physiological and biochemical purposes, principally in redox transformations and the catalytic reactions of metalloproteins and enzymes. The metal affinity of haemocyanin (present in gastropods, cephalopods and, very occasionally, bivalves) has been discussed in the previous section, as have metalloproteins such as α-mannosidase, alkaline phosphatase and carbonic anhydrase. Additionally, there is a variety of complexing compounds (besides metallothionein) which may be involved in modulating the donation of metal to these proteins. Ligands in oysters (which maintain high levels of Zn) include ATP and low molecular weight compounds such as the amino acids taurine, lysine and homarine (often present in large amounts in molluscs and involved in osmotic regulation; Coombs, 1974). This Zn pool presumably acts as a readily available metal source and ensures constant saturation and maximum activity of Zn-dependent enzymes, particularly carbonic anhydrase, which is present in blood at considerably higher levels than in other bivalves. The major low molecular mass Cu moiety in oysters is a nitrogen heterocycle, with smaller amounts as amino acid complexes, including the dipeptide Cys-Gly and glutathione (Fayi and George, 1985).

In the bivalve *Semele solida*, Mn is required for maintenance of the quaternary structure of the enzyme arginase. Although Mn is the main activator of this enzyme, some activation is also produced by Ni and, to a lesser extent, Cd and Co; Zn, in contrast, is inhibitory (Carvajal *et al.*, 1994).

Requirements for Fe are many and varied in molluscs. They encompass redox-based properties of the element – as in the respiratory pigment haemoglobin and the haem-based cytochrome P450 mixed function oxidase system – together with more specific structural assets; for example, the radular teeth of gastropods belonging to the Chitonidae and Patellidae are composed of up to 14% Fe, which confers hardness to these rasping structures. Regulation of Fe is controlled by binding of the metal to regulatory proteins which, through conformational changes, influence interaction with nucleic acids, thus modulating transcription and translation of the major Fe transport and storage proteins, transferrin and ferritin, respectively; in many respects this is analogous to the mechanism of MT induction by metals (Hamer, 1986). Assuming that the mammalian model is relatively ubiquitous, Fe is transported from uptake sites by the plasma glycoprotein transferrin (80 kDa) to sites where haem and other Fe-containing proteins are synthesized. After release of Fe into the cytoplasm, transferrin is recycled. The presence of Fe in cells is regulated by induction of ferritin, the major iron protein, or by inhibition of transferrin. In mussels, ferritin consists of an electron-dense iron core surrounded by a polypeptide shell. It has a mass of 480 kDa (composed of subunits of 18.5 and 24.6 kDa in equal proportions) and is capable of binding 200 Fe atoms per molecule of protein (Bootsma *et al.*, 1988). Lysosomal breakdown of ferritin results in Fe-rich residual bodies (haemosiderin) which are stored principally in the digestive gland. Though Fe storage is its major function, ferritin is also found in association with Zn in oysters. In the absence of Fe, transferrin may bind a number of other metals including Al, Cr, Co, Cu, Mn and V.

Haem compounds isolated from the haemolymph of bivalves such as *Scapharca inaequivalvis* include haemoglobin (Hb) (for oxygen transport) and also haematin, which is involved in an unusual role in the immobilization of sulphide by precipitation with Fe (a possible adaptive advantage for Hb-containing invertebrates from sulphide-rich environments; Vismann, 1993). Various unusual means of voiding excess Fe also exist: mussels, for example, can eliminate Fe and, to some extent, other metals during the process of byssal-thread secretion.

There are a number of other categories of metal-binding protein which have been isolated in molluscs, usually associated with non-essential elements and, therefore, assumed to have a detoxifying role. Some of these resemble MT superficially, but differ to varying degrees in terms of molecular weight, cysteine and aromatic amino acid content; thus, two Cd-, Cu- and Hg-binding glycoproteins of 8 and 13 kDa, containing 52 and 96 aromatic residues, respectively (but few cysteines), have been isolated from the whelk *Buccinum tenuissimum* (Dohi *et al.*, 1986). The 55 kDa Cd-binding protein responsible for binding most of the Cd in the digestive gland of the scallop *Pecten maximus* could conceivably be a multiple form of MT – perhaps the result of disulphide bridging – though sequence homology is still required to establish similarities. Stone *et al.* (1986) concluded that the low cysteine content (8%), non-inducible nature and other characteristics are sufficiently removed from MT to make it appear a novel and constitutive protein, and similar conclusions were drawn by Fowler and Gould (1988) concerning a 45 kDa Cd-binding protein in the kidney of another scallop, *Placopecten magellanicus*. Unspecified high molecular weight proteins which sequester Cd and other metals are a common feature of size-exclusion chromatographs in, for example, gastropods such as *Murex trunculus* and *L. littorea* and bivalves *M. balthica* and *S. plana* (Bouquegneau *et al.*, 1983; Langston and Zhou, 1986, 1987b). A 60 kDa stress protein (hsp60) is produced in the mantle of mussels exposed to Cu (Sanders *et al.*, 1991), though little is known of its contribution in metal metabolism and MT is considered the primary protein involved in Cu detoxification in these bivalves; nevertheless, hsp60 induction occurs in proportion to Cu levels, suggesting some involvement in metal adaptation (albeit minor).

A number of low molecular weight peptides and amino acids are involved in metal sequestration, and in some molluscs (oysters, for example) this represents a major pool, as discussed previously. The thiolic tripeptide glutathione (GSH) may have a more ubiquitous involvement in complexation and detoxification. Metals, particularly those like Cu which are capable of redox cycling, induce lipid peroxidation; subsequently, the production of highly reactive toxic radicals can impair a variety of cellular processes. Reduced GSH may ameliorate these effects – firstly, by sequestration of metal with cysteine (and also amino acid and carboxylic acid groups, and peptide linkages) and secondly, through its antioxidant properties, by removal of superoxide radicals and hydrogen peroxide (reviewed in Mason and Jenkins, 1995). It is possible that GSH acts as the first line of defence in buffering intracellular metal during the lag-phase of MT induction; metal may be transferred subsequently from GSH to the newly synthesized MT. In vertebrates, a wide range of metal–GSH complexes have been determined, involving essential (Cu, Fe, Mn, Zn) and pollutant (Ag, Cd, Hg, Pb) elements, implying a central role in the transport and metabolism of metals, as well as detoxification. To give one example of this basic involvement, ontogenic changes in the ability to excrete Hg in the bile of neonatal rats has been shown to coincide with an increased capacity to secrete GSH (Ballatori and Clarkson, 1982). Because of its instability, the presence of the GSH–metal complex has proved more difficult to confirm in molluscs, but recently Zaroogian and Anderson (1995) have indicated the involvement of GSH in the facilitated diffusion of Cd^{2+} and Ni^{2+} into red gland cells of *Mercenaria mercenaria*.

Depletion of GSH to oxidized forms such as GSSG is symptomatic of intracellular stress. Antioxidant enzymes such as GSH-reductase, which may themselves be metal dependent, are needed to restore ratios. Paradoxically, then, metals are involved in both the generation and removal of reactive oxyradicals. Redox pathways are probably a principal target for metal toxicity in view of the myriad possibilities for metal excess, competition and deficiency.

8.3.3 LYSOSOMES AND GRANULES

Metals may become diverted from MT and other cytosolic modulating ligands towards insoluble and hence less reactive stores, i.e. the various concretions and inclusions that are particularly well documented for molluscan tissues (e.g. Viarengo and Nott, 1993). Association with lysosomes and related vesicle-bound granules further reduces cellular toxicity and may eventually lead to exocytosis and elimination from the body. In essence, this is analogous to the process of intracellular digestion described earlier for bivalves, where absorption is followed by elimination of undigested sedimentary particles in large membrane-bound excretory spheres.

There is a tendency to categorize metal sequestration systems independently because of uncertainties over their origin and mode of operation; consequently, interpretation of their interrelationships remains speculative. However, evidence is emerging that they may often provide complementary strategies for cation homeostasis.

Lysosomes, such as those in mussel kidney and digestive gland, can assimilate metals (e.g. Cu, Cd and Zn) in the insoluble granular matrix (lipofuchsin) or in association with cysteine-rich proteins (George, 1983a,b; Roesijadi *et al.*, 1989; Viarengo *et al.*, 1989). As much as 70% of the Cu in the digestive gland of mussels may be as Cu-MT incorporated into primary lysosomes. This lysosomal MT is typical of molluscan MT in terms of molecular weight and high cysteine content, but may differ somewhat from cytosolic MT, being more acidic, lower in glycine and higher in aspartic and glutamic acids, and with a tendency to oxidize and polymerize (Roesijadi *et al.*, 1989; Viarengo *et al.*, 1989; Viarengo and Nott, 1993). Enzymatic proteolysis of lysosomal MT, and accompanying peroxidation of membrane lipids, results in the formation of insoluble lipofuschin granules, at rates which are metal, tissue and species specific (George and Olsson, 1994). High concentrations of metals are released under the acidic conditions present in these secondary lysosomes, though during the process the intra-lysosomal pH will increase, presumably enabling rebinding of released metals to other ligands, including MT, or to the 'chelating sinks' offered by tertiary lysosomes (containing undigestible remnants and isolated lipofuchsin granules). The latter may contain up to 10% metal, formed as part of the general catabolic pathway. Granule production is stimulated by Cu and results in enhanced exocytosis and elimination, with a relatively short half-time. However, Cd does not appear to stimulate the peroxidation process and since Cd (and Zn) thiolate complexes are less stable than Cu at lysosomal pH, release of Cd (and Zn) to the cytosol is possible, followed by resequestration to newly synthesized MT, with little elimination of metal. Not surprisingly, this differential 'recycling' leads to the variations and disparity between half-lives of MT and associated metals described earlier. Both MT and lysosomes therefore play a fundamental and linked homeostasic role in metal metabolism.

Granules of one sort or another are fairly ubiquitous in molluscs, though they may serve different functions within different cells in relation to the distribution of metals. In bivalve kidneys the presence of high levels of Ag, Hg and Pb, together with Ca, Fe, Zn, S and P, in lysosomal granules and assorted concretions, signifies an important role in pollutant detoxification, as well as in metabolism of essential elements; quantitatively, these inclusions may account for $\sim 20\%$ of the kidney volume following metal exposure. Before they are voided in the urine, the granules may take up metals passively (by adsorption) to add to the original contribution arising from catabolism of proteins (George and Pirie, 1979; Viarengo *et al.*, 1980; Fowler *et al.*, 1981; George, 1983b). The lysosomal origin of these and other types of membrane-limited granule is confirmed by analytical electron microscope studies: in the kidney of scallops, for example, a gradual development from lysosomal membranous vesicles to highly mineralized, membrane-bound excretory granules is observed (George *et al.*, 1980). Derivation of concretions from other organelles, including Golgi or mitochondria, is possible, whilst the presence of further types (outside membranes) may indicate alternative sources for initiation.

Digestive glands of gastropod molluscs are host to several types of inclusion in addition to lysosomal granules. Three of these types have been isolated in *L. littorea* (reviewed in Viarengo and Nott, 1993; Mason and Jenkins, 1995). The first is the virtually pure Ca/Mg carbonate-based inclusion in connective tissue cells, often situated close to blood vessels, thought to be involved primarily in Ca metabolism (particularly shell deposition and growth) and pH regulation. Ca is highly regulated in order to maintain intracellular levels at about 10^{-7} mol l^{-1}, compared with extracellular levels of 10^{-2} mol l^{-1}. The function of the second type of granule, based on Ca and Mg phosphate/pyrophosphate, is more controversial: possibilities include storage and regulation of Ca and PO_4^{3-} and detoxification of intracellular Ca^{2+}. However, the fact that these inclusions are capable of trapping, through oxygen donors, a number of transition and group IIB metals (Mn, Zn, Cu, Fe, Co, Ni and Cd) – by virtue of the insolubility of their phosphates (particularly pyrophosphates) – is consistent with protection against toxicity: once bound, such forms are unlikely to be remobilized, even during digestion by predators (Chapter 12). Furthermore, the location of these metal-rich granules in cells from which they can be extruded in times of metal stress (basophil cells of the digestive gland and in kidney cells), or upon cell death, appears to confirm a detoxifying role (Mason and Jenkins, 1995). A recent demonstration of the association of Cd with S, in an organic form around the periphery of these phosphate granules, is interesting in that it demonstrates a further possible linkage between the sulphur-rich, metal-binding proteins (e.g. MT) and the inorganic granule detoxification system (Nott and Langston, 1989). Their close association points to one mechanism whereby excess metal may be transferred from a rapidly metabolized protein pathway to a more thermodynamically stable and permanent trap within the cell.

The ability to sequester metals in granular form, as insoluble phosphate, follows phylogenetic traits and, among gastropods, appears to be outstanding in some members of the superfamily Cerithiacea (Mesogastropoda). Examples studied include the detritivore *Cerethium vulgatum* whose phosphate granules in the digestive gland contain a suite of metals including Mn, Fe, Co, Ni and Zn (Chapter 12). Carnivorous neogastropods such as *Nucella lapillus* also accumulate significant body burdens (by molluscan standards), partially stored in granule form, even though contamination in their environment may not be evident.

A third type of granule in *L. littorea*, based on Cu-S, arises from the turnover of HCY and is predominant in the digestive gland pore cells. It provides a further example of the involvement of a complex chain of ligands during metal homeostasis; thus, whilst MT appears to be involved in regulating the pool of available Cu for reactivating apo-HCY, the final products of catabolism reside as phagocytosed granules of Cu and S in pore cells. Cu is organically complexed with S in archaeogastropods (*Monodonta crassa*) whilst in mesogastropods (*L. littorea*) and neogastropods (*N. lapillus*, *Murex brandaris*) it is bound both organically and inorganically as CuS (Bouquegneau and Martoja, 1982). This influences subsequent metabolism: in archaeogastropods, Cu in pore cells may be recycled or excreted; in other groups, granular metal is more permanently fixed and thus accumulates with age. Cu-S granules can bind other group B metals by virtue of their sulphur donors, and may be involved in detoxifying not only excess metabolic Cu but also Cu, Cd and Ag arising from environmental contamination (Viarengo and Nott, 1993). Primarily, this type of granule appears to be the product of normal physiological (lysosomal) breakdown of metallo-sulphur proteins; however, the inclusion of cysteine-rich MT (and its associated metals) in this pathway could explain the occurrence of contaminant elements in CuS granules. The storage of Ag in this form, as silver sulphide, appears to be particularly important in marine gastropods and also the brown cells of some bivalves such as *P. maximus* (Martoja *et al.*, 1985). Copper-rich inclusions have been identified in the kidney and digestive glands of marine bivalves (oysters and mussels). These are not derived from HCY metabolism and are probably of lysosomal origin (George *et al.,* 1978; Viarengo *et al.*, 1989): Cu is carried there, as Cu-MT, following exposure – further illustration that different mechanisms of metal complexation are not necessarily exclusive of each other.

Iron-containing granules in epithelial cells from the stomach of *L. littorea* are of lysosomal (dietary) origins (Mason and Jenkins, 1995). On a broader scale, inclusions containing Fe in one form or another appear to be present in a large number of species, indicating a fundamental role in regulating iron requirements. A variety of proteins may be involved in Fe complexation, though the ubiquitous nature of ferritin and transferrin implies that the underlying mechanisms of Fe regulation and metabolism are common to the majority of organisms.

Most of the granular metal forms described here act as kinetic traps for metals (storage and detoxification) and are localized in digestive gland (especially in gastropods), gills, kidneys (notably in bivalves) and mantle edge – tissues from which they can be readily eliminated from the animal, either at the surface or in faeces and urine (reviewed by Nott, 1993). Other forms of granular metal fulfil different roles by virtue of their percolatory nature. Thus, most of the Cu and Zn in phagocytic amoebocytes (haemocytes) of oysters is encased in membrane-bound vesicles. *Ostrea edulis* from Restronguet Creek contains three types of amoebocyte, which may contain Cu or Zn individually or in combination (George *et al.*, 1983). Localization of metals within vesicles enables the oysters to tolerate body burdens of Zn and Cu which are 10 to 100 times background levels: compare, for example, Cu and Zn burdens in *O. edulis* at unpolluted sites (40 μg g^{-1} and 1500 μg g^{-1}, respectively) with those from Restronguet in Figure 8.1. However, in addition to detoxification, it has been proposed that these cells have a defensive role similar to that of mast cells or platelets in vertebrates: release of Cu (and Zn) may enhance clotting at sites of physical damage or under other forms of stress (Fayi and George, 1985).

Phagocytic cells are present in all molluscs to some degree and assist in the compartmentalization, within lysosomes, of several class A and B metals derived from catabolized blood proteins. The granules themselves are not usually extruded, though secretion of the metal-laden amoebocytes (diapedesis) is one mechanism for eliminating metals from bivalves such as oysters and also from some gastropods. Alternatively, MT-bound metals in amoebocytic blood cells may be sequestered in the kidneys, via lysosomal concretions, and processed into residual bodies that can be released through the urine (Carmichael *et al.*, 1979; George, 1983a,b; Fowler and Gould, 1988; Viarengo, 1989). The accumulation of soluble metals in pinocytotic vesicles of blood cells has also been taken as a sign of involvement in metal regulation, additional to their somewhat indiscriminate phagocytic activity (George *et al.*, 1978; Mason and Jenkins, 1995).

The branchial hearts of the more highly evolved, predatory cephalopods *Eledone cirrhosa* and *Sepia officinalis* contain elevated concentrations of Cu and Fe which are not related to contamination, but to the presence of respiratory pigments in atrial cells. Polyhedral cells in these organs are possibly analogous in function to the pore cell in gastropods and contain granules (adenochromes) that are rich in Fe, consistent with their circulatory and excretory involvement. The presence of other metals in adenochromes also indicates a role in storage and detoxification, though this requires verification and may be purely adventitious (Miramand and Bently, 1992).

We should finish this section by emphasizing that elevated metal burdens in aquatic molluscs are not always a sign of pollution but are in many cases a natural phenomenon, reflecting the variety of biological adaptations that have evolved to handle metals. Understanding these metal-handling systems, and the chemical speciation of metals in various mollusc types, would seem to be essential for the prediction of higher-order effects, and for the explanation of different sensitivities to metal contamination. Faced with abnormal metal influxes, there may be a complex interaction of strategies to avoid toxicity. In Cd-exposed *Littorina,* for example, this involves sequestration with MT (especially in digestive gland and kidney), together with: limited excretion (voiding of lysosomal vesicles from digestive cells); partial storage in CuS granules of pore cells; diapedesis of haemocytes from kidney, gill and digestive epithelia; expulsion of granules from nephrocytes; and even re-direction to the shell, following non-discriminatory secretion (in place of Ca) from cells at the mantle edge.

8.3.4 SHELLS

Though the shell is an external structure in most molluscs, its origins are intracellular. Epithelial cells lining the mantle cavity accumulate Ca (and bicarbonate) from water and surrounding tissues and they transport Ca – either as diffusible ions through Ca pumps, or bound to mucoproteins and mineralized spherites – to the extrapallial cavity, where it is secreted along with periostrocum protein. Mineralization of mollusc shells within the tanned organic matrix usually entails the deposition of calcium carbonate in the form of calcite and/or aragonite (involving the Zn-containing enzyme, carbonic anhydrase).

Magnesium and trace elements with ionic radii less than Ca (e.g. Fe^{2+}, Fe^{3+}, Cu^{2+}, Zn^{2+} and Mn^{2+}) accumulate preferentially in calcite. Sr and metal ions larger than Ca (e.g. Pb) tend to become incorporated in aragonite. Interpretation of metals in shells may be complicated by their association with the organic matrix.

Compared with soft tissues, partitioning of metals into mollusc shells is usually low, though there are some exceptions. Thus, whilst Cd, Zn and Hg accumulation in *Macoma balthica* is dominated by soft tissues (> 80%), Ag and Co are assimilated significantly (67% and 61% respectively) into the shell matrix (Luoma and Jenne, 1977; Langston and Zhou, 1987b and unpublished results). Nevertheless, because trace element concentrations are usually low in mollusc shells and are often variable, they have seldom been used as monitoring tools. Furthermore, as it is usually impossible to distinguish the relative proportion of secreted, as opposed to adsorbed, metal (i.e. that which is metabolically involved), the relevance of metals in shells is uncertain. Where they have been tested, as in limpets and mussels, environmental availability of metals is not always reflected in metal concentrations in the shell. Steric interactions between metal ions for sites in the shell matrix (an influx of Sr, for example, would exclude Mn and Fe), together with environmental factors, might explain some of this variability (Foster and Chacko, 1995).

Internal shells of the cuttlefish *Sepia officianalis* are also relatively low in many metals (Ag, Cd, Co, Ni, Pb, V), compared with other tissues, suggesting negligible involvement in metabolism. However, this material does contain 13–17% of the body Fe, Mn and Zn and could represent a storage organ for these essential elements, as appears to be the case for Cu and Zn in squid (Miramand and Bentley, 1992).

8.4 DIAGNOSTIC INDICATORS OF METAL EXPOSURE

8.4.1 BODY BURDENS

Molluscs, particularly bivalves, have assumed a major role in monitoring contaminants worldwide. This arises from several obvious strategic advantages in terms of ease of collection, widespread distribution, relatively sedentary habit, suitable size and, often, economic or ecological importance. The choice of mollusc species used as bioindicators encompasses a number of diverse accumulation strategies and, not surprisingly, it is these species about which we know most in terms of metal metabolism – hence their inclusion in previous sections. Thus, the most commonly employed include, but are not restricted to:

- suspension-feeding bivalves such as oysters (e.g. *O. edulis, C. virginica, C. gigas* and the genus *Saccostrea*) and mussels (*M. edulis, M. galloprovincialis, P. viridis*);
- deposit-feeding clams (e.g. the tellinids *S. plana* and *M. balthica*);
- detrital, herbivorous and carnivorous gastropods (particularly littorinids, patellids and muricids).

The determination of metal concentrations in bioindicators is highly relevant in ecotoxicological terms, since direct measurements are made of bioavailable – and hence potentially deleterious – metal. Analysis of sediments and water, though helpful in terms of defining comparative levels of environmental contamination, usually includes refractory forms which may be of little biological significance: metal 'speciation' techniques, discussed earlier, promise to be more meaningful than 'total' metals in predicting bioavailability, but have yet to be widely adopted. Body burden data, therefore, remain a practical management tool, provided that validation of indicating ability is rigorous. Knowledge of the kinetics and pathways of uptake and loss, metal handling systems, potential metal-regulatory behaviour and other biotic factors are prime requirements in underpinning the selection and application of bioindicators. A full discussion of the topic is not possible here (reviewed by Phillips, 1980, 1990; Bryan *et al.*, 1985; Lauenstein *et al.*, 1990; Rainbow, 1995; Langston and Spence, 1995; and Chapter 7). However, the type of information that might be beneficial in evaluating body burden data is summarized in Table 8.3 for three molluscan species collected extensively around the coastline and

Table 8.3 Examples of molluscs used as diagnostic indicators of metal exposure[1,2]

Species	Habitat	Source/most useful for:	Least useful	Kinetic aspects	Allometric variables
Scrobicularia plana	Estuarine, deposit-feeding bivalve (sediment)	Oxidized surface sediment. Good organism–sediment relationships for Ag, Co, Cd, As, Cr, Hg, Pb and Zn, especially if take into account major sediment binding components[2]. Uptake from water may also be significant for Cd, Co, Zn.	Cu: high concentrations sometimes found in absence of environmental contamination, possibly due to redox-related phenomena. May be some capacity for Zn regulation at high levels leading to potential underestimation of contamination. High levels of sediment Cu may reduce bioavailability of Ag due to competitive effects.	Easily transplanted. Relatively slow equilibration times: 4 months (Cu, As)→1 year (Cd, Pb).	Cd, Co, Cr, Ni, Pb and Zn tend to increase with age. Standardization of sample size recommended (4 cm). Concentration differences between sites of 15–30% are detectable, depending on metal.
Mytilus edulis	Marine/estuarine suspension feeding bivalve (rocks)	Solution, ingested phytoplankton and other suspended particles. Reasonable proportionality with most metals in water (lab). Field results sometimes indicate contradictory behaviour possibly due to competitive effects between metals where present at high levels (Cd, Cu, Zn). Generally acceptable indicators, particularly in view of their widespread distribution. (Cd, Co, Cr, Hg, Pb, Se, Sn)	Partial regulation at moderate levels of contamination may result in some underestimation of Cu, Zn and As. Some forms of Ag may not be bioavailable to mussels. Cd and Hg absorption suppressed by high levels of Zn. Accumulation of Cu enhanced by presence of Ag. High individual variability for Zn.	Equilibration times for metals in transplanted mussels somewhat variable between sites. Most metals accumulated or lost within 3 months though complete exchange may take 12 months, occasionally longer.	Cd and occasionally other metals may vary with size/condition. Attempts to normalize with respect to these parameters have been suggested. Concentration differences between sites of 20–40% are detectable.
Littorina littorea	Marine/estuarine gastropod (rocks/seaweed) sediment)	Algal diet (a reflection of dissolved contamination) is probably the major source for most metals. Based on comparisons of metal levels in *Littorina littorea* and macroalga *Fucus vesiculosus*, this gastropod is a good indicator for Cd, Ag, Pb, As and Hg. Correlations for Cu, Fe and Zn are significant but slopes shallow, perhaps reflecting regulation by the winkle.	Ag concentrations suppressed by elevated levels of dietary Cu (competition for uptake). Relationships to environmental (dietary) levels are of only marginal significance for Cr and Mn, and of no significance for Co and Ni.	Metals in transplanted individuals approach those of native populations in approximately two months	Ag increases and Cd, Cu, Fe, Ni, Pb and Zn usually decrease with size. Analysis of similar sized snails is therefore preferable. Coefficients of variation for individuals range between 10 and 30%.

[1]Major source of references for this table in Bryan *et al.*, 1985; [2]Bryan and Langston, 1992.

estuaries of the UK: the deposit-feeding clam *S. plana*, the suspension-feeding mussel *M. edulis* and the grazing prosobranch *L. littorea*.

Interpretation of concentration ranges for the essential metals (e.g. Cu and Zn) can be complicated by regulation, coupled in some cases with the presence of inherently high tissue burdens. White and Rainbow (1985) calculate that enzymatic requirements for Cu and Zn in molluscs are of the order of 25 and 35 μg g^{-1}, respectively; a further 125 and 65 μg Cu g^{-1} may be needed in HCY-containing gastropods and cephalopods. Though these are only estimates, they illustrate some of the important systematic differences in metal concentrations between mollusc groups. Baseline Cu concentrations in *L. littorea* (150 μg g^{-1}), for example, are an order of magnitude higher than in *S. plana* or *M. edulis*, reflecting the winkle's requirement for Cu-containing HCY. Basal Cu and Zn concentrations in oysters exceed those of clams and mussels by similar proportions but for somewhat different reasons, described above. Thus, in some molluscs, metabolically functional Cu and Zn constitute the dominant pool which, when combined with an effective regulatory mechanism, masks the detection of contamination at all but the most heavily polluted sites.

Homeostatis of non-essential elements is seldom observed in molluscs. Hence, baseline concentrations and responses to contamination tend to be more uniform and predictable for these metals. Nevertheless, the scale of response may vary between species as a result of the numerous biological features that modify uptake and retention. The three species described in Table 8.3, for example, occupy different niches and exhibit different feeding strategies and metal-handling behaviour; consequently, body burdens do not always depict identical patterns of contamination. This is demonstrated for Ag in the maps shown in Figure 8.4, where, despite some superficial similarities, trends in Ag bioavailability display subtle interspecific variations. Concentrations in *S. plana* and *L. littorea* indicate elevated Ag bioavailability at Whitehaven and in the Thames estuary (from industrial and sewage sources) together with the East Looe estuary (from photographic processing, historic mining and sewage). The outstanding degree of Ag enhancement in the deposit feeder *S. plana* from East Looe indicates that bioavailable sediment-bound Ag is especially significant here. In contrast, Ag bioaccumulation in the mussel *M. edulis* is not particularly elevated at East Looe: mussels would appear to be underestimating the potential impact of Ag at this site (although not at Whitehaven and in the Thames Estuary). Such results do not necessarily imply that one species is uniquely preferable to any other as a bioindicator, but rather that various organisms sometimes respond differently to different forms of the metal in the environment. Presumably mussels, being suspension feeders, are not responding to sedimentary Ag in the Looe estuary in this example. In the absence of any 'universal bioindicator' (molluscan or otherwise), monitoring programmes should incorporate analysis of several species as a means of assessing contamination in different phases of the environment (e.g. water, suspended solids and plankton, benthic sediments).

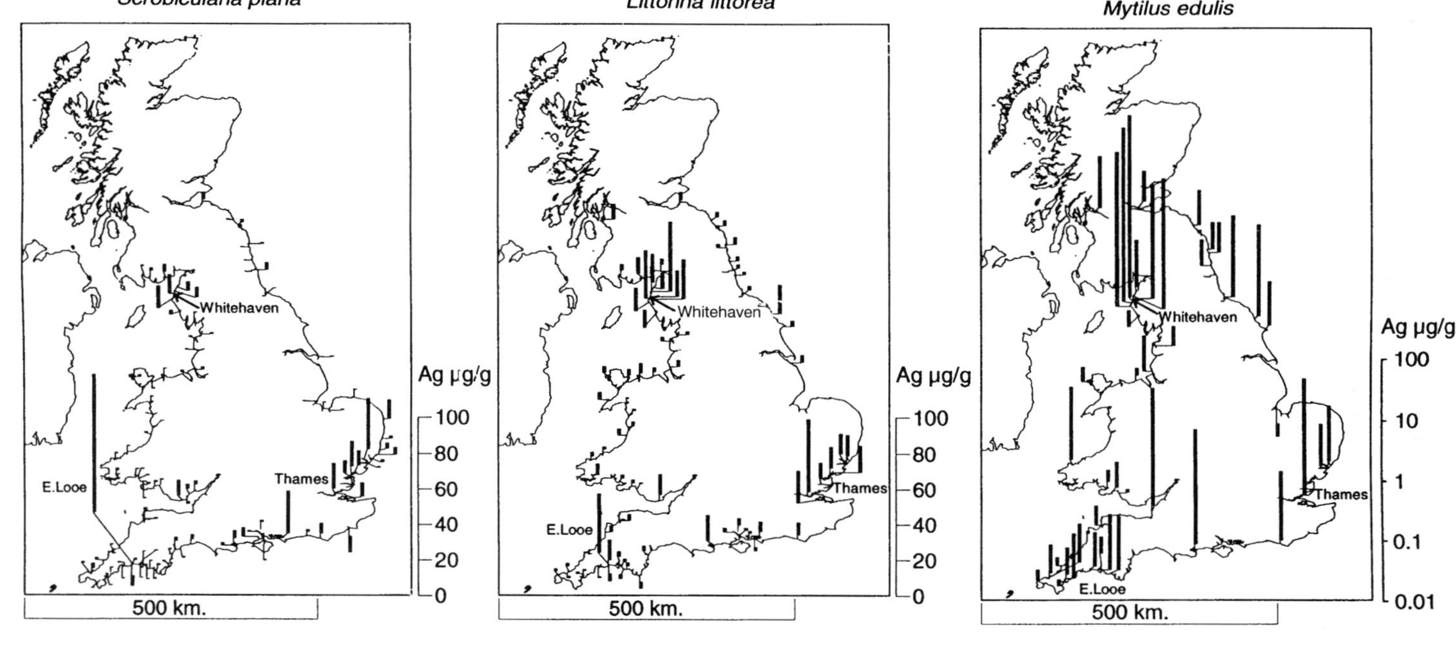

Figure 8.4 Distribution of Ag in *Scrobicularia plana*, *Littorina littorea* and *Mytilus edulis* (μg g^{-1} dw) collected from UK estuaries and coastal waters. (Note logarithmic scale for *M. edulis*.)

Having selected the most appropriate species, the next step towards validating bioindicator potential usually involves minimizing the effect of intrinsic variability on metal burdens. Analysis of individual tissues may be useful: kidney, viscera and muscle are considered as long-term storage sites (reflecting persistent contamination), whilst gills and mantle are capable of more rapid integration and characterization of short-term pollution episodes (e.g. Duquesne and Coll, 1995). In practice, however, constraints on resources often dictate that whole-body analyses are performed, at least where an initial assessment of sites is required.

We are now sufficiently aware of allometric, seasonal and reproductive effects (on trace metal levels) to be able to minimize their influence when monitoring different populations – by correcting for size and avoiding collection of molluscs during breeding seasons, or periods of rapidly changing condition (reviewed by Bryan *et al.*, 1985; Langston and Spence, 1995). The problem of variability within populations can be more problematic: seemingly homogeneous samples of mussels and oysters may display 10-fold variation in metal concentrations, particularly for Zn, with some individuals appearing to regulate the metal and others assimilating unusually high levels (Boyden,1977; Lobel *et al.,* 1982). Comparisons of means for different populations usually require that the data are normally distributed. Where this is not the case, transformation is needed to normalize data (Chapter 7). Optimization of sample numbers, to achieve minimum coefficients of variation, is also an important means of fine-tuning results. Levels of variance typical of *L. littorea*, *S. plana* and *M. edulis* are shown in Table 8.3. Higher coefficients are sometimes a feature of gastropods such as dogwhelk, *Nucella lapillus*, and the limpets *Patella vulgata* (Cd 57–61%; Cu 18–28%; Fe 19–57%; Mn 65–89%; Zn 18–23%), probably due to the diversity in metal-handling systems and diets.

If not controlled, biological parameters such as these may bias estimates of exposure in any one species by up to an order of magnitude (occasionally more). Nevertheless, by standardizing techniques with the most appropriate mollusc species, it is possible to detect concentration differences (attributable to contamination) of the order of 20%, with confidence. Considering that metal concentrations in non-regulating species sometimes exceed background values by several orders of magnitude at grossly contaminated sites, molluscs are clearly capable of revealing the magnitude and spatial extent of anthropogenic contributions of bioavailable metals in the environment, provided that responses are validated. They can be used equally successfully to depict temporal changes in environmental quality which might arise from new pollution sources, or from remediation of existing inputs.

It is worth re-emphasizing that the availability and distribution of internal ligands will partly dictate the kinetics and net balance of metal uptake and loss, and hence body burdens: interspecific variation is therefore to be expected. Paradoxically, the metal detoxifying nature of induced MT or newly formed granules helps to maintain inward diffusion gradients for metals and may enhance accumulation in organisms where their expression is greatest. Some sequestration systems will lead to the presence of extremely high concentrations in certain tissues, and an understanding of their distribution and metal-binding properties can usually help to interpret variable kinetics among different body compartments. For example, rapid (asymptotic) accumulation of both ^{65}Zn and ^{54}Mn occurs in the gills of the clam *Mercenaria campeciensis*, though equilibrium concentrations are relatively higher for Zn because of incorporation into metalloenzymes, which are enriched in this tissue. In the kidney, both elements are accumulated linearly with time, initially by adsorption on to existing granules and subsequently through increased granule formation. For Mn this represents the major accumulation pathway and hence, in contrast to Zn, Mn tissue distributions in clams are dominated by the kidney (Miller *et al.*, 1985). In many other molluscs, particularly in gastropods, interpretation of kinetic data may be more complicated due to the presence of multiple ligand systems within individual tissues.

Information on body/tissue concentrations is clearly important in identifying situations where bioaccumulation of metals is promoted but, *per se*, cannot give a comprehensive insight into biological response. For a more complete assessment, data on metal burdens must be allied to measures of disturbance in metal-handling systems, some of which have been described already. Before addressing the issue in further detail, it is worth examining the evidence for metal-induced impact on native mollusc populations, in order to place into context the relevance and needs of future ecotoxicological research.

8.4.2 ECOLOGICAL EFFECTS

Metal-contaminated water and sediments, and resultant body burdens, have been shown to be harmful to molluscs in numerous laboratory studies but there is controversy surrounding metal toxicity in the field. Thus, although controlled bioassays of effects on growth, respiration, behaviour, reproduction and recruitment are important in identifying vulnerable species and life stages, extrapolation to complex field situations is problematic. This is partly due to the myriad of environmental variables which confound any assessment of change caused by pollutants (Langston, 1990).

Areas exhibiting exceptional metal contamination, usually arising from mining or smelting operations, provide the most unequivocal evidence of impact, since quantities of metals in these locations far outweigh other contaminants. Reduced species diversity, commonly associated with most forms of anthropogenic disturbance, is almost certainly attributable to metals in such cases, though confirmation of the mechanisms that lead to the loss of sensitive species is still needed.

The conspicuous absence of bivalve and gastropod molluscs from highly metal-contaminated sites in the Fal estuary (Restronguet Creek) is a consequence of the long history of metal mining in that area (Bryan *et al.*, 1987): here, for example, Cu and Zn are believed to act by inhibiting the settlement of juvenile bivalves, including *S. plana*, *Cerastoderma edule* and *M. edulis*. Enhanced metal tolerance, which is observed in polychaetes from the same estuary, does not seem to be as effective in ameliorating impact in these molluscs. Oysters may be better adapted, though even they are restricted to the mouth of the estuary. Large reductions in shellfish production in estuaries in Goa can also be linked to impact from metalliferous mine spoils (Parulekar *et al.*, 1986).

In marine environments, evidence for community effects usually pertains to metal-laden sediments. For example, studies in metal-contaminated fjords indicate considerable reductions in faunal diversity at sediment Cu concentrations above 200 μg g^{-1} (Rygg, 1985). Sensitive species, including molluscs, are lost, leaving a high proportion of (tolerant) polychaetes, demonstrating similarities with observations in contaminated estuaries.

Smelting and refining wastes are potentially most hazardous because of their chemical and biological reactivity, as indicated by multivariate techniques in seagrass communities near a Pb smelter in Australia (Ward *et al.*, 1984). Although effects are greatest near the effluence source (where sediment metals are up to 1000 times background levels), poor species richness is observed at sediment concentrations of around 0.7 μg Cd g^{-1}, 10 μg Pb g^{-1} and 92 μg Zn g^{-1} (five times background). Effects are considered to be due to an array of sublethal impacts resulting from the joint actions of various metals.

There is a considerable body of evidence linking metal contamination with biochemical, physiological, reproductive and recruitment anomalies in *M. balthica* populations from San Francisco Bay (Luoma, 1995). Population densities and conditions in *M. balthica* have also been reported to be inversely related to metal levels (0.89 μg Hg g^{-1}, 234 μg Cu g^{-1}, 264 μg Zn g^{-1}) in mudflats of the Fraser River Estuary (McGreer, 1982). In both studies, possible contributions from other stressors cannot be entirely eliminated, despite correlations between metal exposure and effects.

Difficulties in demonstrating the ecological impact of metals on molluscs are perhaps an indication that effects are rare. It is also likely that techniques used in the past to demonstrate impact have seldom been sufficiently sensitive to separate pollutant effects from the variability inherent in natural systems. Multivariate methods that measure responses to metals at the community level are beginning to separate out some of the structural complexities which have previously made definition of sensitive species difficult. Increasingly, attempts will be made in future to link indices of impact based on chemical and biochemical observations, with disturbances at higher organizational level. It may then be possible to trace the chain of events from bioavailability through to reduced competitive ability of the most sensitive organisms and their removal from the community. In the following section we will review, briefly, the application of techniques that, by quantifying abnormalities in metal metabolism, may help to diagnose metal-induced stress, or at least adaptation to stress – which, in turn, may ultimately compromise the ability of molluscs to survive contamination.

8.4.3 BIOCHEMICAL AND CYTOLOGICAL MARKERS OF METAL IMPACT

The metal-sensitive nature of molluscs (compared with, for example, many crustaceans and polychaetes) is partly due to their poor regulatory ability. Nevertheless, metal-detoxifying ligands must confer some resistance and demonstration of the induction (and saturation) of soluble metal-binding proteins (such as MT) or proliferation of inclusions (granules, lysosomes and other concretions) would seem, intuitively, to be a sound basis for examining responses. Systematic investigation of these mechanisms should help to identify vulnerable species and life stages, confirm whether or not they are attempting to adapt to contamination and, hence, define the extent of damage in the field.

We have established that an influx of metals into cells can lead to an elevated level of MT synthesis (in an effort to regulate or detoxify the sudden increase of free metal ions) and detection of this increase in MT protein has often been proposed as a means of demonstrating biological response to metal pollution. Metallothionein is now quantifiable by a number of sensitive methods (chromatography, polarography, metal saturation, immunoassay, MT mRNA) and, as some of the studies reviewed earlier in this chapter show, induction in molluscs is elicited by several metals, often according to exposure levels, at least in the laboratory. Responses are least equivocal in mussels – *Mytilus* spp. (whole animal or gills) – for which numerous isoforms have been sequenced, and adequate knowledge of metal-binding behaviour exists. Metallothionein measurements in mussels are potentially useful as indicators of Cd, Hg, Ag and possibly Cu pollution, though not Zn (this contrasts significantly with fish, where Zn appears to be the most potent inducer of MT; Chapter 10). Further in their favour, interpretation of field data in mussels is not greatly affected by biotic factors (size, condition, age, position in the sexual cycle), temperature or other stressors besides metals. Such endogenous and exogenous factors can be problematic in MT assays using fish and other vertebrates.

Oysters produce MT in response to metals such as Cd and would appear to be possible candidates for sentinel organisms. However, other ligands, particularly very low molecular weight compounds (described above), may be more important in binding high levels of Cu and Zn in some oysters. In addition, since metals do not bind to MT with equal strength, competitive displacement can occur and caution is needed when interpreting data from the field, particularly if contamination with several metals is suspected. Where measurements have been attempted, the increments in MT concentrations in oysters, as in mussels, are relatively small (usually within a factor of three). Comparable increases have also been observed in the clam *Tridacna crocea* (kidney) exposed to metals from large-scale mining operations: MT levels in individuals nearest the source (138 μg g^{-1} ww) are almost double those from more remote sites, and are more closely related to Cd and Zn burdens than to Cu (Duquesne and Coll, 1995).

The value of MT as a 'biomarker' in metal-metabolizing tissues of gastropods, chitons and cephalopods is diminished by the presence of interferences from competing sequestration systems. For example, inherently high levels of MT in *Littorina* digestive gland probably reflect the involvement of this tissue in the turnover of HCY and Cu, and although MT is efficient in Cd detoxification, levels are largely pre-determined by the constitutive role of the protein and are not greatly influenced by external Cd contamination. Several hundred micrograms of Cd per gram may be sequestered before the protein approaches saturation, apparently without harm to the winkle and without inducing a significant net increase in MT production (Table 8.4). Cytosolic metal profiles in the visceral mass of another gastropod, *Nassarius reticulatus*, are likewise independent of contamination (in sediment) and show little evidence of MT induction, though the major Cd-binding protein in this species appears to be of lower thiolic content and exhibits weaker binding than in *Littorina* (Andersen *et al.*, 1989). The involvement of pore-cell granular systems in HCY turnover in gastropod digestive glands could further complicate MT biomarker interpretations. However, in the kidney and gills of *Littorina*, where the protein is less heavily involved in essential metal metabolism, induction of MT by Cd is exposure-related and therefore of potential use as an indicator (Bebianno and Langston, 1995). Significantly, HCY has not been observed in bivalves, other than a small number of primitive protobranchs (Morse *et al.*, 1986). It is perhaps the associated reduced MT involvement in Cu homeostatis and absence of other major competing ligand systems that promote the use of bivalve MTs in a detoxifying and possibly bioindicating capacity.

There are clearly objections to the unqualified adoption of MT assays, even in bivalves. Competitive effects between metals need to be resolved, and

Table 8.4 *Littorina littorea*. Cd and metallothionein (MT) concentrations in digestive gland following exposure to Cd for 65 days (concentrations expressed on a dry weight basis)

Cd exposure (μg l^{-1}) (65 days)	Cd bound to MT pool (μg g^{-1})	MT (mg g^{-1})	MT : Cd molar ratio
0	1.6	10.0	102 : 1
4	8.0	13.5	27 : 1
40	43	8.0	3 : 1
400	358	10.6	1 : 3

there may be reservations regarding the relatively small net increases in MT in molluscs (in contrast to fish MT, where order of magnitude changes have been observed). Indeed some bivalve species, or populations thereof, have a very limited ability to synthesize MT in response to metals. Intraspecific variability has not been studied in many species, but is indicated for geographically separated populations of *M. balthica* (see above), and has also been demonstrated in the oyster *C. virginica*: two molecular weight classes of MT (20 and 10 kDa) are found in *C. virginica* from southern states of the United States, but only the monomeric form is present in individuals collected from north of Chesapeake Bay (Engel and Brouwer, 1982; Fowler *et al.*, 1986; Engel, 1988; Roesijadi *et al.*, 1989). Intensive studies of molluscan MT have undoubtedly improved our understanding of the general biochemical principles governing metal metabolism, but also highlight considerable scope for diversity. Each application of MT protein induction as a biomarker should be accompanied by a critical appraisal of biotic and abiotic factors, turnover times, dose responses and competing ligands/metals.

Apart from limited successes with mussels, and occasionally other species, there are very few examples of field validation. On the basis of current evidence, it seems that contamination in the environment seldom reaches the severity necessary to cause significant net increase in MT protein levels – except perhaps very close to metal-rich outfalls or similar heavily polluted environments. Detection of more subtle pollution events, based on MT protein concentration alone, is therefore likely to prove difficult in molluscs. However, by combining MT determination with observations of associated disturbances in metal partitioning within cells, a more sensitive insight into impact may be gained. Thus, despite the fact that cytosolic metals sometimes represent only a small proportion of the total, they are often more responsive to contamination (i.e. are more representative of bioavailable fractions) than whole-body burdens, the bulk of which may be minimally involved in biological processes. (In such cases, inclusion of this residual non-reactive metal pool in whole-body analysis may mask subtle changes arising from contamination.) In particular, saturation of metallothionein as a result of contamination leads to the appearance of excess metal in 'abnormal' metal pools (the spillover phenomenon) and provides a 'fingerprint' with which to characterize excessive metal influx. Where this involves substitution of non-essential metals, in enzymes and metalloproteins, it is easy to envisage that direct toxic effects will follow (as indicated for oysters by Engel, 1983), though wider confirmation of the consequences of spillover is still required.

Combined observations of MT induction and disturbances in metal partitioning can be used, therefore, to determine whether or not a population is attempting to adapt under metal stress, and to detect the first stages of pollution impact or recovery. Long-term studies of clam populations in San Francisco Bay illustrate the value of this approach since disturbances in metal-binding behaviour (including saturation of MT and spillover) during periods of high Ag and Cu exposure coincide with a decline in biomass production and disruption to the reproductive cycle (Johansson *et al.,* 1986), and thus appear to be markers of higher level disorders. Analogous disturbances in metal partitioning occur in bivalves from contaminated reaches of the Fal Estuary (Restronguet Creek): these include unusually high levels of Cu, Cd and Zn associated with very low molecular weight cytosolic ligands in oysters (Figure 8.1), whilst in transplanted mussels saturation of MT and spillover of Cu to high and low molecular weight pools precede their eventual elimination due to the effects of contamination.

The activities of glutathione (GSH), related GSH-dependent molecules and other antioxidant enzymes have potential as indicators of metal impact since they ameliorate effects, either by sequestration of metal or by reducing oxidative damage. Exposure of the mussel *M. galloprovincialis* to metals, in the laboratory (Cu) or in the field, elicits responses in gills and digestive gland in terms of lowered levels of GSH and raised activities of glyoxalases, allied to damage to cell membranes caused by lipid peroxidation (Regoli and Principato, 1995). Related changes in antioxidative enzymes, including, for example, GSH reductases and peroxidases, have also been detected following transplantation of mussels to sites exposed to multiple pollutants, or acute exposure to Cu in the laboratory. However, activity of the latter group of enzymes was found to be independent of environmental contamination in native mussel populations, suggesting adaptation in chronically polluted organisms, and so limiting their use in determining population responses (Regoli and Principato, 1995). A similar conclusion was drawn over the use of alkaline phosphatase in mussels, despite the enzyme being widely regarded as a general indicator of metabolic condition which is sensitive to metals. A variety of other biomolecules, such as the heat stress protein HSP60 (whose levels increase in mussels following Cu exposure; Sanders *et al.,* 1991), display potential as biomarkers, though their induction may be indicative of a generalized stress response rather than reaction to metal cytotoxicity.

Enhanced accumulation of lipofuchsin in tertiary lysosomes is a quantifiable feature of mussels from metal-polluted environments and is consistent with the notion that metal exposure causes an increase in oxidative damage to membrane lipids (Regoli and Principato, 1995). Disruption of cell membranes in response to metals is further manifested by the destabilization of lysosomes, and measures of lysosomal membrane integrity might also be considered for use in biological effects monitoring. Indices are usually based on the premise that the ability of lysosomes to retain a lipohilic dye is impaired under stress, causing the lysosomal membranes to become leaky and allowing the dye to diffuse back to the cytoplasm. In laboratory trials with the freshwater snail *Viviparus contectus*, for example, neutral red retention time in lysosomes of blood cells declines in proportion to exposure concentrations of Cu in the range 30–1000 μg l^{-1} (Svendsen and Weeks, 1995). As diagnostic tests for metals under complex field conditions, both of the above types of lysosomal assay (lipofuchsin accumulation and neutral red retention) may be compromised by the fact that other forms of stress could inflict identical symptoms. As with most other indices, deployment as one of a suite of tests would be the preferred option in determining impact.

Observations of the formation or extrusion of intracellular granules in response to metals have led to suggestions that they may be used as indicators of metal stress. Granule proliferation occurs, for example, in kidney cells of the scallop *Placopecten magellanicus* exposed to Cu (Fowler and Gould, 1988); in the clam *Mercenaria campechiensis* exposed to Zn and Mn (where there is an increase in both size and numbers of nephrocytic granules) (Miller *et al.*, 1985); and in the bivalve *Donax trunculus* (where extracellular, Mn-rich renal concretions are produced in response to elevated Mn concentrations, both in laboratory exposures and at polluted field sites; Mauri and Orlando, 1982). The high incidence of granular inclusions in digestive glands of the gastropod *L. littorea* from Restronguet Creek is also a characteristic feature attributable to metal pollution (Mason *et al.*, 1984).

Together, this evidence supports the use of granules in a diagnostic capacity, though consideration must be given to the fact that in most cases granule formation and loss is a normal component of metal metabolism; in *D. trunculus*, for example, other factors such as sex and season are influential. Practical applications are therefore currently restricted to descriptive assessments and are likely to be most valuable when tested alongside other response indices. The ultrastructural and biochemical study of Cd- and Cu-binding patterns and damage in kidney cells of *P. magellanicus* by Fowler and Gould (1988) illustrates the potential benefits of integral studies in interpreting toxicological signals. Thus, extrusion of intracellular concretions from the kidney is seen to coincide with displacement of Zn and Cd from cytosolic proteins during Cu exposure. These disturbances to metal partitioning are accompanied by marked cellular degeneration of the nephrocytes, together with reduction in enzyme (renal isocitrate dehydrogenase) activity. Altered binding patterns and granule secretion would appear to signify not only the disruption of normal regulatory functions but may also herald the toxicity of previously sequestered metal (displaced by Cu). In contrast, although Cd exposure results in minor disturbances to cytosolic metals (including induction of MT-like protein), metal-binding granules are retained in scallop kidneys and there is no evidence of cell injury: any changes are presumably within the range of homeostasis offered by Cd-binding granules and cytosolic proteins. Quantitative measures of granule proliferation or excretion therefore represent a feasible means of gauging impact, but they should not be considered in isolation in view of the significance of interactions with other intracellular ligands.

In summary, a number of biochemical and related cellular signals are displayed by molluscs which, together with body burden data, are capable of diagnosing responses to metal contamination. Disturbances in metal partitioning exhibit some consistent features but are modified at the species and tissue level; they must be fully characterized before being deployed in bio-monitoring. Metallothionein concentrations and induction rates are low and somewhat variable among molluscs; their value lies in the specificity to metals, compared with other (particularly higher-level) responses, which are often general symptoms of stress. For the immediate future, potential markers of damage and altered metal metabolism require further validation in the field, especially where interactions between metals may arise from complex inputs. It is obvious that integrated studies, involving several diagnostic components, will be more enlightening than individual chemical or biological assessments. In the long term, more evidence is needed to link these disturbances in metal metabolism with higher order effects, and to define the mechanisms of impact in molluscs. For example, it may be postulated that pollution-induced MT or granule proliferation results in the diversion of energy reserves away from normal requirements. However, there are as yet no satisfactory indications as to the share of the energy budget diverted into detoxification or the related costs of tolerance (e.g. MT-gene multiplication, loss of diversity), though it is clear that these processes could influence the general health of the organism or population.

8.5 CONCLUSIONS

This brief review highlights the diversity in metal metabolism encountered within a small subsample of the phylum Mollusca. As more species are examined it is certain that this diversity will increase: to date only a handful of mollusc species has been studied in any depth, yet there is estimated to be a total of 50 000 marine species alone (Yonge and Thompson, 1976) and perhaps as many as 120 000 molluscs in total. Nevertheless, an understanding of the processes of metal bioaccumulation, transfer, storage and elimination (albeit limited) promotes the use of representative molluscan species as monitors of the state of the aquatic environment. This is an excellent demonstration of the application and benefits of basic science, which needs to be carried forward. We are currently able to define the most suitable properties of molluscan bioindicators and to apply appropriate strategies to ensure that body burden data accurately reflect the bioavailability of contaminants in the environment. The continuing search for biochemical criteria to classify the toxicological status of these burdens should prove rewarding, provided we accept the limitations. For example, metallothionein studies, in isolation, seem unlikely to be of value in predicting specific functional changes in ecosystems, and may even prove difficult at the level of the individual or population. Equally, an ecological study, *per se*, is unlikely to be an acceptable indicator of stress unless accompanied by chemical or biochemical information on causative agents and the mechanisms of effect – damaged communities display similar symptoms (e.g. reduced abundance and species diversity) whatever the reason. Without some indication of specific cause and effect – whether it be manifested in metal accumulation, MT induction, altered enzyme activity or granule production – it would be difficult for the ecotoxicologist to make an appropriate assessment of damage and to propose the correct remedial action.

REFERENCES

Andersen, R.A., Eriksen, K.D.H and Bakke, T. (1989) Evidence of presence of a low molecular weight, non-metallothionein-like metal-binding protein in the marine gastropod *Nassarius reticulatus* L. *Comparative Biochemistry and Physiology* **94B**, 285–291.

Ballatori, N. and Clarkson, T. W. (1982) Developmental changes in the billary excretion of methylmercury and glutathione. *Science* **216**, 61–63.

Balogh, K.V. and Salanki, J. (1984) The dynamics of mercury and cadmium uptake into different organs of *Anodonta cygnea* L. *Water Research* **18**, 1381–1387.

Bebianno, M.J. and Langston, W.J. (1991) Metallothionein induction in *Mytilus edulis* exposed to cadmium. *Marine Biology* **108**, 91–96.

Bebianno, M.J. and Langston W.J. (1992) Cadmium induction of metallothionein synthesis in *Mytilus galloprovincialis*. *Comparative Biochemistry and Physiology* **103C**, 79–85.

Bebianno, M.J. and Langston W.J. (1993) Turnover rate of metallothionein and cadmium in *Mytilus edulis*. *BioMetals* **6**, 239–244.

Bebianno, M.J. and Langston, W.J. (1995) Induction of metallothionein synthesis in the gill and kidney of *Littorina littorea* exposed to cadmium. *Journal Marine Biological Association of the UK* **75**, 173–186.

Bebianno, M.J., Nott, J.A. and Langston, W.J. (1993) Cadmium metabolism in the clam *Ruditapes decussata*: the role of metallothioneins. *Aquatic Toxicology* **27**, 315–334.

Berthet, B. (1990) Influence de la voie de contamination sur les formes physico-chimiques de l'argent chez *Crassostrea gigas* Thurnberg. *Océanis* **16**(5), 249–357.

Bootsma, N., Macey, D.J., Webb, J. and Talbot, V. (1988) Isolation and characterisation of ferritin from the hepatopancreas of the mussel *Mytilus edulis*. *Biology of Metals* **1**, 106–111.

Bouquegneau, J.M. and Martoja, M. (1982) La teneur en cuirve et son degré de complexation chez quatre Gasteropodes marins. Donnèes sur le cadmium et le zinc. *Oceanol. Acta* **5**(2), 219–228.

Bouquegneau, J.M., Martoja, M. and Truchet, M. (1983) Localisation biochimique du cadmium chez *Murex trunculus* L. (Prosobranche Neogasteropode) en mileu naturel non pollué et après intoxication expérimentale, *C.R. Acad. Sc. Paris* **296**, 1121–1124.

Bourgoin, B.P., Risk, M.J., Evans, R.D. and Cornett, R.J. (1991) Relationships between the partitioning of lead in sediments and its accumulation in the marine mussel, *Mytilus edulis* near a lead smelter. *Water, Air and Soil Pollution* **57–58**, 377–386.

Boyden, C.R. (1977) The effect of size upon metal content of shellfish. *Journal Marine Biological Association of the UK* **57**, 675–714.

Brown, S.L. (1986) Faeces of intertidal benthic invertebrates: influence of particle selection in feeding on trace element concentration. *Marine Ecology Progress Series* **28**, 219–231.

Bryan, G.W. and Langston, W.J. (1992) Bioavailability, accumulation and effects of heavy metals in sediments with special reference to United Kingdom estuaries: a review. *Environmental Pollution* **76**, 89–131.

Bryan, G.W., Langston, W.J., Hummerstone, L.G. and Burt, G.R. (1985) A guide to the assessment of heavy-metal contamination in estuaries using biological indicators. *Marine Biological Association of the UK Occasional Publication* **4**.

Bryan, G.W., Gibbs, P.E., Hummerstone, L.G. and Burt, G.R. (1987) Copper, zinc and organotin as long-term factors governing the distribution of organisms in the Fal estuary Southwest England, *Estuaries* **10**, 208–219.

Carmichael, N.G., Squibb, K.S. and Fowler, B.A. (1979) Metals in the molluscan kidney: a comparison of two closely related bivalve species (*Argopecten*) using X-ray microanalysis and atomic absorption spectroscopy, *J. Fish Res. Bd. Can.* **36**, 1149–1155.

Carpenè, E. (1993) Metallothionein in marine molluscs, in *Ecotoxicology of Metals in Invertebrates*, (eds R. Dallinger and P.S. Rainbow), CRC Press, Boca Raton, Florida, pp. 55–72.

Carpenè, E., Cortesi, P., Crisetig, G. and Serrazanetti, G.P. (1980) Cadmium-binding proteins from the mantle of *Mytilus edulis* (L.) after exposure to cadmium, *Thallassia Jugoslavia* **16** (2–4), 317–323.

Carpenè, E., Cattani, O., Hakim, G. and Serrazanetti, G.P. (1983) Metallothionein from the foot and posterior adductor muscle of *Mytilus galloprovincialis*. *Comparative Biochemistry and Physiology* **74C**(2), 331–336.

Carvajal, N., Uribe, E. and Torres, C. (1994) Subcellular localization, metal ion requirement and kinetic properties of arginase from the gill tissue of the bivalve *Semele solida. Comparative Biochemistry and Physiology* **109B**, 683–689.

Cassini, A., Tallindini, L., Favero, N. and Albergoni, V. (1986) Cadmium bioaccumulation studies in the freshwater molluscs *Anodonta cygnea* and *Unio elongatulus. Comparative Biochemistry and Physiology* **84C**, 35–41.

Casterline, J.L. and Yip, G. (1975) The distribution and binding of cadmium in oyster, soybean, and rat liver and kidney. *Archives Environmental Contamination and Toxicology* **3**, 319–329.

Castillo, L.V. and Maita, Y. (1991) Isolation and partial characterization of cadmium binding proteins from the oceanic squid, *Ommastrephes bartrami. Bulletin Faculty Fisheries Hokaido University* **42**(1), 26–34.

Chan, H.M. (1988) Accumulation and tolerance to cadmium, copper, lead and zinc by the green mussel *Perna viridis. Marine Ecology Progress Series* **48**, 295–303.

Coombs, T.L. (1974) The nature of zinc and copper complexes in the oyster *Ostrea edulis. Marine Biology* **28**, 1–10.

Couillard, Y., Campbell, P.G.C. and Tessier, A. (1993) Response of metallothionein concentrations in a freshwater bivalve (*Anodonta grandis*) along an environmental cadmium gradient. *Limnology and Oceanography* **38(2)**, 299–313.

Decho, A.W. and Luoma, S.N. (1991) Time-courses in the retention of food material in the bivalves *Potamocorbula amurensis* and *Macoma balthica*: significance to the absorption of carbon and chromium. *Marine Ecology Progress Series* **78**, 303–314.

Decho, A.W. and Luoma, S.N. (1994) Humic and fulvic acids: sink or source in the availability of metals to the marine bivalves *Macoma balthica* and *Potamocorbula amurensis. Marine Ecology Progress Series* **108**, 133–145.

Decho, A.W. and Luoma, S.N. (1996) Flexible digestion strategies and trace metal assimilation. *Limnology and Oceanography* **41(3)**, 568–572.

Di Toro, D.M., Mahony, J.D., Hansen, D.J. *et al.* (1992) Acid volatile sulfide predicts the acute toxicity of cadmium and nickel in sediments, *Environmental Science and Technology* **26**, 96–101.

Dixon, M. and Webb, E.C. (1979) *Enzymes*, 3rd edn, Longman, London.

Dohi, Y., Kosaka, K., Ohba, K. and Yoneyama, Y. (1986) Cadmium binding proteins of three molluscs and characterization of two cadmium-binding glycoproteins from the hepatopancreas of a whelk *Buccinum tenuissum*. *Environmental Health Perspectives* **65**, 49–55.

Duquesne, S.J. and Coll, J.C. (1995) Metal accumulation in the clam *Tridacna crocea* under natural and experimental conditions. *Aquatic Toxicology* **32**, 239–253.

Duquesne, S.J., Flowers, A.E. and Coll, J.C. (1995) Evidence for a metallothionein-like heavy metal-binding protein in the marine tropical bivalve *Tridacna crocea*. *Comparative Biochemistry and Physiology* **112C**(1), 69–78.

Engel, D.W. (1983) The intracellular partitioning of trace metals in marine shellfish. *Science of Total Environment* **28**, 129–140.

Engel, D.W. (1988) The effect of biological variability on monitoring strategies, metallothioneins as an example. *Water Resources Bulletin* **24**, 981–987.

Engel, D.W. and Brouwer, M. (1982) Detoxification of accumulated metals by the American oyster, *Crassostrea virginica*; laboratory vs environment, in *Physiological Mechanisms of Marine Pollutant Toxicity*, (eds W. B. Vernberg, A. Calabrese, F.P. Thurberg and F.J. Vernberg), Academic Press, New York, pp. 89–107.

Engel, D.W., Sunda, W.G. and Fowler, B.A. (1981) Factors affecting trace metal uptake and toxicity to estuarine organisms. I. Environmental parameters, in *Biological Monitoring of Marine Pollutants*, (eds J. Vernberg, A. Calabrese, F.P. Thurberg and W.B. Verhberg), Academic Press, New York, pp. 127–145.

Evtushenko, Z.S., Belcheva, N.N. and Lukyanova, O.N. (1986) Cadmium accumulation in organs of the scallop *Mizuhopecten yessoensis*-II. Subcellular distribution of the metals and metal-binding proteins. *Comparative Biochemistry and Physiology* **83C**(2), 377–383.

Evtushenko, Z.S., Lukyanova, O.N. and Belcheva, N.N. (1990) Cadmium bioaccumulation in organs of the scallop *Mizuhopecten yessoensis*. *Marine Biology* **104**, 247–250.

Fayi, L. and George, S.G. (1985) Purification of very low molecular weight Cu-complexes from the European oyster, In *Marine Pollution and Physiology: Recent Advances* (eds Vernberg, F.J., Thurberg, F.P., Calabrese, A. and Vernberg, W.B.), CRC Press Inc., Boca Raton, Florida. 123–142.

Finger, J.M. and Smith, J.D. (1987) Molecular association of Cu, Zn, Cd and ^{210}Po in the digestive gland of the squid *Nototodarus gouldi*. *Marine Biology* **95**, 87–91.

Fisher, N.S. and Reinfelder, J.R. (1995) The trophic transfer of metals in marine systems, in *Metal Speciation and Bioavailability in Aquatic Systems* (eds A. Tessier and D.R. Turner), John Wiley, New York, pp. 363–406.

Foster, P. and Chacko, J. (1995) Minor and trace elements in the shell of *Patella vulgata* (L.). *Marine Environmental Research* **40**(1), 55–76.

Fowler, B.A. and Gould, D. (1988) Ultrastructural and biochemical studies of intracellular metal-binding patterns in the kidney tubule of the scallop *Placopecten magellanicus* following prolonged exposure to cadmium and copper. *Marine Biology* **97**, 207–216.

Fowler, B.A., Carmichael, N.G. and Squibb, K.S. (1981) Factors affecting trace-metal uptake and toxicity to estuarine organisms. II. Cellular mechanisms, in *Biological Monitoring of Marine Pollutants* (eds F.J. Vernberg, A. Calabrese, F.P. Thurberg and W.B. Vernberg), Academic Press, New York, pp. 145–163.

Fowler, B.A., Engel, D.W. and Brouwer, M. (1986) Purification and characterization studies of cadmium-binding proteins from the American oyster, *Crassostrea virginica*. *Environmental Health Perspectives* **65**, 63–69.

Frankenne, F., Noel-Lambot, F. and Disteche, A. (1980) Isolation and characterization of metallothioneins from cadmium-loaded mussel *Mytilus edulis*. *Comparative Biochemistry and Physiology* **66C**, 179–182.

Frazier, J.M. and George, S.G. (1983) Cadmium kinetics in oysters – a comparative study of *Crassostrea gigas* and *Ostrea edulis*. *Marine Biology* **76**, 55–61.

George, S.G. (1983a) Heavy metal detoxification in mussel *Mytilus edulis*. Composition of Cd-containing kidney granules (tertiary lysosomes). *Comparative Biochemistry and Physiology* **76C**, 53–57.

George, S.G. (1983b) Heavy metal detoxification in mussel *Mytilus edulis* – an *in vitro* study of Cd- and Zn-binding to isolated tertiary lysosomes. *Comparative Biochemistry and Physiology* **76C**, 59–65.

George, S.G. (1990) Biochemical and cytological assessments of metal toxicity in marine animals, in *Heavy Metals in the Marine Environment* (eds R.W. Furness and P.S. Rainbow), CRC Press Inc., Boca Raton, Florida, pp. 123–142.

George, S.G. and Coombs, T.L. (1977) The effects of chelating agents on the uptake and accumulation of cadmium by *Mytilus edulis*. *Marine Biology* **39**, 261–268.

George, S.G. and Langston W.J. (1994) Metallothionein as an indicator of water quality: assessment of the bioavailability of cadmium, copper, mercury and zinc in aquatic animals at the cellular level, in *Water Quality and Stress Indicators in Marine and Freshwater Systems, Linking Levels of Organisation*, (ed. D.W. Sutcliffe), Freshwater Biological Association, Ambleside, Cumbria, pp. 138–153.

George, S.G. and Olsson, P.-E. (1994) MT as indicators of trace metal pollution, in *Biomonitoring of Coastal Waters and Estuaries*, (ed. K.J.M. Kramer), CRC Press, Boca Raton, Florida, pp. 151–178.

George, S.G. and Pirie, B.J.S. (1979) The occurrence of cadmium in sub-cellular particles in the kidney of the marine mussel *Mytilus edulis*, exposed to cadmium. *Biochimica Biophysica Acta* **580**, 234–244.

George, S.G., Pirie, B.J.S., Cheyne, A.R. *et al.* (1978) Detoxification of metals by marine bivalves: an ultrastructural study of the compartmentation of copper and zinc in the oyster *Ostrea edulis*. *Marine Biology* **45**, 147–156.

George, S.G., Carpenè E., Coombs T.L. *et al.* (1979) Characterization of cadmium-binding proteins from mussels, *Mytilus edulis* (L.) exposed to cadmium. *Biochimica Biophysica Acta* **580**, 225–233.

George, S.G., Pirie, B.J.S. and Coombs, T.L. (1980) Isolation and elemental analysis of metal-rich granules from the kidney of the Scallop, *Pecten Maximus* (L.), *J. Exp. Mar. Biol. Ecol.* **42**, 143–156.

George, S.G., Pirie, B.J.S. and Frazier, J.M. (1983) Effects of cadmium exposure on metal-containing amoebocytes of the oyster *Ostrea edulis*. *Marine Biology* **76**, 63–66.

Gnassia-Barelli, M., Romeo, M. and Puiseux-Dao, S. (1995) Effects of cadmium and copper contamination on calcium content of the bivalve *Ruditapes decussatus*. *Marine Environmental Research* **39**, 325–328.

Gully, J.R. and Mason, A.Z. (1993) Cytosolic redistribution and enhanced accumulation of Cu in gill tissue of *Littorina littorea* as a result of Cd exposure. *Marine Environmental Research* **35**(1/2), 53–57.

Hamer, D.H. (1986) Metallothionein. *Annual Review Biochemistry* **55**, 913–951.

Harrison, F. (1979) Effect of the physiochemical form of trace metals and their accumulation by bivalve molluscs, in *Chemical Models in Aqueous Systems*, (ed. A. Jenne), American Chemical Symposium Series 93, pp. 611–634.

Harrison, F.L. and Lam, J.R. (1985) Partitioning of copper and copper-binding proteins in the mussel *Mytilus edulis* exposed to soluble copper. *Marine Environmental Research* **16**, 151–163.

Harrison, F.L., Watness, K., Nelson, D.A. and Calabrese, A. (1987) Mercury-binding proteins in the slipper limpet, *Crepidula fornicata*, exposed to increased soluble mercury. *Estuaries* **10**(1), 78–83.

Harvey, R.W. and Luoma, S.N. (1985) Separation of solute and particulate vectors of heavy-metal uptake in controlled suspension-feeding experiments with *Macoma balthica*. *Hydrobiologia* **121**, 97–102.

Hennig, H.F-K.O. (1986) Metal-binding proteins as metal pollution indicators. *Environmental Health Perspectives* **65**, 175–187.

Howard, A.G. and Nickless, G. (1977) Heavy metal complexation in polluted molluscs. I. Limpets (*Patella vulgata* and *Patella intermedia*). *Chemical Biological Interactions* **16**, 107–114.

Howard, A.G. and Nickless, G. (1978) Heavy metal complexation in polluted molluscs. III. Periwinkles (*Littorina littorea*), cockles (*Cardium edule*) and scallops (*Chlamys opercularis*). *Chemical Biological Interactions* **23**, 227–231.

Hylland, K., Kaland, T. and Andersen, T. (1994) Subcellular Cd accumulation and Cd-binding proteins in the netted dog welk, *Nassarius reticulatus* L. *Marine Environmental Research* **38**, 169–193.

Imber, B.E., Thompson, J.A.J. and Ward, S. (1987) Metal-binding protein in the Pacific oyster, *Crassostrea gigas*, assessment of the protein as a biochemical environmental indicator. *Bulletin Environmental Contamination Toxicology* **38**, 707–714.

Johansson, C., Cain, D.J. and Luoma, S.N. (1986) Variability in the fractionation of Cu, Ag, and Zn among cytosolic proteins in the bivalve *Macoma balthica*. *Marine Ecology Progress Series* **28**, 87–97.

Julsham, K. and Andersen, K.-J. (1983) Subcellular distribution of major and minor elements in unexposed molluscs of western Norway. II. The distribution and binding of cadmium, zinc, copper, magnesium, manganese and iron in the kidney and digestive system of the common mussel *Mytilus edulis*. *Comparative Biochemistry and Physiology* **75A**, 13–16.

Kojima, Y. and Kagi J.H.R. (1978) Metallothionein. *Trends Biochemical Science* **3**, 90–93.

Langston, W.J. (1980) Arsenic in UK estuarine sediments and its availability to benthic organisms. *Journal Marine Biological Association of the UK* **60**, 869–881.

Langston, W.J. (1982) The distribution of mercury in British estuarine sediments and its availability to deposit-feeding bivalves. *Journal Marine Biological Association of the UK* **62**, 667–684.

Langston, W.J. (1986) Metals in sediments and benthic organisms in the Mersey estuary. *Estuarine Coastal Shelf Science* **23**, 239–261.

Langston, W.J. (1990) Toxic effects of metals and the incidence of metal pollution in marine ecosystems, in *Heavy Metals in the Marine Environment*, (eds R.W. Furness and P.S. Rainbow), CRC Press, Boca Raton, pp. 101–122.

Langston, W.J. (1996) Recent developments in TBT ecotoxicology. *Toxicology and Ecotoxicology News* **3**, 179–187.

Langston, W.J. and Spence, S.K. (1995) Biological factors involved in metal concentrations observed in aquatic organisms, in *Metal Speciation and Bioavailability in Aquatic Systems*, (eds A. Tessier and D.R. Turner), John Wiley, New York, pp. 407–478.

Langston, W.J. and Zhou, M. (1986) Evaluation of the significance of metal-binding proteins in the gastropod *Littorina littorea*. *Marine Biology* **92**, 505– 515.

Langston, W.J. and Zhou, M. (1987a) Cadmium accumulation, distribution and metabolism in the gastropod *Littorina littorea*: the role of metal-binding proteins. *Journal Marine Biological Association of the UK* **67**, 585–601.

Langston, W.J. and Zhou, M. (1987b) Cadmium accumulation, distribution and elimination in the bivalve *Macoma balthica* : neither metallothionein nor metallothionein-like proteins are involved. *Marine Environmental Research* **21**, 225–237.

Langston, W.J., Bryan, G.W., Burt, G.R. and Gibbs, P.E. (1990) Assessing the impact of tin and TBT in estuaries and coastal regions. *Functional Ecology* **4**, 433–443.

Lauenstein, G.G., Robertson, A. and O'Connor, T.P. (1990) Comparison of trace metal data in mussels and oysters from a mussel watch programme of the 1970s with those from a 1980s programme. *Marine Pollution Bulletin* **21**, 440–447.

Lobel, P.B., Mogie, P., Wright, D.A. and Wu, B.L. (1982) Metal accumulation in four molluscs. *Marine Pollution Bulletin* **13**, 170–174.

Lukyanova, O.N., Belcheva, N.N. and Chelomin, V.P. (1993) Cadmium bioaccumulation in the scallop *Mizuhopecten yessoensis* from an unpolluted environment, in *Ecotoxicology of Metals in Invertebrates,,* (eds R. Dallinger and P.S. Rainbow), CRC Press, Boca Raton, Florida, pp. 25–36.

Luoma, S.N. (1983) Bioavailability of trace metals to aquatic organisms – a review. *Science of the Total Environment* **28**, 1–22.

Luoma, S.N. (1995) Prediction of metal toxicity in nature from bioassays: limitations and research needs, in *Metal Speciation and Bioavailability in Aquatic Systems*, (eds A. Tessier and D.R. Turner), John Wiley, New York, pp. 609–659.

Luoma, S.N. and Bryan, G.W. (1978) Factors controlling availability of sediment-bound lead to the estuarine bivalve *Scrobicularia plana*. *Journal Marine Biological Association of the UK* **58**, 793–802.

Luoma, S.N. and Bryan, G.W. (1982) A statistical study of environmental factors controlling concentrations of heavy metals in the burrowing bivalve *Scrobicularia plana* and the polychaete *Nereis diversicolor*. *Estuarine Coastal Shelf Science* **15**, 95–108.

Luoma, S.N. and Jenne, E.A. (1976) Factors affecting the availability of sediment-bound cadmium to the estuarine deposit-feeding clam, *Macoma balthica*, in *Radioecology and Energy Resources: Proceedings of the 4th National Symposium on Radioecology, May 12–14 1976, Cornwallis, Oregon*, (ed. C.E. Cuming Jr), Oregon State University, The Ecological Society of America, Special Publication No. 1, pp. 238–290.

Luoma, S.N. and Jenne, E.A. (1977) The availability of sediment-bound cobalt, silver and zinc to a deposit-feeding clam. Paper presented at the 15th Life Sciences Symposium, Biological Implications of Metals in the Environment, Hanford, Washington, September 29–October 1, 1977, 30 pp.

Luoma, S.N., Johns, C., Fisher, N.S. *et al.* (1992) Determination of selenium bioavailability to a benthic bivalve from particulate and solute pathways. *Environmental Science Technology* **26**, 485–491.

Mackay, E.A., Overnell, J., Dunbar, B. *et al.* (1993) Complete amino acid sequences of five dimeric and four monomeric forms of metallothionein from the edible mussel *Mytilus edulis*. *European Journal of Biochemistry* **218**, 183–194.

Margoshes, M. and Vallee, B.L. (1957) A cadmium protein from equine kidney cortex, *Journal of the American Chemical Society* **79**, 4813–4819.

Marshall, A.T. and Talbot, V. (1979) Accumulation of cadmium and lead in the gills of *Mytilus edulis*: X-ray microanalysis and chemical analysis. *Chemical Biological Interactions* **27**, 111–123.

Martoja, M. and Marcaillou, C. (1993) Localisation cytologique du cuivre et de quelques autres métaux dans la gland digestive de la seiche, *Sepia officinalis* L. (Mollusque céphalopode) *Canadian Journal of Fisheries and Aquatic Science* **50**, 542–550.

Martoja, M., Bouquegneau, J.M., Truchet, M. and Martoja, R. (1985) Recherche de l'argent chez quelques mollusques marins, dulcicoles et terrestres. Formes chimiques et localisation histologique. *Vie Millieu* **35(1)**, 1–13.

Mason, A.Z. and Jenkins, K.D. (1995) Metal detoxification in aquatic organisms, in *Metal Speciation and Bioavailability* (eds A. Tessier and D.R. Turner), John Wiley, New York, pp. 449–608.

Mason, A.Z. and Nott, J.A. (1981) The role of intracellular biomineralised granules in the regulation and detoxification of metals in gastropods with special reference to the marine prosobranch *Littorina littorea*. *Aquatic Toxicology* **1**, 239–256.

Mason, J.W., Cho, J.H. and Anderson, A.C. (1976) Uptake and loss of inorganic mercury in the eastern oyster (*Crassostrea virginica*). *Arch. Environ. Contamin. Toxicol* **4**, 361–376.

Mason, A.Z., Simkiss, K. and Ryan, K.P. (1984) The ultrastructural localisation of metals in specimens of *Littorina littorea* collected from clean and polluted sites. *Journal of the Marine Biological Association of the UK* **64**, 699–720.

Mauri, M. and Orlando, E. (1982) Experimental study on renal concretions in the wedge shell *Donax trunculus* (L). *Journal of Marine Biology and Ecology* **63**, 47–57.

McGreer, E.R. (1982) Factors affecting the distribution of the bivalve, *Macoma balthica* (L) on a mudflat receiving sewage effluent, Fraser River estuary, British Columbia. *Marine Environmental Research* **7**, 131–149.

Miller, W.L., Blake, N.J. and Byrne, R.H. (1985) Uptake of Zn^{65} and Mn^{54} into body tissues and renal concretions by the southern Quahog, *Mercenaria campechiensis* (Gmelin): effects of elevated phosphate and metal concentrations. *Marine Environmental Research* **17**, 167–171.

Miramand, P. and Bentley, D. (1992) Concentration and distribution of heavy metals in tissues of two cephalopods, *Eledone cirrhosa* and *Sepia officinalis*, from the French coast of the English Channel. *Marine Biology* **114**, 407–414.

Morse, M.P., Meyhofer, E., Otto, J.J. and Kuzirian, A.M. (1986) Hemocyanin respiratory pigment in bivalve molluscks. *Science* **231**, 1302–1304.

Noel-Lambot, F. (1976) Distribution of cadmium, zinc and copper in the mussel *Mytilus edulis*. Existence of cadmium-binding proteins similar to metallothioneins. *Experientia* **32**, 324–325.

Noel-Lambot, F., Bouquegneau, J.M., Frankenne, F. and Disteche, A. (1978) Le role des metallothioneines dans le stockage des metaux lourds chez les animaux marins. *Revue International Oceanografie Medical* **XLIX**, 13–20.

Noel-Lambot, F., Bouquegneau, J.M., Frankenne, F. and Disteche, A. (1980) Cadmium, zinc and copper accumulation in limpets (*Patella vulgata*) from the Bristol Channel with special reference to metallothioneins. *Marine Ecology Progress Series* **2**, 81–89.

Nolan, C.V. and Duke, E.J. (1983a) Cadmium-binding proteins in *Mytilus edulis*: relation to mode of administration and significance in tissue retention of cadmium. *Chemosphere* **12**, 65–74.

Nolan, C.V. and Duke, E.J. (1983b) Cadmium accumulation and toxicity in *Mytilus edulis*: involvement of metallothioneins and heavy-molecular weight protein. *Aquatic Toxicology* **4**, 153–163.

Nott, J.A. (1993) X-ray microanalysis in pollution studies, in *X-ray Microanalysis in Biology: Experimental Techniques and Applications*, (eds D.C. Sigee, A.J. Morgan, A.T. Sumner and A. Warley), Cambridge University Press, pp. 257–281.

Nott, J.A. and Langston, W.J. (1989) Cadmium and the phosphate granules in *Littorina littorea*. *Journal Marine Biological Association of the UK* **69**, 219–227.

Nott, J.A. and Langston, W.J. (1993) Effects of cadmium and zinc on the composition of phosphate granules in the marine snail *Littorina littorea*. *Aquatic Toxicology* **25**, 43–54.

Ohtake, H., Suyemitsu, T. and Koga, M. (1983) Sea urchin (*Anthocidaris crassispina*) egg zinc-binding protein. Cellular localization, purification and characterization, *Biochemical Journal* **211**, 109–118.

Olafson, R.W., Sim, R.G. and Boto, K.G. (1979) Isolation and chemical characterization of heavy metal-binding protein metallothionein from marine invertebrates, *Comparative Biochemistry and Physiology* **62B**, 404–416.

Ozaki, K., Terakita, A. , Ozaki, M. *et al.* (1994) Molecular characterization and functional expression of squid retinal-binding protein, *Journal of Biological Chemistry* **269**(5), 3838–3845.

Parulekar, A.M., Ansari, Z.A. and Ingole, B.S. (1986) Effect of mining activities on the clam fisheries and bottom fauna of Goa estuaries. *Proceedings Indian Academy Science (Anim. Sci.)* **95**, 325–339.

Pavicic, J., Raspor, B. and Martincic, M. (1993) Quantitative determination of metallothionein-like proteins in mussels. Methodological approach and field evaluation. *Marine Biology* **15**, 435–444.

Pavicic, J., Skreblin, M., Tusek-Znidaric, M. and Stegnar, P. (1994) Embryo-larval tolerance of *Mytilus galloprovincialis* exposed to elevated sea water metal concentrations. I. Toxic effects of Cd, Zn and Hg in relation to the metallothionein level. *Comparative Biochemistry and Physiology* **107C**, 249–257.

Pavicic, J., Skreblin, M., Kregar, I. *et al.* (1985) Formation of inducible Cd-binding proteins similar to metallothioneins in selected organs and life stages of *Mytilus galloprovincialis*. *Proceedings VII Journés Etudes Pollutions:-CIESM*, pp. 699–705.

Phillips, D.J.H. (1979) Trace metals in the common mussel *Mytilus edulis* (L.) and in the alga *Fucus vesiculosus* (L.) from the region of the Sound (Oresund). *Environmental Pollution* **18**, 31–43.

Phillips, D.J.H. (1980) *Quantitative Aquatic Biological Indicators*, Applied Science Publishers Ltd., London, 488 pp.

Phillips, D.J.H. (1990) The use of macroalgae and invertebrates as monitors of metal levels in estuaries and coastal waters, in *Heavy Metals in the Marine Environment*, (eds Furness, R.W. and Rainbow, P.S.), CRC Press Inc., Boca Raton, Florida, pp. 81–99.

Rainbow, P.S. (1995) Biomonitoring of heavy metal availability in the marine environment. *Marine Pollution Bulletin* **31**, 183–192.

Raspor, B. and Pavicic, J. (1991) Induction of metallothinein-like proteins in the digestive gland of *Mytilus galloprovincialis* after a chronic exposure to the mixture of trace heavy metals. *Chemical Speciation and Bioavailability* **3**(2), 39–46.

Regoli, F and Principato, G. (1995) Glutathione, glutathione-dependent and anti-oxidant enzymes in mussel, *Mytilus galloprovincialis*, exposed to metals under field and laboratory conditions: implications for the use of biochemical biomarkers. *Aquatic Toxicology* **31**, 143–164.

Ridlington, J.M. and Fowler, B.A. (1979) Isolation and partial characterisation of a cadmium-binding protein from the American oyster (*Crassostrea virginica*). *Chemical Biological Interactions* **25**, 127–138.

Riisgard, H.U., Bjornwstad, E. and Mohlenberg, F. (1987) Accumulation of cadmium in the mussel, *Mytilus edulis*: kinetics and importance of uptake via food and sea water, *Marine Biology* **96**, 349–353.

Ringwood, A.H. and Brouwer, M. (1993) Expression of constitutive and metal-inducible metallothioneins in oyster embryos (*Crassostrea virginica*). *Comparative Biochemistry and Physiology* **106B**, 523–529.

Ringwood, A.H. and Brouwer, M. (1995) Patterns of metallothionein expression in oyster embryos. *Marine Environmental Research* **39**, 101–105.

Robinson, W.E., Morse, M.P., Penney, B.A. *et al.* (1985) The eulamellibranch *Mercenaria mercenaria* (L.), a review and current data on metals accumulation and the internal transport of cadmium, in *Marine Pollution and Physiology, Recent Advances*, (eds F.J. Vernberg, F.P. Thurnberg, A. Calabrese and W.B. Vernberg), University of South Carolina Press, Columbia, SC, pp. 83–106.

Roesijadi, G. (1980) Influence of copper on the clam *Protothaca staminea*: effects on gills and occurence of copper-binding proteins. *Biological Bulletin* **158**, 233–247.

Roesijadi, G. (1986) Mercury-binding proteins from the marine mussel, *Mytilus edulis*. *Environmental Health Perspectives* **65**, 45–48.

Roesijadi, G. (1992) Metallothioneins in metal regulation and toxicity in aquatic animals. *Aquatic Toxicology* **22**, 81–114.

Roesijadi, G. (1994a) Behavior of metallothionein-bound metals in a natural population of an estuarine mollusc. *Marine Environmental Research* **38**, 147–168.

Roesijadi, G. (1994b) Metallothionein induction as a measure of response to metal exposure in aquatic animals. *Environmental Health Perspectives* **102** (Suppl. 12), 91–95.

Roesijadi, G. and Fellingham, G.W. (1987) Influence of Cu, Cd and Zn preexposure on Hg toxicity in the mussel *Mytilus edulis*. *Canadian Journal Fisheries Aquatic Science* **44**, 680–684.

Roesijadi, G. and Hall R.E. (1981) Characterisation of mercury-binding proteins from the gills of marine mussels exposed to mercury, *Comparative Biochemistry and Physiology* **70C**, 59-64.

Roesijadi, G. and Klerks, P. L. (1989) Kinetic analysis of cadmium binding to metallothionein and other intracellular ligands in oyster gills. *Journal of Experimental Zoology* **251**, 1–12.

Roesijadi, G. and Unger, M.E. (1993) Cadmium uptake in gills of the mollusc *Crassostrea virginica* and inhibition by calcium channel blockers. *Aquatic Toxicology* **24**, 95–206.

Roesijadi, G. and Robinson, W.E. (1994) Metal regulation in aquatic animals, mechanisms of uptake, accumulation and release, in *Aquatic Toxicology, Molecular, Biochemical and Cellular Perspectives*, (eds D.C. Malins and G.K. Ostrander), Lewis, Boca Raton, Florida, pp. 385–420.

Roesijadi, G., Calabrese, A. and Nelson, D.A. (1982) Mercury-binding proteins of *Mytilus edulis*, in *Physiological Mechasnisms of Marine Pollutant Toxicity*, (eds W.B. Venberg, A. Calabrese, F.P. Thurberg and F.J. Vernberg), Academic Press, New York, pp. 75–87.

Roesijadi, G., Young, J.S., Crecelius, E.A. and Thomas, L.E. (1985) Distribution of trace metals in the hydrothermal vent clam *Calyptogena magnifica*. *Biological Society Washington Bulletin* **6**, 311–324.

Roesijadi, G., Unger, M.E. and Morris J.E. (1988) Immunochemical quantification of metallothioneins of a marine mollusc. *Canadian Journal Fisheries Aquatic Science* **45**, 1257–1263.

Roesijadi, G., Kielland, S.L. and Klerks, P. (1989) Purification and properties of novel molluscan metallothioneins. *Archives Biochemical Biophysics* **273**, 403–413.

Romeo, M. and Gnassia-Barelli, M. (1995) Metal distribution in different tissues and in clam subcellular fractions of the Mediterranean clam *Ruditapes decussatus* treated with cadmium, copper or zinc. *Comparative Biochemistry and Physiology* **111C**, 457–463.

Rygg, B. (1985) Effect of sediment copper on benthic fauna, *Marine Ecology Progress Series* **25**, 83–89.

Sanders, B.M., Martin, L.S., Nelson, W.G. *et al.* (1991) Relationships between accumulation of a 60 kDa stress protein and scope for growth in *Mytilus edulis* exposed to a range of copper concentrations. *Marine Environmental Research* **31**, 81–97.

Scholz, N. (1980) Accumulation, loss and molecular distribution of cadmium in *Mytilus edulis. Helgolander Meeresunters* **33**, 68–79.

Serra, R., Carpenè, E., Marcantonio, A.C. and Isani, G. (1995) Cadmium accumulation and Cd-binding proteins in the bivalve *Scapharca inaequivalvis. Comparative Biochemistry and Physiology* **111C**, 165–174.

Slater, J.D., Mildvan, A.S. and Loeb, L.A. (1971) Zinc in DNA polymerases. *Biochemical and Biophysical Research Communications* **44**, 37–43.

Stone, H.C., Wilson, S.B. and Overnell, J. (1986) Cadmium binding components of scallop (*Pecten maximus*) digestive gland. Partial purification and characterization. *Comparative Biochemistry and Physiology* **85C,** 259–268.

Svendsen, C. and Weeks, J.M. (1995) The use of a lysosome assay for the rapid assessment of cellular stress from copper to the freshwater snail *Viviparus contectus* (Millet). *Marine Pollution Bulletin* **31**, 139–142.

Tallandini, L., Cassini, A., Favero, N. and Albergoni, V. (1986) Regulation and subcellular distribution of copper in the freshwater molluscs *Anodonta cygnea* (L.) and *Unio elongatulus* (Pf.). *Comparative Biochemistry and Physiology* **84C**, 43–49.

Tessier, C. and Blais, J.-S. (1996) Determination of cadmium-metallothioneins in the zebra mussels exposed to subchronic concentrations of Cd^{2+}. *Ecotoxicology and Environmental Safety* **33**, 245–252.

Tessier, A. and Campbell, P.G.C. (1990) Partitioning of trace metals in sediments and its relationship to their accumulation in benthic organisms, in *Nato ASI Series Vol. G23, Metal Speciation in the Environment*, (eds J.A.C. Broekaert, S. Güger and F. Adams), Springer Verlag, Berlin Heidleberg.

Tessier, A., Campbell, P.G.C., Auclair, J.C. and Bisson, M. (1984) Relationships between the partitioning of trace metals in sediments and their accumulation in the tissues of the freshwater mollusc *Elliptio complanata* in a mining area, *Canadian Journal Fisheries Aquatic Science* **41**, 1463–1472.

Tessier, A., Couillard, Y., Campbell, P.G.C. and Auclair, J.C. (1993) Modelling Cd partitioning in oxic lake sediments and Cd concentrations in the freshwater bivalve *Anodonta grandis. Limnology and Oceanography* **38**(1), 1–17.

Tusek-Znidaric, M., Skreblin, M., Pavicic, J. *et al.* (1986) Mercury-binding proteins of the gills and digestive gland of *Mytilus galloprovincialis. FAO/UNEP Meeting on the Toxicity and Bioaccumulation of Selected Substances in Marine Organisms*, Rovinj (Yugoslavia), 5–9 November 1984.

Udom, A.O. and Brady, F.O. (1980) Reactivation *in vitro* of zinc-requiring apoenzymes by rat-liver zinc thionein. *Biochemical Journal* **187**, 329–335.

Unger, M.E. and Roesijadi, G. (1993) Sensitive assay for molluscan metallothionein induction based on ribonuclease protection and molecular titration of metallothionein and actin mRNAs. *Molecular Marine Biology and Biotechnology* **2**(5), 319–324.

Unger, M.E., Chen, T.T., Murphy, C.M. *et al.* (1991) Primary structure of molluscan metallothioneins deduced from PCR-amplified cDNA and mass spectrometry of purified proteins. *Biochimica Biophysica Acta* **1074**, 371–377.

Vallee, B.L. (1991) Introduction to metallothionein, in *Methods in Enzymology 205* (eds J.F. Riordan and B.L. Vallee), Academic Press, New York, pp. 3–7.

Viarengo, A. (1989) Heavy metals in marine invertebrates: mechanisms of regulation and toxicity at the cellular level. *Aquatic Sciences* **1**(2), 295–317.

Viarengo, A. and Nott J.A. (1993) Mechanisms of heavy metal cation homeostatis in marine invertebrates. *Comparative Biochemistry and Physiology* **103C**, 355–372.

Viarengo, A., Pertica, M., Mancinelli, G. *et al.* (1980) Rapid induction of copper-binding proteins in the gills of metal-exposed mussels. *Comparative Biochemistry and Physiology* **67C**, 215– 218.

Viarengo, A., Pertica, M., Mancinelli, G. *et al.* (1981) Synthesis of Cu-binding proteins in different tissues of mussels exposed to the metal. *Marine Pollution Bulletin* **12**(10), 347–350.

Viarengo, A., Moore, M.N., Mancinelli, G. *et al.* (1985a) Significance of metallothioneins and lysosomes in cadmium toxicity. *Marine Environmental Research* **17**, 184–187.

Viarengo, A., Palmero, S., Zanicchi, G. *et al.* (1985b) Role of metallothioneins in Cu and Cd accumulation and elimination in the gill and digestive gland cells of *Mytilus galloprovincialis* Lam. . *Marine Environmental Research* **16**, 23–36.

Viarengo, A., Pertica, M., Canesi, L. *et al.* (1989) Purification and biochemical characterization of a lysosomal copper-rich thionein-like protein involved in metal detoxification in the digestive gland of mussels. *Comparative Biochemistry and Physiology* **93C**(2), 389– 395.

Viarengo, A., Canesi, L., Mazzucotelli, A. and Ponzano, E. (1993) Cu, Zn and Cd content in different tissues of the Antartic scallop *Adamussium colbecki*, role of metallothionein in heavy metal homeostasis and detoxification. *Marine Ecology Progress Series* **95**, 163–168.

Vismann, B. (1993) Hematin and sulfide removal in hemolymph of the hemoglobin-containing bivalve *Scapharca inaequivalvis*. *Marine Ecology Progress Series* **98**, 115–122.

Walsh, A.R. and O'Halloran, J. (1997) The accumulation of chromium by mussels *Mytilus edulis* (L.) as a function of valency, solubility and ligation. *Marine Environmental Research* **43**, 41–54.

Ward, T.J., Warren, L.J. and Tiller, K.G.(1984) The distribution and effects of metals in the marine environment near a lead-zinc smelter, South Australia, in *Environmental Impacts of Smelters*, (ed. J.O. Nriagu), John Wiley, New York, pp. 1–73.

Wen-Xiong Wang, Fisher, N.S. and Luoma, S.N. (1995) Assimilation of trace elements ingested by the mussel *Mytilus edulis*: effects of algal food abundance. *Marine Ecology Progress Series* **129**, 165–176.

White, S.L. and Rainbow, P.S. (1985) On the metabolic requirements for copper and zinc in molluscs and crustaceans. *Marine Environmental Research* **16**, 215–229.

Winsor, M.H., Boese, B.L., Lee, H. *et al.* (1990) Determination of the ventilation rates of interstitial and overlying water by the clam *Macoma nasuta. Environmental Toxicology and Chemistry* **9**, 209–213.

Wright, D.A. and Zamuda, C.D. (1987) Copper accumulation by two bivalves: salinity effect is independent of cupric ion activity, *Marine Environmental Research* **23**, 1–14.

Ying, W., Batley, G.E. and Ashanullah, M. (1992) The ability of sediment extractants to measure the bioavailability of metals to three marine invertebrates. *The Science of the Total Environment* **125**, 67–84.

Yonge, C.M. and Thompson, T.E. (1976) *Living Marine Molluscs*, Collins, London, 288 pp.

Zamuda, C.D. and Sunda, W.G. (1982) Bioavailability of dissolved copper to the American oyster *Crassostrea virginica*. 1. Importance of chemical speciation. *Marine Biology* **66**, 77–82.

Zaroogian, G. and Anderson, S. (1995) Comparison of cadmium, nickel and benzo(a)pyrene uptake into cultured brown cells of the hard shell clam, *Mercenaria mercenaria. Comparative Biochemistry and Physiology* **111C**, 109–116.

Phylogeny of trace metal accumulation in crustaceans

PHILIP S. RAINBOW

9.1 INTRODUCTION

Invertebrates accumulate trace metals in their tissues whether or not these trace metals are essential to their metabolism. Different invertebrates accumulate different trace metals to different degrees and accumulated concentrations vary greatly at tissue, organ and body levels (Eisler, 1981; Depledge and Rainbow, 1990; Rainbow, 1990a, 1993; Rainbow *et al.*, 1990; Phillips and Rainbow, 1993; Chapter 8). Whether an accumulated concentration is high or low, therefore, cannot be assessed on an absolute scale, but the significance of an accumulated concentration depends greatly on the specific tissue and the specific invertebrate (Rainbow, 1987, 1988, 1990a, 1993, 1996; Dallinger, 1993). To take one taxon as an example, tissue and body concentrations of trace metals vary greatly in crustaceans, even in the absence of anthropogenic input of trace metal contaminants (Bryan, 1968, 1976; Moore and Rainbow, 1987; Rainbow, 1987, 1988, 1989, 1990a,b, 1993; White and Rainbow, 1987; Ridout *et al.*, 1989). Table 9.1 highlights this interspecific variation in the body concentrations of zinc, copper and cadmium in crustaceans collected from clean and metal-contaminated sites.

This chapter explores possible reasons behind these interspecific differences in crustacean metal concentrations. The accumulated metal concentration depends on the net difference between rates of uptake and excretion, whether at tissue or body level, and how much metal can be held within a tissue depends on the nature and extent of the metabolic processes of metal detoxification available (Viarengo, 1989; Roesijadi, 1992; Roesijadi and Robinson, 1994). The processes of trace metal uptake, detoxified storage (whether temporary or permanent) and excretion are all strongly affected by other features of the biology of a crustacean – not least its morphology, its physiology, its mode of feeding and its ecophysiological adaptations to the

Metal Metabolism in Aquatic Environments. Edited by William J. Langston and Maria João Bebianno. Published in 1998 by Chapman & Hall, London. ISBN 0 412 80370 4

Table 9.1 A selection of body concentrations (μg g^{-1} dry weight) of three trace metals (Zn, Cu, Cd) in a systematic range of crustaceans from clean and metal-contaminated sites

Species	Location	Zinc	Copper	Cadmium	Reference
Copepoda					
Anomalocera patersoni	Off Monaco	–	41.9	1.5	Polikarpov *et al.* (1979)
Cirripedia					
Capitulum mitella	Hung Hom, Hong Kong (contaminated)	19890	545	10.0	Phillips and Rainbow (1988)
	Cape D'Aguilar, Hong Kong	2852	29.2	5.2	" "
Tetraclita squamosa	Hung Hom, Hong Kong (contaminated)	6963	94.9	2.8	" "
	Tung Chung, Hong Kong	2245	14.9	4.2	" "
Balanus amphitrite	Chai Wan Kok,Hong Kong (contaminated)	9353	3472	7.3	" "
	Lai Chi Chong, Hong Kong	2726	59.3	5.5	" "
Semibalanus balanoides	Dulas Bay, Wales (contaminated)	50280	3750	–	Walker (1977)
	Menai Strait, Wales	19230	170	–	Rainbow (1987)
	Southend, England	27837	232	28	Rainbow *et al.* (1980)
Malacostraca					
Peracarida					
Isopoda					
Oniscus asellus	Haw Wood, Avon, England (contaminated)	524	454	154	Hopkin (1990)
	Wetmoor Wood, Avon, England	62.9	92.9	15.6	"
Porcellio scaber	Haw Wood, Avon, England (contaminated)	897	651	48.9	"
	Wetmoor Wood, Avon, England	186	171	2.4	"

Table 9.1 *Continued*

Species	Location	Zinc	Copper	Cadmium	Reference
Amphipoda					
Orchestia gammarellus	Restronguet Creek, England (contaminated)	392	139	–	Rainbow *et al.* (1989)
	Millport, Scotland	188	77.5	1.6	¨
Talorchestia quoyana	St Kilda, Dunedin, New Zealand	481	31.9	17.2	Rainbow *et al.* (1993a)
	Sandfly Bay, Dunedin,	133	15.6	8.9	¨
Themisto gaudichaudii	Antarctic Ocean	65.7	31.1	52.6	Rainbow (1989)
Mysidacea					
Eucopia unguiculata	N.E. Atlantic	44.2	20.7	2.0	Ridout *et al.* (1989)
Eucarida					
Euphausiacea					
Euphausia superba	Atlantic Ocean	67.5	54.4	0.81	Rainbow (1989)
Meganyctiphanes norvegica	NE Atlantic	102	57.5	0.66	¨
Decapoda					
Dendrobranchiata					
Gennadas valens	NE Atlantic	47.2	40.2	1.9	Ridout *et al.* (1989)
Sergia robustus	NE Atlantic	53.7	25.2	2.3	¨
Metapenaeopsis palmensis	Long Harbour, Hong Kong	63.4	52.2	< 0.3	Rainbow (1990b)
Pleocyemata					
Systellaspis debilis	NE Atlantic	46.9	49.0	8.7	Ridout *et al.* (1989)
Acanthephyra purpurea	NE Atlantic	43.6	29.3	2.2	¨

Table 9.1 *Continued*

Species	Location	Zinc	Copper	Cadmium	Reference
Palaemon elegans	Millport, Scotland	80.6	110	0.9	White and Rainbow (1986a)
Pandalus montagui	Firth of Clyde, Scotland	57.5	57.4	–	Nugegoda and Rainbow (1988)
Crangon allmanni	North Sea	73.7	37.2	3.79	Everaarts *et al.* (1990)
Callianassa australiensis	Western Port, Victoria, Australia	99.8	127	0.26	Ahsanullah *et al.* (1980)
Carcinus maenas	Firth of Clyde, Scotland	96.6	30.2	0.87	Rainbow (1985)

physicochemistry of its environment. For example, an estuarine macrophagous decapod inevitably receives trace metals in a different form in comparison with a microphagous stenohaline marine barnacle, and differences in physiology and morphology almost necessitate that metal handling will differ between the two. This chapter explores the interaction between the biological features of a crustacean allocated by its phylogeny and those features interpretable as adaptations to a habitat or way of life, and considers how that interaction might affect that crustacean's metal biology. Do barnacles and decapods have different trace metal accumulation patterns simply because one is a barnacle and the other a decapod?

9.2 CRUSTACEAN EVOLUTION

There are as many views on the details of the evolution of the Crustacea as there are crustacean biologists, but few carcinologists would dispute that the evolution of crustaceans has seen the transition from a small marine epibenthic microphagous filter feeder to larger benthic walking macrophagous crustaceans, the acme of such a trend being represented by the crabs (Brachyura). Figure 9.1 is a visual representation of one possible classification of the Crustacea. This figure is not intended to indicate phylogenetic relationships but does provide a reference framework for the many crustacean subtaxa introduced hereafter.

The first crustaceans may have been of the order of a few millimetres long with a distinct head and with a long series of undifferentiated body segments, each bearing a pair of paddles or phyllopodia. The phyllopodia would have carried out a variety of functions, including locomotion, filter feeding and gaseous exchange. Metachronal waves of beating of the phyllopodia would have propelled the crustacean through the water. Phyllopodia also provided a means of filter feeding on small particles, possibly of stirred-up deposited detritus in the manner of epibenthic cephalocarids today, collected food particles being passed forward to the mouth along the ventral body surface. The high ratio of surface area to volume of the phyllopodia would have been ideal for the exchange of oxygen and carbon dioxide with the medium, although the cuticle over the whole body would have been permeable to gaseous exchange.

The drainage system of crustaceans today is based on the remnants of an ancestrally segmental arrangement of paired coelomoducts (perhaps with a hint of combination with nephridia distally) (Barrington, 1967). The only segments with such coelomoducts fully developed in living crustaceans are those associated with the second antennae and/or second maxillae, although small podocytic thoracic limb glands found in many crustaceans are probably homologous (Hessler and Elofsson, 1995). In the Branchiopoda, Ostracoda, Copepoda, Branchiura, Cirripedia and 'lower' Malacostraca the paired antennary glands are present in the larva and the maxillary glands in the adult; in the Amphipoda, Euphausiacea and Decapoda the antennary glands persist in the adult but the maxillary glands disappear or fail to develop (Barrington,

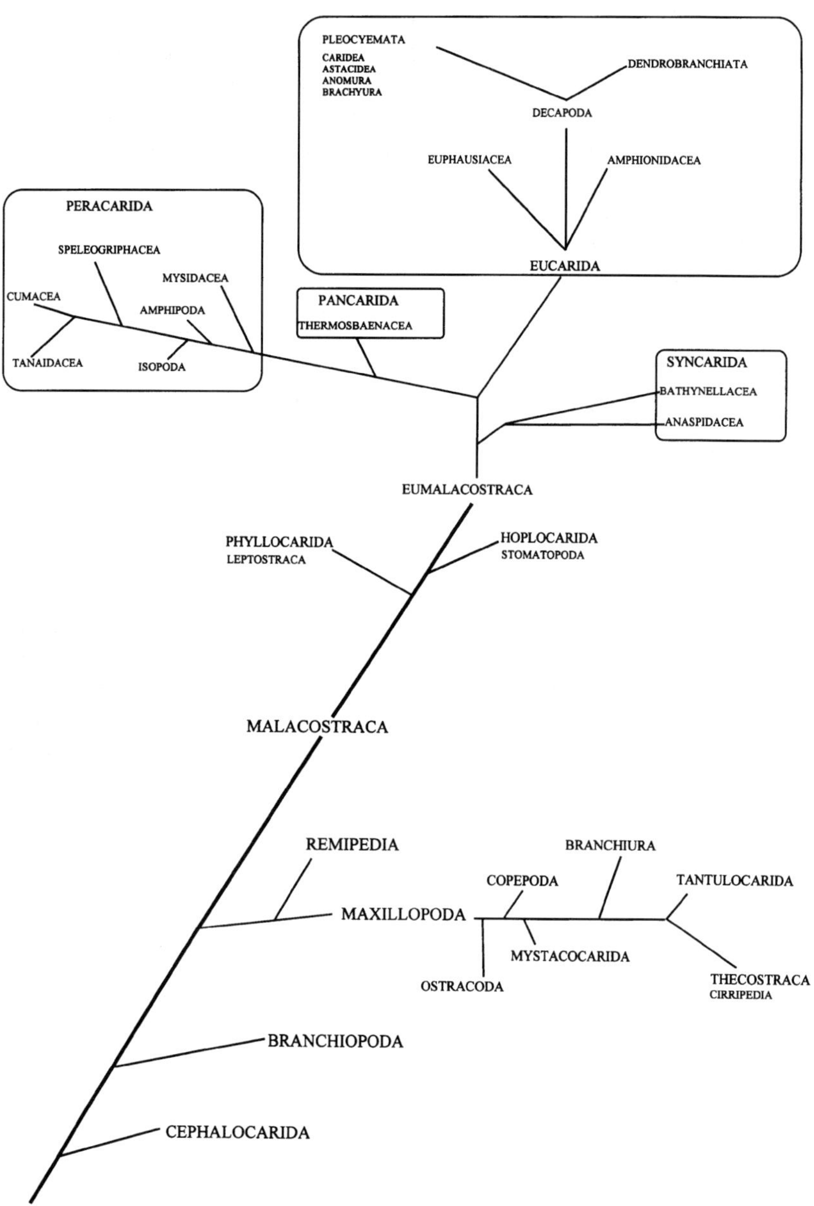

Figure 9.1 A visual representation of crustacean classification. This is not intended to indicate phylogenetic relationships and should not be so interpreted. (After Bowman and Abele, 1982.)

1967). In the Mysidacea both pairs of glands may be functional in the adult (Barrington, 1967). Although often referred to as excretory glands, the antennary and/or maxillary glands are particularly important in ionic and osmotic regulation, being well developed in freshwater crustaceans.

Many of the crustaceans alive today fall in the class Malacostraca (Figure 9.2). In comparison with the possible ancestral model described above, malacostracans are larger, many of them being benthic walkers. Body segments are divided into thorax and abdomen, primitively with six and eight segments, respectively, and phyllopodia are typically replaced by appendages with more specific functions. Many malacostracans walk and swim, walking being executed by thoracic limbs and swimming by abdominal appendages. The originally biramous appendages have become modified. Up to three anterior pairs of thoracic appendages are maxillipeds, assisting the five pairs of head appendages in the handling of food. The inner branches (endopodites) of the remaining thoracic appendages have become strutlike stenopodia and are used for walking. Between three and five pairs of abdominal appendages may be modified as biramous paddles (pleopods) for swimming, the remaining pair or pairs being uropods, possibly as part of a tail fan.

Given the increase in the size of the crustacean and the loss of flat phyllopodia with high surface areas for gaseous exchange, malocostracans may develop limbs or parts of limbs as gills. Thus walking isopod crustaceans that have lost the ability to swim may use abdominal pleopods as gills, while decapods typically develop new growths at the base of thoracic legs (epipodites) as gills, often housed in an expanded cavity beneath a carapace. Also associated with the increase in size has been the development of a more

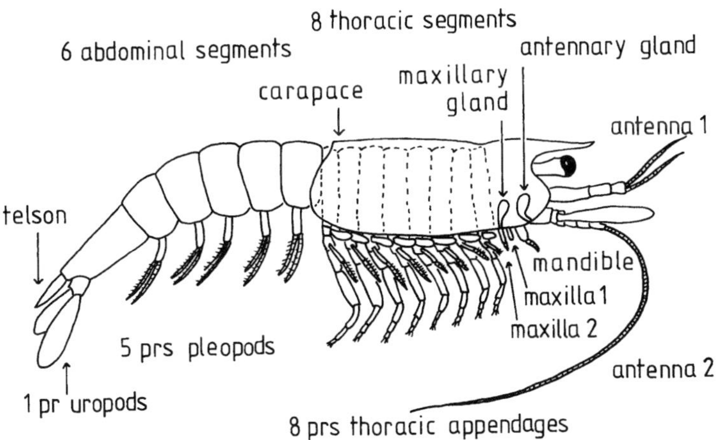

Figure 9.2 A generalized malacostracan crustacean. Appendages and excretory glands are paired.

robust exoskeleton, for both support and protection. Lobsters and crabs, for example, have a hard carapace in which the cuticle has been tanned and calcified, the latter process being the laying down of calcium carbonate in the cuticle. The calcium is typically resorbed at the time of moulting. The processes of tanning and calcification which harden the cuticle also render it impermeable, and are therefore associated with the coevolution of new, extensive permeable areas of cuticle – in effect the gills. The restriction of gaseous and indeed ionic exchange over the cuticle to a single region (the gills) has consequences for the physiology of malacostracans, not least with respect to ionic and osmotic balance and regulation.

Most decapod crustaceans are macrophagous, as predators or scavengers, feeding on large pieces of animal (or more rarely vegetable) matter cut off with claws and macerated by the array of mouthparts and maxillipeds. Large items of food are received by the decapod stomach, a development of the cuticle-lined foregut with regions for further maceration of food particles and ultimately their filtration before the filtrate is passed to a new evolutionary development – the hepatopancreas. The decapod hepatopancreas is a complex tubular organ derived from the midgut, the functions of which include absorption and storage of nutrients, synthesis of digestive enzymes and detoxification of trace metals and xenobiotics (Gibson and Barker, 1979; Icely and Nott, 1992; Vogt, 1994). Other crustaceans, particularly other malacostracans, have a variety of midgut caeca with a subset of similar functions (Icely and Nott, 1992).

Other features of crustacean phylogeny include the evolution of the Cirripedia – the barnacles. Barnacles are sessile crustaceans which have six pairs of thoracic appendages modified to filter food particles from suspension in the relatively large volumes of water passed across the body. Two peracarid groups, the isopods and amphipods, have fully terrestrial representatives: the woodlice (or sowbugs) and landhoppers, respectively. Woodlice use the abdominal pleopods for aerial respiration and both isopods and amphipods brood their young, obviating the need to return to the sea for larval development. The latter requirement remains as a restriction on the otherwise terrestrial land hermit crabs (Anomura) and land crabs (Brachyura).

Crustaceans are also found in fresh water. The term Entomostraca encompasses the crustaceans that are not malacostracans (Figure 9.1), though the grouping is disparate and polyphyletic. Entomostracans have long been resident in fresh water but malacostracans have also invaded rivers via estuaries – not least the decapods, in the shape of astacidean crayfish. Many decapods are euryhaline and straddle the estuarine boundary between sea and fresh water, as exemplified by palaemonid carideans, and some brachyuran crabs. Mysids are also common inhabitants of estuaries. Euryhaline crustaceans have characteristics interpreted as physiological adaptations to their estuarine or littoral habitats, not least strong powers of ionic and osmotic regulation and the ability to vary epidermal permeability (Mantel and Farmer, 1983).

9.3 PREADAPTATIONS TO TRACE METAL BIOLOGY

Tables 9.2 and 9.3 attempt to relate biological features highlighted above to the patterns of trace metal accumulation seen in crustaceans. These tables will form the basis of the subsequent interpretation of the significance of the inter-specific differences in accumulated metal concentrations. Thus Tables 9.2 and 9.3 summarize some of the characteristic features of the biology of particular subtaxa of crustaceans that would affect their patterns of trace metal accumulation. In effect these are preadaptations of crustaceans to a particular trace metal accumulation pattern, resulting from either their phylogeny (Table 9.2) or their ecology (Table 9.3).

In the case of metal uptake from solution (Table 9.2), decapods could be expected to have a low rate of metal uptake per unit body weight because most of the body is covered by an impermeable cuticle, permeability being restricted to the gills. Entomostracans in general would be covered in permeable cuticle and filter feeding requires the passage of large volumes of water across such surfaces; they might therefore be expected to have high specific

Table 9.2 Phylogenetic preadaptations of crustaceans to a trace metal accumulation pattern

Process	Examples	
Metal uptake	*Low uptake*	*High uptake*
From solution		
Cuticle impermeability (tanning, calcification)	Decapods	Entomostracans
Volume of feeding/respiratory current	Decapods	Barnacles
From food		
Macrophagy/microphagy	Decapods	Barnacles
Food type: animal/plant/detritus	Carnivorous crabs	Barnacles
Metal excretion	*High excretion*	*Low excretion*
Antennary/maxillary glands	Decapods	?
High potential urine flow	Mysids	
Gills	Decapods?	?
Gut epithelium turnover, (e.g. caeca, hepatopancreas)	?	?
Moulting	?	?
Detoxified storage capacity		
Hepatopancreas, caeca	Malacostracans	
Body parenchyma	Barnacles	
Cuticle	Brachyurans?	
Blood haemocyanin	Most malacostracans	
Longevity/growth rate		
Long life: accumulation	Lobsters?	
Rapid diluting growth	Penaeids	

Table 9.3 Ecological preadaptations of crustaceans to a trace metal accumulation pattern

Process	Examples	
Metal uptake	*Low uptake*	*High uptake*
From solution		
Cuticle impermeability (tanning, calcification)	Freshwater/estuarine	Oceanic?
Physiological response: permeability change	Estuarine/littoral	
Volume of feeding/respiratory current		Filter feeders Burrow irrigators
From food		
Macrophagy/microphagy	Macrophages	Microphages
Food type: animal/plant	Carnivores	Detritivores
	Terrestrial herbivores	Specific metal-rich diets
Metal excretion	*High excretion*	*Low excretion*
Antennary/maxillary glands (high potential urine flow)	Estuarine/freshwater	Stenohaline Terrestrial
Gills	?	?
Gut epithelium turnover, e.g. caeca, hepatopancreas	?	?
Moulting	Juveniles	Long final intermoult
Longevity/growth rate		
Long life/slow growth	Deep sea, polar	
Rapid diluting growth	Tropical, coastal	

rates of metal uptake from solution. Barnacles, as an example, have a large permeable surface area represented not least by the thoracic legs (cirri) and they move large volumes of water across these cirri during suspension feeding. Furthermore the food of barnacles typically consists of small particles, including in many cases detritus which characteristically is organically rich, with the subsequent potential for strong trace metal adsorption. Macrophagous feeders such as carnivorous predatory or scavenging crabs take in large pieces of animal tissue. Animal tissue, especially muscle, is not particularly metal-rich, although a predator feeding selectively on an invertebrate storage organ equivalent to the crustacean hepatopancreas may ingest considerable amounts of trace metals in a detoxified form. Even in this latter case, however, such detoxified metal may be chemically bound so tightly that it is not available to the predator (Nott and Nicolaidou, 1990; Chapter 12). Crabs feeding macrophagously on plant material tend to be grapsids feeding on fallen leaves on the top of tropical mangrove-covered shores, and the leaves of essentially terrestrial plants are low in trace metal concentrations in comparison with their marine ecological counterparts, the macrophytic algae (Moore and Rainbow, 1987; Rainbow, 1992).

From an ecological standpoint (Table 9.3), freshwater and estuarine malacostracans have low integumental permeabilities compared with their marine relatives (Mantel and Farmer, 1983), a difference reflected in trace metal uptake rates (Nugegoda and Rainbow, 1989; Rainbow, 1995a). Moreover, a feature of many euryhaline crustaceans is their ability, in response to changes in medium osmotic pressure, to vary their integumental permeability, often described as the apparent water permeability (Mantel and Farmer, 1983; Campbell and Jones, 1990). Such changes may have consequences on rates of trace metal uptake from solution (Rainbow *et al.*, 1993c; Rainbow, 1995a).

Ecological specialization will also affect potential trace metal uptake from food (Table 9.3). Not all decapods are macrophagous; some are microphagous filter feeders – for example, thalassinidean mud lobsters and hippid crabs. Both the latter would be expected to have a more significant input of trace metals from food (and from solution via extensive feeding currents) than their strictly macrophagous relatives. Burrowing malacostracans also have the potential for high trace metal uptake from irrigatory currents.

It is much more difficult to highlight examples of potential predaptations of crustaceans to the excretion of trace metals (Tables 9.2, 9.3). Certain crustacean subtaxa do have well developed antennary and/or maxillary glands, not least the mysids, and the specific distribution of the development of these coelomoducts probably follows ecological as opposed to systematic divisions. Furthermore, the role of antennary glands (for example) in metal excretion by decapods is by no means universal, for trace metal excretion may also be significant from the gills and the gut epithelium, with relative proportions varying between species (Bryan, 1966, 1971). Excretion of trace metals in detoxified granular form from the epithelium of the midgut or a midgut derivative like the hepatopancreas is not uncommon in crustaceans (Hopkin and Nott, 1979; Icely and Nott, 1980; Hopkin and Martin, 1984; Moore and Rainbow, 1984; Vogt and Quinitio, 1994). The significance of metal excretion by this route is difficult to interpret in terms of systematic relationships or ecological specialization. The rate of epithelium turnover (Al-Mohanna and Nott, 1989; Vogt, 1994) and the variability of this turnover rate as well as the maximum capacity of epithelial cells to hold detoxified granules will affect the potential for trace metal excretion from the midgut epithelium of a crustacean.

The incorporation of trace metals into detoxified granules in the epithelium of the midgut (or a derivative thereof) represents detoxified storage, which is temporary if the epithelial cell is part of a cell cycle with subsequent release of the granules into the gut lumen and ultimately the faeces. Storage can also be permanent – as in the case of barnacles, which store accumulated trace metals in trace metal detoxification granules in the body cells beneath the midgut (Rainbow, 1987; Pullen and Rainbow, 1991). There is no access for these granules to any route for the excretion of solid material such as the lumen of the gut or maxillary gland, and so they accumulate over the lifetime of the barnacle with consequent increases of body concentrations of trace metals like zinc to staggeringly high levels (Table 9.1) (Rainbow, 1987,

1990a, 1993). Although barnacles do have extensions of the midgut into midgut caeca, the caecal epithelia are morphologically similar to that of the tubular midgut itself, with no apparent specific 'hepatic' function (such as enhanced trace metal detoxification) over and above digestion and absorption (Rainbow and Walker, 1977), as would be the case for the decapod hepatopancreas. The *stratum perintestinale* beneath the midgut epithelium carries out some such functions with storage of detoxified metal-rich granules extending to the surrounding parenchyma cells (Walker *et al.*, 1975a; Rainbow and Walker, 1977; White and Walker, 1981). In effect the lack of a hepatopancreas with lumen confluent with that of the rest of the midgut prevents the barnacle from excreting metal-rich granules, and perhaps commits it to a pattern of strong permanent net accumulation of trace metals.

Crustaceans transfer accumulated trace metals, including zinc and cadmium, into the cuticle, resorbing at least a portion of this metal upon ecdysis (White and Rainbow, 1984a,b, 1986b; Chan and Rainbow, 1993a,b). Some metal is lost, however, and so the cuticle can act as both a storage site and an excretion route, with variation in relative proportions between species. Moulting also exposes permeable cuticle to the medium, at least until tanning and calcification are complete, exposing the crustacean to temporarily increased rates of trace metal uptake (White and Rainbow, 1984a,b, 1986b). Does a high frequency of moulting provide an avenue for increased excretion of trace metals via the cast and therefore bring about reduced net metal accumulation, or does it cause a temporarily increased rate of trace metal uptake so often that net accumulation increases?

The blood of many malacostracans, including the decapods, contains the respiratory pigment haemocyanin. As a large soluble protein, haemocyanin offers binding sites for trace metals, particularly zinc (Bryan, 1966; Zatta, 1984), and the blood may therefore act as a storage site for accumulated metal (Chan and Rainbow, 1993a). Such storage is not available to entomostracans and those malacostracans lacking haemocyanin.

If a particular trace metal is accumulated without significant net excretion by a crustacean, the body concentration of that metal will be greatly affected by the growth rate of the crustacean (Tables 9.2, 9.3). If the elaboration of new body tissue is taking place at a much greater relative rate than the rate of net accumulation of a trace metal, growth will dilute the body concentration of that metal. Rapid growth rate may well cause body concentrations of accumulated cadmium to be low in the tropical coastal penaeid prawn *Metapenaeopsis palmensis*, as opposed to those accumulated by their temperate caridean ecological equivalents (Rainbow, 1990b). Conversely the apparently high body cadmium concentrations of some mesopelagic caridean decapods of the northeast Atlantic Ocean (for example, *Systellaspis debilis*) (Ridout *et al.*, 1985, 1989; White and Rainbow, 1987) may be the result of a slow growth rate in the cold midwater environment relative to cadmium accumulation rate. The rate of cadmium uptake is potentially high, given the nutrient-type profile of dissolved cadmium concentrations in the open ocean with raised concentrations at depth (Bruland, 1983), and the probably low integumental permeability of a stenohaline oceanic crustacean.

9.4 CRUSTACEAN EXAMPLES OF TRACE METAL ACCUMULATION PATTERNS

The previous section draws attention to the many biological characteristics of crustaceans that have the potential to affect, or perhaps predetermine, the accumulation pattern exhibited by a crustacean for particular trace metals. This section incorporates these characteristics into possible explanations of trace metal accumulation patterns in specific crustacean examples.

9.4.1 CIRRIPEDIA

The first crustacean subtaxon considered is that of the Cirripedia – the barnacles. Barnacles are notoriously strong accumulators of trace metals (Rainbow, 1985, 1987; Rainbow and White, 1989, 1990). As detailed for zinc and cadmium in Table 9.4, barnacles have very high rates of uptake of trace metals from solution in comparison with their malacostracan relatives, in this case an amphipod and a decapod (Rainbow and White, 1989, 1990). Relatively high rates of dissolved trace metal uptake are to be expected from the biological arguments put forward earlier. Given the mode of feeding of barnacles, uptake from food will also be very significant. Excretion of trace metals accumulated in the body by barnacles is insignificant, certainly in the short term (Rainbow, 1987; Rainbow and White, 1989). Some long-term excretion may occur (Walker and Foster, 1979; White and Walker, 1981) but most reductions in body metal concentrations are probably attributable to growth dilution (Powell and White, 1990) or loss of gametes or moulted casts.

The net effect of high uptake and no (or very low) excretion of trace metals by barnacles is their strong net accumulation in the body over time (Rainbow, 1985; Rainbow and White, 1989). Figures 9.3, 9.4 and 9.5, for example, illustrate how the net accumulation of zinc, copper and cadmium from solution by a barnacle (*Elminius modestus*) in each case is much higher than that of an amphipod (*Echinogammarus pirloti*) and a decapod (*Palaemon elegans*). Such strong net accumulation is necessarily accompanied by detoxification. The dominant detoxification system involves the incorporation of accumulated trace metals into granules based on pyrophosphate (Walker *et al.*, 1975a,b; Rainbow, 1987; Pullen and Rainbow, 1991). There are many trace metals accumulated in these granules, the dominant ones being zinc and iron (Walker *et al.*, 1975a,b; White and Walker, 1981; Pullen and Rainbow, 1991). When barnacles are exposed to an extremely high availability of copper, a different organically rich intracellular deposit is found that is rich in copper (Walker, 1977). This deposit appears to be a lysosomal residual body

Table 9.4 Zinc and cadmium uptake from solution by a decapod crustacean (*Palaemon elegans*), an amphipod crustacean (*Echinogammarus pirloti*) and a barnacle (*Elminius modestus*) under identical physiochemical conditions (artificial seawater, 33 ppt salinity, 10°C), after Rainbow and White (1989, 1990)

	Zinc		Cadmium	
Dissolved concentration				
μg l^{-1}	31.6	100	31.6	100
μmol l^{-1}	0.48	1.53	0.28	0.89
Palaemon elegans				
Uptake rate				
μg g^{-1} per day	0.62	1.52	0.074	0.23
nmol g^{-1} per day	9.43	23.3	0.66	2.05
Uptake index	0.020	0.015	0.0023	0.0023
Echinogammarus pirloti				
Uptake rate				
μg g^{-1} per day	0.51	2.68	0.72	1.70
nmol g^{-1} per day	7.80	41.0	6.41	15.1
Uptake index	0.016	0.027	0.023	0.017
Elminius modestus				
Uptake rate				
μg g^{-1} per day	11.7	26.9	2.75	12.5
nmol g^{-1} per day	179	411	24.5	111
Uptake index	0.373	0.269	0.088	0.125

The uptake index is calculated as the molar rate of metal uptake (mol g^{-1} per day) per molar dissolved metal concentration (mol l^{-1})

derived from the breakdown of copper-rich metallothionein induced in the barnacle by copper exposure (Rainbow, 1987). Metallothioneins are present in barnacles (Pullen, 1988) and under raised but less than severe availabilities of copper and cadmium, these trace metals appear to be incorporated into the pyrophosphate granules subsequent to their being bound by metallothionein (Pullen, 1988; Pullen and Rainbow, 1991).

The detoxified granules lie below the midgut in the *stratum perintestinale* and the parenchyma tissue of the body. The rate of production of the metabolically inert granules matches the high uptake rate of trace metals; they are not excreted and build up over the lifetime of the barnacle, committing it to a pattern of strong net trace metal accumulation.

9.4.2 BRANCHIOPODA

Not least because of their small size, relatively little work has been carried out on the accumulation of trace metals by other entomostracans. Species of the brine shrimp *Artemia* (Branchiopoda: Anostraca) (Figure 9.1) are often used

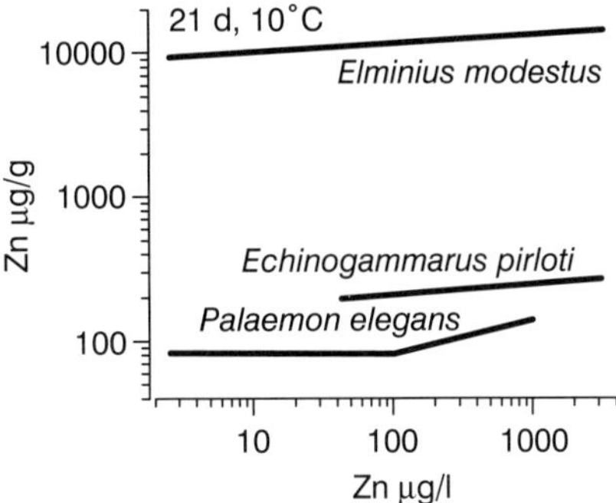

Figure 9.3 Comparative patterns of zinc accumulation by three crustaceans – the decapod *Palaemon elegans*, the amphipod *Echinogammarus pirloti* and bodies of the barnacle *Elminius modestus* – exposed to a range of dissolved zinc concentrations under identical physicochemical conditions (artificial seawater, 33 ppt salinity, 10°C) for 21 days. (After Rainbow and White, 1989.)

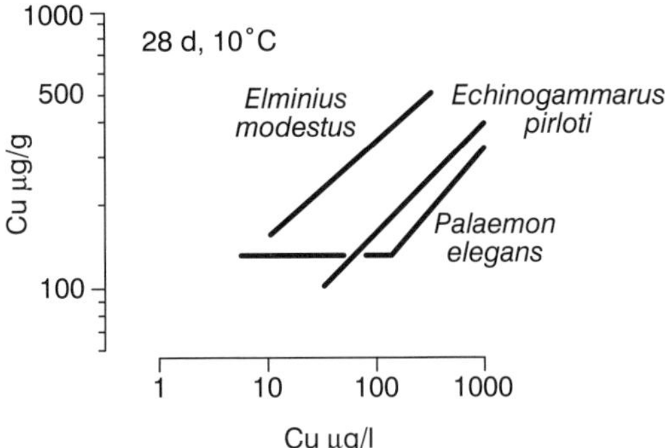

Figure 9.4 Comparative patterns of copper accumulation by three crustaceans exposed to a range of dissolved copper concentrations for 28 days. Other details as for Figure 9.3. (After Rainbow and White, 1989.)

as models for marine crustaceans, particularly in uptake studies (Blust *et al.*, 1992, 1995). Brine shrimp typically live in the high salt concentrations of brine lakes of salinities higher than that of seawater (*c.* 35 ppt). *Artemia salina* has a very low permeability and in high salinity it drinks the medium to replace the water lost from the body of osmosis; the water is taken up in the gut and excess salt is excreted from the body (Mantel and Farmer, 1983). It is important to realize that brine shrimp may not be an ideal model for a marine crustacean in that most of trace metal uptake from solution occurs in the gut, after drinking. There is no reason to believe, however, that different physico-chemical and physiological processes control metal uptake at the gut epithelium than at permeable ectodermal surfaces.

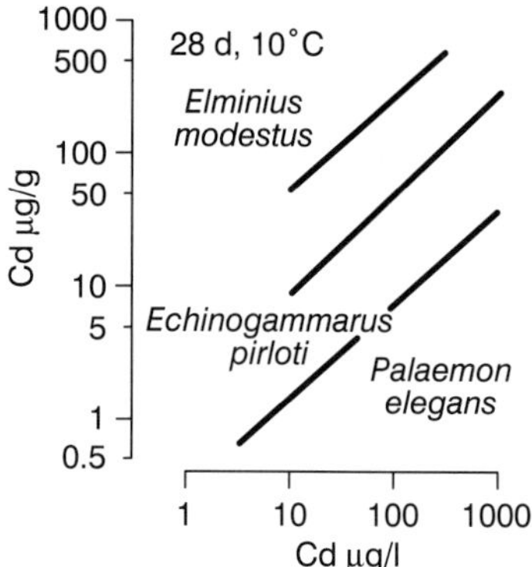

Figure 9.5 Comparative patterns of cadmium accumulation by three crustaceans exposed to a range of dissolved cadmium concentrations for 28 days. Other details as for Figure 9.3. (After Rainbow and White, 1989.)

Artemia salina accumulates cadmium from solution and from food, and after allowance for desorption of surface-adsorbed cadmium and for defaecation there appears to be no excretion (Jennings and Rainbow, 1979b). At least for cadmium, therefore, *A. salina* is a net accumulator. A related anostracan, the fairy shrimp *Branchinecta longiantenna*, is also a net accumulator of cadmium, as it is of lead and zinc (Mizutani *et al.*, 1991).

For approximate comparison with the figures given in Table 9.4, the uptake rate of cadmium by *A. salina* from solution at 100 µg Cd l⁻¹ is 3.5 µg Cd g⁻¹ per day (after allowance for adsorption) (Jennings and Rainbow, 1979b), though differences in the physicochemistries of the seawater used in the experiments require caution to be exercised. The cadmium uptake rate of *Artemia salina* is lower than that of a barnacle at the same concentration, but higher than those of the amphipod and decapod (Table 9.4). It should be remembered that this uptake, albeit from solution, is probably across the gut as opposed to across the ectoderm, given the low integumental permeability of *Artemia*.

Members of another subtaxon of the Branchiopoda (Figure 9.1), in this case the Cladocera, also appear to be strong net accumulators of trace metals, as for example in the case of zinc by *Daphnia magna* (Memmert, 1987). Generally, however, information is lacking on the physiological detoxification of accumulated trace metals in branchiopods and most small entomostracans.

9.4.3 ISOPODA

More information is available on the trace metal accumulation patterns of malacostracans. The first malacostracan order considered here is the Isopoda. Paradoxically, given the title of this book, the first isopods considered here are the terrestrial woodlice. Most isopod research relevant to the arguments put forward here has been carried out on woodlice (Hopkin and Martin, 1984; Hopkin, 1989), and the smaller body of aquatic isopod work is best interpreted in the light of the woodlice studies.

Since they are terrestrial, woodlice accumulate trace metals from the food, absorption taking place in each of four so-called hepatopancreatic caeca. The epithelia of these caeca are differentiated into two types of cells: B cells, which are secretory and absorptive; and S cells, the main function of which appears to be storage of accumulated trace metals (Hopkin and Martin, 1984). Woodlice incorporate trace metals into two types of intracellular granules. In the S cells there are dense spherical homogeneous granules which routinely contain copper, sulphur and calcium, but in woodlice from metal-contaminated sites there are potentially also zinc, cadmium and lead (Hopkin and Martin, 1984). The B cells also play a role in trace metal detoxification for they contain another granule type – a more loosely bound deposit of flocculent material containing iron; these granules may additionally contain zinc and lead in woodlice from metal-contaminated sites (Hopkin and Martin, 1984).

Thus woodlice are carrying out storage of detoxified trace metals in the hepatopancreatic caeca. Whether this storage is temporary or permanent depends on the turnover of cells in the hepatopancreatic caeca and the possible release (excretion) of metal-rich granules into the gut lumen. Differences between species in rates of turnover of each cell type, or indeed of the whole hepatopancreatic caecal epithelium, will cause differences in the patterns of accumulation of a trace metal. For example, there are clear differences in the accumulated trace metal concentrations of the two woodlice species *Oniscus asellus* and *Porcellio scaber* from the same sites (Table 9.5) (Hopkin, 1990). *O. asellus* has higher hepatopancreatic caecal and total body cadmium and lead concentrations than *P. scaber* but lower zinc and copper (and to a lesser extent iron) concentrations (Table 9.5). Interspecific differences may be caused by different trace metal assimilation rates, but not apparently in the case of zinc (Hopkin, 1990). A clear interspecific difference is the ability of *O. asellus* but not *P. scaber* to excrete zinc deposited in hepatopancreatic caeca, possibly because the two species have different proportions of accumulated zinc stored in the two cell types in the heptopancreatic caeca (Hopkin *et al.*, 1989; Hopkin, 1990). S cells have a long residence time in the hepatopancreatic caeca, and cadmium and copper (found only in S cell granules) are not lost readily therefrom (Hopkin, 1990; Hames and Hopkin, 1991). Most of the B cells, on the other hand, void their apical cytoplasm into the hepatopancreatic caecal lumen at the end of a 24-hour digestive cycle, and the iron-rich granules are expelled therewith (Hames and Hopkin, 1991). *Oniscus asellus* stores the majority of assimilated zinc in the granules in the B cells; *Porcellio scaber* stores the majority in the granules in S cells (Hopkin *et al.*, 1989). Thus zinc may be lost from *O. asellus* but not *P. scaber* because of the rapid turnover of the B cells (Hopkin, 1990).

Table 9.5 Concentrations (μg g^{-1}, mean $\pm$ SE, $n = 12$) of accumulated trace metals in two species of woodlice (*Oniscus asellus* and *Porcellio scaber*) collected from Wetmoor Wood (uncontaminated) and Haw Wood (metal-contaminated), Avon, England (after Hopkin, 1990)

Site and species	Zn	Cd	Pb	Cu	Fe
Wetmoor Wood (uncontaminated)					
Oniscus asellus					
Hepatopancreas	324 $\pm$ 52	329 $\pm$ 44	275 $\pm$ 74	837 $\pm$ 93	331 $\pm$ 41
Total	62.9 $\pm$ 3.2	15.6 $\pm$ 2.5	21.1 $\pm$ 5.6	92.2 $\pm$ 4.3	97.3 $\pm$ 10.9
Porcellio scaber					
Hepatopancreas	2140 $\pm$ 460	32.6 $\pm$ 4.2	27.6 $\pm$ 6.1	1780 $\pm$ 380	426 $\pm$ 58
Total	186 $\pm$ 21	2.44 $\pm$ 0.31	2.57 $\pm$ 0.40	171 $\pm$ 13	110 $\pm$ 12
Haw Wood (contaminated)					
Oniscus asellus					
Hepatopancreas	8250 $\pm$ 1050	2940 $\pm$ 230	8570 $\pm$ 1240	7920 $\pm$ 700	388 $\pm$ 69
Total	524 $\pm$ 42	154 $\pm$ 10	508 $\pm$ 59	454 $\pm$ 26	164 $\pm$ 18
Porcellio scaber					
Hepatopancreas	12000 $\pm$ 1600	706 $\pm$ 52	927 $\pm$ 199	9150 $\pm$ 1340	508 $\pm$ 58
Total	897 $\pm$ 111	48.9 $\pm$ 6.3	116 $\pm$ 13	651 $\pm$ 31	184 $\pm$ 22

In summary, woodlice accumulate trace metals in detoxified granules in the hepatopancreatic caeca (Hopkin and Martin, 1984; Hopkin, 1989; Dallinger, 1993). The degree of permanence of this detoxified storage depends on the rate of turnover of cells containing the granules, and indeed on the relative distribution of accumulated metal between granule type and therefore cell types. The generalized pattern appears to be the same for all woodlice but rates of turnover vary so much, even between woodlice species, that the pattern of trace metal accumulation can have very different kinetics, with or without significant excretion of accumulated body load.

Asellus aquaticus and *Asellus (Proasellus) meridianus* are freshwater isopods. Trace metal accumulation patterns of *A. aquaticus* are comparable to those of the woodlouse *Oniscus asellus* (Hopkin, 1989); these freshwater isopods show rapid elimination of accumulated zinc and lead but usually no significant elimination of accumulated copper and cadmium (Van Hattum *et al.*, 1993). It remains to be clarified whether the same copper-containing granules in S cells and iron-containing granules in B cells of the hepatopancreatic caeca are in operation (Van Hattum *et al.*, 1993), although preliminary microscopy on copper- and lead-tolerant populations of the related *A. meridianus* suggests that this is likely (Brown, 1977, 1978).

9.4.4 AMPHIPODA

The second malacostracan order to be discussed is the Amphipoda. Data for the littoral amphipod *Echinogammarus pirloti* (Table 9.4; Figures 9.3, 9.4 and 9.5) are directly comparable to those for the barnacle *Elminus modestus* and the decapod *Palaemon elegans*, for all work has been carried out under identical physicochemical conditions (Rainbow and White, 1989, 1990).

The accumulation patterns of *Echinogammarus pirloti* for zinc, copper and cadmium are intermediate between those of the barnacle and the decapod (Figures 9.3, 9.4 and 9.5), with weak net accumulation in each case (Rainbow and White, 1989). The uptake rate of cadmium from solution by *E. pirloti* is similarly intermediate between the cadmium uptake rates of the barnacle and the decapod (Table 9.4), as is also the case for cobalt uptake (Rainbow and White, 1990). The zinc uptake rate of *E. pirloti* is very low, matching that of *P. elegans* (Tables 9.4, 9.6). *E. pirloti* did not excrete any accumulated zinc and cadmium over 21 and 28 days, respectively, the amphipod accumulating all metal taken up from solution. The net accumulation of the amphipod is weaker than that of the barnacle for each metal, simply as a result of the lower absolute uptake rates of the two metals. The lack of a convenient radioactive tracer for copper has prevented the detection of the presence or absence of copper excretion but it remains likely that the weak accumulation of copper results from a low uptake rate from solution with no excretion of metal taken up by this route.

The comparative study of uptake and accumulation of trace metals from solution has been extended to two closely related talitrid amphipods,

Table 9.6 Mean rates of zinc uptake (μg g^{-1} per day $\pm$ SE) from solution of five crustaceans – barnacle *Elminius modestus*, amphipods *Echinogammarus pirloti, Orchestia gammarellus* and *Orchestia mediterranea*, and decapod *Palaemon elegans* – under identical physiochemical conditions (31.6 μg Zn l^{-1} in artificial seawater, 33 ppt salinity, 10°C) (after Rainbow and White, 1989; Weeks and Rainbow, 1991)

Crustacean	Zinc uptake rate
Barnacle	
Elminius modestus	11.69 ± 276
Amphipod	
Echinogammarus pirloti	0.509 ± 0.052
Orchestia gammarellus	0.430 ± 0.073
Orchestia mediterranea	0.408 ± 0.052
Decapod	
Palaemon elegans	0.617 ± 0.034

Orchestia gammarellus and *O. mediterranea* (Weeks and Rainbow, 1991). Under identical physicochemical conditions these two amphipods show zinc uptake rates from solution very similar to those of *Echinogammarus pirloti*, and lower than or matching that of *Palaemon elegans* (Table 9.6), suggesting that a low rate of zinc uptake from solution is an amphipod characteristic. Accumulation patterns for the two *Orchestia* species of both zinc and copper taken up from solution are remarkably similar to those for *E. pirloti* (Weeks and Rainbow, 1991), even though the genera *Orchestia* and *Echinogammarus* are in the families Talitridae and Gammaridae, respectively.

Accumulation patterns of trace metals taken up from solution are not consistent across all amphipods. The freshwater amphipod *Hyatella azteca* is a net accumulator of cadmium, lead and (weakly) zinc but appears to regulate body concentrations of copper (Borgmann *et al.*, 1993). *Gammarus pulex*, another freshwater amphipod, can eliminate zinc taken up from solution, at least at high experimental exposures (Xu and Pascoe, 1993). In the case of estuarine amphipods, *Gammarus duebeni* regulates body concentrations of zinc taken up from dissolved concentrations up to 200 μg Zn l^{-1} (Johnson and Jones, 1989) and *Gammarus zaddachi* also regulates body concentrations of zinc, but not copper (Amiard *et al.*, 1987). The littoral talitrid *Allorchestes compressa* appears to regulate body concentrations of both zinc and copper (Ahsanullah and Williams, 1991). Differences observed between accumulation patterns of amphipods taking up trace metals from solution may be down to differences in experimental protocols, not least duration and temperature (Borgmann *et al.*, 1993), but they may be real consequences of differing physiological adaptations to freshwater, estuarine and littoral environments.

Amphipods – for example, *Orchestia gammarellus* and *O. mediterranea* (Weeks and Rainbow, 1993) – accumulate trace metals such as zinc and copper from food as well as solution. In the case of these amphipods collected from the Firth of Clyde, Scotland, the high-littoral *O. gammarellus* is the more dependent on its decaying seaweed food source to meet both copper and zinc requirements, whereas the mid-littoral *O. mediterranea* obtains sufficient copper from solution but not from the food source alone (Weeks and Rainbow, 1991, 1993). Copper from the diet accumulates over the long term in copper-rich granules in cells of the ventral caeca (a midgut derivative), with more granules present in these cells in amphipods from copper-contaminated sites (Weeks, 1992). Another amphipod, *Corophium volutator*, from copper-contaminated sites has similar copper-rich granules in cells of the same ventral caeca (termed hepatopancreatic caeca in this case) (Icely and Nott, 1980). The degree of turnover of the ventral caecal cells and therefore the potential release of the metal-rich granules into the gut lumen and faeces await elucidation. Nevertheless the ventral caeca are the sites of storage of detoxified copper, temporary or permanent, and in this way parallel the role of the hepatopancreatic caecal cells of the isopods, their peracarid relatives.

The incorporation of trace metals accumulated from the diet into detoxified granules that are held temporarily before expulsion provides the basis for a particular trace metal accumulation pattern. Amphipods (or indeed isopods such as woodlice) exposed anew to a metal-rich food source would accumulate metal in the caecal cells, with the consequence that the metal concentrations of the caeca and of the whole body would rise. Eventually caecal cells laden with metal-rich granules reach the end of the cell cycle and metal is excreted as granules are expelled in the faeces. A steady state may be reached where excretion matches uptake. Exposure to a higher metal challenge would cause the production of more metal-rich granules per caecal cell. Initial metal accumulation would be at an increased rate but eventually a new steady state is reached, albeit at a raised caecal and body metal concentration. Metal excretion again might balance uptake from the diet, but now there are more metal-rich granules held in the caecal cells passing through their respective cell cycles. Even if excretion does not match uptake and net accumulation continues, the same principles hold and amphipods or isopods receiving different metal inputs in the diet would have different accumulated metal concentrations. Amphipods and isopods can therefore be used in metal biomonitoring programmes (Hopkin *et al.* 1986; Rainbow *et al.*, 1989, 1993a; Dallinger *et al.*, 1992; Shutes *et al.*, 1993), as can barnacles with no net metal excretion (Rainbow, 1987; Phillips and Rainbow, 1988; Rainbow *et al.*, 1993b).

Amphipods of one family, the Stegocephalidae, have an equivalent detoxificatory system for iron taken up from their diet of cnidarians (Moore and Rainbow, 1984, 1989, 1992). Iron is detoxified in the form of large crystals of ferritin in the R/F cells of the ventral caeca, these crystals eventually being expelled in the faeces. The storage of high concentrations of iron, albeit temporarily, in the ventral caeca causes the total body concentration of iron in stegocephalid amphipods to be raised in comparison with that in other amphipods (Moore and Rainbow, 1984, 1992) but without toxicological danger, as the iron is not metabolically available. The adoption of this particular iron detoxification system involving crystals of ferritin appears to be a specific feature of stegocephalids, probably in response to an iron challenge presented by their characteristic cnidarian prey (Moore and Rainbow, 1984).

9.4.5 DECAPODA

The final crustacean examples to be considered are decapods (Figure 9.1). Much comparative quantitative experimental work has been carried out (White and Rainbow, 1982, 1984a,b, 1986a,b; Rainbow and White, 1989, 1990) on the pleocyemate caridean decapod *Palaemon elegans*, variously referred to as a prawn or shrimp. As regards uptake from solution *P. elegans* has at least as low an uptake rate of zinc as amphipods (Tables 9.4, 9.6), much lower than that of a barnacle (Table 9.4) (Rainbow and White, 1989). The rate of dissolved cadmium uptake by *P. elegans* is much lower than those of the amphipod *Echinogammarus pirloti* and the barnacle *Elminius modestus* (Table 9.4) (Rainbow and White, 1989), as is the rate of cobalt uptake from solution (Rainbow and White, 1990). Such a low uptake rate might be expected, given the morphology of decapods: the majority of the cuticle is impermeable, and metal entry across the ectoderm is restricted to the gills. The comparative order of uptake rates by these three crustaceans is reflected in the comparative patterns of accumulation of trace metals taken up from solution (Figures 9.3, 9.4 and 9.5) (Rainbow and White, 1989, 1990). In the cases of both cadmium and cobalt, all metal taken up from solution over 28 days by *Palaemon elegans* is added sequentially to the existing body load of metal, with no excretion (White and Rainbow, 1986b; Rainbow and White, 1989, 1990). The accumulation pattern in each case is that of net accumulation (see Figure 9.5 for cadmium).

The pattern of accumulation of zinc taken up from solution by *Palaemon elegans* is strikingly different (White and Rainbow, 1982, 1984a,b; Rainbow and White, 1989). Even though the rate of zinc uptake from solution by the decapod increases with exposed zinc concentration, the body concentration of zinc does not show any change until a threshold external zinc concentration is reached (Figure 9.3). At lower exposed zinc concentrations new zinc is still entering the decapod (via the gills) in significant amounts but an equivalent amount of accumulated zinc is excreted to match the rate of zinc uptake (Figure 9.6) (White and Rainbow, 1984a,b; Rainbow and White, 1989). This pattern is interpreted as the regulation of the body zinc concentration (White and Rainbow, 1982, 1984a,b; Rainbow, 1988; Rainbow and White, 1989). When the decapod is exposed to increasing dissolved zinc concentrations, the

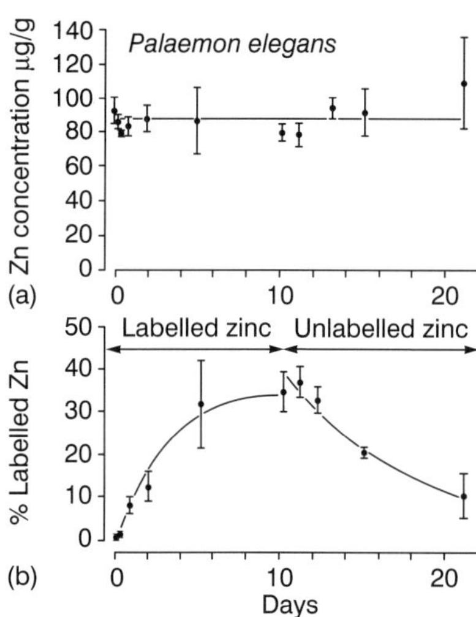

Figure 9.6 Regulation of body concentration of zinc by the decapod *Palaemon elegans* exposed to 100 µg Zn l⁻¹ for 20 days at 10°C in artificial seawater (days 0–10 in radioactively labelled zinc; thereafter in unlabelled zinc). (a) Total body concentrations of zinc; (b) changes in body concentrations of labelled zinc over time. (After White and Rainbow, 1984a; Phillips and Rainbow, 1993.)

rate of uptake of zinc increases until, at the point of regulation breakdown, the increasing rate of zinc excretion can no longer match the rate of zinc uptake, and net accumulation begins (Figure 9.3) (White and Rainbow, 1982, 1984a,b; Rainbow and White, 1989). The body zinc accumulated after regulation breakdown appears to remain in metabolically available form, for death ensues when the total body concentration reaches a value only about twice that of the regulated body concentration (White and Rainbow, 1982; Rainbow, 1988).

The situation concerning copper is less clear given the lack of a suitable radiotracer to allow investigation of the kinetics of copper accumulation. Nevertheless the accumulation pattern of copper taken up from solution by *Palaemon elegans* has similarities to that for zinc (Figure 9.4) (White and Rainbow, 1982). There appears to be a pattern of regulation up to a threshold external copper concentration, when net accumulation begins (Figure 9.4). Net accumulation of copper proceeds further from the 'regulated' body concentration, reaching 700 µg Cu g^{-1} before death (White and Rainbow, 1982). It appears, therefore, that much of this accumulated copper is in detoxified form. Indeed *P. elegans* accumulating copper from high external dissolved copper concentrations shows the presence of copper-rich granules in cells of the hepatopancreas (Rainbow, unpublished). These granules appear very similar to the copper-rich granules identified above in barnacles (Walker, 1977) and amphipods (Icely and Nott, 1980; Weeks, 1992) and probably represent residual bodies from the lysosomal breakdown of metallothionein binding copper.

Other caridean decapods have similar patterns of accumulation of trace metals taken up from solution. There is no regulation but net accumulation of body concentrations of cadmium in *Palaemon serratus* larvae (Devineau and Amiard-Triquet, 1985), *Crangon crangon* (Dethlefsen, 1978) and *Pandalus montagui* (Ray *et al.*, 1980). There is, however, regulation of body zinc concentrations in *P. serratus* larvae (Devineau and Amiard-Triquet, 1985), *C. crangon* (Amiard-Triquet *et al.*, 1983; Amiard *et al.*, 1985) and *Palaemonetes varians* (Nugegoda and Rainbow, 1988). The zinc uptake rates of three of the carideans exposed to dissolved zinc, investigated under otherwise identical physicochemical conditions, fall along an ecological gradient of decreasing salinity. As Table 9.7 shows, the uptake rate of the stenohaline sublittoral *Pandalus montagui* exceeds that of the littoral *Palaemon elegans*, which is greater than that of the brackish water inhabitant *Palaemonetes varians* (Nugegoda and Rainbow, 1988, 1989). The reductions in zinc uptake rate correlate with expected differences in exoskeleton permeability as an adaptation to life in low salinity (Mantel and Farmer, 1983; Campbell and Jones, 1990).

Detailed studies of cellular detoxification of accumulated trace metals in prawn and shrimps are more extensive for dendrobranchiate penaeids than for pleocyemate carideans (see Figure 9.1). Copper and zinc are stored in two types of granule in the R cells of the hepatopancreas of *Panaeus semisulcatus*, and granule abundance and hepatopancreas copper and zinc concentra-

Table 9.7 Comparative zinc uptake rates (μg Zn g^{-1} per day $\pm$ S.D where shown) of three caridean decapods under identical physicochemical conditions of exposure to dissolved zinc (after Nugegoda and Rainbow, 1988, 1989)

Conditions	Crustacean	Zinc uptake rate per day
20 μg Zn l^{-1}, 33 ppt salinity, 10°C	*Pandalus montagui* (sublittoral)	0.931 μg Zn g^{-1}
	Palaemon elegans (littoral pools)	0.582 $\pm$ 0.155 μg Zn g^{-1}
100 μg Zn l^{-1} 16.5 ppt salinity, 10°C	*Palaemon elegans* (littoral pools)	5.27 $\pm$ 3.67 μg Zn g^{-1}
	Palaemonetes varians (brackish water)	1.80 $\pm$ 0.61 μg Zn g^{-1}

tions vary with the moult cycle (Al-Mohanna and Nott, 1985). In *P. semisulcatus* this stored copper is predominantly found in sulphur-rich granules which also contain zinc (Al-Mohanna and Nott, 1985). These copper- and sulphur-rich granules are similar to those described earlier in the amphipods *Orchestia gammarellus* (Weeks, 1992) and *Corophium volutator* (Icely and Nott, 1980), and the isopod *Asellus meridianus* (Brown, 1977). Such copper-rich granules are also found in carideans such as *Palaemon elegans* (Rainbow, unpublished) and *Crangon crangon* (as *C. vulgaris*) (Djangmah and Grove, 1970). In *P. semisulcatus* zinc is also found (at higher relative concentration) in phosphorus-rich granules which are present in the hepatopancreas R cells during the pre-moult stage (Al-Mohanna and Nott, 1985). Decapods therefore employ storage detoxification of the essential trace metals copper and zinc in the hepatopancreas as the two metals are mobilized or immobilized at different stages of the moult cycle (Al-Mohanna and Nott, 1985, 1989), in the absence of any external metal challenge. These same detoxificatory processes are available when copper and zinc (and probably other metals such as cadmium) are accumulated from metal-contaminated conditions. For example Vogt and Quinitio (1994) have confirmed that in another species of *Penaeus* (*P. monodon*) copper-rich granules are accumulated in the hepatopancreas after exposure of the prawns to 1000 μg Cu l^{-1} for 10 days. The copper-rich granules increase in size and number along the cells of the hepatopancreas tubule in accordance with the age of the cells, and are released into the faeces upon discharge of senescent hepatopancreas cells. As in the amphipods and isopods, details of the patterns and kinetics of copper and zinc accumulation by different prawns and shrimps will depend on the nature of cell cycles, in this case in the hepatopancreas proper.

The site of detoxification of lead accumulated by *Penaeus monodon* after exposure to 1000 μg Pb l^{-1} for 10 days is different, the lead being primarily accumulated in granules in the antennary glands (Vogt and Quinitio, 1994). These lead-rich granules are discharged individually into the antennary gland lumen by apocrine secretion and excreted with the urine; the lead granules do not therefore require cell death before release (Vogt and Quinitio, 1994).

At the other evolutionary end of the decapods lie the brachyuran crabs. The accumulation of trace metals has received particular attention in the common shore crab *Carcinus maenas* (Bryan, 1966, 1971; Wright and Brewer, 1979; Rainbow, 1985; Chan and Rainbow, 1993a,b).

The accumulation pattern of zinc taken up from solution by *C. maenas* (Bryan, 1966, 1971; Rainbow, 1985) is superficially similar to that for *Palaemon elegans* (Figure 9.3) (White and Rainbow, 1982; Rainbow and White, 1989). There is very little, if any, significant increase in the body zinc concentration of crabs exposed to a range of dissolved zinc concentrations. This lack of increase is not the result of rapid and significant turnover of accumulated zinc in the body of the crustacean as for *P. elegans*, for all zinc taken up from solution by *C. maenas* is accumulated (Table 9.8) (Chan and Rainbow, 1993a,b). At very high zinc exposures there is net accumulation of zinc in spite of some zinc excretion across the gills (Bryan, 1966, 1971). In fact the rate of uptake of zinc by adult intermoult crabs is so low that under most environmental conditions newly entering zinc makes no significant contribution to the total body content of the crab (Table 9.8) (Chan and Rainbow, 1993a). Under the physicochemical conditions described in Table 9.4 the uptake rate of zinc by *C. maenas* exposed to 100 μg Zn l^{-1} is only 0.475 $\pm$ 0.069 μg Zn g^{-1} per day (Chan and Rainbow, 1993a) – less than a third of that of *P. elegans*, itself already low in comparison with other crustaceans (Table 9.4). The apparent zinc regulation of *C. maenas* therefore results from a very low absolute zinc uptake rate. This low uptake rate is to be expected from the systematic position of *C. maenas* as a brachyuran crab but ecology may also intervene, for *C. maenas* is a euryhaline inhabitant of seashores and estuaries as well as being found in fully marine sublittoral habitats. To extend the ecological argument, the Chinese mitten crab *Eriocheir sinensis*, living in dilute brackish water and in fresh water, has even lower uptake rates of zinc (and also cadmium) than *C. maenas* (W.H. Black, unpublished).

The accumulation pattern of copper taken up from solution by *Carcinus maenas* is similar to that for zinc with no significant increase in accumulated copper concentration over a wide range of ambient dissolved copper exposures (Rainbow, 1985), but lack of a radioactive tracer prevents explanation of copper accumulation kinetics. Cadmium, on the other hand, is clearly accumulated by the crabs in proportion to dissolved cadmium exposure (Jennings and Rainbow, 1979a; Rainbow, 1985).

As regards detoxificatory processes, metallothioneins binding cadmium, copper and zinc have been identified in *Carcinus maenas* (Wong and Rainbow, 1986a,b; Pedersen *et al.*, 1994) and in its fellow members of the family Portunidae – *Scylla serrata* (Otvos *et al.*, 1982) and *Callinectes sapidus* (Engel and Brouwer, 1984; Schlenk *et al.*, 1993; Narula *et al.*, 1995) – and also in other crabs such as *Cancer pagurus* (Overnell, 1986).

Table 9.8 Accumulated content of zinc (mean μg Zn $\pm$ SD ($n = 5$) in a standardized 10 g dry weight crab) taken up from solution by the crab *Carcinus maenas* exposed to a range of concentrations of radioactively labelled zinc for 21 days in artificial seawater (33 ppt salinity, 10°C) (after Chan and Rainbow, 1993a)

Exposure	Total Zn content	Labelled Zn content	Increase in total Zn content
Control	945 $\pm$ 100		
23 μg Zn l^{-1}	1048 $\pm$ 50	21.1 $\pm$ 2.6	103 $\pm$ 112
100 μg Zn l^{-1}	1064 $\pm$ 110	99.8 $\pm$ 15.0	119 $\pm$ 149
3162 μg Zn l^{-1}	2720 $\pm$ 320	1392 $\pm$ 307	1775 $\pm$ 335

Note that the accumulated content of labelled (new) zinc at all exposures is not significantly different from the observed increase in total zinc content, indicating that no newly taken up zinc is excreted.

Metallothioneins are not only induced in response to an external trace metal challenge, but are also prominent in the 'normal' biology of copper and zinc, as these metals are mobilized and immobilized during the moult cycle (Engel and Brouwer, 1991, 1993). The formation of the copper- and sulphur-rich granules identified above in penaeid decapods, for example (Al-Mohanna and Nott, 1985, 1989; Vogt and Quinitio, 1994), may be associated with the regular breakdown of such metallothioneins. Here then is further exemplification of a physiological trace metal detoxificatory system required in the absence of metal contamination, but with the potential to be employed at times of atypically high external metal challenge (see also Chapter 8).

Granules also play a role in the detoxification of trace metals in crabs. *Carcinus maenas* has concentric granules based on calcium phosphate in the R cells of the hepatopancreas, and these can immobilize lead (Hopkin and Nott, 1979). *Carcinus maenas* also has a high concentration of haemocyanin in the blood; the haemocyanin stores accumulated zinc (Zatta, 1984; Chan and Rainbow, 1993a,b) and acts as a more rapid transfer vehicle for accumulated cadmium (Wright and Brewer, 1979).

As for other decapods, lobsters regulate the body concentration of zinc; excretion takes place via the urine as opposed to the gills (Bryan, 1964, 1966, 1967; Bryan *et al.*, 1986). Freshwater crayfish also appear to regulate body concentrations of zinc but these crustaceans are so impermeable that there is insignificant uptake and loss of zinc across the body surface, all uptake and loss taking place in the gut (Bryan, 1966, 1967). Thus the ecological trend already noted is again in action here: adaptation to the low osmotic pressure of fresh water is associated with the development of very low cuticle permeability.

9.5 COMPARATIVE SUMMARY OF CRUSTACEAN ACCUMULATION PATTERNS FOR ZINC

Table 9.9 summarizes some of the examples discussed above. It is clear that crustaceans fall along a gradient of metal accumulation patterns from strong net accumulation to the regulation of body trace metal concentration. The separate processes of uptake, detoxification and excretion come together in different relative contributions in different crustaceans to produce the final integrated trace metal accumulation pattern. Such patterns vary for different trace metals within a single species, but nevertheless it can be concluded that an understanding of the biology of a crustacean, not least its phylogeny and ecology, does cast some light on why and how accumulation patterns do differ between crustacean taxa.

This chapter has also highlighted the features that make some crustaceans more valuable than others in metal ecotoxicological research. Trace metal biomonitoring programmes set up to investigate temporal or spatial variation in the bioavailabilities of toxic metals (in short, metal pollution) need to use organisms that are net accumulators of the metal in question (Phillips and Rainbow, 1993; Rainbow and Phillips, 1993; Rainbow, 1995b). Barnacles, which are

Table 9.9 A summary of patterns of accumulation of zinc shown by crustaceans

Species		Uptake	Turnover	Detoxified storage	Excretion	Accumulation pattern
From solution						
Elminius modestus (barnacle)		High	None	High	None	Strong net accumulation
Echinogammarus pirloti (amphipod)		Low	None	Low	None	Weak net accumulation
Orchestia gammarellus (amphipod)		Low	None	Low	None	Weak net accumulation
Orchestia mediterranea (amphipod)		Low	None	Low	None	Weak net accumulation
Palaemon elegans (decapod)	increased zinc ↓	Low / Increase / High	Low / Increase / High	None / None / Limited?	Low / Increase / Maximum	Regulation / Regulation / Regulation breakdown = accumulation
Carcinus maenas (decapod)	increased zinc ↓	V. Low / Increase	None / Some	Blood/ Exoskeleton	None / Some but < uptake	Negligible accumulation = regulation? / Net accumulation
From food						
Porcellio scaber (isopod woodlouse)		High?	Low	High	Low	Strong net accumulation
Oniscus asellus (isopod woodlouse)		High?	Rapid	Low	High	Weak net accumulation
Orchestia gammarellus (amphipod)		High	Low	High?	Low?	Net accumulation
Orchestia mediterranea (amphipod)		High?	Low	High?	Low?	Net accumulation

strong accumulators of trace metals, are particularly suitable as biomonitors given their sessile nature, but talitrid amphipods also have potential, certainly for copper and zinc (Rainbow and Phillips, 1993; Rainbow, 1995b). Decapods regulating body concentrations of copper and zinc would not be suitable as biomonitors of at least these two metals. Other relevant metal ecotoxicological research involving crustaceans includes the use of biomarkers such as the induction of metallothionein by specific trace metals – for example, in crabs. Metallothioneins do have potential as biomarkers of a toxic metal challenge but further research is needed to elucidate the background physiological variation in metallothionein synthesis – for example, with the interaction of the moult cycle and copper metabolism (Engel and Brouwer, 1991, 1993).

REFERENCES

Ahsanullah, M. and Williams, A.R. (1991) Sublethal effects and bioaccumulation of cadmium, chromium, copper and zinc in the marine amphipod *Allorchestes compressa*. *Marine Biology* **108**, 59–65.

Ahsanullah, M., Negilski, D.S. and Tawfik, F. (1980). Heavy metal content of *Callianassa* spp. in Western Port. *Australian Journal of Marine and Freshwater Research* **31**, 847–850.

Al-Mohanna, S.Y. and Nott, J.A. (1985) The accumulation of metals in the hepatopancreas of the shrimp *Penaeus semisulcatus* de Haan (Crustacea: Decapoda) during the moult cycle, in *Marine Environment and Pollution, Proceedings of the First Arabian Gulf Conference on Environment and Pollution*, (eds. R. Halwagy, D. Clayton and M. Behbehani), Kuwait University, Kuwait, pp. 195–207.

Al-Mohanna, S.Y. and Nott, J.A. (1989) Functional cytology of the hepatopancreas of *Penaeus semisulcatus* (Crustacea: Decapoda) during the moult cycle. *Marine Biology* **101**, 535–544.

Amiard-Triquet, C., Amiard, J.C., Robert, J.M. *et al.* (1983) Etude comparative de l'accumulation biologique de quelques oligo-elements metalliques dans l'estuarine interne de la Loire et les zones neritiques voisines (Baie de Bourgneuf). *Cahiers de Biologie Marine* **24**, 105–118.

Amiard, J.C., Amiard-Triquet, C. and Metayer, C. (1985) Experimental study of bioaccumulation, toxicity and regulation of some trace metals in various estuarine and coastal organisms. *Symposia Bioilogica Hungarica* **29**, 313–323.

Amiard, J.C., Amiard-Triquet, C., Berthet, C. and Metayer, C. (1987) Comparative study of the patterns of bioaccumulation of essential (Cu, Zn) and non-essential (Cd, Pb) trace metals in various estuarine and coastal organisms. *Journal of Experimental Marine Biology and Ecology* **106**, 73–89.

Barrington, E.J.W. (1967) *Invertebrate Structure and Function*, Thomas Nelson and Sons, London.

Blust, R., Kockelberg, E. and Baillieul, M. (1992) Effect of salinity on the uptake of cadmium by the brine shrimp *Artemia franciscana*. *Marine Ecology Progress Series* **84**, 245–254.

Blust, R., Bailleul, M. and Decleir, W. (1995) Effect of total cadmium and organic complexing on the uptake of cadium by the brine shrimp *Artemia franciscana*. *Marine Biology* **123**, 65–73.

Borgmann, U., Norwood, W.P. and Clarke, C. (1993) Accumulation, regulation and toxicity of copper, zinc, lead and mercury in *Hyalella azteca*. *Hydrobiologia* **25**, 79–89.

Bowman, T.E. and Abele, L.G. (1982) Classification of the recent Crustacea, in *The Biology of Crustacea, Vol. 1, Systematics, the Fossil Record, and Biogeography*, (ed. L.G. Abele), Academic Press, New York, pp. 1–27.

Brown, B.E. (1977) Uptake of copper and lead by a metal-tolerant isopod *Asellus meridianus* Rac. *Freshwater Biology* **7**, 235–244.

Brown, B.E. (1978) Lead detoxification by a copper-tolerant isopod. *Nature* **276**, 388–390.

Bruland, K.W. (1983) Trace elements in seawater, in *Chemical Oceanography*, (eds. J.P. Riley and R. Chester), Academic Press, London, pp. 157–220.

Bryan, G.W. (1964) Zinc regulation in the lobster *Homarus vulgaris*. I. Tissue zinc and copper concentrations. *Journal of the Marine Biological Association of the UK* **44**, 549–563.

Bryan, G.W. (1966) The metabolism of Zn and ^{65}Zn in crabs, lobsters and freshwater crayfish, in *Radioecological Concentration Processes*, (eds. B. Aberg and F.P. Hungate), Pergamon Press, Oxford, pp. 1005–1016.

Bryan, G.W. (1967) Zinc regulation in the freshwater crayfish (including some comparative copper analyses). *Journal of Experimental Biology* **46**, 281–296.

Bryan, G.W. (1968) Concentrations of zinc and copper in the tissues of decapod crustaceans. *Journal of the Marine Biological Association of the UK* **48**, 303–321.

Bryan, G.W. (1971) The effects of heavy metals (other than mercury) on marine and estuarine organisms. *Proceedings of the Royal Society of London, Series B* **177**, 389–410.

Bryan, G.W. (1976) Heavy metal contamination in the sea, in *Marine Pollution*, (ed. R. Johnston), Academic Press, London, pp. 185–302.

Bryan, G.W., Hummerstone, L.G. and Ward, E. (1986) Zinc regulation in the lobster *Homarus gammarus*: importance of different pathways of absorption and excretion. *Journal of the Marine Biological Association of the UK*, **66**, 175–199.

Campbell, R.J. and Jones, M.B. (1990) Water permeability of *Palaemon longirostris* and other euryhaline caridean prawns. *Journal of Experimental Biology* **150**, 145–158.

Chan, H.M. and Rainbow, P.S. (1993a) The accumulation of dissolved zinc by the shore crab *Carcinus maenas* (L.). *Ophelia* **38**, 13–30.

Chan, H.M. and Rainbow, P.S. (1993b) On the excretion of zinc by the shore crab *Carcinus maenas* (L.). *Ophelia* **38**, 31–45.

Dallinger, R. (1993) Strategies of metal detoxification in terrestrial invertebrates, in *Ecotoxicology of Metals in Invertebrates*, (eds. R. Dallinger and P.S. Rainbow), Lewis Publishers, CRC Press, Boca Raton, Florida, pp. 245–289.

Dallinger, R., Berger, B. and Birkel, S. (1992) Terrestrial isopods as biological indicators of urban metal pollution. *Oecologia* **89**, 32–41.

Depledge, M.H. and Rainbow, P.S. (1990) Models of regulation and accumulation of trace metals in marine invertebrates. *Comparative Biochemistry and Physiology* **97C**, 1–7.

Dethlefsen, V. (1978) Uptake, retention and loss of cadmium by brown shrimp *Crangon crangon*. *Meeresforschung* **26**, 137–152.

Devineau, J. and Amiard-Triquet, C. (1985) Patterns of bioaccumulation of an essential trace element (zinc) and a pollutant metal (cadmium) in larvae of the prawn *Palaemon serratus*. *Marine Biology* **86**, 139–143.

Djangmah, J.S. and Grove, D.J. (1970). Blood and hepatopancreas copper in *Crangon vulgaris* (Fabricius). *Comparative Biochemistry and Physiology* **32**, 733–745.

Eisler, R. (1981) *Trace Metal Concentrations in Marine Organisms*, Pergamon Press, Oxford.

Engel, D.W. and Brouwer, M. (1984) Cadmium-binding proteins in the blue crab, *Callinectes sapidus*: laboratory–field comparison. *Marine Environmental Research* **14**, 139–151.

Engel, D.W. and Brouwer, M. (1991) Short-term metallothionein and copper changes in blue crabs at ecdysis. *Biological Bulletin* **180**, 447–452.

Engel, D.W. and Brouwer, M. (1993) Crustaceans as models for metal metabolism: I. Effects of the molt cycle on blue crab metal metabolism and metallothionein. *Marine Environmental Research* **35**, 1–5.

Everaarts, J.M., Otter, E. and Fischer, C.V. (1990) Cadmium and polychlorinated biphenyls: different distribution pattern in North Sea benthic biota. *Netherlands Journal of Sea Research* **26**, 75–82.

Gibson, R. and Barker, P.L. (1979) The decapod hepatopancreas. *Oceanography Marine Biology Annual Review* **17**, 285–346.

Hames, C.A.C. and Hopkin, S.P. (1991) A daily cycle of apocrine secretion by the B cells in the hepatopancreas of terrestrial isopods. *Canadian Journal of Zoology* **69**, 1931–1937.

Hessler, R.R. and Elofsson, R. (1995) Segmental podocytic excretory glands in the thorax of *Hutchinsoniella macracantha* (Cephalocarida). *Journal of Crustacean Biology* **51**, 61–69.

Hopkin, S.P. (1989) *Ecophysiology of Metals in Terrestrial Invertebrates*, Elsevier Applied Science, Barking, UK.

Hopkin, S.P. (1990) Species-specific differences in the net assimilation of zinc, cadmium, lead, copper and iron by the terrestrial isopods *Oniscus asellus* and *Porcellio scaber*. *Journal of Applied Ecology* **27**, 460–474.

Hopkin, S.P. and Martin, M.H. (1984) Heavy metals in woodlice. *Symposia of the Zoological Society of London* **53**, 143–166.

Hopkin, S.P. and Nott, J.A. (1979) Some observations on concentrically structured intracellular granules in the hepatopancreas of the shore crab *Carcinus maenas* (L.) *Journal of the Marine Biological Association of the UK* **59**, 867–877.

Hopkin, S.P., Hames, C.A.C. and Dray, A. (1989) X-ray microanalytical mapping of the intracellular distribution of pollutant metals. *Microscopy and Analysis* **14**, 23–27.

Hopkin, S.P., Hardisty, G. and Martin, M.H. (1986) The woodlouse *Porcellio scaber* as a 'biological indicator' of zinc, cadmium, lead and copper pollution. *Environmental Pollution* **11B**, 271–290.

Icely, J.D. and Nott, J.A. (1980) Accumulation of copper within the 'hepatopancreatic' caeca of *Corophium volutator*. *Marine Biology* **57**, 193–199.

Icely, J.D. and Nott, J.A. (1992) Digestion and absorption: digestive system and associated organs, in *Microscopic Anatomy of Invertebrates, Vol. 10: Decapod Crustacea*, (eds F.W. Harrison and A.G. Humes), Wiley-Liss, New York, pp. 147–201.

Jennings, J.R. and Rainbow, P.S. (1979a) Studies on the uptake of cadmium by the crab *Carcinus maenas* in the laboratory. I. Accumulation from seawater and a food source. *Marine Biology* **50**, 131–139.

Jennings, J.R. and Rainbow, P.S. (1979b) Accumulation of cadmium by *Artemia salina*. *Marine Biology* **51**, 47–53.

Johnson, I. and Jones, M.B. (1989) Effects of zinc/salinity combinations on zinc regulation in *Gammarus duebeni* from the estuary and the sewage treatment works at Looe, Cornwall. *Journal of the Marine Biological Assocation of the UK* **69**, 249–260.

Mantel, L.H. and Farmer, L.L. (1983) Osmotic and ionic regulation in *The Biology of Crustacea, Vol. 5, Internal Anatomy and Physiological Regulation*, (ed. L.H. Mantel), Academic Press, New York, pp. 54–161.

Memmert, U. (1987) Bioaccumulation of zinc in two freshwater organisms (*Daphnia magna*, Crustacea and *Brachydanio rerio*, Pisces). *Water Research* **21**, 99–106.

Mizutani, A., Ifune, E., Zanella, A. and Eriksen, C. (1991) Uptake of lead, cadmium and zinc by the fairy shrimp, *Branchinecta longiantenna* (Crustacea: Anostraca). *Hydrobiologia* **212**, 145–149.

Moore, P.G. and Rainbow, P.S. (1984) Ferritin crystals in the gut caeca of *Stegocephaloides christianiensis* Boeck and other Stegocephalidae (Amphipoda: Gammaridea): a functional interpretation. *Philosophical Transactions of the Royal Society, Series B* **306**, 219–245.

Moore, P.G. and Rainbow, P.S. (1987) Copper and zinc in an ecological series of talitroidean Amphipoda (Crustacea). *Oecologia* **73**, 120–126.

Moore, P.G. and Rainbow, P.S. (1989) Feeding of the mesopelagic gammaridean amphipod *Parandania boecki* (Stebbing, 1888) (Crustacea: Amphipoda: Stegocephalidae) from the Atlantic Ocean. *Ophelia* **30**, 1–19.

Moore, P.G. and Rainbow, P.S. (1992) Aspects of the biology of iron, copper and other metals in relation to feeding in *Andaniexis abyssi*, with notes on *Andaniopsis nordlandica* and *Stegocephalus inflatus* (Amphipoda: Stegocephalidae), from Norwegian waters. *Sarsia* **78**, 215–225.

Narula, S.S., Brouwer, M., Hua, Y. and Armitage, I.M. (1995) Three-dimensional solution structure of *Callinectes sapidus* metallothionein-1 determined by homonuclear and heteronuclear magnetic resonance spectroscopy. *Biochemistry* **34**, 620–631.

Nott, J.A. and Nicolaidou, A. (1990) Transfer of metal detoxification along marine food chains. *Journal of the Marine Biological Association of the UK* **70**, 905–912.

Nugegoda, D. and Rainbow, P.S. (1988) Zinc uptake and regulation by the sublittoral prawn *Pandalus montagui* (Crustacea: Decapoda). *Estuarine Coastal and Shelf Science* **26**, 619–632.

Nugegoda, D. and Rainbow, P.S. (1989). Effects of salinity changes on zinc uptake and regulation by the decapod crustaceans *Palaemon elegans* and *Palaemonetes varians*. *Marine Ecology Progress Series* **51**, 57–75.

Otvos, J.D., Olafson, R.W. and Armitage, J.M. (1982) Structure of an invertebrate metallothionein from *Scylla serrata*. *Journal of Biological Chemistry* **257**, 2427–2431.

Overnell, J. (1986) Occurrence of cadmium in crabs (*Cancer pagurus*) and the isolation and properties of cadmium metallothionein. *Environmental Health Perspectives* **65**, 101–105.

Pedersen, K.L., Pedersen, S.N., Hojrup, P. *et al.* (1994) Purification and characterisation of a cadmium-induced metallothionein from the shore crab *Carcinus maenas* (L.). *Biochemical Journal* **297**, 609–614.

Phillips, D.J.H. and Rainbow, P.S. (1988) Barnacles and mussels as biomonitors of trace elements: a comparative study. *Marine Ecology Progress Series* **49**, 83–93.

Phillips, D.J.H. and Rainbow, P.S. (1993) *Biomonitoring of Trace Aquatic Contaminants*, Chapman & Hall, London.

Polikarpov, G.G., Oregioni, B., Parschevskaya, D.S., and Benayoun, G. (1979) Body burden of chromium, copper, cadmium and lead in the neustonic copepod *Anomalocera patersoni* (Pontellidae) collected from the Mediterranean sea. *Marine Biology* **53**, 79–82.

Powell, M.I. and White, K.N. (1990) Heavy metal accumulation by barnacles and its implications for their use as biological monitors. *Marine Environmental Research* **30**, 91–118.

Pullen, J.S.H. (1988) Aspects of the biochemical detoxification of accumulated heavy metals by barnacles with reference to metallothioneins and pyrophosphate granules. PhD thesis, University of London.

Pullen, J.S.H. and Rainbow, P.S. (1991) The composition of pyrophosphate heavy metal detoxification granules in barnacles. *Journal of Experimental Marine Biology and Ecology* **150**, 249–266.

Rainbow, P.S. (1985) Accumulation of Zn, Cu and Cd by crabs and barnacles. *Estuarine, Coastal and Shelf Science* **21**, 669–686.

Rainbow, P.S. (1987) Heavy metals in barnacles, in *Barnacle Biology*, (ed. A.J. Southward), A.A. Balkema, Rotterdam, pp. 405–17.

Rainbow, P.S. (1988) The significance of trace metal concentrations in decapods. *Symposia of the Zoological Society of London* **59**, 291–313.

Rainbow, P.S. (1989) Copper, cadmium and zinc concentrations in oceanic amphipod and euphausiid crustaceans, as a source of heavy metals to pelagic seabirds. *Marine Biology* **103**, 503–518.

Rainbow, P.S. (1990a) Heavy metal levels in invertebrates, in *Heavy Metals in the Marine Environment*, (eds. R.W. Furness and P.S. Rainbow), CRC Press, Boca Raton, Florida, pp. 67–79.

Rainbow, P.S. (1990b) Trace metal concentrations in a Hong Kong penaeid prawn. *Metapenaeopsis palmensis* (Haswell), in *Proceedings of the Second International Marine Biological Workshop: The Marine Flora and Fauna of Hong Kong and Southern China, Hong Kong, 1986*, (ed. B. Morton), Hong Kong University Press, Hong Kong, pp. 1221–1228.

Rainbow, P.S. (1992) The talitrid amphipod *Platorchestia platensis* as a potential biomonitor of copper and zinc in Hong Kong: laboratory and field studies, in *Proceedings of the Fourth International Marine Biological Workshop: The Marine Flora and Fauna of Hong Kong and Southern China, 1989*, (ed. B. Morton), Hong Kong University Press, Hong Kong, pp. 599–610.

Rainbow, P.S. (1993) The significance of trace metal concentrations in marine invertebrates, in *Ecotoxicology of Metals in Invertebrates*, (eds. R. Dallinger and P.S. Rainbow), Lewis Publishers, CRC Press, Boca Raton, Florida, pp. 3–23.

Rainbow, P.S. (1995a) Physiology, physicochemistry and metal uptake – a crustacean perspective. *Marine Pollution Bulletin* **31**, 55–59.

Rainbow, P.S. (1995b) Biomonitoring of heavy metal availability in the marine environment. *Marine Pollution Bulletin* **31**, 183–192.

Rainbow, P.S. (1996) Heavy metals in aquatic invertebrates, in *Environmental Contaminants in Wildlife Tissues: Interpreting Tissue Concentrations*, (eds W.N. Beyer, G.H. Heinz and A.W. Redmon-Norwood), CRC Press, Boca Raton, Florida, pp. 405–425.

Rainbow, P.S. and Phillips, D.J.H. (1993) Cosmopolitan biomonitors of trace metals. *Marine Pollution Bulletin* **26**, 593–601.

Rainbow, P.S. and Walker, G. (1977) The functional morphology of the alimentary tract of barnacles (Cirripedia: Thoracica). *Journal of Experimental Marine Biology and Ecology* **28**, 183–206.

Rainbow, P.S. and White, S.L. (1989) Comparative strategies of heavy metal accumulation by crustaceans: zinc, copper and cadmium in a decapod, an amphipod and a barnacle. *Hydrobiologia* **174**, 245–262.

Rainbow, P.S. and White, S.L. (1990) Comparative accumulation of cobalt by three crustaceans: a decapod, an amphipod and a barnacle. *Aquatic Toxicology* **16**, 113–126.

Rainbow, P.S., Scott A.G., Wiggins, E.A. and Jackson, R.W. (1980) Effect of chelating agents on the accumulation of cadmium by the barnacle *Semibalanus balanoides*, and complexation of soluble Cd, Zn and Cu. *Marine Ecology Progress Series* **2**, 143–152.

Rainbow, P.S., Moore, P.G. and Watson, D. (1989) Talitrid amphipods as biomonitors for copper and zinc. *Estuarine, Coastal and Shelf Science* **28**, 567–582.

Rainbow, P.S., Phillips, D.J.H. and Depledge, M.H. (1990) The significance of trace metal concentrations in marine invertebrates: a need for laboratory investigation of accumulation strategies. *Marine Pollution Bulletin* **21**, 321–324.

Rainbow, P.S., Emson, R.H., Smith, B.D. *et al.* (1993a) Talitrid amphipods as biomonitors of trace metals near Dunedin, New Zealand. *New Zealand Journal of Marine and Freshwater Research* **27,** 201–207.

Rainbow, P.S., Huang, Z.G., Yan, S.K. and Smith, B.D. (1993b) Barnacles as biomonitors of trace metals in the coastal waters near Xiamen, P.R. China. *Asian Marine Biology* **10**, 109–121.

Rainbow, P.S., Malik, I. and O'Brien, P. (1993c) Physicochemical and physiological effects on the uptake of dissolved zinc and cadmium by the amphipod crustacean *Orchestia gammarellus*. *Aquatic Toxicology* **25**, 15–30.

Ray, S., McLeese, D.W., Waiwood, B.A. and Pezzack, D. (1980) The disposition of cadmium and zinc in *Pandalus montagui*. *Archives of Environmental Contamination and Toxicology* **9**, 675–681.

Ridout, P.S., Willcocks, D.A., Morris, R.J. *et al.* (1985) Concentrations of Mn, Fe, Cu, Zn and Cd in the mesopelagic decapod *Systellaspis debilis* from the East Atlantic Ocean. *Marine Biology* **87**, 285–288.

Ridout, P.S., Rainbow, P.S., Roe, H.S.J. and Jones, H.R. (1989) Concentrations of V, Cr, Mn, Fe, Ni, Co, Cu, Zn, As, Cd in mesopelagic crustaceans from the north east Atlantic Ocean. *Marine Biology* **100**, 465–71.

Roesijadi, G. (1992) Metallothioneins in metal regulation and toxicity in aquatic animals. *Aquatic Toxicology* **22**, 81–114.

Roesijadi, G. and Robinson, W.E. (1994) Metal regulation in aquatic animals: mechanisms of uptake, accumulation and release, in *Aquatic Toxicology. Molecular Biochemical and Cellular Perspectives*, (eds. D.C. Malins and G.K. Ostrander), Lewis Publishers, Chelsea, Michigan, pp. 387–420.

Schlenk, D., Ringwood, A.H., Brouwer-Hoexum, T. and Brouwer, M. (1993) Crustaceans as models for metal metabolism. II. Induction and characterisation of metallothionein isoforms from the blue crab (*Callinectes sapidus*). *Marine Environmental Research* **35**, 7–11.

Shutes, B., Ellis, B., Revitt, M. and Bascombe, A. (1993) The use of freshwater invertebrates for the assessment of metal pollution in urban receiving waters, in *Ecotoxicology of Metals in Invertebrates*, (eds. R. Dallinger and P.S. Rainbow), Lewis Publishers, CRC Press, Boca Raton, Florida, pp. 201–222.

Van Hattum, B., Korthals, G., Van Straalen, N.M. *et al.* (1993) Accumulation patterns of trace metals in freshwater isopods in sediment bioassays – influence of substrate characteristics, temperature and pH. *Water Research* **27**, 669–684.

Viarengo, A. (1989) Heavy metals in marine invertebrates: mechanisms of regulation and toxicity at the cellular level. *Reviews in Aquatic Sciences* **1**, 295–317.

Vogt, G. (1994) Life-cycle and functional cytology of the hepatopancreatic cells of *Astacus astacus* (Crustacea, Decapoda). *Zoomorphology* **114**, 83–101.

Vogt, G. and Quinitio, E.T. (1994) Accumulation and excretion of metal granules in the prawn, *Penaeus monodon*, exposed to water-borne copper, lead, iron and cadmium. *Aquatic Toxicology* **28**, 223–241.

Walker, G. (1977) 'Copper' granules in the barnacle *Balanus balanoides*. *Marine Biology* **39**, 343–349.

Walker, G. and Foster, P. (1979) Seasonal variation of zinc in the barnacle *Balanus balanoides* (L.) maintained on a raft in the Menai Strait. *Marine Environmental Research* **2**, 209–222.

Walker, G., Rainbow, P.S., Foster, P. and Crisp, D.J. (1975a) Barnacles: possible indicators of zinc pollution? *Marine Biology* **30**, 57–75.

Walker, G., Rainbow, P.S., Foster, P. and Holland, D.L. (1975b) Zinc phosphate granules in tissues surrounding the midgut of the barnacle *Balanus balanoides*. *Marine Biology* **33**, 162–166.

Weeks, J.M. (1992) Copper-rich granules in the ventral caeca of talitrid amphipods (Crustacea; Amphipoda; Talitridae). *Ophelia* **36**, 119–133.

Weeks, J.M. and Rainbow, P.S. (1991) The uptake and accumulation of zinc and copper from solution by two species of talitrid amphipods (Crustacea). *Journal of the Marine Biological Association of the UK* **71**, 811–26.

Weeks, J.M. and Rainbow, P.S. (1993) The relative importance of food and seawater as sources of copper and zinc to talitrid amphipods (Crustacea; Amphipoda; Talitridae). *Journal of Applied Ecology* **30**, 722–735.

White, S.L. and Rainbow, P.S. (1982) Regulation and accumulation of copper, zinc and cadmium by the shrimp *Palaemon elegans*. *Marine Ecology Progress Series* **8**, 95–101.

White, S.L. and Rainbow, P.S. (1984a) Regulation of zinc concentration by *Palaemon elegans* (Crustacea: Decapoda): zinc flux and effects of temperature, zinc concentration and moulting. *Marine Ecology Progress Series* **16**, 135–147.

White, S.L. and Rainbow, P.S. (1984a) Zinc flux in *Palaemon elegans* (Crustacea: Decapoda): moulting, individual variation and tissue distribution. *Marine Ecology Progress Series* **19**, 153–166.

White, S.L. and Rainbow, P.S. (1986a) A preliminary study of Cu-, Cd- and Zn-binding components in the hepotopancreas of *Palaemon elegens* (Crustacea: Decapoda). *Comparative Biochemistry and Physiology* **83C**, 111–116.

White, S.L. and Rainbow, P.S. (1986b) Accumulation of cadmium by *Palaemon elegans* (Crustacea: Decapoda). *Marine Ecology Progress Series* **32**, 17–25.

White, S.L. and Rainbow, P.S. (1987) Heavy metal concentrations and size effects in the mesopelagic decapod crustacean *Systellaspis debilis*. *Marine Ecology Progress Series* **37**, 147–151.

White, K.N. and Walker, G. (1981) Uptake, accumulation and excretion of zinc by the barnacle *Balanus balanoides* (L.) *Journal of Experimental Marine Biology and Ecology* **51**, 285–298.

Wong, V.W.T. and Rainbow, P.S. (1986a) Two metallothioneins in the shore crab *Carcinus maenas*. *Comparative Biochemistry and Physiology* **83A**, 149–156.

Wong, V.W.T. and Rainbow, P.S. (1986b) Apparent and real variability in the presence and metal contents of metallothioneins in the crab *Carcinus maenas* including the effects of isolation procedure and metal induction. *Comparative Biochemistry and Physiology* **83A**, 157–177.

Wright, D.A. and Brewer, C.C. (1979) Cadmium turnover in the shore crab *Carcinus maenas*. *Marine Biology* **50**, 151–156.

Xu, Q. and Pascoe, D. (1993) The bioconcentration of zinc by *Gammarus pulex* (L.) and the application of a kinetic model to determine bioconcentration factors. *Water Research* **27**, 1683–1688.

Zatta, P. (1984) Zinc transport in the haemolymph of *Carcinus maenas* (Crustacea: Decapoda). *Journal of the Marine Biological Association of the UK* **64**, 801–807.

10 *Mechanisms of heavy metal accumulation and toxicity in fish*

PER-ERIK OLSSON, PETER KLING AND CHRISTER HOGSTRAND

10.1 INTRODUCTION

As a result of mining, forestry, waste disposal and fuel combustion, our environment is becoming increasingly contaminated with heavy metals. The aquatic environment receives waste products from such activities and may be the final depository for these anthropogenically remobilized heavy metals. In order to understand the impact of heavy metals on aquatic biota it is important to characterize the mechanisms available for aquatic life to transport, immobilize and excrete heavy metals.

Relatively little is known about the specific mechanisms of uptake of metals in cells of non-mammalian vertebrates. This chapter attempts to assemble available information on fish. In general, metals absorbed across the gills or the intestinal wall are distributed via the circulation, bound to transport proteins, to different tissues of the body. Within the tissues metals can participate in the essential life functions, but can also exert toxic actions, or be detoxified by binding to the protein, metallothionein (MT). The intracellular levels of essential metal are regulated by transporters (which translocate metals across the plasma membrane) as well as by MT and other metal-binding proteins. Metals themselves cannot be metabolized (using the strict definition of the term) and can only be eliminated from the body by excretion.

Metals that enter the body will react with different components of the cell (Chapter 1). The heavy metals are soft donors and will therefore readily bind to soft acceptors, such as sulfhydryl-groups. MT is a low molecular weight cytosolic, cysteine-rich protein that binds group 1B and 2B heavy metals (Olsson and Haux, 1985). Within the cells of the body, MT is the major heavy metal-binding protein (Dunn *et al.*, 1987). Environmental exposure of fish to heavy metals has been shown to result in primarily renal and hepatic accu-

Metal Metabolism in Aquatic Environments. Edited by William J. Langston and Maria João Bebianno. Published in 1998 by Chapman & Hall, London. ISBN 0 412 80370 4

mulation of the metal in MT (Olsson and Haux, 1986; Norey *et al.*, 1990; Hogstrand *et al.*, 1991). Heavy metals will further activate biosynthesis of new MT (Olsson, 1993). Exposure of fish to heavy metals has been shown to result in protection against subsequent higher levels of these metals and induction of MT during the initial exposure may be a mechanism that confers acquired tolerance (Goering and Klaassen, 1984; Hamilton and Mehrle, 1986). Thus, MT detoxifies metals by binding the free ions and thereby making them less available for interaction with sensitive biomolecules.

This chapter is divided into five main sections, which deal (in order) with the uptake, distribution, accumulation and storage, physiological and biochemical effects, and excretion of heavy metals. Implications for environmental hazard assessment are summarized at the end of the chapter.

10.2 UPTAKE

10.2.1 ABSORPTION ACROSS THE GILLS

Metal uptake in fish differs fundamentally from that in terrestrial animals because of the fact that fish have gills that are constantly submersed in a solution of metal ions. At least in freshwater fish (which do not drink), the gills are the main point of entry for dissolved metals. As we will see later, gills are also a major target for metal toxicity. The most important physiological roles of the gills are to take up oxygen from the water and excrete carbon dioxide. Because water contains much less dissolved oxygen than air, a fish has to move approximately 20 litres of water across the respiratory surfaces to extract the same amount of oxygen as a mammal can obtain from 1 litre of air. This physical constraint means that large amounts of metals are passed over the gills at any time, which enhances metal uptake.

The gills of teleost fish are central for the ion regulation of the animal and they possess at least one type of specialized ion-transporting cell. These are usually referred to as chloride cells because they were first identified as cells involved in chloride transport (Figure 10.1). In contrast to the flat respiratory pavement cells, the chloride cells have a columnar shape; they are rich in mitochondria, and the basolateral surfaces (i.e. serosal side) are deeply invaginated. At the apical (i.e. mucosal) side most of the chloride cell is covered by pavement cells and only a minor portion of their apical membrane is in contact with the water. However, this portion is under dynamic control; the gill surface area covered by chloride cells can change markedly within a few hours (Goss *et al.*, 1992; Perry *et al.*, 1993) and this change in surface area correlates to changes in ion fluxes across the gills (Goss *et al.*, 1992; Perry *et al.*, 1993). There is now strong evidence that at least the divalent Co^{2+}, Zn^{2+} and Cd^{2+} ions traverse the gill epithelium through the chloride cells (Spry and Wood, 1989; Verbost *et al.*, 1989; Wicklund Glynn *et al.*, 1994; Hogstrand *et al.*, 1994, 1996a; Comhaire *et al.*, submitted). These metals all seem to enter the apical membrane of the chloride cells through the same transporter as Ca^{2+} (Spry and Wood, 1989; Verbost *et al.*, 1989; Hogstrand *et al.*, 1994, 1996a;

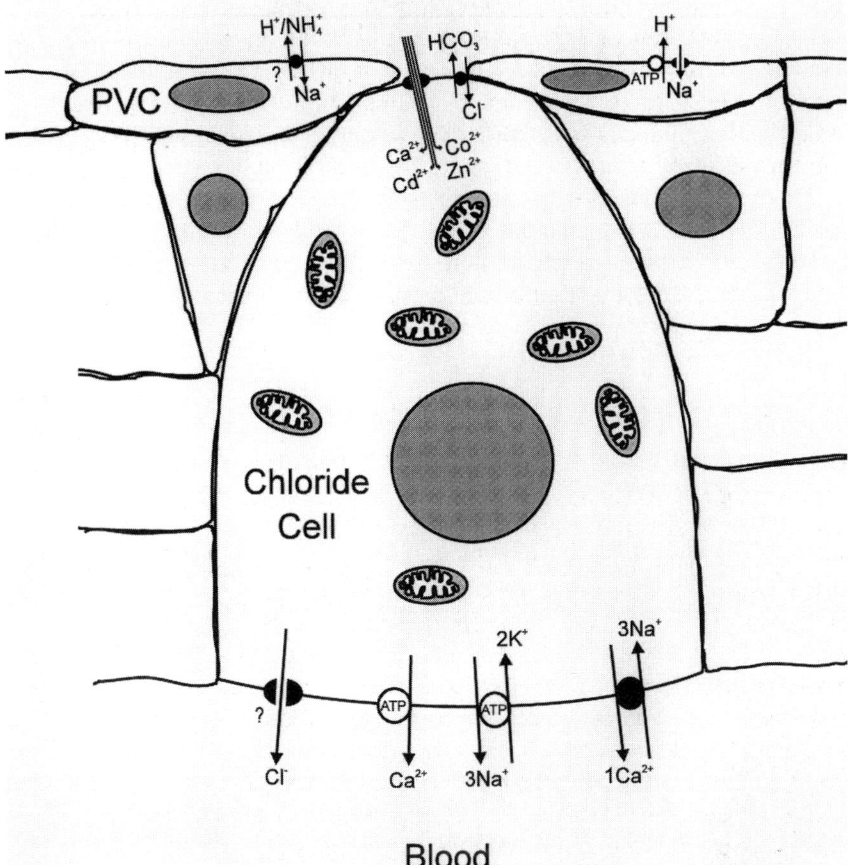

Figure 10.1 Schematic presentation of a freshwater fish chloride cell and a hypothetical model of mechanisms involved in unidirectional ion transport across the gill epithelium. Most of the apical membrane of the chloride cell is covered by pavement cells (PVC). These PVC may regulate the apical uptake of ions from the water into the chloride cell by modifying the area of the chloride cell that is in direct contact with the water. The apical membrane of the chloride cell contains a La^{3+}-sensitive cation transporter (channel?) which is permeable to divalent ions, such as Ca^{2+}, Co^{2+}, Zn^{2+} and Cd^{2+}. In the apical membrane, there is also an electroneutral Cl^-/HCO_3^- exchanger that is responsible for the Cl^- uptake into the chloride cell. The PVC (and possibly the chloride cells) have an apical H^+ pump, which drives the uptake of Na^+ into the gill epithelium through an amiloride-sensitive channel. Earlier theories of apical Na^+/H^+ and Na^+/NH_4^+ electroneutral exchangers now appear less likely, but are not completely disproven. On the basolateral side, Cl^- may be transported into the blood through a Cl^--channel or possibly a Cl^-/HCO_3^- exchange mechanism. The movement of Na^+, as well as Cl^-, across the basolateral membrane is driven by the Na^+/K^+ ATPase. Basolateral transfer of Ca^{2+} is mediated by a high-affinity Ca^{2+} ATPase and a low-affinity Na^+/Ca^{2+} exchange mechanism, which uses the transmembrane Na^+ gradient for the uphill transport of Ca^{2+} from the chloride cell cytosol to the blood.

Comhaire *et al.*, 1994). Thus, Ca^{2+}, Co^{2+}, Zn^{2+} and Cd^{2+} compete for the same apical uptake sites. This competition explains why water hardness, which is mostly made up by the calcium concentration, reduces the uptake of these elements and protects against toxicity (Bradley and Sprague, 1985a,b; Wicklund and Runn, 1988; Comhaire *et al.*, 1994, submitted). Furthermore, during long-term exposure to waterborne Zn, rainbow trout seems to be capable of reducing the branchial Zn^{2+} influx by reducing the affinity of this carrier (Hogstrand *et al.*, 1994).

It is not clear how any of the metals Co^{2+}, Zn^{2+} and Cd^{2+} are transferred across the basolateral membrane of the gill epithelium. Studies using isolated intestinal basolateral membranes from tilapia (*Oreochromis mossambicus*) suggest that Cd^{2+} can be transported by the Na^+/Ca^{2+} exchanger, operating in a Cd^{2+}/Ca^{2+} exchange mode (Schoenmakers *et al.*, 1992). The importance of the Na^+/Ca^{2+} exchanger in basolateral transfer of Cd^{2+} in chloride cells as well as enterocytes is uncertain. Influx of Cs^+ across the gills of rainbow trout follows saturation kinetics, which suggests that the uptake takes place through a transcellular, carrier mediated pathway (Morgan *et al.*, 1994).

All metals discussed above seem to cross the gill epithelium through specific pathways. This is apparently not the case for all metals. Wicklund Glynn *et al.* (1994) studied the localization of radioactive Cd and inorganic Hg after a short exposure to either of these metals in the water. Whereas Cd staining was found in discrete epithelial cells, tentatively identified as chloride cells, radioactive Hg was present in all cell types.

10.2.2 GASTROINTESTINAL ABSORPTION

As in mammals, the gastrointestinal tract is an important route for metal absorption in fish. In general, the relative importance of the intestine for metal uptake increases with increasing hardness and salinity of the water. These factors reduce the branchial uptake of many metals. An additional factor is that marine fish drink considerable quantities of water, which means that metals may be taken up not only from the food, by the gastrointestinal tract, but also from the water. Heavy metals in their inorganic form are believed to enter the cell through facilitated diffusion while methyl-Hg, due to its high lipophilicity, may diffuse directly across the membrane (Chapter 5). Since only a limited number of studies have dealt with fish, comparisons will be made with mammalian systems.

In general, most of the Cd that reaches the gastrointestinal tract is not taken up by the body. Exposure of epithelial cell surfaces to heavy metals results in the formation of mucus, which traps metals and decreases intestinal metal uptake (Sorensen, 1991). It has been shown that approximately 95% of the ingested cadmium binds to mucus in the fish gut (Pärt and Lock, 1983). It has been suggested that Cd binds to the luminal surface of the mucosal cells and is taken up by the cell by membrane fluidity processes (Handy, 1996). In fish, only a fraction (1–2%) of the ingested Cd is thereby absorbed by the mucosal cells (Harrison and Klaverkamp, 1989). The situation is similar to that in other animal groups. In rats only 0.3 % of the Cd is accumulated and about 7% in poultry (Kello and Kostial, 1977a; Sell, 1977). In humans the absorbed fraction of Cd is about 10% (Flanagan *et al.*, 1978). Intestinal metal retention may be further reduced by trapping of the metals in the epithelial cell layer, which is removed by desquamation (Bremner, 1979). Neither Cd nor Hg appears to depend on reactive sulfhydryl groups or on oxidative metabolism for their transfer across the intestinal mucosa (Foulkes and Bergman, 1993), but one of the most important chelating molecules for Cd in the enterocytes is the sulfhydryl-rich MT. MT induction thus appears to reduce the Cd transfer from the luminal to the serosal side of the intestinal epithelium (Bremner, 1979). Cd is not taken up against its concentration gradient in any segment of the intestinal wall, which would indicate that there may be no active transport mechanisms for Cd uptake through the intestinal mucosa (Gruden, 1981). Schoenmakers *et al.* (1992) found that Cd^{2+} could be transported by the Na^+/Ca^{2+} exchanger in isolated basolateral membranes from tilapia enterocytes, but the importance of such a transport mechanism *in vivo* remains to be verified. It has been observed that Cd absorption was greatly enhanced in mice during a one month dietary aluminium exposure experiment (Sugawara and Sugawara, 1992), suggesting that an induced abnormality of the gut wall, caused by aluminium, enhanced Cd absorption. When comparing the toxicity of $CdCl_2$ and Cd-MT in intestinal primary cultures and cell lines it has been observed that the toxicity of $CdCl_2$ is much higher than that of Cd-MT (Groten *et al.*, 1992). It has been proposed that the intestinal uptake mechanisms for Cd, as $CdCl_2$ or Cd-MT, are different, and that deficiencies of metals such as Fe and Zn increase the deposition of Cd in the kidney of rats (Ohta and Cherian, 1995).

Hg mainly accumulates as methyl-Hg in fish, often regardless of their position in the food chain (Handy, 1996; Chapter 5). Most of the methyl-Hg is then stored in muscle tissue, subsequently leading to biomagnification through the food chain. The uptake efficiency of methyl-Hg is approximately 20% while only about 6% of the ingested inorganic Hg will be absorbed (Boudou and Ribeyre, 1985). Due to its high lipophilicity, methyl-Hg may diffuse directly across the cellular membrane. Like Cd, inorganic Hg is trapped in the mucus, which reduces its bioavailability (Pärt and Lock, 1983). Active transport of inorganic Hg is deemed unlikely due to its inhibitory effect on ATP-ases and ion channels (Baatrup *et al.*, 1990; Gill *et al.*, 1990).

Uptake of Cu occurs mainly as Cu^{2+} at neutral pH. The mechanism of Cu uptake over the gastrointestinal epithelium is open for investigation. Due to its complexation with mucosa, Cu bioavailability (uptake efficiency) has been calculated to be approximately 3% in fish (Handy, 1992). In mammals, the bioavailability of Cu is higher than this, ranging from 10% in cattle up to 75% in humans. The regulation of intracellular Cu levels depends, to a large extent, on specific Cu transport proteins that actively remove Cu from the cells. These will be discussed more in detail in section 10.3.

Information on the nature of intestinal Zn uptake in fish is scarce and is limited to marine species. The absorption of Zn in the winter flounder (*Pseudopleuronectes americanus*) and plaice (*Pleuronectes platessa*) seems to be highest in the anterior part of the intestine (Pentreath, 1976; Renfro *et al.*, 1975). The uptake of Zn by the enterocytes becomes saturated at high Zn loads, suggesting that binding of Zn to more or less specific sites on the surface of, or inside, the enterocytes is involved (Shears and Fletcher, 1983). The entry step through the brush-border membrane (e.g. the membrane facing the intestinal lumen) is inhibited by the presence of Cu, Cd, Co, Cr, Ni, Mg and Hg. The uptake of Zn is unaffected by the Ca content of the lumen, which would indicate that the mechanism of Zn uptake in winter flounder may differ from that in mammals, where high Ca reduces the absorption of Zn. The kinetics of basolateral Zn transfer in winter flounder enterocytes indicate a passive diffusion process, superimposed on a saturable component (Shears and Fletcher, 1983). There is no available information on the kinetics of intestinal Zn uptake in freshwater fish.

In mammals, Zn is absorbed mainly from the duodenum, ileum and jejunum (Lee *et al.*,1989; Lönnerdal, 1988). Despite a number of studies on the topic, the underlying mechanisms of Zn uptake in mammals are poorly understood. As in marine fish, the kinetics of intestinal transepithelial Zn transport in mammals suggests the presence of both saturable and non-saturable components (Lönnerdal, 1988). The bioavailability of dietary Zn, expressed as uptake efficiency, depends on the nutritional needs of the individual and can be as little as 20%, or as much as 70% during Zn deficiency, in both mammals and fish (Matsusaka *et al.*, 1985, Hardy *et al.*, 1987).

10.3 DISTRIBUTION

Following absorption, heavy metals such as Zn, Cu, Cd or Hg are distributed by the blood to different tissues in the body, where they may cause toxic effects before they are excreted via urine or faeces, or across the gills.

10.3.1 CADMIUM

When Cd has been absorbed by the body it is transported primarily to liver and kidneys. When Cd first enters the blood it is present in the plasma pool, but is then gradually transferred to the erythrocytes, until equilibrium is reached between the erythrocyte pool and the plasma pool. Following injection of Cd in mice, Perry and Erlanger (1971) found that, initially, almost all Cd remained in the plasma pool, but after 4 hours about 50% of injected Cd had been transferred from the plasma to the erythrocytes. Cd in the blood is either distributed to different organs or excreted. There are several molecules that bind Cd in the plasma pool; in mammalian systems both albumin and α2-macroglobulin have been shown to bind and transport any Cd that remains for prolonged times in the serum (Giroux and Henkin, 1972; Watkins *et al* ., 1977).

The distribution and toxicity of Cd can be altered by a variety of different substances. In rainbow trout it has been shown that fish fed ascorbic acid-deficient food have elevated hepatic Cd levels (Palace *et al.*, 1993). This effect has also been observed in ascorbic acid-deficient mice (Rambeck *et al.*, 1988). Diplock (1976) showed that Se injections protect against Cd-induced testicular damage and cause Cd to accumulate in the testes at higher concentrations than in animals exposed to Cd without Se. Selenium appears to divert the binding of Cd from low to high molecular weight proteins in the cytosol of these testicular cells. The altered binding pattern for Cd caused by Se supplements is thought to be responsible for the protective effect (Chen, 1974). Diets containing Se have also been shown to decrease the accumulation of Cd in liver and kidney, but tend to increase Cd deposition in the testes.

As discussed above, Cd may be transported within erythrocytes. MT is one of the major Cd-binding proteins within the erythrocytes (Tanaka *et al.*, 1985). It is thought that MT is primarily an intracellular protein, but it may be present in high levels in plasma during certain situations (Bremner and Morrison, 1986). The mechanism of the MT release from the cell is unclear but MT-containing exocytotic vesicles have been observed in parenchymal liver cells of Cd-exposed rainbow trout (Morka, 1991). It has been shown that MT secretion is inhibited following Cd exposure of rat liver parenchymal cells (Mitane *et al.*, 1987). The presence of MT in extracellular compartments may increase the circulating pool of metals. For example, rats injected with Cd-MT have been shown to develop higher levels of circulating Cd than rats injected with $CdCl_2$ (Goyer, 1995; Chan *et al.*, 1992).

The redistribution of Cd from the blood to different tissues may, to some extent, depend on the affinity for detoxifying systems. Certain freshwater fish, including rainbow trout, are very sensitive to Cd in their environment, whilst other species, such as stone loach (*Noemachelius barbatulus*) and pike (*Esox lucius*), are more tolerant to Cd (Norey *et al.*, 1990; Norey, 1991). Comparison of the organ distribution of Cd between rainbow trout and stone loach has shown that rainbow trout preferentially accumulate Cd in the gills and kidney, while stone loach tends to accumulate Cd primarily in the gills and the liver (Olsson *et al.*, 1989a; Norey *et al.*, 1990). Functional promoter analysis of the MT genes from rainbow trout and stone loach show that MT is equally inducible in both species (Olsson and Kille, 1997). In rainbow trout the MT-A gene has six metal response elements (MREs) whereas the stone loach MT gene has seven MREs (Kille *et al.*, 1993; Olsson *et al.*, 1995a). When these gene promoters have been tested in cell transfection experiments they give similar reporter gene activities (Table 10.1). One difference between these two species is that Cu levels are one order of magnitude higher in rainbow trout liver (Norey, 1991). Cu has a high affinity for MT and it is possible that the high hepatic Cu levels in rainbow trout may in part be responsible for the preferential distribution of Cd to the kidney in this species. In stone loach, on the other hand, the hepatic Cu levels are much lower and in this species Cd is primarily distributed to the liver. Another Cd-tolerant species, the pike, shows a very different tissue distribution pattern (Norey *et al.*, 1990; Norey, 1991). In the pike, the Cd level of the gills is low and very little Cd is accumulated by this species compared with rainbow trout and stone loach. When analysing the MT promoter from pike and comparing it with rainbow trout and stone loach MT, two major differences can be observed. Firstly, pike has only four MREs; secondly, the pike MT gene has a TTTA box instead of a TATA box (Kille *et al.*, 1993). Both these differences result in reduced metal inducibility of the pike MT gene (Olsson and Kille, 1997). It is tempting to speculate that the reduced inducibility of MT is involved in the lower Cd retention observed for this species (as postulated for molluscs in Chapter 8). From the above analysis of the MT gene promoters, it is clear that a multitude of parameters may be involved in determining the metal sensitivity of a species. While the difference in Cd sensitivity between rainbow trout and stone loach appears to be due to factors other than MT, it is possible that impaired MT regulation is involved in the lowered Cd retention in pike.

10.3.2 ZINC

Absorbed Zn is bound to albumin and α-globulin in the blood and transported to the liver, where it accumulates before being distributed to other organs (Giroux, 1975): Zn is primarily distributed to bone, muscles and skin. Altered Zn distribution can result in deleterious effects. For example, urethane induction of MT synthesis in pregnant rats can dramatically alter the distribution of Zn in the fetus. This altered Zn distribution is believed to be responsible for the diminished development observed in urethane-treated rats (Daston *et al.*, 1991).

Mechanisms of Zn transport into and out of cells have received significant attention during recent years, although few Zn transporters have been particularly well characterized (Reyes, 1996). Zn can enter erythrocytes as a $ZnCO_3Cl^-$ or $Zn(HCO_3)Cl.OH^-$ complex, through the DIDS (4,4′-diisothiocyanatostilbene-2,2′-disulfonic acid)-sensitive Cl^-/HCO_3^- transporter, or as a $Zn(SCN)_2$ complex through a DIDS-insensitive mechanism (Kalfakakou and Simons, 1990). Eel erythrocytes seem to transport Zn through the Cl^-/HCO_3^- exchanger and also through an unidentified DIDS-insensitive pathway (Mandolfino *et al.*, 1994). Under physiological conditions, transport mechanisms other than the Cl^-/HCO_3^- exchanger may be more important for uptake of Zn into erythrocytes. Van Woewe *et al.* (1990) reported the presence of a high-affinity Zn^{2+} transporter with a K_M of 0.2 nM. Incidentally, this concentration is very similar to the activity of free Zn^{2+} in the plasma of mammals (Magneson *et al.*, 1987), which makes the high-affinity Zn^{2+}-transporter a likely pathway for Zn uptake in erythrocytes *in vivo*. Yet another Zn uptake mechanism seems to be present in mammalian erythrocytes. Zn can be transported into erythrocytes bound to L-histidine $[Zn(His)_2]$ by a Na-dependent amino acid transport system. The efflux of Zn from human erythrocytes seems to be mediated by a Zn^{2+}/Ca^{2+} exchange mechanism that uses the

Table 10.1 Comparison of rainbow trout-MTA (RT-MTA), stone loach-MT (SL-MT) and pike-MT (P-MT) gene promoter activity

Promoter	Zinc	Copper	Cadmium	H_2O_2
RT-MTA	32	6	16	12
SL-MT	34	5	19	nd
P-MT	23	4	9	nd

Experiments were performed on RTH-149 cells, cotransfected with plasmids containing luciferase or β-galactocidase reporter genes, driven by the RT-MTA and SV40 promoters, respectively (Olsson *et al.*, 1996, Olsson and Kille, 1997). Following induction with 150 μm Zn, 100 μM Cu, 10 μM Cd or 100 μM H_2O_2 the cells were harvested and the protein concentration was determined by the Bradford assay. Equal amounts of sample were used for luciferase and β-galactosidase assays. The luciferase activities were standardized for transfection efficiency by measurement of β-galactosidase activity. The activity of the different constructs are shown as fold induction with the activity of the pGL-2 basic vector subtracted. nd, not determined.

inward Ca^{2+} gradient to catalyse the uphill outward movement of Zn^{2+} (Simons, 1991).

Zn uptake in hepatocytes from rat or longjaw squirrelfish (*Holocentrus marianus*) is characterized by a rapid uptake phase followed by a slower phase (Pattison and Cousins, 1986; Hogstrand *et al.*, 1996b). The same uptake pattern was observed in liver slices from puffer fish (*Tetraodon hispidus*) (Saltman and Boroughs, 1960). In rat hepatocytes, both phases consist of saturable and non-saturable components (Pattison and Cousins, 1986; Taylor and Simons, 1994). The saturable pathway of the rapid phase is probably the uptake route that is of importance *in vivo* because of its high affinity for Zn^{2+} (K_M = 2–50 μM total Zn). The rapid-phase non-saturable pathway, possibly an ion channel, is present in squirrelfish hepatocytes (Hogstrand *et al.*, 1996b). Since the plasma Zn concentration is 10 times higher in fish than in mammals (Hogstrand and Wood, 1996), this low-affinity pathway may be of importance in fish. It is unknown if fish hepatocytes also have a high-affinity Zn^{2+} uptake system, such as that in the rat. The kinetics of Zn efflux in squirrelfish hepatocytes suggest that the outward transport of Zn is carrier mediated in this species, but the nature of the efflux mechanism has not been identified (Hogstrand *et al.*, 1996b). Palmiter and Findley (1995) cloned a novel transmembrane protein (ZnT-1) from a rat kidney library that seemed to mediate transport of Zn out of cells. The ZnT-1 transporter does not appear to transport Cd since no change in Zn efflux can be observed following Cd exposure of cells. Mutation of the ZnT-1 gene leads to elevated intracellular Zn levels and increased Zn toxicity. It remains to be investigated whether or not ZnT-1 and/or the Ca^{2+}-dependent Zn^{2+} efflux system, found in human erythrocytes, are present in cells from fish.

10.3.3 COPPER

Cu is primarily bound to albumin and ceruloplasmin in the blood. Cu has also been shown to bind to α-fetoprotein, suggesting its important role as a Cu-transporting protein during fetal development (Sarkar, 1989). Experiments on $CuCl_2$-injected rats have shown that Cu, taken up by the liver, is bound to MT, suggesting that the uptake of Cu by the liver depends on the induction of MT synthesis (Suzuki *et al.*, 1989). Among transepithelial proteins for Cu, the Cu-binding protein CTR-1 in *Saccaromyces cerevisiae* has been localized within the cell membrane (Dancis *et al.*, 1994). The CTR-1 protein is regulated by Cu availability, being induced by Cu deprivation and repressed by Cu excess. Mutations on the CTR-1 gene resulted in a change in response to Cu exposure, suggesting a physiological role for CTR-1 as a transmembrane transporter for Cu (Dancis *et al.*, 1994). In mammals there are at least two types of Cu-transport protein. A gene for one of these is found in most tissues and mutations in this gene results in Menkes disease in humans, whilst defects in the other gene, that is primarily located in the liver, leads to Wilson's disease. Both of these transporters belong to the P-type ion motive ATPases (Bull *et al.*, 1993; Chelly *et al.*, 1993). As discussed above, the rainbow trout has high hepatic Cu levels and is very sensitive to Cd exposure. The situation in rainbow trout is reminiscent of the circumstances that have been observed in Wilson's disease and LEC (Long-Evans Cinnamon) rats (Lee *et al.,* 1989; Friedman and Yarze, 1993). In Wilson's disease the mutated hepatic Cu-transport protein is responsible for elevated hepatic Cu levels, which cause toxicity. Pretreatment of rainbow trout and LEC mice with Zn increases the tolerance to toxic heavy metals (Sugawara and Sugawara, 1994; Norey *et al.*, 1990). Elucidation of the Cu-transport proteins from fish would allow determination of their functionality in different fish species in relation to Cu sensitivity.

10.3.4 MERCURY

Uptake of inorganic Hg by primary rat renal cortical epithelial cells indicates that a small amount of inorganic Hg is transported by an active mechanism, but most transport of inorganic Hg occurs through passive diffusion (Endo and Shaikh, 1995). Hg is predominantly distributed to muscles when it occurs as methyl-Hg, while inorganic Hg also accumulates in the kidneys (McKim *et al.*, 1976). Sublethal exposure of *Tilapia nilotica* to inorganic Hg, for a period of 60 days, showed that accumulation of Hg was highest in muscles, liver and gills in increasing order (Canli and Erdem, 1994). Comparison of intermittent and continuous exposure to $HgCl_2$ in rainbow trout, goldfish (*Carassius auratus*) and fathead minnow (*Pimephales promelas*) has shown that Hg concentrations in the intermittently exposed groups were lower than in the continuously exposed groups, as would be expected. This was not because of Hg excretion, but rather due to reduced accumulation during the recovery periods (Handy, 1995). As pointed out earlier, heavy metals tend to cause mucus formation (Varanasi *et al.*, 1975), which can cause Hg^{2+} to be trapped in the mucus layer, while methyl-Hg appears to pass easily through the mucus layer (Lock *et al.*, 1981). Even if Hg is given as an ion, it almost always tends to become methylated, resulting in accumulation within body organs. This is in conflict with Handy's (1995) observations, where Hg was not accumulated in fish during recovery periods following each Hg exposure. Experiments on rats showed that the toxic action of dimethyl-Hg could be prevented by administering the chemical spironolactone. It is proposed that spironolactone can bind dimethyl-Hg, reducing its toxicity and facilitating its excretion (Kourounakis and Rekka, 1994).

10.4 ACCUMULATION AND STORAGE

10.4.1 INTRACELLULAR COMPARTMENTALIZATION

Metals are distributed throughout the cell, but some compartments are particularly important for metal storage. Typically, 50–80% of the metal stored in cells is found in the cytosol and during chronic exposure to metals the portion of cytosolic metals increases in relation to the total metal load of the cell (Bouquegneau *et al.*, 1975; Bunton *et al.*, 1987; Olsson and Hogstrand, 1987; Hogstrand *et al.*, 1991; Wicklund Glynn and Olsson, 1991; Wicklund Glynn *et al.*, 1992). The primary metal-binding protein in the cytosol is the low molecular weight protein, MT, which will be specifically discussed in section 10.4.2.

The lysosomes constitute another major compartment for metal accumulation. A variety of elements have been localized in lysosomes, including Cu, Zn, Cd and Hg (Sternlieb and Goldfisher, 1976; George, 1982; Fowler, 1987). Cu in lysosomes is typically found together with sulphur-rich compounds (Bunton *et al.*, 1987; Lanno *et al.*, 1987; see also Chapter 8 for molluscs). Studies on dogs with inherited abnormal Cu accumulation have shown that at least some of these sulphur-rich compounds are MT (Johnson *et al.*, 1981). Metals, in particular Zn, are accumulated in fish scales (Abdullah *et al.*, 1976; Sauer and Watabe, 1989a) and this accumulation is increased during exposure (Sauer and Watabe, 1989b). There is good evidence that Zn, incorporated in scales, originates from lysosomes in the osteoblasts (Saur and Watabe, 1989b). Zinc in the numerous osteoblast lysosomes coexists with sulphur-rich compounds.

Cu and Zn are abundant in the nucleus (Baatrup, 1989; Julshamn *et al.*, 1988; Hogstrand *et al.*, 1996b). The level of MT is increased in the nuclei of hepatocytes and pancreatic cells from Cd-treated carp (Kito *et al.*, 1986). Since MT binds most of the Cd in the cell, it is very possible that Cd-MT accumulates in the nucleus. In squirrelfish (*Holocentridae*) females, which utilize massive amounts of Zn during reproduction, half of the Zn in the liver is present in the nucleus and most of this Zn is bound to MT (Hogstrand and Haux, 1996; Hogstrand *et al.*, 1996b). The physiological significance of nuclear Zn is related, at least partially, to its presence in a large number of DNA-binding proteins (Coleman, 1992).

10.4.2 METALLOTHIONEIN

For the metals in groups 1B and 2B of the periodic table, MT is quantitatively the single most important intracellular binding site. The binding affinity for cysteinyl residues in MT is highest for Hg, followed in decreasing order by Cu, Cd and Zn. Upon translation, MT folds into two domains with a capacity to bind three and four divalent metal ions, respectively (Furey *et al.*, 1986). Determination of the primary structures of teleost MTs has shown that the cysteinyl residues are perfectly conserved with the displaced antepenultimate cysteine always located at position 55, instead of position 57 as observed in mammalian and avian MTs. In the dessert pupfish (Fam. *Cyprinodon*) there is an extra cysteinyl residue in position 27 where all other teleosts have a serine residue (Kille *et al.*, 1996). This change creates a superfluous cysteine residue, possibly providing an additional ligand for metal coordination. Changes in the cysteine content and arrangement may have consequences for both the stoichiometry and affinity of metal binding.

Administration of heavy metals, such as Zn, Cu, Cd and Hg, has been shown to induce MT synthesis. However, in contrast to the binding affinity to MT, the ability to activate transcription is highest for Zn, while Cd is a less effective inducer and Cu is a poor inducer of MT (Palmiter, 1994). A transcription factor responsible for the induction of MT transcription via the metal response elements (MREs) has been cloned from mouse (Radke *et al.*, 1993). The MTF-1 has six Zn fingers and it has been suggested that Zn may be the only metal that can activate this transcription factor (Heuchel *et al.*, 1994; Palmiter, 1994). This would indicate that the other heavy metals induce MT gene transcription via other transactivators or via secondary mechanisms and this could in turn explain the lower inducibility via these other metals. In fish, the promoter regions of four MT genes have been characterized so far. These include the two rainbow trout MT-A and MT-B genes, the pike MT gene and the stone loach MT gene (Zafarullah *et al.*, 1988; Murphy *et al.*, 1990; Kille *et al.*, 1993; Olsson *et al.*, 1995a). Common to all these genes is the presence of both proximal and distal MREs (Figure 10.2). Both the proximal and distal MREs have been shown to participate in metal induction when tested in transient transfection systems (Zafarullah *et al.*, 1988; Olsson *et al.*, 1995; Olsson and Kille, 1997).

MT can be induced in fish by substances other than metals. It has been shown, in cultured fish cells, that glucocorticoid, progesterone, noradrenaline and hydrogen peroxide treatment results in increased biosynthesis of MT (Hyllner *et al.*, 1989; Olsson *et al.*, 1990; George *et al.*, 1992; Kling *et al.*, 1996). An involvement of MT in stress and inflammation reactions has been indicated in fish (Baer and Thomas, 1990; Maage *et al.*, 1990). In Atlantic salmon (*Salmo salar*) injected with a carrier-hapten antigen (NIP$_{11}$-LPH) it was observed that the MT levels increased significantly and that Zn increased in correlation with MT (Maage *et al.*, 1990). In another study it was found that Zn induction of MT resulted in a reduction of antibodies in carp (*Cyprino*

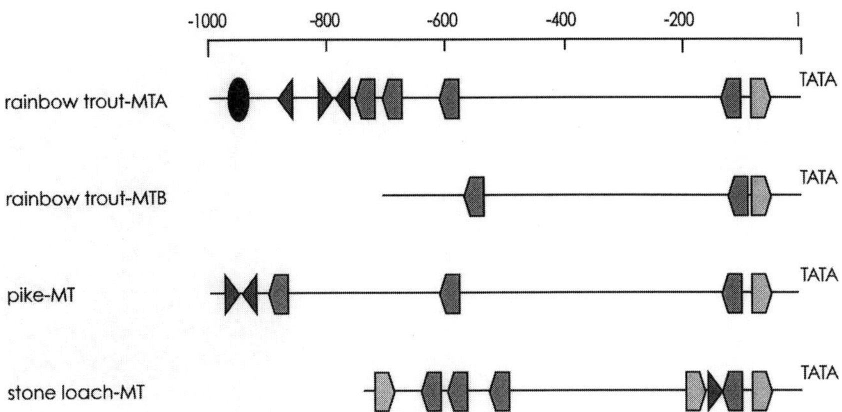

Figure 10.2 The localization and orientation of regulatory elements in the different fish MT genes. Pentagons, metal response elements; triangles, activator protein-1 (AP-1) sites; oval, nuclear factor interleukin-6 (NF-IL6) site. In all fish species there are both distal and proximal MREs. The distal MREs in rainbow trout-MTA, stone loach-MT and pike-MT have all been shown to contribute to metal inducibility. Furthermore, the four AP-1 sites in rainbow trout-MTA have been shown to respond to hydrogen peroxide treatment. Induction levels in response to metals and hydrogen peroxide are shown in Table 10.1.

carpio L.) (O'Neil, 1981). Antibody production has furthermore been shown to be inhibited in cunners (*Tautogolabrus adspesus*) and enhanced in striped bass (*Morone saxatilis*) following Cd exposure (Robohm, 1986). The exact mechanism of MT induction in response to inflammation is still unclear. The observed inhibition of the immune system in certain teleost species, and the inducibility of MT via AP-1 elements, may indicate that trace metals do not directly induce MT in response to inflammation. While no functional *cis*-acting elements have been shown to mediate the induction via hormone treatment in fish, several activator protein-I (AP-I) sites have been demonstrated in rainbow trout, pike and stone loach MT genes (Kille *et al.*, 1993; Olsson *et al.*, 1995a). In rainbow trout these AP-I elements have been shown to be actively regulated by hydrogen peroxide (Olsson *et al.*, 1995a). In transient transfection experiments, H_2O_2 has been shown to be a good inducer of MT (Table 10.1). However, determination of cellular MT mRNA induction has shown that free radicals are not as strong inducers of MT as some metals (Kling and Olsson, unpublished results). Besides being induced by free radicals, fish MT has been indicated to be involved in the protection against cell damage induced by free radicals (Kling *et al.*, 1996). Specifically, free radicals can replace the metals in MT (Sato and Bremner, 1993). In a study on RTG-2 (rainbow trout gonadal cell line) cells it was found that pre-exposure to Zn resulted in increased cell survival during a subsequent challenge with different concentrations of hydrogen peroxide (Kling *et al.*, 1996). Further support for the involvement of MT in this protection against free radical damage stems from the observation that Zn pre-treatment of CHSE-214 (a chinook salmon embryonic cell line) did not result in increased protection against free radicals (Kling and Olsson, unpublished results): it has been suggested that the MT genes in the CHSE-214 cell line may be methylated and therefore not expressed (Price-Haughey *et al.*, 1987; Zafarullah *et al.*, 1989). In a recent study it was found that the CHSE-214 cell line does express MT mRNA, although at a very low level (Kling and Olsson, unpublished results).

Metals and MT levels in different organs of fish have been indicated to be influenced by endogenous stimuli. Thus, the basal levels of MT have been shown to be modulated by gender, reproductive state, developmental stage, time of year and water temperature (Fletcher and King, 1978; Olsson *et al.*, 1987, 1995a, 1996; Overnell *et al.*, 1987; Hogstrand *et al.*, 1996b). This natural fluctuation in MT levels has implications for the use of piscine MT as a biomarker for the determination of the toxicity of heavy metals (reviewed by George and Olsson, 1994); it also indicates that, although MT may be involved in detoxification, the binding of non-essential heavy metals to MT may lead to disturbances in normal trace metal regulation.

Female squirrelfish (Fam. Holocentridae) have perhaps the highest MT levels recorded for any animal: hepatic MT contents of up to 30 mg g^{-1} have been observed (Hogstrand and Haux, 1990, 1996; Hogstrand *et al.*, 1996b). The liver of female squirrelfish has up to 100 times more MT than that of an average male squirrelfish. The MT from female squirrelfish contains almost exclusively Zn, which appears to be used for reproductive purposes (Hogstrand *et al.*, 1996b). Curiously, virtually all MT in female squirrelfish liver is present in the nucleus, whereas the males have the typical cytosolic MT location. Nuclear MT is usually associated with rapid cell division (Cherian, 1994) but the physiological significance of nuclear MT in squirrelfish is unclear.

MT synthesis is induced at the end of the vitellogenic period in female rainbow trout (Olsson *et al.*, 1987). This increase has been shown to occur once the estradiol-induced synthesis of vitellogenin has ceased (Olsson *et al.*, 1989b). During vitellogenesis the Zn levels increase in the liver, due to the increased metabolic activity. It appears that a redistribution of Zn at the end of the vitellogenic period is responsible for this induction of hepatic MT (Olsson *et al.*, 1989b). Rainbow trout co-injected with oestradiol and Cd are more sensitive to Cd than fish not injected with oestradiol. It is unlikely that oestradiol would induce MT synthesis in rainbow trout. In fact, oestradiol inhibits Cd-mediated induction of MT (Olsson *et al.*, 1995b). Cd has been shown to inhibit vitellogenesis in flounder (*Platychtys flesus* L.) and this inhibition correlates with reduced RNA/DNA ratios in cells (Faaborg Povlsen *et al.*, 1990). In a study on rainbow trout it was found that the Cd inhibition of vitellogenesis correlated with an inability of Cd to bind to or induce MT (Olsson *et al.*, 1995b). In addition, it appears that oestradiol alters the distribution of Cd between different tissues of rainbow trout (Mason and Olsson, unpublished results). Following Cd injection, most of the Cd is found in the liver; following co-injection of Cd and oestradiol, an increasing amount of Cd is found in the kidney and in the blood. This inhibition and redistribution could explain the increased sensitivity of female fish for Cd during the vitellogenic period.

When Cd enters the fish through the gills at sublethal levels, about half of the branchial Cd binds to MT (Olsson and Hogstrand, 1987; Wicklund Glynn and Olsson, 1991; Wicklund Glynn *et al.*, 1992). From the gills, Cd is primarily distributed to kidneys and liver (Olsson and Hogstrand, 1987). With time, an increasing portion of the total renal and hepatic Cd binds to MT (Olsson and Hogstrand, 1987; Wicklund Glynn and Olsson, 1991; Wicklund Glynn *et al.*, 1992). During chronic exposure to sublethal levels of Cd, 60–90% of all Cd in these organs is sequestered by MT (Olsson and Haux, 1986; Hogstrand *et al.*, 1991).

Studies of the binding behaviour of inorganic Hg to piscine MT have produced rather ambiguous results. During exposure to low levels of waterborne inorganic Hg, it was found that a significant amount of the metal was bound to MT (Bouquegneau *et al.*, 1975). However, when the fish were exposed to high levels of Hg, MT bound little of the accumulated Hg. Interestingly, in a more recent study on isolated primary hepatocytes, it was found that inorganic Hg induced synthesis of MT, but did not bind to the protein (Gagné *et al.*, 1990).

MT is an important factor for the binding of Cu and Zn in fish but the importance of MT for the sequestration of these elements seems to differ between fish species and depends on the levels of Cu and Zn in the tissues. In the liver of unexposed fish, 10–30% of the Cu and 3–25% of the Zn is typically bound to MT, depending on species (Hogstrand and Haux, 1990; Olsson *et al.*, 1990; Hogstrand *et al.*, 1991). In Zn-exposed fish, the corresponding figures for Cu and Zn are 20–50% and 17–40%, respectively (Kito *et al.*, 1982a,b; Hogstrand and Haux, 1990; Hogstrand *et al.*, 1991). The overall trend is that the importance of MT as a compartment for Cu and Zn is increased during exposure, indicating that MT is binding the excess Cu and Zn that enters the cell. In environmentally exposed fish, which presumably have been accumulating Cu and Zn over a long time period, this trend is more clear than in laboratory studies (Hogstrand and Haux, 1990; Hogstrand *et al.*, 1991).

10.5 PHYSIOLOGICAL AND BIOCHEMICAL EFFECTS

In general, fish are less tolerant to metals than terrestrial animals and the reason for this is that the gills are very susceptible to metal toxicity. The teleost gill is a complex organ with functions equivalent to those performed in mammals by lungs (i.e. ventilation), intestine and kidneys (i.e. mineral balance and osmoregulation). Many waterborne metals obstruct one or more of these functions, which is often the direct cause of acute metal toxicity. At 'industrial' concentrations, most metals affect the gills in a similar manner; metals typically cause oedema, inflammation, and sloughing of epithelial cells in the gill epithelium, and 'clubbing' and fusion of the secondary lamellae. The result of such effects is a reduced diffusion distance for gases and the primary cause of mortality is usually the resulting hypoxia and/or acidosis. Such morphological effects of metals have been reviewed in depth in two papers (Mallat, 1985; McDonald and Wood, 1993) and will not be discussed in detail here. We will instead focus on the effects of metals on fish at lower concentrations.

Since many toxic metals are soft donors (Lewis acids), they readily bind to soft acceptors (Lewis bases), such as sulphur. They have an especially strong preference for sulfhydryl groups and, in fact, most metabolic alterations that are associated with metal toxicity involve interactions with sulfhydryl groups (Goering *et al.*, 1987). One of the most important destructive properties of toxic metals is their ability to alter the effects of enzymes. The mode of action includes binding to sulfhydryl groups in the active site and interference with regulatory mechanisms. An example of the latter is the binding of Cd to calmodulin (Flik *et al.*, 1987).

In the gills, metals often target a specific class of enzymes, namely ion transporters (Figure 10.1) (McDonald and Wood, 1993). Cu and Ag are both powerful inhibitors of the Na^+/K^+ ATPase in the basolateral membrane of gills (Stagg and Shuttleworth, 1982; Laurén and McDonald, 1986; Morgan *et al.*, 1995; Ferguson *et al.*, 1996). Exposure to Cu^{2+} or Ag^+ in fresh water results in a dramatic decrease in the concentrations of Na^+ and Cl^- in the blood plasma, which subsequently can lead to cardiovascular collapse (Laurén and McDonald, 1986; Wood *et al.*, 1996a). Speciation has a remarkable influence on silver toxicity. In a recent study, silver thiosulphate ($Ag[S_2O_3]_n$), which is the major silver species discharged into natural water systems, had no effect on plasma Na^+ and Cl^- levels up to a concentration of 0.3 mM in spite of the fact that high levels of silver accumulated in gills and other tissues (Wood *et al.*, 1996b). Aluminium is another metal that affects the plasma Na^+ and Cl^- levels, but unlike the other metals discussed here it can also cause respiratory problems, due to inflammation of the gill epithelium, at environmentally realistic concentrations (Muniz and Leivestad, 1980; Playle *et al.*, 1989; Playle and Wood, 1989, 1991). The Al-induced losses of plasma Na^+ and Cl^- are brought about by an inhibition of the branchial Na^+/K^+ ATPase (and thereby a reduction of the influx of Na^+ and Cl^-), and by stimulating passive losses of Na^+ and Cl^- across paracellular pathways (Staurnes *et al.*, 1984; Playle *et al.*, 1989; Playle and Wood, 1989, 1991). Although Al is by no means as toxic as Cu or Ag, it is a significant environmental concern in acidified lakes where the low pH can elevate total dissolved Al to concentrations as high as 40 μM (Wright *et al.*, 1980). In addition to the increased leaching of Al from sediment and soil, a low pH augments the effect of Al by reducing the uptake, and stimulating the passive loss, of Na^+ and Cl^- across the gills.

Zn and Cd have little effect on Na^+ and Cl^- regulation (Larsson *et al.*, 1981; Spry and Wood, 1985). Instead, waterborne Zn^{2+} and Cd^{2+} both inhibit branchial Ca^{2+} uptake, which can lead to fatal hypocalcaemia (Verbost *et al.*, 1987, 1989; Spry *et al.* 1988; Spry and Wood, 1989). The primary toxic action of Cd^{2+} is to block the basolateral Ca^{2+} ATPase, which effectively inhibits the transfer of Ca^{2+} across the basolateral membrane of the chloride cells (Verbost *et al.*, 1988). The free Zn^{2+} ion is also a very strong inhibitor of the basolateral Ca^{2+} ATPase, but at sublethal concentrations of waterborne Zn^{2+} a competition for the apical Ca^{2+} transporter (Figure 10.1) is probably a more important mechanism of toxicity (Hogstrand *et al.*, 1994, 1996a).

As discussed earlier, metals that are absorbed by gills or intestine are distributed by the blood to various parts of the body. The pattern of accumulation in the body is different for different metals and the target accumulatory tissue for a specific metal is often the target for chronic toxicity of that metal. Such long-term effects are often very similar to those described in humans (Chapter 13) and other mammals. For example, chronic toxicity to Cu includes haemolytic anaemia (Gwozdzinski *et al.*, 1992), Cd causes renal failure and demineralization of the skeleton (Larsson *et al.*, 1981) and lead exposure can result in inhibition of an enzyme in the haemoglobin synthesis pathway (i.e. δ-aminolevulinic acid dehydratase, ALA-D) (Hodson *et al.*, 1977; Haux *et al.*, 1986) and in neurotoxicity (Hodson *et al.*, 1984).

Another very important effect of many metals in fish is the resulting disturbance in reproductive success (Birge *et al.*, 1981). Impairment in the capacity to reproduce probably represents the most sensitive gross effect of long-term Zn exposure (Spear, 1981; Eisler, 1993). At the molecular level this effect may be related to the tendency of Zn to disrupt calcium homeostasis (Hogstrand and Wood, 1996). Exposure of female fish to very low concentrations of Cd markedly reduces the viability of their eggs (Birge *et al.*, 1981). Cd reduces vitellogenin production in female fish (Faaborg Povlsen *et al.*, 1990; Olsson *et al.*, 1995b). In addition, a recent study suggests that Cd readily binds to vitellogenin and that Cdvitellogenin is rapidly taken up by oocytes (Ghosh and Thomas, 1995). These effects on yolk formation are likely to contribute to the observed reduction of egg viability. Cu is another metal that has severe effects on fish reproduction. Mount (1968) found that the egg production of fathead minnows was reduced at Cu concentrations that had no effects on eggs, fry or adults. In contrast to most metals, Cu seems to have more severe effects on larval and juvenile life stages than on embryos (McKim *et al.*, 1978). Hg is one of the most toxic metals to early life stages of fish (Birge *et al.*, 1979). It produces high rates of embryonic and larval mortalities, and it is also extremely teratogenic.

10.6 EXCRETION

Since metals cannot be degraded metabolically, the organism must use other ways to inactivate these potential toxicants. Metals may be excreted through the gills, skin, intestine, gall bladder or kidney. Excretion is one of the possible ways for an organism to protect itself against the toxic actions of heavy metals.

Once Cd has been distributed to the liver and kidney it is bound to MT (Olsson and Hogstrand, 1987), which may explain the long half-life of Cd in fish (Haux and Larsson, 1984). Cd elimination can occur through the gills, by the hepatic-biliary route and via the intestine (Varanasi and Markey, 1978). In fish, secretion of mucus also provides an important means of reducing the body burden of Cd (Varanasi and Masrkey, 1978). Quantitatively, the primary routes of Cd excretion are by faecal loss and efflux across the gills, whilst very little is lost via urine (Harrison and Klaverkamp, 1989; Giles, 1988). It has been shown that Cd has a very high affinity for the intestinal mucosa, and that it can be secreted directly into the lumen of the small and large intestine (Gruden,1981). The accumulated Cd in the intestinal mucosa is thought to be present as Cd-MT (Valberg *et al.*, 1977). In mammals, small amounts of Cd are excreted via the bile as Cd-glutathione (Cherian and Vostal, 1977).

Excretory pathways for Zn have not been particularly well investigated. Little Zn seems to be excreted via the kidneys (Cunnane, 1988). However, Nakatani (1966) proposed that excretion of ingested Zn occurs through the urine, the gastrointestinal tract and the gills. In rainbow trout it has been suggested that dietary Zn is excreted through the gills (Hardy *et al.*, 1987).

Even less is known about excretion of Cu in fish. The gills are certainly involved in Cu homeostasis and Cu excretion may therefore occur across the gill surface (Handy, 1992; Laurén and MacDonald, 1986). In marine species renal excretion of Cu is improbable due to the low filtration rate (Handy, 1996). Because of the concentration of Cu granules in the liver, bile has been suggested to be an important excretory organ in fish (Lanno *et al.*, 1987), but other studies have shown that, during depuration, very little Cu is present in the gut, arguing against biliary excretion of this metal (Handy, 1992).

Hg is methylated in the liver, resulting in a non-polar organic complex. Because of the lipophilic character of methyl-Hg, it is not readily excreted and it accounts for up to 100% of the total Hg in fish muscle. In fish there is virtually no elimination of methyl-Hg (Gottofrey, 1990); on the other hand, Hg^{2+} is soluble in water and easier to excrete (Chapter 5).

10.7 SIGNIFICANCE OF BIOCHEMICAL AND PHYSIOLOGICAL RESPONSES IN TERMS OF ENVIRONMENTAL HAZARD ASSESSMENT

The greatest advantage of mechanistic studies, at the molecular, cellular and organism levels of biological organization, is that they enable us to explain effects of environmental contaminants based on interactions between toxicants and biomolecules. Understanding of such basic principles helps us to explain how changes in water chemistry, physiological status and other factors modulate toxicity. For example, there is strong experimental evidence that both Zn^{2+} and Cd^{2+} cross the apical membrane of the gill epithelium using the same transporter as Ca^{2+} (Verbost *et al.*, 1989; Hogstrand *et al.*, 1996a). The competition between Ca^{2+}, Zn^{2+} and Cd^{2+} for the same uptake sites explains the well known phenomenon that high water hardness protects against toxicity from these metals (Bradley and Sprague, 1985a,b). Recently, the 'gill-binding model' (Playle *et al.*, 1993a,b; Janes and Playle, 1995; MacRae *et al.*, 1996) has been discussed as a possible future aid in determining water quality standards in the United States (Wood *et al.* 1996a,b). The approach is to determine stability constants for metal interactions with biological surfaces, such as the fish gill (Playle *et al.* 1993a,b; Janes and Playle, 1995). These values are inserted into aquatic chemistry modelling programs and used to calculate predicted metal uptake and toxicity in waters with differing chemistry. Using conditional stability constants for Cu–gill interactions, established in defined media, Playle *et al.* (1993a) were able to show that Cu-binding to the gills of minnows could be predicted in various lake and river waters. Furthermore, there appears to be a direct link to survival, because Zn- and Cu-binding to rainbow trout gills was found to correlate with acute toxicity (Bradley *et al.* 1985; MacRae *et al.*, 1996). Hence, basic mechanistic studies are very important for the development of models that can be used to predict ecotoxicological problems before they occur, rather than to confirm that an ecosystem is already damaged.

While biochemical and physiological research is needed for the understanding of toxicant–organism interactions, studies of impact on community structure are of highest relevance. After all, it is populations and ecosystems that we want to protect, not biomolecules. Therefore, a concern with biochemical and physiological approaches to aquatic toxicology is that it is not always easy to interpret such endpoints in terms of population balance and ecosystem quality. In a few cases, serious attempts have been made to integrate endpoints at multiple levels of biological organization (Roch *et al.*, 1985; Bayne *et al.*, 1988; Addison and Clarke, 1990; Phillips and Lipton, 1995). Such studies are intrinsically difficult to manage and require thorough logistic and experimental planning. Nevertheless, they are necessary to demonstrate connections between biochemical, physiological and ecological parameters. Past investigations suggest that such relationships do exist (Roch *et al.*, 1985; Bayne *et al.*, 1988; Addison and Clarke, 1990; Phillips and Lipton, 1995), but more research is needed in this area.

ACKNOWLEDGEMENTS

This study was supported by a grant from the Swedish National Research Council to P.E.O. and a grant from the Silver coalition to C.H.

REFERENCES

Abdullah, M.I., Banks. W., Miles, D.L. and O'Grady, K.T. (1976) Environmental dependence of manganese and zinc in the scales of Atlantic salmon, *Salmo salar* (L), and brown trout, *Salmo trutta* (L). *Freshwater Biology* **6**, 161–166.

Addison, R.F. and Clarke, K.R. (1990) Introduction: the IOC/GEEP Bermuda workshop. *J. Exp. Mar. Biol. Ecol.* **138**, 1–8.

Baatrup, E. (1989). Selenium-induced autometallographic demonstration of endogenous zinc in organs of the rainbow trout, *Salmo gairdneri*. *Histochemistry* **90**, 417–425.

Baatrup, E., Doving, K.B. and Winberg, S. (1990) Differential effects of mercurial compounds on the electroolfactogram (EOG) of salmon (*Salmo salar*). *Ecotox. Environ. Safety* **20**, 269–276.

Baerm K.N. and Thomas, P. (1990) Influence of capture stress, salinity and reproductive status on zinc associated with metallothionein-like proteins in the liver of three marine teleost species. *Mar. Environ. Res.* **28**, 157–161.

Bayne, B.L., Clarke, K.R. and Gray, J.S. (1988) Background and rationale to a practical workshop on biological effects of pollutants. *Mar. Ecol. Prog. Ser.* **46**, 1–5.

Birge, W.J., Black, J.A., Westerman, A.G. and Hudson, J.A. (1979) The effects of mercury on reproduction of fish and amphibians, in *The Biogeochemistry of Mercury in the Environment* (ed. J.O. Nrigau), Elsevier/North Holland Biomedical Press, pp. 629–655.

Birge, W.J., Black, J.A. and Ramey, B.A. (1981) The reproductive toxicology of aquatic contaminants. *Haz. Assess. Chem.* **1**, 59–115.

Boudou, A. and Ribeyre, F. (1985) Experimental study of trophic contamination of *Salmo gairdneri* by two mercury compounds – $HgCl_2$ and CH_3HgCl – analysis at the organism and organ level. *Water Air Soil Polln* **26**, 137–148.

Bouquegneau, J.M., Gerday, C. and Disteche, A. (1975) Fish mercury-binding thionein related to adaptation mechanisms. *Febs Lett.* **55**, 173–177.

Bradley, R.W. and Sprague, J.B. (1985a) Accumulation of zinc by rainbow trout as influenced by pH, water hardness and fish size. *Environ. Toxicol. Chem.* **4**, 685–694.

Bradley, R.W. and Sprague, J.B. (1985b) The influence of pH, water hardness, and alkalinity on the acute lethality of zinc to rainbow trout (*Salmo gairdneri*). *Can. J. Fish. Aquat. Sci.* **42**, 731–736.

Bradley, R.W., DuQuesnay, C. and Sprague, J.B. (1985) Acclimation of rainbow trout to zinc: kinetics and mechanism of enhanced tolerance induction. *J. Fish. Biol.* **27**, 367–379.

Bremner, I. (1979) Mammalian absorption, transport and excretion of cadmium, in *The Chemistry, Biochemistry and Biology of Cadmium* (ed. Webb), Elsevier/Northholland.

Bremner, I. and Morrison, J.N. (1986) Assessment of zinc, copper and cadmium status in animals by assay of extracellular metallothionein. *Acta Pharmacol. Toxicol. Copenh.* **59**, 502–509.

Bull, P.C., Thomas, G.R., Rommens, J. *et al.* (1993) The Wilson disease gene is a putative copper transporting P-type ATPase similar to the Menkes disease. *Nature Genet* **5**, 327–337.

Bunton, T.E., Baksi, S.M., George, S.G. and Frazier, J.M. (1987) Abnormal hepatic copper storage in a teleost fish (*Morone americana*). *Vet. Pathol.* **24**, 515–524.

Canli, M. and Erdem, C. (1994) Mercury toxicity and effects of exposure concentration and period on mercury accumulation in tissues of a tropical fish, *Tilapia nilotica*. *Turk. J. Zool.* **18**, 233–239.

Chan, H.M., Bjerregard, P., Rainbow, P.S. and Depledge, M.H. (1992) Uptake of zinc and cadmium by two populations of shore crabs *Carcinus maenas* at different salinities. *Mar. Ecol. Prog. Ser.* **86**, 91–97.

Chelly, J., Timer, Z., Tonneson, T. *et al.* (1993) Isolation of a candidate gene for Menkes disease that encodes a potential heavy metal binding protein. *Nature Genet.* **5**, 344–350

Chen, R.W. (1974) Affinity labelling studies with [109]Cd in cadmium-induced testicular injury in rats. *J. Reprod. Fertil.* **38**, 293–306.

Cherian, M.G. (1994) The significance of the nuclear and cytoplasmic localization of metallothionein in human liver and tumor cells. *Environ. Health Perspect.* **102**, 131–135.

Cherian, M.G. and Vostal, J.J. (1977) Biliary excretion of cadmium in rat. I: Dose-dependent biliary excretion and the form of cadmium in the bile. *J. Toxicol. Environ. Health* **2**, 945–954.

Coleman, J.E. (1992) Zinc proteins: enzymes, storage proteins, transcription factors, and replication proteins. *Annu. Rev. Biochem.* **61**, 897–947.

Comhaire, S., Blust, R., Van Ginneken, L. and Vanderborght, O.L.J. (1994) Cobalt uptake across the gills of the common carp, *Cyprinus carpio*, as a function of calcium concentration in the water of acclimation. *Comp. Biochem. Physiol.* **109C**, 63–76.

Comhaire, S., Blust, R., Van Ginneken, L. *et al.* (1996) Branchial cobalt uptake by the carp, *Cyprinus carpio*: partial involvement of the calcium uptake system. (submitted).

Cunnane, S.C. (1988) *Zinc: Clinical and Biochemical Significance*, CRC Press. Boca Raton, Florida.

Dancis, A., Haile, D., Yuan, D.S. and Klausner R.D. (1994) The *Saccharomyces cerevisiae* copper transport protein (Ctr 1p). Biochemical characterization, regulation by copper, and physiologic role in copper uptake. *J. Biol. Chem.* **269**, 25660–25667.

Daston, G.P., Overmann, G.J., Taubenekk, M.W. *et al.* (1991) The role of metallothionein induction and altered zinc status in maternally mediated developmental toxicity: comparison of the effects of urethane and styrene in rats. *Toxicol. Appl. Pharmacol.* **110**, 450–463.

Diplock, A.T. (1976) Metabolic aspects of selenium action and toxicity. *CRC Crit. Rev. Toxicol.* **4**, 271–329.

Dunn, M.A., Blalock, T.L. and Cousins, R.J. (1987) Metallothionein. *Proc. Soc. Exp. Biol. Med.* **185**, 107–119.

Eisler, R. (1993) Zinc hazards to fish, wildlife, and invertebrates: a synoptic review. *Contaminant Hazard Reviews*, Report 26, Fish and Wildlife Service/US DPA, Washington, DC, 106 pp.

Endo, T. and Shaikh, Z.A. (1995) Cadmium uptake by primary cultures of rat renal cortical epithelial cells: influence of cell density and other metal ions. *Tox. Appl. Pharmacol.* **121**, 203–209.

Faaborg Povlsen, A., Korsgaard, B. and Bjerregaard, P. (1990) The effect of cadmium on vitellogenin metabolism in estradiol-induced flounder (*Platichtys flesus* (L.)) males and females. *Aquat. Toxicol.* **17**, 253–262.

Ferguson, E.A., Leach, D.A. and Hogstrand, C. (1996) Metallothionein protects against silver blockage of the Na$^+$/K$^+$-ATPase, in *Proceedings of the 4th International Conference of Transport, Fate and Effects of Silver in the Environment* (eds A.W. Andrén and T.W. Bober), University of Wisconsin, Wisconsin (accepted for publication).

Flanagan, P.R., McLellan, J.S., Haist, J. *et al.* (1978) Increased dietary cadmium absorption in mice and human subjects with iron deficiency. *Gastroenterology* **74**, 841–846.

Fletcher, G.L. and King, M.J. (1978) Seasonal dynamics of Cu^{2+}, Zn^{2+}, Ca^{2+} and Mg^{2+} in gonads and liver of winter flounder (*Pseudopleuronectes americanus*): evidence for summer storage of Zn^{2+} for winter gonad development in females. *Can. J. Zool.* **56**, 284–290.

Flik, G., van de Winkel J.G.J., Psrt, P. *et al.* (1987) Calmodulin-mediated cadmium inhibition of phosphodiesterase activity, *in vitro. Arch. Toxicol.* **59**, 353–359.

Foulkes, E.C. and Bergman, D. (1993) Inorganic mercury absorption in mature and immature rat jejunum: transcellular and intercellular pathways *in vivo* and in everted sacs. *Toxicol. Appl. Pharmacol.* **120**, 89–95.

Fowler, B.A. (1987). Intracellular compartmentation of metals in aquatic organisms: roles in mechanisms of cell injury. *Environ. Health Perspect.* **71**, 121–128.

Friedman, L.S. and Yarze, J.C. (1993) Zinc treatment of Wilson's disease: how it works. *Gastroenterology* **104**, 1566–1568.

Furey, W.F.D., Robbins, A.H., Clancy, L.L. *et al.* (1986) Crystal structure of Cd,Zn metallothionein. *Science* **231**, 704–710.

Gagné, F., Marion, M. and Denizeau, F. (1990) Metallothionein induction and metal homeostasis in rainbow trout hepatocytes exposed to mercury. *Toxicol. Letts* **51**, 99–107.

George, S.G. (1982) Subcellular accumulation and detoxification of metals in aquatic animals, in *Physiological Mechanisms of Marine Pollutant Toxicity*, (eds. W.B. Vernberg, A. Calabrese, F.P. Thurberg and F.J. Vernberg), Academic Press, New York, pp. 3–52.

George, S. G. and Olsson, P.-E. (1994) Metallothioneins as indicators of trace metal pollution, in *Biomonitoring of Coastal Waters and Estuaries*, (ed. K.J.M. Kramer), CRC Press, Boca Raton, Florida, pp. 151–178.

George S., Burgess, D., Leaver, M. and Frerich N. (1992) Metallothionein induction in cultured fibroblasts and liver of a marine flatfish, the turbot, *Scopothalmus maximus. Fish. Physiol. Biochem.* **10**, 43–54.

Ghosh, P. and Thomas, P. (1995) Binding of metals to Red drum vitellogenin and incorporation into oocytes. *Mar Environ. Res.* **39**, 165–168.

Giles M.A. (1988) Accumulation of cadmium by rainbow trout, *Salmo gairdneri*, during extended exposure. *Can. J. Fish. Aq. Sci.* **45**, 1045–1053.

Gill, T.S., Tewari, H. and Pande, J. (1990) Use of fish enzyme system in monitoring water quality: effects of mercury on tissue enzymes. *Comp. Biochem. Phys.* **97C**, 287–292.

Giroux, E.L. (1975) Determination of zinc distribution between albumin and $\alpha 2$-macroglobulin in human serum. *Biochem. Med.* **12**, 258–266.

Giroux, E.L. and Henkin, R.I. (1972) Competition for zinc among serum albumin and amino acids. *Biochim. Biophys. Acta* **273**, 64–72.

Goering, P.L. and Klaassen, C.D. (1984) Tolerance to cadmium-induced toxicity depends on presynthesized metallothionein in liver. *J. Toxicol. Environ. Health* **14**, 803–812.

Goering P.L., Mistry P. and Fowler B.A. (1987) Mechanisms of metal-induced cell injury, in *Handbook of Toxicology*, (eds. T.J. Haley and W.O. Berndt), Hemisphere, New York, pp. 384–425.

Goss G.G., Perry S.F., Wood C.M. and Laurent P. (1992) Mechanisms of ion and acid–base regulation at the gills of freshwater fish. *J. Exp. Zool.* **263**, 143–159.

Gottofrey, J. (1990) The disposition of cadmium, nickel, mercury and methylmercury in fish and effects of lipophilic metal chelation. PhD thesis, Swedish University of Agricultural Sciences, Uppsala, Sweden, 71 pp.

Goyer, R.A. (1995) Nutrition and metal toxicity. *Am. J. Clin. Nutr.* **61**, 646–650.

Groten, J.P., Luten, J.B., Bruggemann, I.M. *et al.* (1992) Comparative toxicity and accumulation of cadmium-chloride and cadmium-metallothionein in primary cells and cell lines of rat intestine, liver and kidney. *Toxicol. in vitro* **6**, 509–517.

Gruden, N. (1981) Influence of cadmium on calcium transfer through the intestinal wall in rats. *Arch. Toxicol.* **37**, 149–154.

Gwozdzinski, K., Roche, M. and Peres, G. (1992) The comparison of the effects of heavy metal ions on the antioxidant enzyme activities in human and fish *Dicentrarchus labrax* erythrocytes. *Comp. Biochem. Physiol.* **102C**, 57–60.

Hamilton, S.J. and Mehrle, P.M. (1986) Evaluation of metallothionein measurement as a biological indicator of stress from cadmium in brook trout. *Trans. Am. Fish. Soc.* **116**, 551–560.

Handy, R.D. (1992) The assessment of episodic pollution. II. The effects of cadmium and copper enriched diets on tissue contaminant analysis in rainbow trout (*Oncorhyncus mykiss*). *Arch. Environ. Contam. Toxicol.* **22**, 82–87.

Handy, R. D. (1995) Comparison of intermittent and continuous exposure to mercuric chloride in rainbow trout (*Oncorhyncus mykiss*), goldfish (*Carassius auratus*) and the fathead minnow (Pimephales promelas). *Can. J. Fish. Aqua. Sci.* **52**, 13–22.

Handy, R.D. (1996) Toxic metals in the diet, in *Toxicology and Aquatic Pollution – Physiological, Cellular and Molecular Approaches*, (ed. E.W. Taylor), Society for Experimental Biology Seminar Series, Cambridge University Press, Cambridge, UK, pp. 29–60.

Hardy, R.W., Craig, V.S. and Koziol, A.M. (1987) Absorption, body distribution and excretion of dietary zinc by rainbow trout (*Salmo gairdneri*). *Fish. Physiol. Biochem.* **3**, 133–143.

Harrison, S.E. and Klaverkamp, J.F. (1989) Uptake, elimination and tissue distribution, and excretion of dietary zinc by rainbow trout (*Salmo gairdneri* Richardson) and lake whitefish (*Coregonus clupeaformis* Mitchill). *Environ. Toxicol. Chem.* **8**, 87–97.

Haux, C. and Larsson, Å. (1984) Long term sublethal physiological effects on rainbow trout, *Salmo gairdneri*, during exposure to cadmium and after subsequent recovery. *Aquat. Toxicol.* **5**, 129–142.

Haux C., Larsson Å., Lithner G. and Sjöbeck M.-L. (1986) A field study of physiological effects on fish in lead-contaminated lakes. *Environ. Toxicol. Chem.* **5**, 283–288.

Heuchel, R., Radtke, F., Georiev, O. *et al.* (1994) The transcription factor MTF-1 is essential for basal and heavy metal-induced metallothionein gene expression. *EMBO J.* **13**, 2870–2875.

Hodson, P.V., Blunt, D.J., Spry, D.J. and Austen, K. (1977) Evaluation of erythrocyte delta-amino-levulinic acid dehydratase activity as short-term indicator in fish of harmful exposure to lead. *J. Fish. Res. Bd. Can.* **34**, 501–508.

Hodson, P.V., Blunt, B.R. and Whittle, D.M. (1984) Monitoring lead exposure of fish, in *Contaminant Effects of Fisheries*, (eds V.W. Cairns, P.V. Hodson and J.O. Nriagu), John Wiley, New York, pp. 87–98.

Hogstrand, C. and Haux, C. (1990) A radioimmunoassay for perch (*Perca fluviatilis*) metallothionein. *Toxicol. Appl. Pharmacol.* **103**, 56–65.

Hogstrand, C. and Haux, C. (1996) Naturally high levels of zinc and metallothionein in liver of several species of the squirrelfish family from Queensland, Australia. *Mar. Biol.* **125**, 23–31.

Hogstrand, C. and Wood, C. (1996) The physiology and toxicology of zinc in fish, in *Toxicology of Environmental Pollution – Physiological, Molecular, and Cellular Approaches* (ed. E.W. Taylor), Society for Experimental Biology Seminar Series, Cambridge University Press, Cambridge, UK.

Hogstrand, C., Lithner, G. and Haux, C. (1991) The importance of metallothionein for the accumulation of copper, zinc and cadmium in environmentally exposed perch, *Perca fluviatilis*. *Pharmacol. Toxicol.* **68**, 492–501.

Hogstrand, C., Wilson, R.W., Polgar, D. and Wood, C.M. (1994) Effects of zinc on the branchial calcium uptake in freshwater rainbow trout during adaptation to waterborne zinc. *J. Exp. Biol.* **186**, 55–73.

Hogstrand, C., Verbost, P.M., Wendelaar Bonga, S.E. and Wood, C.M. (1996a) Mechanisms of zinc uptake in gills of freshwater rainbow trout: interplay with calcium transport. *Am. J. Physiol.* **270**, R1141–1147.

Hogstrand, C., Gassman, N.J., Popova, B. *et al.* (1996b) The physiology of massive zinc accumulation in the liver of female squirrelfish and its relation to reproduction. *J. Exp. Biol.* **199**, 2543–2554.

Hyllner, S.J., Anderson, T., Haux, C. and Olsson, P.-E. (1989) Cortisol induction of metallothionein in primary culture of rainbow trout hepatocytes. *J. Cell. Physiol.* **139**, 24–28.

Janes, N. and Playle, R.C. (1995) Modeling silver binding to gills of rainbow trout (*Oncorhynchus mykiss*). *Environ. Toxicol. Chem.* **14**, 1847–1858.

Johnson, G.F., Morell, A.G., Stockert, R.J. and Sternlieb, I. (1981) Hepatic lysosomal copper protein in dogs with an inherited copper toxicosis. *Hepatology* **1**, 243–248.

Julshamn, K., Andersen, K.-J., Ringdal, O. and Brenna, J. (1988). Effect of dietary copper on the hepatic concentration and subcellular distribution of copper and zinc in the rainbow trout (*Salmo gairdneri*). *Aquaculture* **73**, 143–155.

Kalfakakou, V. and Simons, T.J. (1990) Anionic mechanisms of zinc uptake across the human red blood cell membrane. *J. Physiol.* **421**, 485–497.

Kello, D. and Kostial, K. (1977) Influence of age and milk diet on cadmium absorption from the gut. *Toxicol. Appl. Pharmacol.* **40**, 277–282.

Kille, P., Kay, J. and Sweeney, G.E. (1993) Analysis of regulatory elements flanking metallothionein genes in Cd-tolerant fish (pike and stone loach). *Biochim. Biophys. Acta* **1216**, 55–64.

Kille, P., Soltz, D., Cosson, R. *et al.* (1996) The primary structure of metallothionein from teleosts inhabiting diverse environments. (Submitted.)

Kito, H., Ose, Y. and Sato, T. (1986). Cadmium-binding protein (metallothionein) in carp. *Environ. Health Perspect.* **65**, 117–124.

Kito, H., Tazawa, T., Osa, Y. *et al.* (1982a). Formation of metallothionein in fish. *Comp. Biochem. Physiol.* **73C**, 129–134.

Kito, H., Tazawa, T., Ose, Y. *et al.* (1982b) Protection by metallothionein against cadmium toxicity. *Comp. Biochem. Physiol.* **73C**, 135–139.

Kling, P., Erkell, L.J., Kille, P. and Olsson, P.-E. (1996) Metallothionein induction in rainbow trout gonadal (RTG-2) cells during free radical exposure. *Mar. Environ. Res.* **42**, 33–36.

Kourounakis, P.N. and Rekka, E. (1994) Effect of spironolactone on dimethyl mercury toxicity: a possible molecular mechanism. *Arzneimittel-Forschung* **44**, 1150–1153.

Lanno, R.P., Hicks, B. and Hilton, J.W. (1987). Histological observations on intra hepatocytic copper-containing granules in rainbow trout reared on diets containing elevated levels of copper. *Aquatic Toxicol.* **10**, 251–263.

Larsson, Å., Bengtsson, B.-E. and Haux, C. (1981). Disturbed ion balance in flounder, *Platichtys flesus* L., exposed to sublethal levels of cadmium. *Aquat. Toxicol.* **1**, 19–36.

Laurén, D.J. and McDonald, D.G. (1985). Effects of copper on branchial ionoregulation in the rainbow trout, *Salmo gairdneri* Richardson. *J. Comp. Physiol. B* **155**, 635–644.

Laurén, D.J. and McDonald, D.G. (1986) Influence of water hardness, pH, and alkalinity on the mechanisms of copper toxicity in juvenile rainbow trout, *Salmo gairdneri. Can. J. Fish. Aq. Sci.* **43**, 1488–1496

Lee, D.Y., Brewer, G.J. and Wang, Y.X. (1989) Treatment of Wilson's disease with zinc. VII. Protection of the liver from copper toxicity by zinc-induced metallothionein in a rat model. *J. Lab. Clin. Med.* **114**, 639–646.

Lock, R.A.C., Cruijsen, P.M.J.M. and van Overbeeke, A.P. (1981) Effects of mercuric chloride and methylmercuric chloride on the osmoregulatory function of the gills in rainbow trout, *Salmo gairdneri* Richardson. *Comp. Biochem. Physiol.* **68C**, 151–159.

Lönnerdal, B. (1988) Intestinal absoption of zinc, in *Zinc in Human Biology*, Springer-Verlag, Berlin, pp. 33–53.

Maage, A., Waagbo, R., Olsson, P.-E. *et al.* (1990) Ascorbate-2-sulfate as a dietary vitamin C source for Atlantic salmon (*Salmo salar*). 2. Effects of dietary levels and immunization on the metabolism of trace elements. *Fish Physiol. Biochem.* **8**, 429–436.

MacRae, R.K., Smith, D.E., Swoboda-Colberg, N. *et al.* (1996) Copper binding affinity of rainbow trout (*Onchorhynchus mykiss*) and brook trout (*Salvelinus fontinalis*) gills. *Environ. Toxicol. Chem.* (in press).

Magneson, G.R., Puvathingal, J.M. and Ray, W.J. Jr (1987) The concentrations of free Mg^{2+} and free Zn^{2+} in equine blood plasma. *J. Biol. Chem.* **262**, 11140–11148.

Mallatt, J. (1985) Fish gill structural changes induced by toxicants and other irritants: a statistical review. *Can. J. Fish. Aquat. Sci.* **42**, 630–648.

Mandolfino, M., Cimino, G., Scuteri, A. *et al.* (1994) Some aspects of zinc transport across eel (*Anguilla anguilla*) red blood cell membranes. *Cell. Biol. Int.* **18**, 279–288.

Matsusaka, N., Berg, D. and Kollmer, W.E. (1985) Influence of changing Zn supply on ^{65}Zn absorption and retention in rats, in *Trace Elements in Man and Animals*, (eds. C.F. Mills, I. Bremner and J.K. Chester), Commonwealth Agricultural Bureau, Slough, pp. 394–397.

McDonald, D.G. and Wood, C.M. (1993) Branchial mechanisms of acclimation to metals in Freshwater fish, in *Fish Ecophysiology*, (eds J.C. Rankin and F.B. Jensen), Chapman & Hall, London, pp. 297–321.

McKim, J.M. and Benoit, D.A. (1971) Effects of long-term exposure to copper on survival, growth and reproduction of brook trout (*Salvelinus fontinalis*). *J. Fish. Res. Bd. Can.* **28**, 655–662.

McKim, J.M., Olson, G.F., Holcombe, G.W. and Hunt, E.P. (1976) Long-term effects of methyl-mercuric chloride on three generations of brook trout (*Salvelinus fontinalis*): toxicity, accumulation, distribution and elimination. *J. Fish. Res. Bd. Can.* **33**, 2726–2739.

McKim, J.M., Eaton, J.G. and Holcombe, G.W. (1978) Metal toxicity to embryos and larvae of eight species of freshwater fish – II: Copper. *Bull. Environ. Contam. Toxicol.* **19**, 608–616.

Mitane, Y., Aoki, Y. and Suzuki, K.T. (1987) Cadmium inhibits protein secretion from cultured rat liver parenchymal cells. *Biochem. Pharmacol.* **15**, 2647–2652.

Morgan, I.J., Tytler, P. and Bell, M.V. (1994) The use of a perfused, whole body preparation to measure the branchial and intestinal influx of 137-caesium in the rainbow trout. *J. Fish Biol.* **45**, 247–256.

Morgan, I.J., Galvez, F., Munger, R.S. *et al.* (1995) The physiological effects of acute silver exposure in rainbow trout (*Oncorhynchus mykiss*), in *Proceedings of the 3rd International Conference of Transport, Fate and Effects of Silver in the Environment*, (eds A.W. Andren and T.W. Bober), University of Wisconsin, Wisconsin, pp. 303–306.

Morka, H. (1991) Immunocytochemisk lokalisering av metallothionein i lever og nyre hos regnueorret (*Oncorhynchus mykiss*). MSc thesis, Department of Molecular and Cellular Biology, Biological Institute, University of Oslo, Norway, 80 pp.

Mount, D.I. (1968) Chronic toxicity of copper to fathead minnows. *Wat. Res.* **2**, 215–223.

Muniz, I.P. and Leivestad, H. (1980) Toxic effects of aluminium on the brown trout, *Salmo trutta* L., in *Proceedings of the International Conference on Acid Precipitation*, (eds D. Drablos and A. Tollan), SNSF, Oslo, pp. 84–92.

Murphy, M.F., Collier, P., Koutz, P. and Howard, B. (1990) Nucleotide sequence of the trout metallothionein A gene 5O flanking region. *Nucl. Acid Res.* **18**, 4622.

Nakatani, R.E. (1966) Biological responses of rainbow trout (*Salmo gairdneri*) ingesting ^{65}Zn, in *Proceeding on the Disposal of Radioactive Wastes at Seas, Oceans, and Surface Waters*, (ed. A. Guillon), International Atomic Energy Agency, Vienna. pp 809–823.

Norey, C.G. (1991) Cadmium poisoning and metallothionein gene expression in freshwater fish. PhD thesis, University of Wales, College of Cardiff, Wales.

Norey, C.G., Brown, M.W., Cryer, A. and Kay, J. (1990) A comparison of the accumulation, tissue distribution and secretion of cadmium in different species of freshwater fish. *Comp. Biochem. Physiol.* **96C**, 181–184.

Ohta, H. and Cherian, M. G. (1995) The influence of nutritional deficiencies on gastrointestinal uptake of cadmium and cadmium-metallothionein in rats. *Toxicology* **97**, (1–3), 71–80.

Olsson, P.-E. (1993) Metallothionein gene expression and regulation in fish, in *Biochemistry and Molecular Biology of Fishes: Molecular Biology Frontiers*, (eds P.W. Hochachka and T.P. Mommsen), Elsevier Science Publishers B.V., Amsterdam, pp. 259–278.

Olsson, P.-E. and Haux, C. (1985) Rainbow trout metallothionein. *Inorg. Chim. Acta* **107**, 67–71.

Olsson, P.-E. and Haux, C. (1986) Increased hepatic metallothionein content correlates to cadmium accumulation in environmentally exposed perch (*Perca fluviatilis*). *Aquat. Toxicol.* **9**, 231–242.

Olsson, P.-E. and Hogstrand, C. (1987) Subcellular distribution and binding of cadmium to metallothionein in tissues of rainbow trout after exposure to ^{109}Cd in water. *Environ. Toxicol. Chem.* **6**, 867–874.

Olsson, P.-E. and Kille, P. (1997) Functional comparison of the metal-regulated transcriptional control regions of metallothionein genes from cadmium sensitive and tolerant fish species. *Biochim. Biophys. Acta* **1350**, 325–334.

Olsson, P.-E., Haux, C. and Förlin, L. (1987) Variation in hepatic metallothionen, zinc and copper levels during an annual reproductive cycle in rainbow trout, *Salmo gairdneri*. *Fish Physiol. Biochem.* **3**, 39–47.

Olsson, P.-E., Larsson, Å., Maage, A. *et al.* (1989a) Induction of metallothionein synthesis in rainbow trout, *Salmo gairdneri*, during long term exposure to waterborne cadmium. *Fish. Physiol. Biochem.* **6**, 221–229.

Olsson, P.-E., Zafarullah, M. and Gedamu, L. (1989b) A role of metallothionein in zinc regulation after oestradiol induction of vitellogenin synthesis in rainbow trout, *Salmo gairdneri. Biochem. J.* **257**, 555–559.

Olsson, P.-E., Hyllner, S.J., Zafarullah, M. *et al.* (1990) Differences in metallothionein gene expression in primary cultures of rainbow trout hepatocytes and the RTH-149 cell line. *Biochim. Biophys. Acta* **1049**, 78–82.

Olsson, P.-E., Zafarullah, M., Foster, R. *et al.* (1995a) Developmental regulation of metallothionein mRNA, zinc and copper levels in rainbow trout, *Salmo gairdneri. Eur. J. Biochem.* **193**, 229–235.

Olsson, P.-E., Kling, P., Pettersson, C. and Silversand, C. (1995b) Interaction of cadmium and oestradiol-17á on metallothionein and vitellogenin synthesis in rainbow trout (*Oncorhynchus mykiss*). *Biochem. J.* **307**, 197–203.

Olsson, P.-E., Larsson, Å. and Haux, C. (1996) Influence of seasonal changes in water temperature on cadmium inducibility of hepatic and renal metallothionein in rainbow trout.*Mar. Environ. Res.* **42**, 41–44.

O'Neil, J.G. (1981) The humoral immune response of *Salmo trutta* L. and *Cyprinus carpio* L. exposed to heavy metals. *J. Fish Biol.* **19**, 297–306.

Overnell, J., McIntoch, R. and Fletcher, T.C. (1987) The levels of liver metallothionein and zinc in plaice, *Pleuronectes platessa* L., during the breeding season, and the effect of oestradiol injection. *J. Fish Biol.* **30**, 539–546.

Palace, V.P., Majewski, H.S. and Klaverkamp, J.F. (1993) Interactions among antioxidant defences in liver of rainbow trout (*Oncorhyncus mykiss*) exposed to cadmium. *Can. J. Fish. Aquat. Sci.* **50**, 156–162.

Palmiter, R.D. (1994) Regulation of metallothionein genes by heavy metals appear to be mediated by a zinc-sensitive inhibitor that interacts with a constitutively active transcription factor, MTF-1. *Proc. Natl Acad. Sci.* **91**, 1219–1223.

Palmiter, R.D. and Findley, F. (1995) Cloning and functional characterization of a mammalian zinc transporter that confers resistance to zinc. *EMBO J.* **14**, 639–649.

Pärt, P. and Lock, R.A.C. (1983) Diffusion of calcium, cadmium and mercury in mucous solution from rainbow trout. *Comp. Biochem. Physiol.* **76C**, 259–263.

Pattison, S.E. and Cousins, R.J. (1986) Zinc uptake and metabolism in hepatocytes. *Federation Proc.* **45**, 2805–2809.

Pentreath, R.J. (1976) Some further studies on the accumulation of ^{65}Zn and ^{54}Mn by the plaice, *Pleuronectes platessa. J. Exp. Mar. Bio. Ecol.* **21**, 179–189.

Perry, S.F., Goss, G.G. and Fenwick , J.C. (1993) Interrelationships between gill chloride cell morphology and calcium uptake in freshwater teleosts. *Fish Physiol. Biochem.* **10**, 327–337.

Perry, H.M. and Erlanger, M.W. (1971) Hypertension and tissue metal levels after intraperitoneal cadmium, mercury and zinc. *Am. J. Physiol.* **220**, 808–811.

Phillips, G. and Lipton, J. (1995) Injury to aquatic resources caused by metals in Montana's Clark Fork River basin: historic perspective and overview. *Can. J. Fish. Aquat. Sci.* **52**, 1990–1993.

Playle, R.C. and Wood, C.M. (1989) Water pH and aluminum chemistry in the gill micro-environment of rainbow trout during acid and aluminum exposures. *J. Comp. Physiol.* **159B**, 539–550.

Playle, R.C. and Wood, C.M. (1991) Mechanisms of aluminum extraction and accumulation at the gills of rainbow trout, *Oncorhynchus mykiss* (Walbaum), in acidic soft water. *J. Fish. Biol.* **38**, 791–805.

Playle, R.C., Goss, G.G. and Wood, C.M. (1989) Physiological disturbances in rainbow trout (*Salmo gairdneri*) during acid and aluminum exposures in soft water of two calcium concentrations. *Can J. Zool* **63**, 314–324.

Playle, R.C., Dixon, D.G. and Burnison, K. (1993a) Copper and cadmium binding to fish gills: estimates of metal-gill stability constants and modeling of metal accumulation. *Can. J. Fish. Aquat. Sci.* **50**, 2678–2687.

Playle, R.C., Dixon, D.G. and Burnison, K. (1993b) Copper and cadmium binding to fish gills: modification by dissolved organic carbon and synthetic ligands. *Can. J. Fish. Aquat. Sci.* **50**, 2667–2677.

Price-Haughey, J., Bonham, K. and Gedamu, L. (1987) Metallothionein gene expression in fish cell lines: its activation in embryonic cells by 5-azacytidine. *Biochim. Biophys. Acta* **908**, 158–168.

Radke, F., Heuchel, R., Georgiev, O. *et al.* (1993) Cloned transcription factor MTF-1 activates the mouse metallothionein I promoter. *EMBO J.* **12**, 1355–1362.

Rambeck, W.A., Bruckner, C., Meier, S. *et al.* (1988) Cadmium bioavailibility and the influence of feed components in chickens, in *Trace Element Analytical Chemisty in Medicine and Biology*, (ed. P. Brattler), Walter de Gruyter and Co., New York.

Renfro, W.C., Fowler, S.W., Heyraud, M. and La Rosa, J. (1975). Relative importance of food and water in long-term zinc^{-65} accumulation by marine biota. *J. Fish. Res. Bd. Can.* **32**, 1339–1345.

Reyes, J.G. (1996) Zinc transport in mammalian cells. *Am. J. Physiol.* **270**, C401–410.

Robohm, R.A. (1986) Paradoxical effects of cadmium exposure on antibacterial antibody responses in two fish species: inhibition in cunners (*Tautogolabrus adspersus*) and enhancement in striped bass (*Morone saxatilis*). *Vet. Immunol. Immunopathol.* **12**, 251–262.

Roch, M., Nordin, R.N., Austin, A. *et al.* (1985) The effects of heavy metal contamination on the aquatic biota of Buttle Lake and the Campbell River drainage (Canada). *Arch. Environ. Contam. Toxicol.* **14**, 347–362.

Saltman, P. and Boroughs, H. (1960) The accumulation of zinc by fish liver slices. *Arch. Biochem. Biophys.* **86**, 169–174.

Sarkar, B. (1989) Metal–protein interactions in transport, accumulation and excretion of metals. *Biol. Trace Elem. Res.* **21**, 137–144.

Sato, M. and Bremner, I. (1993) Oxygen free radicals and metallothionein. *Free Radical Biol. Med.* **14**, 325–337.

Sauer, G.R. and Watabe, N. (1989a) Temporal and metal-specific patterns in the accumulation of heavy metals by the scales of *Fundulus heteroclitus. Aquat. Toxicol.* **14**, 233–248.

Sauer, G.R. and Watabe, N. (1989b) Ultrastructural and histochemical aspects of zinc accumulation by fish scales. *Tiss. Cell* **21**, 935–943.

Schoenmakers, Th.J.M., Klaren, P.H.M., Flik, G. *et al.* (1992) Actions of cadmium on basolateral plasma membrane proteins involved in calcium uptake by fish intestine. *J. Membrane Biol.* **127**, 161–172.

Sell, J.L. (1977) Cadmium and the laying hen: apparent absorption, tissue distribution and virtual absence of transfer into eggs. *Poultry Sci.* **54**, 1674–1678.

Shears, M.A. and Fletcher, G.L. (1983) Regulation of Zn^{2+} uptake from the gastrointestinal tract of a marine teleost, the Winter flounder (*Pseudopleuronectes americanus*). *Comp. Biochem. Physiol.* **64B**, 297–299.

Simons, T.J.B. (1991) Calcium-dependent zinc efflux in human red blood cells. *J. Membr. Biol.* **123**, 73–82.

Sorensen, E.M. (1991) *Metal Poisoning in Fish*, CRC Press, Boca Raton, 374 pp.

Spear, P.A. (1981) *Zinc in the Aquatic Environment: Chemistry, Distribution, and Toxicology*, National Research Council of Canada, Environmental Secretariat publication no. 17589, Publications NRCC/CNRC, Ottawa.

Spry, D.J. and Wood, C.M. (1985) Ion flux rates, acid–base status, and blood gases in rainbow trout, *Salmo gairdneri*, exposed to toxic zinc in natural soft water. *Can. J. Fish. Aquat. Sci.* **42**, 1332–1341.

Spry, D.J. and Wood, C.M. (1989) A kinetic method for the measurement of zinc influx *in vivo* in the rainbow trout, and the effects of waterborne calcium on flux rates. *J. Exp. Biol.* **142**, 425–446.

Spry, D.J., Hodson, P.V. and Wood, C.M. (1988) Relative contributions of dietary and waterborne zinc in the rainbow trout, *Salmo gairdneri. Can. J. Fish. Aquat. Sci.* **45**, 32–41.

Stagg, R.M. and Shuttleworth, T.J. (1982). The effects of copper on ionic regulation by the gills of the seawater-adapted flounder (*Platichthys flesus* L.). *J. Comp. Physiol.* **149**, 83–90.

Staurnes, M., Sigholt, T. and Reite, O.B. (1984) Reduced carbonic anhydrase and Na-K-ATPase activity in gills of salmonids exposed to aluminium-containing acid water. *Experentia* **40**, 226–227.

Sternlieb, I. and Goldfischer, S. (1976). Heavy metals and lysosomes, in *Lysosomes in Biology and Pathology*, (eds. J.T. Dingle and R.T. Dean), North Holland/American Elsevier, pp. 185–200.

Sugawara, N. and Sugawara, C. (1992) The effects of aluminium ingestion on intestinal cadmium absorption, lipid peroxidation and alkaline phosphatase activity. *Toxicol. Letts* **61**, 225–232.

Sugawara, N. and Sugawara, C. (1994) A copper deficient diet prevents hepatic copper accumulation and dysfunction in Long-Evans Cinnamon (LEC) rats with an abnormal copper metabolism and hereditary hepatitis. *Arch. Toxicol.* **69**, 137–140.

Suzuki, K.T., Karazawa, A., Sunaga, H. *et al.* (1989) Uptake of copper from the bloodstream and its relation to induction of metallothionein synthesis in the rat. *Comp. Biochem. Physiol.* **94**, 93–97.

Tanaka, K., Min, K.S., Onosaka, S. *et al.* (1985) The origin of metallothionein in red blood cells. *Toxicol. Appl. Pharmacol.* **78**, 63–68.

Taylor, J.A. and Simons, T.J.B. (1994) The mechanism of zinc uptake by cultivated rat liver cells. *J. Physiol.* **474**, 55–64.

Valberg, L.S., Haist, J. and Cherian, M.G. (1977) Cadmium-induced enteropathy: comparative toxicity of cadmium chloride and cadmium-thionein. *J. Toxicol. Environ. Health* **2**, 963–975.

Van Woewe, J.P., Veldhuizen, M., De Groij, J.J. and Van den Hamer, C.J. (1990) *In vitro* exchangeable erythrocytic zinc. *Biol. Trace Elem. Res.* **25**, 57–69.

Varanasi, U. and Markey, D. (1978) Uptake and release of lead and cadmium in skin and mucus of coho salmon (*Oncorhyncus kisutch*). *Comp. Biochem. Physiol.* **60C**, 187–191.

Varanasi, U., Robisch, P.A. and Malins, D.C. (1975) Structural alterations in fish epidermal mucus produced by waterborne lead and mercury. *Nature* **258**, 431–432.

Verbost, P.M., Flik, G., Lock, R.A.C. and Wendelaar Bonga, S.E. (1987) Cadmium inhibition of Ca^{2+} uptake in rainbow trout gills. *Am. J. Physiol.* **253**, R216–221.

Verbost, P.M., Flik, G., Lock, R.A.C. and Wendelaar Bonga, S.E. (1988) Cadmium inhibits plasma membrane calcium transport. *J. Membrane Biol.* **102**, 97–104.

Verbost, P.M., van Rooit, J., Flik, G. *et al.* (1989) The movement of cadmium through freshwater branchial epithelium and its interference with calcium transport. *J. Exp. Biol.* **145**, 185–197.

Watkins, S.R., Hodge, R.M., Cowman, D.C. and Wickham, P.P. (1977) Cadmium-binding serum protein. *Biochem. Biophys. Res. Commun.* **74**, 1403–1410.

Wicklund, A. and Runn, P. (1988) Calcium effects on cadmium uptake, redistribution and elimination in minnows, *Phoxinus phoxinus*, acclimated to different calcium concentrations. *Aquat. Toxicol.* **13**, 109–122.

Wicklund Glynn, A. and Olsson, P.-E. (1991) Cadmium turnover in minnows, *Phoxinus phoxinus*, pre-exposed to water-borne cadmium. *Environ. Toxicol. Chem.* **10**, 383–394.

Wicklund Glynn, A., Haux, C. and Hogstrand, C. (1992) Chronic toxicity and metabolism of Cd and Zn in juvenile minnows (*Phoxinus phoxinus*) exposed to a Cd and Zn mixture. *Can. J. Fish. Aquat. Sci.* **49**, 2070–2079.

Wicklund Glynn, A., Norrgren, L. and Mussener, A. (1994) Differences in uptake of inorganic mercury and cadmium in the gills of the zebrafish *Brachydanio rerio*. *Aquat. Toxicol.* **30**, 13–26.

Wood C.M., Adams W.J., Ankley G.T. *et al.* (1996) Environmental toxicology of metals, in *Reassessment of Metals Criteria for Aquatic Life Protection: Priorities for Research and Implementation*, SETAC Technical Publication Series, SETAC Press, Pensacola, Florida, Chapter 4.

Wood, C.M., Hogstrand, C., Galvez, F. and Munger, R.S. (1996a) The physiology of waterborne silver toxicity in freshwater rainbow trout (*Oncorhynchus mykiss*): 1. The effects of ionic Ag⁺. *Aquat. Toxicol.* (in press).

Wood, C.M., Hogstrand, C., Galvez, F. and Munger, R.S. (1996b) The physiology of waterborne silver toxicity in freshwater rainbow trout (Oncorhynchus mykiss): 2. The effects of silver thiosulfate. *Aquat. Toxicol.* (in press).

Wright, R.F., Conroy, N., Dickson, W.T. *et al.* (1980) Acidified lake districts of the world: a comparison of water chemistry of lakes in southern Norway, southern Sweden, southwestern Scotland, the Adirondack Mountains of New York, and southeastern Ontario, in *Proceedings of the International Conference on Acid Precipitation*, (eds D. Drablos and A. Tollan), SNSF, Oslo, pp. 377–379.

Zafarullah, M., Bonham, K. and Gedamu, L. (1988) Structure of the rainbow trout metallothionein B gene and characterization of its metal-responsive region. *Mol. Cell. Biol.* **8**, 4469–4476.

Zafarullah, M., Olsson P.-E. and Gedamu, L. (1989) Endogenous and heavy-metal-induced metallothionein gene expression in salmonid tissues and cell lines. *Gene* **83**, 85–93.

11   *Influence of ecological factors on accumulation of metal mixtures*

CLAUDE AMIARD-TRIQUET AND JEAN-CLAUDE AMIARD

11.1 INTRODUCTION

A very large number of experimental studies in aquatic ecotoxicology have been devoted to the assessment of metal uptake in different classes of living organisms. Most of them are based on analysis of single toxicants, despite the fact that under field conditions the biota is generally exposed to multicontaminant pollution. In addition to producing a direct response in biota, a toxic metal can also act by interfering with other metal ions or elements. Thus, the concurrent presence of two or more metallic contaminants in the organism often yields metal uptake that deviates from those currently observed for individual metals. If the effect is an enhancement of bioaccumulation, it is said to be synergistic. If the effect is a decrease of metal incorporation, it is said to be antagonistic.

Most of the knowledge about metal–metal interactions has been acquired in mammalian species. In such species, the major source of metals is the diet and metal uptake depends on the physicochemical forms of storage of metals in food organisms (Amiard-Triquet *et al.*, 1993; Dallinger, 1993). In these conditions, interactions occur inside the organism and changes may be expected at the level of toxicant uptake, distribution, storage and excretion. Biochemical processes such as competitive affinities for intracellular ligands or co-accumulation as mineral granules are probably predominant. In the case of aquatic organisms, interactions can occur both inside organisms (as in mammals) as well as outside through changes of metal speciation, distribution between soluble and insoluble phases and finally bioavailability. Great differences have been detected in metal body burdens when different species (or even different populations within the same species) were compared (Rainbow, 1993). If the species or populations that accumulate the most are

Metal Metabolism in Aquatic Environments. Edited by William J. Langston and Maria João Bebianno. Published in 1998 by Chapman & Hall, London. ISBN 0 412 80370 4

not necessarily the most sensitive to intoxication, nevertheless the incorporated metal is the most liable to produce a toxic effect. Thus indirect evidence of metal interactions can be derived from data concerning the biological effects of metal mixtures. Most studies on interactions between contaminants are based on acute or sublethal toxicity experiments (Westernhagen *et al.*, 1979; Wang, 1987; Sorensen, 1991). Very few studies have been based on metal uptake, with most work in this field dealing with the combined storage of Se and Hg (see for example Pelletier, 1985). When both types of studies are considered, the most striking feature is the heterogeneity of the results for similar combinations of metals. Nevertheless, this 'similarity' is often limited to the nature of the constituents of metal mixtures whereas the experimental conditions differ in metal concentration, direct or trophic exposure pathways, species and stages of development studied, and organs or tissues analysed. The experimental procedures and the methods used for data treatment are an important and difficult judgement to make.

11.2 METHODOLOGIES

Most studies of interactions between contaminants are based on experiments. However, in certain conditions, field studies may be sometimes tentatively interpreted to assess potential interactions.

11.2.1 FIELD STUDIES

In organisms exposed to metal mixtures in their environment, metal concentrations may be determined individually in different organs. A correlation matrix of metals versus metals is then established and significant correlations are noted (e.g. Mackay *et al.*, 1975; Norheim, 1987; Barghigiani *et al.*, 1993; Cosson and Métayer, 1993; Caurant *et al.*, 1994). These correlations may reveal similarities in the mechanisms leading to metal uptake in the organisms or affinities for metabolic pathways. On the other hand, they can reflect concomitant variations due to some factors independent from the pair of metals, such as the influence of age in the case of cumulative pollutants (Figure 11.1) or the phenomenon of biological 'dilution' or 'concentration' associated with sexual maturation in bivalves (Amiard and Berthet, 1996). To go further in the matter, it is necessary to determine the biological or ecological process that is responsible for the observed correlation (Caurant *et al.*, 1994).

Independently of its high ecological relevance, a field approach is the only available method for the study of species as large as dolphins.

11.2.2 LABORATORY STUDIES

(a) Contamination level

Many early ecotoxicological studies used unrealistically high concentrations and short-term exposure and therefore reported only on acute effects. At very

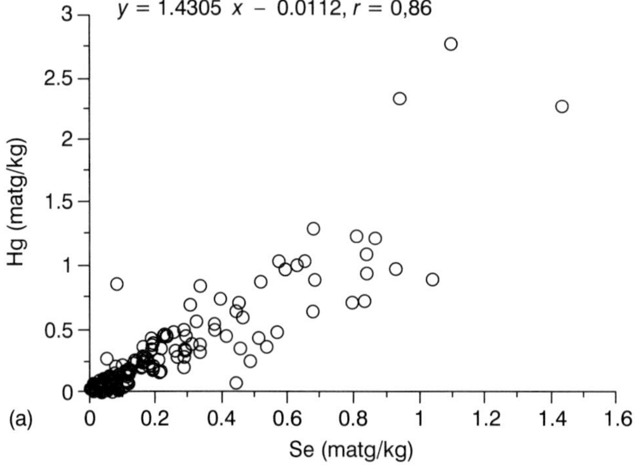

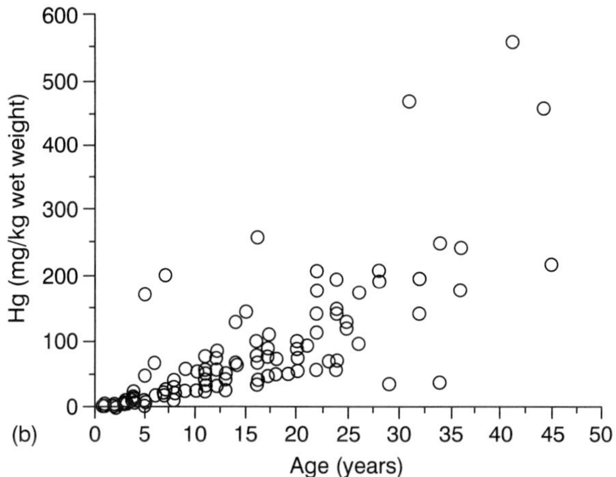

Figure 11.1 Bioaccumulation of mercury and selenium in the liver of pilot whale *Globicephala melas* caught off the Faroe Islands. Relationship between (a) trace element concentrations and (b) mercury concentrations and age of individuals. The trend was similar for selenium concentrations. (After Caurant, 1994.)

low and environmentally relevant levels of contaminants, the biological response may differ. At high concentrations, elements such as Fe are partly insoluble and can smother the gills of aquatic organisms (Martin, 1973). Metal uptake is thus limited but interactions at the level of the innermost biological processes – which are the more liable to occur in environmental conditions – are concealed. In connection with this, interactions between metals and biological macromolecules depend on the nature and severity of the stress. In the plaice *Pleuronectes platessa* injected intraperitoneally with Cd, good correlations between Cd and metallothionein (MT) levels were obtained but at the highest doses (0.5–1 mg kg^{-1}), this relationship deviated from linearity (George and Olsson, 1994). Similarly, in the seabass *Dicentrarchus labrax*, intraperitoneal injection of 0.5 mg Cu kg^{-1} induced a lowering of hepatic MT level compared with controls instead of the increase theoretically expected (Risso *et al.*, 1996). Cd or Cu doses were so high that instead of inducing metal binding to newly synthesized MT, they probably disturbed MT synthesis.

(b) Exposure to single contaminants versus mixtures

Several authors have mentioned that data from experimental groups exposed to mixtures do not conform to patterns determined for single elements (e.g. Eisler and Gardner, 1973; Pelgrom *et al.*, 1994). Such observations suggest two types of interactions: inside the organism, between a metal M1 already present in the body and another metal M2 taken up from the experimental medium; or during the uptake of both metals M1 and M2 as a result of co-exposure. In the first situation, biochemical mechanisms may be predominant, whereas physiological and behavioural mechanisms may be involved in the second.

The difficulties inherent in quantifying interactions due to mixtures may be linked with the methodological choices for devising protocols and the methods used for the data treatment. Experimental factorial designs associated with variance analysis or multiple regression can provide an important contribution to research development in this area (Ribeyre *et al.*, 1995).

(c) Exposure regimes

The majority of experimental studies in aquatic ecotoxicology choose the direct (aqueous) route as the contamination vector, as this generally involves less unwieldy protocols and induces more rapid bioaccumulation in the short term, compared with the trophic (dietary) route. With a view to extrapolating laboratory data to the field situation, it is necessary to examine the representativeness of direct contamination. At both ends of the food chains, the situation is clear since water is the single source of metals for primary producers whereas the diet is the major source for marine mammals or water birds. In other groups (for example, worms, bivalves, crustaceans, fish) the relative importance of food versus water is still a topic of controversy (Chapters 8, 9, 10). However, except in the cases of Hg and radioactive Cs, biomagnification along food chains has not been universally demonstrated (Amiard-Triquet, 1989).

(d) Choice of biological model

It has been hypothesized that metal–metal interactions could result from competition for binding sites at the level of biological macromolecules such as transport proteins in the membranes and metabolic sites within the cell. Cytosolic ligands (glutathione and phytochelatins in plants, transport proteins, enzymes and metallothioneins in animals) and mineral structures (polyphosphate bodies in plants, metal-containing 'granules' in animals) involved in such processes are highly variable among different taxonomic groups. With a view to a comprehensive survey of potential interactions, it is thus necessary to consider a sufficient range of biological models to cover most of the accumulation strategies adopted in aquatic organisms.

(e) Selection of the criteria and levels of analysis

Bioaccumulation may be quantified at different levels of biological organization: organism, organ, subcellular fractions. The choice depends on different factors from material constraint (e.g. size of the studied species) to the aim of the study (mechanistic approach versus environmental quality concern) or the theories to be tested (metal interaction outside or inside the organism; within the soft tissues versus the structural tissues; in intracellular granules versus inducible proteins).

The analysis of metal–metal interaction at the level of the whole organism has a limited interest from a mechanistic point of view since the different processes that may be involved act concomitantly and through indistinguishable ways. However, this procedure may be envisaged to assess the global changes that may occur in some species considered as key species due to their role in the ecosystem or to their commercial value. It may be also used as a preliminary approach before a more in-depth analysis of bioaccumulation mechanisms, but if the studied interaction takes place in a specific organ or tissue, the analysis at the general level of the organism may concealed the phenomenon.

If the analysis is carried out at the level of organs or tissues, different criteria may dictate the choice:

- the relationship of the studied organ to the source of contaminants;
- the physiological role of the organ (uptake, metabolization, storage, elimination);
- the ability of the organ to accumulate preferentially one or several metals;
- the relative abundance of a known metal ligand.

11.3 CASE STUDIES

11.3.1 PLANTS

Metal–metal interactions have been examined in freshwater macrophytes (*Spirodela polyrhiza*), euryhaline macroalgae (*Enteromorpha prolifera*) and different species of unicellular algae (Table 11.1).

Copper seems generally responsible for an enhancement of the accumulated levels of other elements, but for the species *Tetraselmis suecica* and *Haslea ostrearia* the effect of Cu on Ag accumulation was, respectively, synergistic

Table 11.1 Literature sources dealing with metal–metal (M_1/M_2) interaction in plants (the effect studied is of metal M_1 on accumulation of metal M_2)

Species	Conditions of exposure	Effect on bioaccumulation (M_1/M_2)
Spirodela polyrhiza[a]	Cu up to 0.5 mol m^{-3}	Cu/Fe, Cu/Zn, Cu/Ca Synergy at the highest doses
Chlorella ellipsoidea[b,c]	Cd 0–20 µeq l^{-1} + Cr 0–50 µeq l^{-1} Continuous culture Batch culture	Cr/Cd: no effect Cd/Cr: synergy Synergy in both cases
Scenedesmus acutiformis[d]	Ni 0–3 mg l^{-1} + Cu 0–1 mg l^{-1}	Synergy in both cases
Phaedactylum tricornutum[e]	Zn 0.1–2 mg l^{-1} + Cd 0.1–2 mg l^{-1}	Antagonism in both cases
Skeletonema costatum[e]	Zn 50–100 µg l^{-1} + Cd 100 µg l^{-1}	Antagonism in both cases or no effect in different clones (see text)
Tetraselmis suecica[f]	Cu 50 µg l^{-1} Ag 20 µg l^{-1}	Enhancement of Cu, Ag, Pb at the global and ultracellular level in both conditions
Haslea ostrearia[g]	Ag 20 µg l^{-1} + Cu 50 µg l^{-1}	Cu/Ag antagonism
Enteromorpha prolifera[h,i]	Cd 400 nM	Cd/Fe Cd/Cu } synergy Cd/Pb Cd/Zn: antagonism
	Cu 800 nM	Cu/Fe } synergy Cu/Pb

[a]Schreinemakers and Dorhout, 1985; [b]Aoyama and Okamura, 1993; [c]Okamura and Aoyama, 1994; [d]Stokes, 1975; [e]Braek *et al.*, 1980; [f]Ballan-Dufrançais *et al.*, 1991; [g]Ettajani and Pirastru, 1992; [h]Rijstenbil *et al.*, 1993; [i]Haritonidis *et al.*, 1994.

and antagonistic. In *T. suecica* exposed to Cu, Ag and Pb (which were present at very low natural levels in the culture medium), the metals were co-accumulated in osmiophilic vesicles located mainly in the posterior part of the cell (Figure 11.2). Interactions can be different according to the conditions of exposure, as shown in fronds of the duckweed *Spirodela polyrhiza* exposed to Cu up to 0.5 mol m^{-3} (Schreinemakers and Dorhout, 1985). The rates of absorption of Cu, Fe, Zn and Ca were enhanced at the highest doses tested. The authors underlined the fact that Cu concentrations larger than 0.31 mol m^{-3} resulted in free ions in the culture medium.

A synergistic effect of Cd was observed on Cr, Fe and Pb concentrations whereas an antagonistic effect on Zn accumulation was shown in different species. A reciprocal antagonism was generally observed in microalgae exposed to Zn. For *Skeletonema costatum*, two clones were used: Skel-0 was Zn resistant and Skel-5 was Cd resistant. Whatever the combinations of both metals, the interactive effect was still antagonistic in *S. costatum* Skel-0, whereas no appreciable changes were shown in *S. costatum* Skel-5 (Braek *et al.*, 1980). Interactive toxic effects and bioconcentration between Cd and Cr were studied in the freshwater green alga *Chlorella ellipsoidea* by using continuous algal cultures (Aoyama and Okamura, 1993) or batch cultures (Okamura and Aoyama, 1994). In the first case, it was found that Cd stimulated the uptake of Cr by algal cells but Cr showed no effect on the uptake of Cd by the cells. The toxicity of Cr – assessed through the growth rate – seemed to be enhanced as a consequence of the presence of Cd in the medium raising the amount of Cr in the cells. On the other hand, in batch cultures, the amount of Cd as well as Cr in the cells was increased due to the presence of the other metal. Accordingly, the growth inhibition rate also increased.

Few data exist about other pairs of elements. Those mentioned in Table 11.1 (Ag/Cu, Ag/Pb, Ni/Cu) show synergistic effects. In Ag-exposed *T. suecica*, co-accumulation of Cu and Pb was observed in osmiophilic vesicles (Figure 11.2), but quantitatively the synergistic effect was not so important as observed in Cu-exposed algae.

11.3.2 MOLLUSCS

Metal–metal interactions have been examined in gastropods, bivalves and cephalopods. In molluscs exposed to Cd, Zn body burden is generally not affected (Table 11.2). In *Mytilus edulis planulatus*, no interactive effect was observed after exposure to a mixture containing 10 µg Cd l^{-1}, 100 µg Zn l^{-1}, plus Pb and Cu (Ritz *et al.*, 1982). On the contrary, a synergistic effect was observed in the same species when molecular ratios were different (Elliott *et al.*, 1986). In *M. edulis* exposed to Cd, Zn concentrations in soft tissues were not modified except in the gills following the longest exposure (16 days) (Berthet *et al.*, 1985).

In contrast, exposure to Zn induces generally an antagonistic effect on Cd bioaccumulation in bivalves (Table 11.2). Studies by Fowler and Benayoun

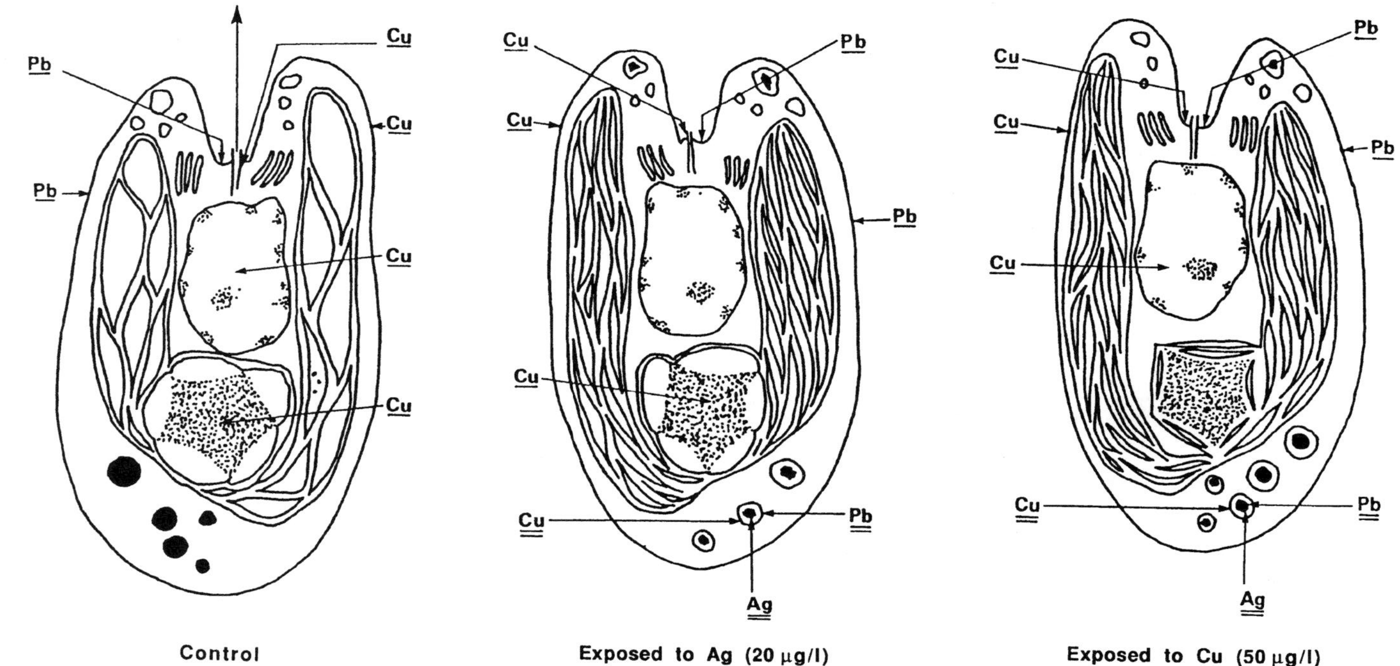

Figure 11.2 Intracellular distribution of metals in marine unicellular algae *Tetraselmis suecica* determined by electron probe microanalysis. Single underlining: 5–500 counts per 100 seconds; double underlining: 500 to 5000 counts per second. (Ballan-Dufrançais, personal communication.)

Table 11.2 Literature sources dealing with metal–metal (M_1/M_2) interaction in molluscs (the effect studied is of metal M_1 on accumulation of metal M_2)

Species	Conditions of exposure	Effect on bioaccumulation (M_1/M_2)
Littorina littorea[a]	Cd 0.005–5 mg l^{-1} Zn 0.01–1 mg l^{-1}	Cd/Zn: no effect Cd/Cu: no effect Cd/Pb: antagonism at intermediate levels Zn/Cd: variable according to exposure (see text) Zn/Pb: variable according to exposure (see text) Zn/Cu: no effect
Littorina littorea[b]	Cd 200 µg l^{-1} + Se 141 µg l^{-1}	Se/Cd: no effect on accumulation in soft parts and shell
Gibbula umbilicalis[a]	Zn 0.01–1 mg l^{-1}	Zn/Cd: variable according to exposure (see text) Zn/Pb: synergy if Zn $\geq$ 0.25 mg l^{-1} Zn/Cu: no effect
Mytilus galloprovincialis[c]	Cd 9.5 nM + Pb 5.1 nM	No effect in both cases
Mytilus edulis planulatus[d]	Cu 10 µg l^{-1} or Cu 20 µg l^{-1} Cd 0–20 µg l^{-1} + Cu 2-20 µg l^{-1} + Zn 11.5–200 µg l^{-1}	Cu/Cd: no interaction Cu/Cd: antagonism Zn/Cd. Cu/Cd: antagonism Zn/Cu, Cu/Zn, Cd/Zn: synergy Cd/Cu: no effect in absence of zinc
Mytilus edulis planulatus[e]	Cd 50 µg l^{-1} + Pb 100 µg l^{-1} Cd 10 µg l^{-1} + Pb 50 µg l^{-1} + Cu 10 µg l^{-1} + Zn 100 µg l^{-1}	Cd and Pb: no interaction in both cases Cd/Cu and Cd/Zn: no effect
Mytilus edulis[a]	Cu 0.01–1 mg l^{-1} Cd 0.0025–2.5 mg l^{-1}	Cu/Cd: no effect in visceral mass; for gills, see text Cu/Zn: no effect in visceral mass; for gills, see text Cd/Zn: no effect, except visceral mass (see text)

Table 11.2 *Continued*

Species	Conditions of exposure	Effect on bioaccumulation (M_1/M_2)
Mytilus edulis[f]	Cd 0.01–60 mg l^{-1} + Cu, Fe, Hg, Pb or Zn 5 mg l^{-1}	No interaction on Cd uptake in gills
Mytilus edulis[g]	Cd 5 µg l^{-1} + Zn 500 or 1000 µg l^{-1}	Zn/Cd: antagonism
Mulinia lateralis[g]	Cd 5 µg l^{-1} + Zn 500 or 1000 µg l^{-1}	Zn/Cd: antagonism
Mytilus edulis[b]	Cd 200 µg l^{-1} + Se 141 µg l^{-1} Cd 200 µg l^{-1} + Se 703 µg l^{-1}	Se/Cd: no effect Se/Cd: synergy
Crassostrea gigas[h]	Ag 10 µg l^{-1} Cu 30 µg l^{-1}	No noticeable effect on Cu concentration Cu/Ag synergy depending on the source of contamination (see text)
Crassostrea gigas[a]	Cu 0.05 mg l^{-1} + Pb 0.20 mg l^{-1} Direct way, or direct + food	Cu + Pb/Cd: antagonism after 20 and 28 days Cu + Pb/Zn: no effect
Macoma balthica[i]	Cd 45 µg l^{-1} + Zn 0–540 µg l^{-1}	Zn/Cd: antagonism
Mizuhopecten yessoensis[j]	Cd 0.5 µg l^{-1}	Cd/Zn: no effect Cd/Cu: antagonism
Anadara granosa[k]	Cd 2.0 µg ml^{-1} + Cu 0.5 µg ml^{-1} Cd 2.0 µg ml^{-1} + Zn 4.0 µg ml^{-1}	Cd and Cu: antagonism in both cases Cd/Zn: no effect Zn/Cd: antagonism
Anodonta cygnea[l]	Cd 122 µg l^{-1} + Cu 139 µg l^{-1}	Reciprocal antagonism
Eledone cirrhosa[m]	*In situ*	Correlated variations of Cd and Se concentrations

[a]Berthet *et al.*, 1985; [b]Bjerregaard, 1988a; [c]Odzak *et al.*, 1994; [d]Elliot *et al.*, 1986; [e]Ritz *et al.*, 1982; [f]Carpene and George, 1981; [g]Jackim *et al.*, 1977; [h]Ettajani *et al.*, 1992; [i]McLeese and Ray, 1984; [j]Evtushenko *et al.*, 1986a; [k]Patel and Anthony, 1991; [l]Holwerta, 1991; [m]Barghigiani *et al.*, 1993.

(1974) showed a trend toward decreased Cd uptake at 100 µg Zn l^{-1} by *Mytilus galloprovincialis* but this was not statistically significant. In the freshwater bivalve *Anodonta cygnea* co-exposure to Zn decreased both *in vivo* accumulation of Cd in the whole animal and the *in vitro* uptake in isolated gills (Hemelraad *et al.*, 1987; Holwerta *et al.*, 1989). In gastropods, no universal trend was detected (Figure 11.3) The influence of Zn on Cd accumulation varied between species (*L. littorea* and *Gibbula umbilicalis*), according to the length of exposure and the Zn concentrations in the experimental medium (Berthet *et al.*, 1985).

The data obtained with other pairs of elements are more heterogeneous (Table 11.2). Some sources of variation have been found. In *M. edulis*

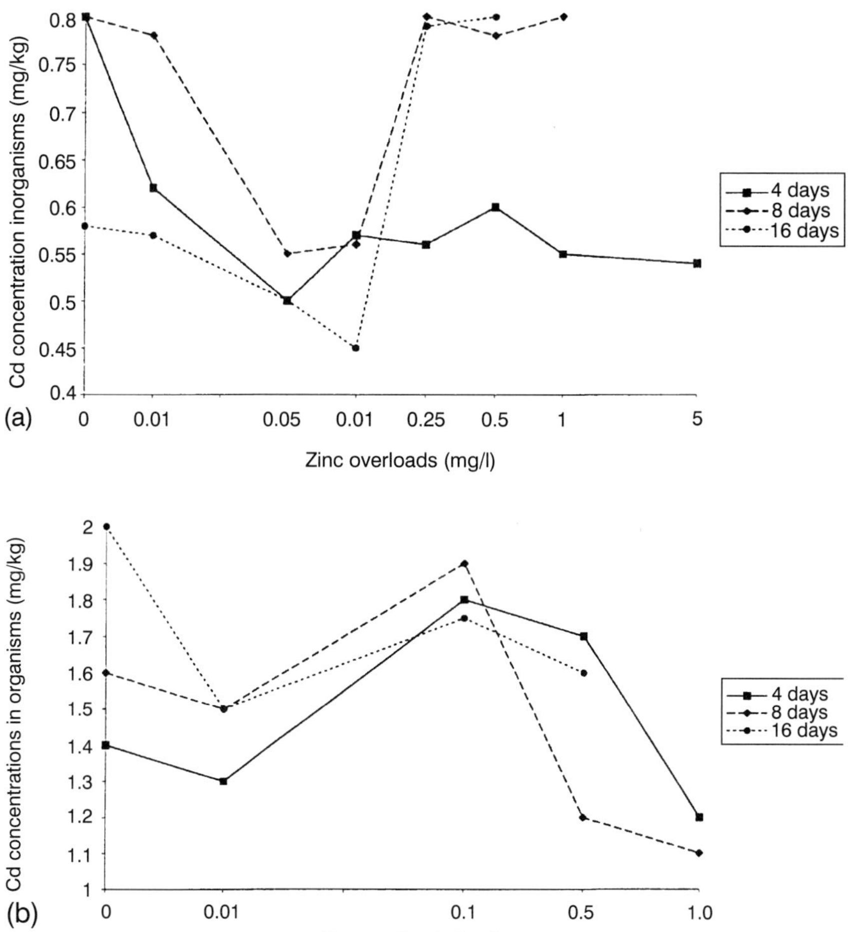

Figure 11.3 Effect of zinc exposure on cadmium accumulation in soft tissues of marine gastropods (a) *Gibbula umbilicalis* and (b) *Littorina littorea*. Influence of doses and length of exposure. (After Berthet *et al.*, 1985.)

exposed to Cu (Berthet *et al.*, 1985), effects on the bioaccumulation of Cd and Zn were examined in different organs or tissues: visceral mass, gills plus palps and remaining tissues. No effect was found in the visceral mass. In other tissues, Cd and Zn levels were not noticeably modified after 4 days. After 8 days of exposure to Cu, Cd concentrations decreased in the remaining tissues whatever the experimental overload (10 μg–1 mg Cu l^{-1}). In the gills, in contrast, a synergistic effect was observed at the lowest dose whereas no effect occurred at 1 mg Cu l^{-1}. This result is in agreement with the study by Carpene and George (1981) showing no effect of 5 mg Cu l^{-1} on the *in vitro* Cd uptake in the gill of *M. edulis*. The influence of the length of exposure has been demonstrated also in young oysters (*Crassostrea gigas*) exposed simultaneously to Cu and Pb: a decrease of Cd level in the soft tissues was observed after 20 and 28 days whereas no effect was appreciable after 12 days (Berthet *et al.*, 1985). In *M. edulis* exposed to Cu (10 or 20 μg l^{-1}), Cd accumulation also depended on the dose (Elliott *et al.*, 1986) (Table 11.2). In mussels exposed to Cd alone (200 μg l^{-1}) or to equimolar concentrations of Cd and Se (141 μg l^{-1}), Cd uptake was similar. Exposure to a five-fold concentration of Se exerted a synergistic effect on Cd uptake in gills, mantle and foot (Bjerregaard, 1988a).

The exposure route has been investigated as a source of variability in metal interactions. Cd and Zn contents of soft tissues were examined in oysters exposed to Cu and Pb introduced in seawater or present in both water and food. No effect of metal sources was shown (Berthet *et al.*, 1985). In contrast, in oysters exposed to Cu in seawater (30 μg l^{-1}) or bound to phytoplanktonic food or sediment, Ag concentrations in soft tissues were significantly increased, the major enhancement being shown with particulate sources (Figure 11.4). The exposure to Ag induced only small changes in Cu concentrations in the soft tissues of oysters (Ettajani *et al.*, 1992).

Selenium and Cd were quantified in mantle muscle tissue of 174 specimens of the cephalopod *Eledone cirrhosa* collected from the northern Tyrrhenian Sea (Barghigiani *et al.*, 1993). Se concentration related both to that of Cd and to the length of specimens.

11.3.3 CRUSTACEANS

Metal–metal interactions in freshwater (*Asellus meridianus*, *Daphnia magna*) and marine crustaceans are listed in Table 11.3. The number of data is limited and in these conditions no clear trend appears.

In molluscs, as mentioned above, Cd exposure had generally no effect on Zn bioaccumulation, a feature that has also been observed in *Pandalus montagui* (Ray *et al.*, 1980). Similarly, dietary Cd had no effect on Zn concentration in the tail muscle of lobster *Homarus americanus* whereas Zn concentration was enhanced in the hepatopancreas (Chou *et al.*, 1987). In the shrimp *Callianassa australiensis*, the relationship between Cd, Zn and also Cu was

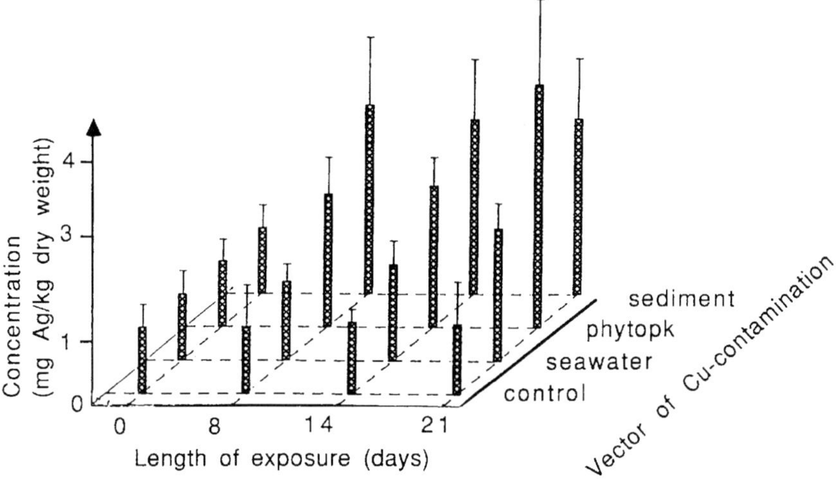

Figure 11.4 Effect of copper exposure on silver accumulation in the soft tissues of oyster *Crassostrea gigas*. Influence of exposure route. (After Ettajani *et al.*, 1992.)

very complicated, as shown by multiple regression analysis (Ahsanullah *et al.*, 1981). In the following models, metal concentrations in the shrimp (μg g^{-1}) are expressed as function of metal concentrations in the experimental medium (μg l^{-1}):

$$Zn_{(shrimp)} = 44 + 58\ Zn + 41\ 834\ Cu + 10\ 452\ Cu^2 - 34\ 621\ ZnCu - 1751\ Cd + 2008\ ZnCd - 8209\ ZnCdCu$$
$$Cd_{(shrimp)} = 4 + 71\ Cd - 36\ Zn + 149\ CdZn - 382\ ZnCu$$
$$Cu_{(shrimp)} = 198 + 3659\ Cu - 5164\ Cu^2 + 322\ CdZn - 1584\ CuZn - 6487\ CuCd$$

In the shrimp *Pandalus montagui*, Cd incorporation was compared after exposure to Cd alone or a mixture of Cd and Zn. In the presence of Zn, Cd concentration was doubled in the hepatopancreas, depressed by one-third in the carcass and unaffected in the other tissues. At highest Zn concentrations (105 or 410 μg Zn l^{-1}) the level of Cd in the hepatopancreas was still greater than after exposure to Cd alone but the synergistic effect was limited compared with the situation observed with the lower Zn overload (70 μg Zn l^{-1}) (Ray *et al.*, 1980). Interactions between Se and Cd accumulation have been studied in the shore crab *Carcinus maenas* (Bjerregaard, 1982, 1985, 1988b). In gills and haemolymph, a synergistic effect on Cd accumulation was observed with some selenite concentrations whereas no effects were observed in other organs.

The history of experimental specimens is important to consider. Interactions between Cu and Pb were examined in two populations of the freshwater isopod *Asellus meridianus*, one of them showing both Cu and Pb

Table 11.3 Literature sources dealing with metal–metal (M_1/M_2) interaction in crustaceans (the effect studied is of metal M_1 on accumulation of metal M_2)

Species	Conditions of exposure	Effect on bioaccumulation
Daphnia magna[a]	Cd 30 µg l^{-1} + Cr 150 µg l^{-1}	Antagonistic effect compared to Cd alone
Asellus meridianus[b]	Cu 0.25 mg l^{-1} + Pb 0–0.4 mg l^{-1}	Antagonism or no effect, according to the origin of the population
	Pb 0.25 mg l^{-1} + Cu 0–0.4 mg l^{-1}	Antagonism or no effect, according to the origin of the population
Xantho hydrophilus[c]	Cd 1–2 mg l^{-1}	Cd/Fe: synergy
Palaemon elegans[c]	Cd 0.01–1 mg l^{-1}	Cd/Cu: antagonism
Pandalus montagui[d]	Cd 40 µg l^{-1} + Zn 70–410 µg l^{-1}	Zn/Cd: variable responses according to organs or doses (see text) Cd/Zn: no effect.
Callianassa australiensis[e]	Zn + Cd Zn + Cu Cd + Cu Zn + Cd + Cu	↑ Zn and Cd accumulation ↑ Zn and ↓ Cu accumulation ↓ Cu accumulation, Cd: no effect ↑ Zn, ↓ Cu accumulation, Cd: no effect.
Homarus americanus[f]	Cd added to food (45 mg Cd kg^{-1})	Hepatopancreas: ↑ Ag, Cu, Zn concentrations Tail muscle: no effect
Carcinus maenas[g]	Cd 1 mg l^{-1} + Se 0.063–1 mg l^{-1}	Gills: reciprocal synergy Cd and Se concentrations correlated in hepatopancreas and carapace
Carcinus maenas[h]	Cd 1 mg l^{-1} + Se in food (200 µg g^{-1})	Se/Cd: synergy in gills; no effect in other organs (hepatopancreas, chela muscle and carapace)

	Cd 1 mg l^{-1} + Se 1 mg l^{-1}	Se/Cd: synergy in haemolymph and gills; no effect in other organs (hepatopancreas, muscle, carapace, heart, testes and hypodermis)
Carcinus maenas[i]	Cd 200 μg l^{-1} + Se > 281 μg l^{-1}	Se/Cd: synergy in haemolymph
	Cd 200 μg l^{-1} + Se 35–2248 μg l^{-1}	Se/Cd: synergy in gills

[a]Kungolos and Aoyama, 1993; [b]Brown, 1978; [c]Skwarzec *et al.*, 1984; [d]Ray *et al.*, 1980; [e]Ahsanullah *et al.*, 1981; [f]Chou *et al.*, 1987; [g]Bjerregaard, 1982; [h]Bjerregaard, 1985; [i]Bjerregaard, 1988b

tolerance, the other being tolerant to Pb (Brown, 1978). In the first case, reciprocal antagonism was observed between Cu and Pb, whereas no interaction was observed in the latter population.

11.3.4 FISH

In fish, most of the studies dealing with elemental interactions include Se as one of the members of coupled elements (Table 11.4). Synergistic effects of Se on bioaccumulation of other elements have been shown in numerous cases: Se vs. Ag in *Brachydanio rerio* (Ribeyre *et al.*, 1995); Se vs. Cu in *Salmo gairdneri* (Hilton and Hodson, 1983); and in the black marlin positive correlations were shown between Se and Cu concentrations in liver (Mackay *et al.*, 1975).

Some heterogeneous data have been published dealing with Se–Hg interactions. The effects of waterborne Se on Hg tissue concentrations were studied in a freshwater fish, the northern pike (*Esox lucius*) (Klaverkamp *et al.*, 1983). At the highest concentrations tested (100 µg Se l⁻¹), no effect was observed either on Hg concentrations actually found in carcass + liver or on $CH_3{}^{203}Hg$ uptake in muscle + skin. A trend of decreased Hg contamination was observed at lower Se concentrations (1 or 10 µg Se l⁻¹). This trend was not confirmed by Turner and Swick (1983) studying the same species with similar concentrations (4.5–6.4 µg Se l⁻¹) but they have shown that, when elevated in food, Se decreased both the body burden of ²⁰³Hg accumulated in pike and the proportion in muscle. In the northern creek chubb (*Semotilus atromaculatus*) Se pre-treatment (3 mg Se l⁻¹ for 48 hours) appears to favour Hg accumulation at the lowest doses tested (0.01–0.07 mg Hg l⁻¹) or to limit Hg accumulation at higher doses (0.1–0.16 mg Hg l⁻¹) (Kim *et al.*, 1977). However, under field conditions, positive correlations between Se and Hg were established in both muscle and liver of black marlin (Mackay *et al.*, 1975).

The interaction between accumulation of waterborne Cd and selenite was studied in juvenile turbot *Scophthalmus maximus* (Nielsen and Bjerregaard, 1991). Concurrent exposure to selenite enhanced Cd uptake in gills, kidney and liver, and reduced accumulation in intestine and erythrocytes; no effect was observed in spleen, skin, muscle and plasma. Concurrent exposure to Cd depleted Se in erythrocytes, and to some extent skin, and reduced Se accumulation in kidney and plasma, but no effect was shown in other organs and tissues. Under field conditions, positive correlations were shown between Cd and Se in the liver of black marlin but not in muscle (Mackay *et al.*, 1975).

The relationship between Cu and Cd depends on organs examined, doses of metals in the mixture and consequent metal ratio, and feeding conditions. Exposure of plaice (*Pleuronectes platessa*) to Cd (2 mg l⁻¹) induced a depletion of Cu in serum, blood cells and gills whereas the concentrations of these essential metals were enhanced in spleen, liver and kidney (Coombs, 1976). Related biochemical changes were observed such as the production

Table 11.4 Literature sources dealing with metal–metal (M_1/M_2) interaction in fish (the effect studied is of metal M_1 on accumulation of metal M_2)

Species	Conditions of exposure	Effect on bioaccumulation
Esox lucius[a]	Se 100 µg l^{-1} Se 1 or 10 µg l^{-1}	No effect on Hg Synergistic effect on Hg
Esox lucius[b]	Se 4.5–6.4 µg l^{-1} Se in food	No effect on Hg Decreased body burden of Hg
Semotilus atromaculatus[c]	Se 3 mg l^{-1} + Hg 0.01–0.07 mg l^{-1} Se 3 mg l^{-1} + Hg 0.1–16 mg l^{-1}	Se/Hg: synergy Se/Hg: antagonism
Salmo gairdneri[d]	Dietary Se	Synergistic effect on liver Cu concentration
Brachydanio rerio[e]	Ag 0–2 µg l^{-1} + MeHg 0–1 µg l^{-1} + Cu 0–20 µg l^{-1} + Zn 0–200 µg l^{-1} + Se 0–50 µg l^{-1}	Ag/MeHg: antagonism Zn, Cu, Se/Ag: synergy
Scophthalmus maximus[f]	Cd 150 µg l^{-1} and/or Se-SeO$_3^{2-}$ 105 µg l^{-1}	Synergy, antagonism or no effect, depending on different tissues
Oreochromis mossambicus[g]	Cd 0–150 µg l^{-1} and/or Cu 0–400 µg l^{-1}	Synergy, antagonism or no effect, depending on metal ratio and feeding conditions
Fundulus heteroclitus[h]	Zn 0–60 mg l^{-1} and/or Cu 0–8 mg l^{-1} and/or Cd 0–10 mg l^{-1}	Cd/Zn: antagonism at all doses Zn/Cd } synergy or antagonism, Cu/Cd } depending on metal ratios
Anguilla rostrata[i]	Cd 75 µg l^{-1} Cd 150 µg l^{-1}	↑ Cu concentration in kidney ↓ Cu concentration in kidney
Makaira indica[j]	Field study	Positive correlations in muscle: Hg-Se, Cu-Pb, Cu-Zn, Zn-Pb, Zn-Cd, Cd-Pb

Table 11.4 *Continued*

Species	Conditions of exposure	Effect on bioaccumulation
		Positive correlations in liver: Hg-Se, Hg-Cd, Hg-Pb, Hg-Cu, Cu-Cd, Cu-Se, Cd-Pb, Cd-Se
Herring eggs[k]	Cd 0.56–5.0 mg l^{-1} + Cu 0.0167–0.15 mg l^{-1} + Pb 0.56–5 mg l^{-1}	Cd/Pb, Cu/Pb: synergy Cd/Cu, Pb/Cu } synergy or antagonism Cu/Cd, Pb/Cd } depending on metal ratios.

[a]Klaverkamp *et al*, 1983; [b]Turner and Swick, 1983; [c]Kim *et al.*, 1977; [d]Hilton and Hodson, 1983; [e]Ribeyre *et al.*, 1995; [f]Nielsen and Bjerregaard, 1991; [g]Pelgrom *et al*, 1994; [h]Eisler and Gardner, 1973; [i]Gill *et al.*, 1992; [j]Mackay *et al.*, 1975; [k]Westernhagen *et al.*, 1979;

of a metallothionein-like protein in liver and kidney. In mummichog (*Fundulus heteroclitus*) surviving exposures to Zn, Cu and Cd in brackish water, interpretation of whole-body metal residues is complicated by the observation that data from groups held in mixtures does not conform to patterns determined for single elements (Eisler and Gardner, 1973). Since the tested doses were very high (up to 60 mg Zn l⁻¹, 8 mg Cu l⁻¹, 10 mg Cd l⁻¹), variations in solubilities or bioavailabilities of individual metals when mixed with other metal salts could account at least partially for several phenomena observed in this study. Interactions between Cu and Cd during single and combined exposure were examined in fed and non-fed juvenile tilapia (*Oreochromis mossambicus*) (Pelgrom *et al.*, 1994). Exposure to one metal (0–150 μg Cd l⁻¹ or 0–400 μg Cu l⁻¹) influenced the concentration of the other metal in the whole body. The effects observed on Cd content after exposure to Cu + Cd were not simply comparable to results obtained after single Cu exposure. In fish exposed to a single contaminant (Cd or Cu) the absence of interactive effect is more frequent in non-fed than in fed specimens. This tendency is not so marked in fish exposed to both contaminants (Table 11.5). Under field conditions, i.e. in fish exposed to both waterborne and dietary trace elements, Mackay *et al.* (1975) have shown a positive correlation between Cd and Cu concentrations in liver.

In oysters, as mentioned above, the effect of the essential metal Cu (and also Zn and Se) on Ag uptake in *Brachydanio rerio* was synergistic whereas no reciprocal effect was evidenced (Ribeyre *et al.*, 1995).

11.3.5 OTHER PHYLA

(a) Invertebrates

In the worms *Nereis virens* and *N. diversicolor*, an antagonistic effect of Zn on Cd bioaccumulation has been observed (Bryan and Hummerstone, 1973; Ray *et al.*, 1979). In the worm *Arenicola marina* exposed to a mixture of Cd (200 μg l⁻¹) and Se (141 μg l⁻¹), no interaction was shown, whereas at the same doses a synergistic effect of Se on Cd uptake was observed in the seastar *Asterias rubens* (Bjerregaard, 1988a).

(b) Water birds

Samples of liver and kidney from 92 birds of ten species collected on Spitsbergen and in the Antarctic were analysed for their content of Cu, Zn, Cd, Pb, Hg and Se (Norheim, 1987). Significant correlations were observed between levels of several of the metals studied, especially between Cd and Zn and between Hg and Se. These correlations could reflect protective mechanisms since the literature provides examples where Se and Zn reduced the toxic effect of Cd and Hg (Underwood, 1977; Magos and Webb, 1980).

Table 11.5 Interactions between copper and cadmium during single and combined exposure in fed and non-fed fish, *Oreochromis mossambicus* (after Pelgrom *et al.*, 1994)

Metals in water (μg l^{-1})	Feeding condition	Number of studied concentrations	Effect on whole body metal-content		
			Cases of synergy	Cases of antagonism	No effect
Effect of Cd exposure on Cu content					
Cd 0–70	Fed	3		2	1
Cd 0–70 + Cu 0–400	Fed	7	1	1	5
Cd 0–150	Non-fed	6	1		5
Cd 0–150 + Cu 0–400	Non-fed	11	1	3	7
Effect of Cu exposure on Cd content					
Cu 0–400	Fed	6	6		
Cu 0–400 + Cd 0–70	Fed	6		5	1
Cu 0–400	Non-fed	9		4	5
Cu 0–400 + Cd 0–150	Non-fed	11		9	2

Trace elements (Cd, Cu, Hg, Pb, Se and Zn) were measured in seven organs or tissues of greater flamingos (*Phoenicopterus ruber*) from the Camargue, France (Cosson and Métayer, 1993). Positive elemental correlations between organs or within the same organ have been interpreted according to the role of organs and the existence of exchange surface with the external medium. The data support the hypothesis of an atmospheric contamination for Cd, Cu, Pb and Hg, plus a trophic contamination for Cd and Cu. The amounts of renal and hepatic metallothionein-like proteins (MTLP) were related to single and combined amounts of Zn and Cd (Cosson, 1989).

(c) Marine mammals

Trace elements (As, Cd, Cu, Hg, Pb, Se and Zn) were measured in liver and kidney of pilot whales (*Globicephala melas*) caught off the Faroe Islands (Caurant *et al.*, 1994). The high correlation between Hg and Se found in pilot whales (Table 11.6) was noticed in earlier studies on both dolphins and seals (references in Caurant *et al.*, 1994) and it seems relevant to relate these data to detoxification mechanisms proposed by Martoja and Berry (1980) and Cuvin-Aralar and Furness (1991).

Significant correlations between levels of As and other elements were shown only in the kidney of pilot whale (Table 11.6). When toxic effects were considered, interactions were evidenced between As-Cd and As-Se (Haguenoer and Furon, 1982); concerning the latter pair, interferences may be due to similarities between metabolism of As and Se (Chappuis, 1991).

In both pilot whale tissues, positive correlations were shown between Cd and Cu and most of the other elements analysed (Table 11.6). In numerous

Table 11.6 *Globicephala melas*: correlation matrix between trace element levels in liver (below diagonal) and kidney (above diagonal) (after Caurant *et al.*, 1994)

	As	Cd	Cu	Hg	Se	Zn
As	–	0.33	0.28	0.09	0.43	0.25
Cd	–0.07	–	0.64[b]	0.68	0.72	0.76[b]
Cu	–0.11	0.58[a]	–	0.20	0.53	0.56[b]
Hg	–0.09	0.24*	0.52	–	0.70	0.57
Se	–0.06	0.37*	0.69	0.86	–	0.58
Zn	0.03	0.65[a]	0.43[a]	–0.04	0.05	–

Significant correlation coefficients ($P < 0.05$) are underlined.
*Deviation from linear trend
Values for liver calculated from 105 individual data except [a] = 164 individuals.
Values for kidney calculated from 54 individual data except [b] = 93 individuals.

marine mammals (Helle, 1981; Honda and Tatsukawa, 1983; Julshamn *et al.*, 1987), as in terrestrial mammals or in humans (Schroeder *et al.*, 1967), Zn concentrations increase with Cd concentrations in liver or kidney.

11.4 TENTATIVE EXPLANATIONS

From case studies described above, several factors influencing interactions have been frequently noted (Table 11.7):

- **Specific differences** occur but we will not develop this point as it is not characteristic of interactive effects.
- More relevant to the topic of interactions is the **tolerance of a population**, within a species, to one or another element (Braek *et al.* 1980; Brown, 1978).
- Different responses have been noted for several pairs of elements in various **tissues or organs** in bivalves, crustaceans and fish (Table 11.7). Field studies have also shown that correlations between pairs of elements were characteristic of each tissue or organ (Mackay *et al.*, 1975; Cosson and Métayer, 1993; Caurant *et al.*, 1994).
- The **conditions of exposure** are the most frequent source of variation (Table 11.7) – more precisely, the length of exposure and the external concentrations of each contaminant, the variations of which determine the metal ratio. This is true even if aquatic organisms are exposed to a range of concentrations of a single contaminant since other elements are naturally present in both the experimental medium and organisms. In terms of competition for binding sites, the importance of metal ratio in metal–metal interaction is evident.
- The influence of the **exposure route** has been examined only in a few cases (Figure 11.4; Table 11.7). A tentative explanation may be built considering that different metabolic pathways are probably involved according to the source of metal, whereas in individuals deprived of food the general metabolism may be affected.
- The **biological role of trace elements**, whether essential (such as Cu, Se and Zn) or toxic (such as Cd, Hg or Ag), seems to be a characteristic that is able to control interactive effects, at least partly. A synergistic effect of essential elements on Ag accumulation has been shown in fish (Ribeyre *et al.*, 1995) and oysters (Ettajani *et al.*, 1992) whereas no or limited reciprocal effects were observed. In numerous invertebrates, the exposure to non-essential Cd has no effect on Zn accumulation but exposure to essential Zn has often an antagonistic effect on Cd accumulation. In vertebrates, Zn and Cd concentrations were correlated in liver and kidney.

Metal–metal interactions may occur at various levels:

- In the external medium, biogeochemical interactions are involved in the control of metal bioavailability.

Table 11.7 Influence of biological and ecological factors on elemental interactions

Species	Contaminant	Bioaccumulated element	Factors having influence	Reference
Skeletonema costatum	Cd + Zn	Zn, Cd	Metal tolerance in different clones	Braek *et al.*, 1978
Asellus meridianus	Cu + Pb	Cu, Pb	Metal tolerance in different populations	Brown, 1978
Merceneria merceneria	Cd	Cu		Robinson and Ryan, 1986
Mytilus edulis	Cu	Cd, Zn	Different responses	Berthet *et al.*, 1985
Mytilus edulis	Cd	Zn	according	Berthet *et al.*, 1985
Pandalus montagui	Cd + Zn	Cd, Zn	to organs or tissues	Ray *et al.*, 1980
Homarus americanus	Cd	Ag, Cu, Zn		Chou *et al.*, 1987
Scophtalmus maximus	Cd and/or Se	Cd, Se		Nielsen and Bjerregaard, 1991
Chlorella ellipsoidea	Cd+ Cr	Cd	Mode of culture (batch, chemostat)	Aoyama and Okamura, 1993; Okamura and Aoyama, 1994
Spirodela polyrhiza	Cu	Fe , Zn, Ca	Level and speciation of Cu	Schreinemakers and Dorhout, 1985
Mytilus edulis	Cd + Se	Cd	Molar ratio	Bjerregaard, 1988a
M. edulis planulatus	Cu	Cd	Level of Cu exposure	Elliot *et al.*, 1986
Crassostrea gigas	Cu + Pb	Cd, Zn	Length of exposure	Berthet *et al.*, 1985
Pandalus montagui	Cd + Zn	Cd	Level of Zn exposure	Ray *et al.*, 1980
Herring eggs	Cd + Cu + Pb	Cd, Cu	Metal ratio	Westernhagen *et al.*, 1979
Oreochromis mossambicus	Cd and/or Cu	Cd, Cu	Metal ratio	Pelgrom *et al.*, 1994
Fundulus heteroclitus	Cd + Cu + Zn	Cd	Metal ratio	Eisler and Gardner, 1973
Semotilus atromaculatus	Se + Hg	Hg	Level of Hg exposure	Kim *et al.*, 1983
Esox lucius	Se	Hg	Level of Se exposure	Klaverkamp *et al.*, 1983
Anguilla rostrata	Cd	Cu	Level of Cd exposure	Gill *et al.*, 1992
Crassostrea gigas	Cu	Ag	Exposure route (water, suspended matter, phytoplankton)	Ettajani *et al.*, 1992
Esox lucius	Se	Hg	Exposure route (waterborne or dietary Se)	Turner and Swick, 1983.

- The competition for absorption sites at the level of biological membranes.
- The competition for metabolic sites within the cell.

11.4.1 ENVIRONMENTAL FACTORS INFLUENCING METAL–METAL INTERACTIONS

Complexation with a variety of naturally occurring ligands can significantly affect the availability and toxicity of trace metals to aquatic organisms. In many cases, uptake and toxicity of trace metals appear to be related to the free metal-ion activity rather than the total concentration of trace metal (Rainbow *et al.*, 1993; Roméo and Gnassia-Barelli, 1993; and references therein) (see also various chapters, this volume). The complexing capacity of natural waters depends on the relative concentrations of metal complexing ligands and of free ions of different metals. Increasing concentrations tested in numerous experimental studies correspond to both quantitative and qualitative changes since the complexing capacity is often exceeded at the highest doses. When metal mixtures are used, the situation is particularly complicated as a consequence of different affinities of each element for each type of ligand.

Another important factor controlling metal speciation is pH. Actively photosynthesizing algae create increased pH around them. At higher pH levels, it has been suggested (Braek *et al.*, 1980) that a precipitation of heavy metals as hydroxides on or around the cells may occur and be confused with metal-binding by the test alga. A rise in pH over more than two units has been observed in batch cultures, whereas the conditions remain almost constant in continuous cultures. This hypothesis is in full agreement with a synergistic effect of Cr on Cd accumulation in *Chlorella ellipsoidea* in batch culture compared with the absence of effect in continuous culture (Aoyama and Okamura, 1993; Okamura and Aoyama, 1994).

Estuaries have been suggested as sinks for dissolved metals of riverine origin due to co-precipitation with Fe oxyhydroxides but in some complex natural systems, such as the anthropogenically contaminated Elbe and Weser estuaries, there is no indication of metal co-precipitation with colloidal Fe oxides (Schoer, 1985, and references therein).

11.4.2 INTERACTIVE EFFECTS ON INFLUX

Possible mechanisms involved in the uptake of metals by aquatic organisms have been reviewed (Simkiss and Taylor, 1989 ; Rainbow, 1993) (Chapter 1). Trace metals have a high affinity for proteins, suggesting that dissolved metals in the external medium may bind passively to transport protein in the membranes of permeable surfaces of the organisms. 'The absence of any clear understanding of what carrier molecules are involved in this system of membrane transport is clearly a major problem in understanding metal toxicity in aquatic organisms' – and more particularly the tendencies and importance of metal–metal interactions (Simkiss and Taylor, 1989). In fish gills, metallic cations induce ionoregulatory failure with at least two discrete sites: at low external concentrations, ion influx is inhibited due in large part to the binding of metal ions to sulfhydryl groups on transport proteins (ATPases), while at higher concentrations ion efflux is stimulated (Lauren, 1991). On the other hand, ecological factors such as salinity (Rainbow *et al.*, 1993) or factors of contamination such as Cu concentration (Hassal, 1963, cited by Stokes, 1975) may influence membrane permeability.

11.4.3 INTERACTIONS WITHIN THE CELL

The competition for metabolic sites within the cell is the hypothesis most frequently proposed to explain metal–metal interactions. Thus we have to consider the distribution of metals at the cellular and tissue level, and metal speciation (protein binding versus co-deposition as mineral compounds) in biological compartments.

Working on Cd–Cr interactions in unicellular freshwater algae *Chlorella ellipsoidea*, Okamura and Aoyama (1994) assume that synergistic toxic effects between the two metals occurred because one metal stimulated the uptake of the other but also influenced its cellular distribution. In the absence of Cr, Cd tended to be accumulated preferentially in the soluble fraction of algal cells (50% vs. 20% in the membrane fraction) whereas the presence of Cr changed the Cd distribution in both fractions to 40%. The amount of Cr accumulated in each fraction was almost the same in the absence of Cd. In the presence of Cd, 90% of Cr was associated to the cell wall fraction after three days of exposure and this percentage stayed as high as 50% even at the end of the study (six days).

It is generally considered that heavy metals exert their noxious effects on organisms mainly by substituting for essential metals in certain macromolecules, such as enzymes. The opposite situation was observed in seawater with low Zn concentration: Cd (known only for its toxic effects) and Co (considered as essential for its role in vitamin B_{12}) stimulate the growth of the diatom *Thalassiosira weissflogii* by substituting for Zn in certain macromolecules (Price and Morel, 1990).

On the other hand, in microalgae and cyanobacteria, numerous studies have shown that toxic metallic ions are being sequestered by polyphosphate or dense core vesicles (Twiss *et al.*, 1989; Ballan-Dufrançais *et al.*, 1991; and references therein). Such intracellular precipitations may be interpreted as specific mechanisms that provide resistance to toxic metals. Moreover, as shown in *Tetraselmis suecica* exposed to Cu or Ag, the 'activation' of such intracellular sinks favours a co-storage of other trace elements present in the culture medium at very low natural levels (Figure 11.2).

In animals, detoxification is achieved according to different kinds of processes. In vertebrates the major process of detoxification is through the binding of metals to particular metalloproteins, namely metallothioneins (Chapter 10). In invertebrates, numerous processes are involved in detoxification including insolubilization of metals as mineral concretions (Brown, 1982) as well as MT (Chapters 8,9).

The metal accumulation strategies vary in different species and may be modified in some populations as a consequence of chronic exposure to metallic contamination. In these populations, metal metabolism may be different according to the nature of metals responsible for acquired tolerance. In specimens of *Asellus meridianus* resistant to Cu and Pb, it has been demonstrated that both metals compete for sites of storage (cuprosomes) in the hepatopancreas whereas the Pb-resistant population was able to restrict the uptake of both Cu and Pb (Brown, 1978). Thus it was not surprising that Cu-Pb interaction was also different between these populations (Table 11.3).

Interactive effects of Se have been shown with several trace metals. Cuvin-Aralar and Furness (1991) hypothesize that the protective effect of Se against Hg may be due to the redistribution of Hg in the presence of Se, competition for binding sites between both elements, formation of Hg–Se complex, conversion of toxic forms of Hg to other forms, and prevention of oxidative damage. In *Ziphius cavirostris* and *Tursiops truncatus*, granules of mercuric selenide have been found in the connective tissue of the liver. The number of these granules is higher in old individuals and they are not eliminated over the life span (Martoja and Berry, 1980). The formation of this non-biodegradable Hg–Se complex (molar ratio 1 : 1) leads to a 'fossilization' of both elements and probably explains the relationship between their concentrations in liver as well as the age-related increase and protective effect mentioned above in marine mammals. The equation shown in Figure 11.1a indicates a Hg/Se molar ratio higher than 1 in the liver of pilot whales and the remaining variation of Hg concentration may be explained through storage as methylmercury or equilibria with other trace metals such as those able to bind to metallothionein, as suggested by correlations (Table 11.6). The formation of a protein-stabilized Cd–Se complex in rats with a molar ratio close to 1 proposed by Gasiewics and Smith (1976, 1978, cited in Bjerregaard, 1988a, and Barghigiani *et al.*, 1993) seems in agreement with data obtained in cephalopods (Barghigiani *et al.*, 1993) but hardly explains all the Cd–Se interactions described in marine invertebrates (Bjerregaard, 1988a). Substitution of Se for sulphur in metallothionein and subsequent interferences with Cu transport or excretion have been suggested as an explanation for Se–Cu interactions (Hilton and Hodson, 1983).

A model for *in vivo* displacement of metallic ions within the MT molecule has been proposed (Funk *et al.*, 1987) and competition for MT binding sites has been evoked to explain some observed interactions (Ray *et al.*, 1980; Skwarzec *et al.*, 1984; Weiss *et al.*, 1986; Pelgrom *et al.*, 1994). Toxic elements (Ag, Hg, Cd) show a greater affinity for MT than essential metals (Otvos *et al.*, 1993). However, Zn is the only metal for which a primary induction has been established and then factors modifying Zn balance may also modify MT levels. In the fish *Gambusia affinis*, it has been assumed that exposure to Hg induced a shift of Zn from native hepatic (Cu–Zn) MT, resulting in *de novo* synthesis of MT induced by free Zn ions (Cosson *et al.*, 1991). The idea that such a substitution process may be at work in most cases of contamination is in full agreement with numerous data showing limited or no interactive effect of toxic metals on bioaccumulation of essential metals. Moreover, the ratios between the quantities of different metal ions interfere with competitive affinities. Both outside and inside the organisms, the concentrations of toxic elements (Ag, Hg, Cd) are several orders of magnitude lower than those determined for essential metals even in contaminated areas. However, in experimental procedures, some authors have used such high concentrations that the normal ratios between essential and toxic metals were deeply modified, generating aberrant data which are at least partly responsible for the heterogeneity observed among interactive effects.

The question of metal ratio is important whatever the ligand involved in metal binding and potentially in metal–metal interaction. The subcellular distribution of metals is governed by numerous factors such as the nature and level of the studied metal and the organ in which the compartmentalization is determined, as shown in fish (Lauren, 1991) or in the scallop *Mizuhopecten yessoensis* (Figure 11.5). In controls, Cd concentrated in hepatopancreas was equally distributed in microsomes and cytoplasm whereas most gill Cd was cytosolic. Exposure to Cd increased the portion of this metal bound to cytosol in both organs. Cadmium also affected Zn distribution. In the hepatopancreas, nuclear Zn increased and cytosolic Zn was depleted. Opposite trends were observed in gills. The length of exposure is another important factor. At the beginning of exposure to 0.5 mg Cd l^{-1}, the greater part of Cd was bound rapidly to high molecular weight proteins (HMWP). Only after 60 days had the metal begun penetrating low molecular weight metallothionein-like proteins. The same species was exposed at length to 0.5 µg Cd l^{-1} (Evtushenko *et al.*, 1984). In this case, Cd was present only in MT-like proteins and was not bound to HMWP. The same pattern of Cd penetrating HMWP at the beginning of the exposure was shown in mussels (Köhler and Riisgard, 1982; Nolan and Duke, 1983), short-necked clams (Ishiguro *et al.*, 1982), oysters (Mouneyrac *et al.*, 1995) and freshwater bivalves (Couillard *et al.*, 1995).

In contrast, Siewicki *et al.* (1983) found that Cd was bound to metallothionein when acute Cd doses were fed to oyster *Crassostrea virginica*, whereas in New Zealand oysters cultured with chronic environmental Cd most of the metal was bound to HMWP (Nordberg *et al.*, 1986).

The influence of the length of exposure on metal–metal interactions may be linked at least partly to changes in metal distribution, but physiological and behavioural processes may also be involved in the short term. Some authors have evoked the role of ventilation and filtration in bivalves. Holwerta (1991) showed that Cu did not influence the *in vitro* uptake of Cd by the excised gills

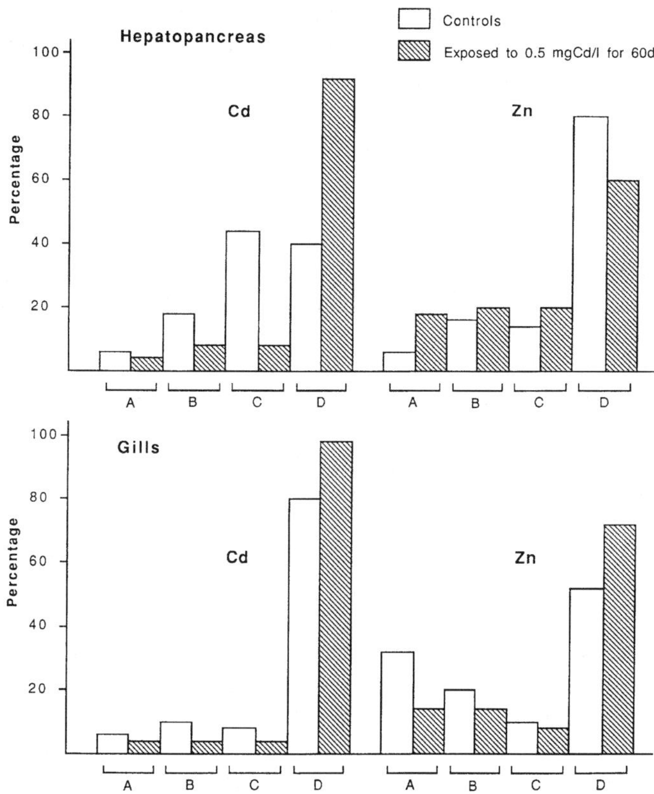

Figure 11.5 Subcellular distribution of cadmium and zinc in scallop *Mizuhopecten yessoensis*. A, nucleus; B, mitochondria; C, microsomes; D, cytosol. (After Evtushenko *et al.*, 1986b.)

of the freshwater bivalve *Anodonta cygnea* but increased the elimination of previously accumulated Cd. Decrease of the ventilation rate probably accounted for the greater part of the Cu effect on Cd levels in tissues. In the marine bivalve *Anadara granosa*, Cd exposure increased the rate of filtration over the first 24 hours but a reverse phenomenon was observed after 96 hours (Patel and Anthony, 1991). On the contrary Zn exposure, or co-exposure to Zn and Cd, resulted in an initial depression of filtration activity which subsequently increased to regain control levels. Copper alone, or in combination with Cd and Zn, influenced filtration negatively. Furthermore, impact of all three metals on both filtration rate and respiratory activity changed from synergistic in the very short term to antagonistic after exposure for 96 hours. Such variations, as well as valve closure (described as an escape mechanism of bivalves in response to contaminants), probably contribute to changing interactive effects according to the length of exposure. Another explanation may reside in the fact that accumulation kinetics are different among organs according to their biological role. If an interaction takes place preferentially in an organ accumulating the contaminants in the long term, interactive effects may be deferred.

11.5 CONCLUSIONS

Considering deleterious effects, the interactions between heavy metals appear to be without pattern. For the most studied metal couple, Cd–Zn, interactive effects varied from antagonism to synergy as observed in different test organisms (Table 11.8) (Wang, 1987). However, it is difficult to compare Table 11.8 with the bioaccumulation data collected in this chapter. Wang (1987) does not distinguish between the effect of Cd on Zn toxicity or the reciprocal effect of Zn on Cd toxicity, but the biological response in terms of bioaccumulation was clearly different: no interactive effects were shown in the first case, and antagonism was the predominant feature in the latter. Moreover, the most numerous references dealing with accumulation of Zn and/or Cd have used bivalves as test organisms (Table 11.2) whereas the literature cited by Wang (1987) covered species belonging to various taxonomic groups. In plants, when toxicity of pairs or mixtures including Cu were studied, synergistic effects were much more frequent than antagonistic ones (Table 11.8). These observations are in agreement with the synergistic effect of Cu on bioaccumulation of other trace elements (Table 11.1).

To summarize, the major factors that have been recognized in the present review as affecting metal–metal interactions are:

- belonging to a taxonomic group (this may be biased due to the fact that authors have preferentially studied certain pairs of elements in certain groups, e.g. combinations including Cu in plants, Cd–Zn in molluscs, Hg–Se in fish and mammals), or to a definite species or a particular population;
- the fact that metal analyses were performed on one or another organ or tissue;
- the conditions of exposure (length, level of exposure and resulting metal ratios);
- the biological role – essential or toxic – of elements involved in potential interactions.

Most of these factors are also involved in absorption and metabolic processes that may be proposed as the most probable origin of interactions. More studies on metal interactions are required before a degree of predictability can be achieved.

Several recommendations are made for future research. Contamination levels should remain realistic in relation to concentrations found in the environment. Since competitions are involved in the complexation of metals with ligands, outside and inside the organisms, defined ionic ratios in experimen-

Table 11.8 Metal–metal interactions which have been shown to influence toxicity (after Wang, 1987)

Pairs of elements	Test organisms	Interactions
Cd-Zn	Freshwater shrimp	Less than additive, or additive, depending on metal level
Cd-Zn	Aquatic plants	Synergistic
Cd-Zn	Freshwater fish	Antagonistic
Cd-Zn	Fiddler crab	Antagonistic
Cd-Zn	Freshwater zooplankton	Synergistic
Cd-Zn	Invertebrate	Antagonistic
Cd-Zn	Freshwater algae	Antagonistic
Cd-Zn	Aquatic plants	Antagonistic
Cu-Fe	Marine algae	Antagonistic
Cu-Mn	Freshwater algae	Synergistic
Cu-Ni	Freshwater algae	Synergistic
Cu-Zn	Marine phytoplankton	Synergistic or antagonistic, depending on species
Cd, Cu, Hg, Pb	Aquatic plants	Synergistic
Cd, Cu, Pb, Zn	Aquatic plants	Synergistic
As, Cd, Cr, Cu, Fe, Hg, Ni, Pb, Se, Zn	Freshwater algae	Synergistic

tal media should be relevant to the known data in natural waters, effluents, sediments or prey and, from a technical point of view, it is necessary to express the results in micromolars rather than weight units (which are generally used in the field of ecotoxicology).

The length of exposure is also an important methodological choice to make: on the one hand, early disturbances can disappear after a short period of acclimation (e.g. Patel and Anthony, 1991); on the other hand, biochemical processes involved in co-storage of metals have been shown to occur over a relatively long period (e.g. Evtushenko *et al.*, 1986b; Couillard *et al.*, 1995; Mouneyrac *et al.*, 1995).

Since different mechanisms probably contribute to metal–metal interactions, it is interesting to compare the effect of single contaminants versus mixtures, hypothesizing (Pelgrom *et al.*, 1994) that the first situation corresponds primarily to intracellular competition and the second to interactions during uptake.

Many early studies using unrealistically high concentrations and short-term exposures have produced responses that were highly heterogeneous in terms of bioaccumulation and noxious effects. The question of interactive effects has thus appeared very confusing and has been practically abandoned by researchers, as illustrated by the small number of recent references quoted in the present chapter. Based on recent advances in the biochemistry of trace elements, relevant hypotheses have now been formed. It is necessary to test them in order to improve the regulations aiming at environmental protection, which generally ignore the multi-element character of metal contamination.

REFERENCES

Ahsanullah, M., Negilski, D.S. and Mobley, M.C. (1981) Toxicity of zinc, cadmium and copper to the shrimp *Callianassa australiensis*. III. Accumulation of metals. *Mar. Biol.* **64**, 311–316.

Amiard, J.-C. and Berthet, B. (1996) Fluctuations of cadmium, copper, lead and zinc concentrations in field populations of the Pacific oyster *Crassostrea gigas* in the Bay of Bourgneuf (France). *Ann. Inst. Oceanogr.* **72(2)**, 195–207.

Amiard-Triquet, C. (1989) Bioaccumulation et nocivité relatives de quelques polluants métalliques à l'égard des espèces marines. *Bull. Ecol.* **20** (2), 129–151.

Amiard-Triquet, C., Jeantet, A.-Y. and Berthet, B. (1993) Metal transfer in marine food chains: bioaccumulation and toxicity. *Acta Biol. Hung.* **44** (4), 387–409.

Aoyama, I. and Okamura, H. (1993) Interactive toxic effect and bioconcentration between cadmium and chromium using continuous algal culture. *Environ. Toxicol. Wat. Qual.* **8** (3), 255–269.

Ballan-Dufrançais, C., Marcaillou, C. and Amiard-Triquet, C. (1991) Response of the phytoplanktonic alga *Tetraselmis suecica* to copper and silver exposure: vesicular metal bioaccumulation and lack of starch bodies. *Biol. Cell* **72**, 103–112.

Barghigiani, C., D'Ulivo, A., Zamboni, R. and Lampugnani, L. (1993) Interaction between selenium and cadmium in *Eledone cirrhosa* of the Northern Tyrrhenian Sea. *Mar. Pollut. Bull.* **26** (4), 212–216.

Berthet, B., Amiard, J.-C., Amiard-Triquet, C. and Métayer, C. (1985) Accumulation de quatre métaux (Cd, Pb, Cu, Zn) chez les animaux marins côtiers et leurs interactions mutuelles, in *Actes du 1er Colloque d'Océanologie côtière* (ed. ADERMA), Bordeaux, pp. 287–299.

Bjerregaard, P. (1982) Accumulation of cadmium and selenium and their mutual interaction in the shore crab *Carcinus maenas* (L.). *Aquat. Toxicol.* **2**, 113–125.

Bjerregaard, P. (1985) Effect of selenium on cadmium uptake in the shore crab *Carcinus maenas* (L.). *Aquat. Toxicol.* **7**, 177–189.

Bjerregaard, P. (1988a) Effect of selenium on cadmium uptake in selected benthic invertebrates. *Mar. Ecol. Prog. Ser.* **48**, 17–28.

Bjerregaard, P. (1988b) Interaction between selenium and cadmium in the hemolymph of the shore crab *Carcinus maenas* (L.). *Aquat. Toxicol.* **13**, 1–11.

Braek, G.S., Malnes, D. and Jensen, A. (1980) Heavy metal tolerance of marine phytoplankton. IV. Combined effect of zinc and cadmium on growth and uptake in some marine diatoms. *J. Exp. Mar. Biol. Ecol.* **42**, 39–54.

Brown, B.E. (1978) Lead detoxification by a copper-tolerant isopod. *Nature* **276**, 388–390.

Brown, B.E. (1982) The form and function of metal-containing granules in invertebrate tissues. *Biol. Res.* **57**, 621–667.

Bryan, G.W. and Hummerstone, L.G. (1973) Adaptation of the polychaete *Nereis diversicolor* to estuarine sediments containing high concentrations of zinc and cadmium. *J. Mar. Biol. Ass. UK* **53**, 839–857.

Carpene, E. and George, S.G. (1981) Absorption of cadmium by gills of *Mytilus edulis* (L.). *Molec. Physiol.* **1**, 23–24.

Caurant, F. (1994) Bioaccumulation de quelques éléments traces (As, Cd, Cu, Hg, Se, Zn) chez le Globicéphale noir (*Globicephala melas*, Delphinidé), pêché au large des îles Féroé. Thèse de Doctorat, Université de Nantes, 206 pp. + Annexes.

Caurant, F., Amiard, J.-C., Amiard-Triquet, C. and Sauriau, P.G. (1994) Ecological and biological factors controlling the concentrations of trace elements (As, Cd, Cu, Hg, Se, Zn) in delphinids *Globicephala melas* from the North Atlantic Ocean. *Mar. Ecol. Prog. Ser.* **103**, 207–219.

Chappuis, P. (1991) *Les oligo-éléments en médecine et biologie*, Lavoisier, Paris.

Chou, C.L., Uthe, J.F., Castell, J.D. and Kean, J.C. (1987) Effect of dietary cadmium on growth, survival, and tissue concentrations of cadmium, zinc, copper, and silver in juvenile American lobster (*Homarus americanus*). *Can. J. Fish Aquat. Sci.* **44** (8), 1443–1450.

Coombs, T.L. (1976) The significance of multielement analyses in metal pollution studies, in *Ecological Toxicology Research*, (eds A.D. McIntyre and C.F. Mills), Plenum Publ. Corp., New York, pp. 187–195.

Cosson, R.P. (1989) Relationship between heavy metal and metallothionein-like protein levels in the liver and kidney of two birds: the greater flamingo and the little egret. *Comp. Biochem. Physiol.* **96C** (1), 243–248.

Cosson, R.P. and Métayer, C. (1993) Etude de la contamination des flamants de Camargue par quelques éléments traces: Cd, Cu, Hg, Pb, Se et Zn. *Bull. Ecol.* **24** (1), 17–30.

Cosson, R.P., Amiard-Triquet, C. and Amiard, J.-C. (1991) Metallothioneins and detoxification. Is the use of detoxication protein for Mts a language abuse? *Water, Air, Soil Pollut.* **57–58**, 555–567.

Couillard, Y., Campbell, P.G.C., Pellerin-Massicotte, J. and Auclair, J.C. (1995) Field transplantation of a freshwater bivalve, *Pyganodon grandis*, across a metal contamination gradient. II. Metallothionein response to Cd and Zn exposure, evidence for cytotoxicity, and links to effects at higher levels of biological organization. *Can. J. Fish. Aquat. Sci.* **52**, 703–715.

Cuvin-Aralar, M.L. and Furness, R. (1991) Mercury and selenium interaction: a review. *Ecotoxicol. Environ. Saf.* **21**, 348–364.

Dallinger, R. (1993) Strategies of metal detoxification in terrestrial invertebrates, in *Ecotoxicology of Metals in Invertebrates*, (eds R. Dallinger and P.S. Rainbow), Lewis Publishers, Boca Raton, pp. 245–289.

Eisler, R. and Gardner, G.R. (1973) Acute toxicity to an estuarine teleost of mixtures of cadmium, copper and zinc salts. *J. Fish. Biol.* **5**, 131–142.

Elliott, N.G., Swain, R. and Ritz, D.A. (1986) Metal interaction during the accumulation by the mussel *Mytilus edulis planulatus*. *Mar. Biol.* **93** (3), 395–399.

Ettajani, H., Amiard-Triquet, C. and Amiard, J.-C. (1992) Etude expérimentale du transfert de deux éléments traces (Ag, Cu) dans une chaîne trophique marine: eau – particules (sédiment naturel, microalgue) – Mollusques filtreurs (*Crassostrea gigas* Thunberg). *Wat. Air Soil Polln* **65**, 215–236.

Ettajani, H. and Pirastru, L. (1992) Méthodologie pour prévoir le transfert des métaux lourds dans les chaînes trophiques marines incluant des Mollusques filtreurs. *Hydroécol. Appl.* **4** (2), 79–90.

Evtushenko, Z.S., Lukyanova, O.N. and Khristoforova, N.N. (1984) Biochemical changes in selected body tissues of the scallop *Patinopecten yessoensis* under long-term exposure to low Cd concentrations. *Mar. Ecol. Prog. Ser.* **20**, 165–170.

Evtushenko, Z.S., Belcheva, N.N. and Lukyanora, O.N. (1986a) Cadmium accumulation in organs of the scallop *Mizuhopecten yessoensis*. I. Activities of phosphatases and composition and amount of lipids. *Comp. Biochem. Physiol.* **83C**, 371–376.

Evtushenko, Z.S., Belcheva, N.N. and Lukyanora, O.N. (1986b) Cadmium accumulation in organs of the scallop *Mizuhopecten yessoensis*. II. Subcellular distribution of metals and metal-binding proteins. *Comp. Biochem. Physiol.* **83C**, 377–383.

Fowler, S.W. and Benayoun, G. (1974) Experimental studies on cadmium flux through marine biota, in *Comparative Studies of Food and Environmental Contamination*, (ed. IAEA, Vienna), *Radioactivity in the Sea* **44**, 159–178.

Funk, A.E., Day, F.A. and Brady, F.O. (1987) Displacement of zinc and copper from copper-induced metallothionein by cadmium and by mercury: *in vivo* and *ex vivo* studies. *Comp. Biochem. Physiol.* **86C** (1), 1–6.

George, S.G. and Olsson, P.-E. (1994) Metallothioneins as indicators of trace metal pollution, in *Biomonitoring of Coastal Waters and Estuaries*, (ed. K.J.M. Kramer), CRC Press, Boca Raton, pp. 151–178.

Gill, T.S., Bianchi, C.P. and Epple, A. (1992) Trace metal (Cu and Zn) adaptation of organ systems of the American eel, *Anguilla rostrata*, to external concentrations of cadmium. *Comp. Biochem. Physiol.* **100C**, 361–371.

Haguenoer, J.M. and Furon, D. (1982) *Toxicologie et hygiène industrielle. Vol. II. Les dérivés minéraux.* Techniques et Documentation, Paris.

Haritonidis, S., Rijstenbil, J.W., Malea, P. *et al.* (1994) Trace metal interactions in the macroalga *Enteromorpha prolifera* (O.F. Müller)J.Ag., grown in water of the Scheldt estuary (Belgium and SW Netherlands), in response to cadmium exposure. *BioMetals* **7**, 61–66.

Helle, E. (1981) Reproductive trends and occurrence of organochlorines and heavy metals in the Baltic seal populations. *Comm. Meet. int. Coun. Explor. Sea C.M.-ICES* **E:37**.

Hemelraad, J., Kleinveld, H.A., De Roos, A.M. *et al.* (1987) Cadmium kinetics in freshwater clams. III. Effects of zinc on uptake and distribution of cadmium in *Anodonta cygnea*. *Arch. Environ. Contam. Toxicol.* **16**, 95–101.

Hilton, J.W. and Hodson, P.V. (1983) Effect of increased dietary carbohydrate on selenium metabolism and toxicity in rainbow trout (*Salmo gairdneri*). *J. Nutr.* **113** (6), 1241–1248.

Holwerta, D.A. (1991) Cadmium kinetics in freshwater clams. V. Cadmium–copper interaction in metal accumulation by *Anodonta cygnea* and characterization of the metal-binding protein. *Arch. Environ. Contam. Toxicol.* **21**, 432–437.

Holwerta, D.A., De Knecht, J.A., Hemelraad, J. and Veenof, P.R. (1989) Cadmium kinetics in freshwater clams. Uptake of cadmium by the excised gill of *Anodonta cygnea*. *Bull. Environ. Contam. Toxicol.* **42**, 382–388.

Honda, K. and Tasukawa, R. (1983) Distribution of cadmium and zinc in tissues and organs and their age-related changes in striped dolphins, *Stenella coeruleoalba*. *Arch. Environ. Contam. Toxicol.* **12**, 543–550.

Ishiguro, T., Kitajama, K., Chiba, M. *et al.* (1982) Studies on Cd-binding proteins in short-necked clam. *Bull. Jap. Soc. Sci. Fish.* **48**, 793–798.

Jackim, E., Morrison, G. and Steele, R. (1977) Effects of environmental factors on radiocadmium uptake by four species of marine bivalves. *Mar. Biol.* **40** (4), 303–308.

Julshamn, K., Andersen, A., Ringdal, O. and Morkore, J. (1987) Trace elements intake in the Faroe Islands. I. Element levels in edible parts of pilot whales (*Globicephalus melaenus*). *Sci. Tot. Environ.* **65**, 53–62.

Kim, J.H., Birks, E. and Heisinger, J.F. (1977) Protective action of selenium against mercury in Northern Creek chubs. *Bull. Environ. Contam. Toxicol.* **17** (2), 132–136.

Klaverkamp, J.F., Hodgins, D.A. and Lutz, A. (1983) Selenite toxicity and mercury–selenium interactions in juvenile fish. *Arch. Environ. Contam. Toxicol.* **12**, 405–413.

Köhler, K. and Rijsgard, H.U. (1982) Formation of metallothioneins in relation to accumulation of cadmium in the common mussel *Mytilus edulis*. *Mar. Biol.* **66**, 53–58.

Kungolos, A. and Aoyama, I. (1993) Interactive effect, food effect, and bioaccumulation of cadmium and chromium for the system *Daphnia magna–Chlorella ellipsoidea*. *Environ. Toxicol. Wat. Qual.* **8** (4), 351–369.

Lauren, D.J. (1991) The fish gill: a sensitive target for waterborne pollutants. *Aquat. Toxicol. Risk Assessment: 14th Vol., ASTM STP 1124*, (eds M.A. Mayes and M.G. Barron), American Society for Testing and Materials, pp. 223–244.

Mackay, N.J., Kazacos, M.N., Williams, R.J. and Leedow, M.I. (1975) Selenium and heavy metals in black marlin. *Mar. Pollut. Bull.* **6** (4), 57–61.

Magos, L. and Webb, M. (1980) The interactions of selenium with cadmium and mercury. *CRC Crit. Rev. Toxicol.* **8**, 1–42.

Magos, L., Clarkson, T.W., Sparrow, S. and Hudson, A.R. (1987) Comparison of the protection given by selenite, selenomethionine and biological selenium against the renotoxicity of mercury. *Arch. Toxicol.* **60**, 422–426.

Martin, J.-L. (1973) Iron metabolism in *Cancer irroratus* (Crustacea Decapoda) during the intermoult cycle, with special reference to iron in the gills. *Comp. Biochem. Physiol.* **46A**, 123–129.

Martoja, R. and Berry, J.P. (1980) Identification of tienmannite as a probable product of demethylation of mercury by selenium in cetaceans. A complement to the scheme of the biological cycle of mercury. *Vie Milieu* **30** (1), 7–10.

McLeese, D.W. and Ray, S. (1984) Uptake and excretion of cadmium, CdEDTA, and zinc by *Macoma balthica*. *Bull. Environ. Contam. Toxicol.* **32**, 85–92.

Mouneyrac, C., Berthet, B., Amiard, J.-C. and Cosson, R.P. (1995) Caractérisation des composés moléculaires fixant le Cd chez des huîtres (*Crassostrea gigas*) provenant d'un site contaminé, in *Colloque International 'Marqueurs biologiques de pollution'*, ANPP, Chinon, pp. 47–54.

Nielsen, G. and Bjerregaard, P. (1991) Interaction between accumulation of cadmium and selenium in the tissues of turbot *Scophthalmus maximus*. *Aquat. Toxicol.* **20** (4), 253–266.

Nolan, C.V. and Duke, E.J. (1983) Cd-binding proteins in *Mytilus edulis*: relation to mode of administration and significance in tissue retention of Cd. *Chemosphere* **12**, 65–74.

Nordberg, M.I., Nuottaniemi, M.I., Cherian, M.G. *et al.* (1986) Characterization studies on the cadmium-binding proteins from two species of New Zealand oysters. *Environ. Health Perspec.* **65**, 57–62.

Norheim, G. (1987) Levels and interactions of heavy metals in sea birds from Svalbard and the Antarctic. *Environ. Pollut.* **47**, 83–94.

Odzak, N., Martincic, D., Zvonaric, T. and Branica, M. (1994) Bioaccumulation rate of Cd and Pb in *Mytilus galloprovincialis* foot and gills. *Mar. Chem.* **46**, 119–131.

Okamura, H. and Aoyama, I. (1994) Interactive toxic effect and distribution of heavy metals in phytoplankton. *Environ. Toxicol. Wat. Qual.* **9** (1), 7–15.

Otvos, J.D., Liu, X., Li, H. *et al.* (1993) Dynamic aspects of metallothionein structure, in *Metallothionein III: Biological Roles and Medical Implications*, (eds K.T. Suzuki, N. Imura, and M. Kimura), Birkhäuser Verlag, Basel, pp. 57–74.

Patel, B. and Anthony, K. (1991) Uptake of cadmium in tropical marine lamellibranchs, and effects on physiological behaviour. *Mar. Biol.* **108** (2), 457–470.

Pelgrom, S.M.G.J., Lamers, L.P.M., Garritsen, J.A.M. *et al.* (1994) Interactions between copper and cadmium during single and combined exposure in juvenile tilapia *Oreochromis mossambicus* : Influence of feeding condition on whole body metal accumulation and the effect of the metals on tissue water and ion content. *Aquat. Toxicol.* **30** (2), 117–135.

Pelletier, E. (1985) Mercury–selenium interactions in aquatic organisms: a review. *Mar. Env. Res.* **18**, 111–132.

Price, N.M. and Morel, F.M.M. (1990) Cadmium and cobalt substitution for zinc in a marine diatom. *Nature* **344**, 658–660.

Rainbow, P.S. (1993) The significance of trace metal concentrations in marine invertebrates, in *Ecotoxicology of Metals in Invertebrates*, (eds R. Dallinger and P.S. Rainbow) Lewis Publishers, Boca Raton, pp. 3–23.

Rainbow, P.S., Malik, I. and O'Brien, P. (1993) Physico-chemical and physiological effects on the uptake of dissolved zinc and cadmium by the amphipod crustacean *Orchestria gammarellus. Aquat. Toxicol.* **25**, 15–30.

Ray, S., McLeese, D.W. and Pezzack, D. (1979) Chelation and interelemental effects on the bioaccumulation of heavy metals by marine invertebrates, in *International Conference Management and Control of Heavy Metals in the Environment* (ed. CCE, OMS, IWPC, IPHE, IWES), pp. 35–38.

Ray, S., McLeese, D.W., Waiwood, B.A. and Pezzack, D. (1980) The disposition of cadmium and zinc in *Pandalus montagui. Arch. Environ. Contam. Toxicol.* **9**, 675–681.

Ribeyre, F., Amiard-Triquet, C., Boudou, A. and Amiard, J.-C. (1995) Experimental study of interactions between five trace elements – Cu, Ag, Se, Zn, and Hg – toward their bioaccumulation by fish (*Brachydanio rerio*) from the direct route. *Ecotox. Environ. Saf.* **32**, 1–11.

Rijstenbil, J.W., Haritonidis, S., Drie, J.v. *et al.* (1993) Interactions of copper with trace metals and thiols in the macro-algae *Enteromorpha prolifera* (O.F. Müll)J.Ag., grown in water of the Scheldt Estuary (Belgium and SW Netherlands), in response to cadmium exposure. *Sci. Tot. Environ.* **Suppl. 1992**, 539–549.

Risso, C., Gnassia-Barelli, M., Cosson, R. *et al.* (1996) Evaluation du contenu en MT dans le foie du Loup *Dicentrachus labrax*, intoxiqué par un mélange de polluants: comparaison de deux méthodes de dosage, in *La qualité de l'eau* (eds C. Amiard-Triquet and T. Hamon), Université de Nantes, Nantes pp. 45–48.

Ritz, D.A., Swain, R. and Elliott, N.G. (1982) Use of the *Mussel Mytilus edulis planulatus* (Lamarck) in monitoring heavy metal levels in seawater. *Aust. J. Mar. Freshwater Res.* **33**, 491–506.

Robinson, W.E. and Ryan, D.K. (1986) Metal interactions within the kidney, gill, and digestive gland of the hard clam, *Mercenaria mercenaria*, following laboratory exposure to cadmium. *Arch. Environ. Contam. Toxicol.* **15**, 23–30.

Roméo, M. and Gnassia-Barelli, M. (1993) Organic ligands and their role in complexation and transfer of trace metals (micronutrients) in marine algae, in *Macroalgae, Eutrophication and Trace Metal Cycling in Estuaries and Lagoons,*

Proceedings of the Cost-48 Symposium, Thessaloniki, 24–26 September 1993, (eds J.W. Rijstenbil and S. Haritonidis).

Schoer, J. (1985) Iron-hydroxydes and their significance to the behavior of heavy metals in estuaries, in *Heavy Metals in the Environment*, (ed. T.D. Lekkas), CEP Consultants Ltd, Edinburgh, Vol. 1, pp. 384–388.

Schreinemakers, W.A.C. and Dorhout, R. (1985) Effects of copper ions on growth and ion absorption by *Spirodela polyrhiza* (L.) Schleiden. *J. Plant Physiol.* **121**, 343–351.

Schroeder, H.A., Nason A.P., Tipton, I.H. and Balassa, J.J. (1967) Essential trace metals in man: zinc. Relation to environmental cadmium. *J. Chron. Dis.* **20**, 179–210.

Siewicki, T.C., Sydlowski, J.S. and Webb, E.S. (1983) The nature of cadmium binding in commercial eastern oysters (*Crassostrea virginica*). *Arch. Environm. Contam. Toxicol.* **12**, 299–304.

Simkiss, K. and Taylor, M.G. (1989) Metal fluxes across the membranes of aquatic organisms. *CRC Crit. Rev. Aquat. Sci.* **1**, 173–187.

Skwarzec, B., Kentzer-Baczewska, A., Styczynska-Jurewicz, E. and Neugebauer, E. (1984) Influence of accumulation of cadmium on the content of other microelements of two species of Black Sea decapods. *Bull. Environ. Contam. Toxicol.* **32**, 93–101.

Sorensen, E.M.B. (1991) Interactions, in *Metal Poisoning in Fish*, (ed. E.M.B. Sorensen), CRC Press, Boca Raton, pp. 333–358.

Stokes, P. (1975) Uptake and accumulation of copper and nickel by metal-tolerant strains of *Scenedesmus*. *Verh. Internat. Verein. Limnol.* **19** (3), 2128–2137.

Turner, M.A. and Swick, A.L. (1983) The English–Wabigoon river system: IV. Interaction between mercury and selenium accumulated from waterborne and dietary sources by northern pike (*Esox lucius*). *Can. J. Fish. Aquat. Sci.* **40** (12), 2241–2250.

Twiss, M.R., Nalewajko, C. and Stockes, P.M. (1989) The influence of phosphorus nutrition on copper tolerance in *Scenedesmus* spp., in *Heavy Metals in the Environment* (ed. J.-P. Vernet), CEP Consultants Ltd, Edinburgh, Vol. 2, pp. 174–177.

Underwood, E.J. (1977) *Trace Elements in Human and Animal Nutrition*, Academic Press, New York.

Wang, W. (1987) Factors affecting metal toxicity to (and accumulation by) aquatic organisms. Overview. *Env. Intern.* **13** (6), 437–457.

Weiss, P., Bodgen, J.D. and Enslee, E.C. (1986) Hg- and Cu-induced hepatocellular changes in the munnichog *Fundulus heteroclitus*. *Environ. Health Perspect.* **65**, 167–173.

Westernhagen, H.V., Dethlefsen, V. and Rosenthal, H. (1979) Combined effects of cadmium, copper and lead on developing herring eggs and larvae. *Helgoländer wiss. Meerest.* **32**, 257–278.

12   *Metals and marine food chains*

JAMES A. NOTT

12.1 INTRODUCTION

Food chains consist of variable numbers of trophic levels linked in successive prey and predator relationships. Networks of these chains form complex food webs that route the supply, transfer and disposal of potentially toxic metals within ecological systems. Metals are transferred along the chains but they differ from the organic constituents of food in two main respects. Firstly, they can enter food chains directly from solution in seawater by crossing permeable membranes of marine organisms. Within tissues they are compartmentalized in membrane-limited vacuoles and bound to ligands to reduce the toxic reactivity (Mason and Nott, 1981; Brown, 1982; George, 1982; Taylor and Simkiss, 1984; Roesijadi, 1992; Viarengo and Nott, 1993). These processes remove metals from tissue fluids, and diffusion gradients inwards from seawater are maintained; concentration factors in excess of a thousand-fold can be attained. Concentration factors represent the net balance of continuous uptake and excretion, which can result in negligible or excessive accumulation according to the metal and its availability and the species of organism (Baudo, 1981, 1985; Bryan, 1976b, 1979; Fowler, 1982; Suedel *et al.*, 1994). Uptake by a particular species can be modified by stage of life history, physiological acclimation and genetic adaptation (Chapter 7).

Secondly, metals are indestructible: they are retained and accumulated within ecosystems because reactive forms bind to sediments and non-reactive forms occur as insoluble oxides and salts. Thus, ecotoxicological effects of metals can persist for decades after pollution incidents and can be damaging when sediments and mine wastes are disturbed. Organic compounds are destructible: they are lost from ecosystems as carbon dioxide, water and gaseous and soluble forms of nitrogen, sulphur and halogens.

Metal Metabolism in Aquatic Environments. Edited by William J. Langston and Maria João Bebianno. Published in 1998 by Chapman & Hall, London. ISBN 0 412 80370 4

12.2 CONCENTRATION FACTORS

The relationship between the concentration of a metal in the environment and the concentration in an organism is a complex phenomenon (Baudo, 1981) which is affected by speciation, active and passive uptake, modalities of uptake, and transformation, transport and distribution between and within tissues, and elimination. Metal bioavailability is a direct result of reactivity and thermodynamic equilibrium, and biomagnification occurs at stages in a food chain when proportionally more metal is retained than energy, in the form of weight gain.

Concentration factors for metals within organisms and the retention of stable metal species within ecosystems both contribute to the transfer of metals along food chains and the toxic effects. Thus, primary producers can accumulate high concentrations and these are consumed by organisms on secondary trophic levels in the food chain. If the diet of the secondary consumer contains biochemically reactive metals, they will be absorbed and accumulated but if they are insoluble they will pass through the gut and be excreted in the faeces. Table 12.1 shows a number of food chains that have been investigated and it emphasizes the variable characteristics of the trophic levels. The results in the quoted papers are assessed and marked subjectively for the relative efficiency with which the metals are taken up, transferred and in some cases excreted.

12.3 HEPATOSOMATIC INDEX

Food chain transfer is affected by the distribution of metals between different tissues of the prey and by the marked degree to which this compartmentalization can vary at different trophic levels (Figure 12.1). The trophic levels range from phytoplankton and macrophytes to zooplankton, invertebrates, fish and mammals.

Phytoplankton has a large ratio of surface area to volume, and metals dissolved in seawater have access to the entire surface of each cell (Baudo, 1981; Bryan, 1976a; Chapter 4). Once within the outer cell membrane, metal distribution is determined by the location of reactive molecules, which is similar in all cells of the same type. Macrophytes also lack specialized regions of uptake and, as a result, older parts of seaweed fronds contain the highest concentrations of metals. Also, they do not excrete metals. Therefore, primary producers are vulnerable to the effects of excess metals and they show some of the highest levels of accumulation in food chains (Preston *et al.*, 1972; Laumond *et al.*, 1973; Hetherington *et al.*, 1975; Guary and Frazier, 1977; Sanders *et al.*, 1989, 1990; Lindsay and Sanders, 1990). This initial accumulation from seawater provides much of the momentum for subsequent transfer along food chains (Preston *et al.*, 1972).

However, situations are far from straightforward. For example, arsenic is incorporated by algae and transformed into reduced and methylated forms which are non-toxic to phytoplankton but may be toxic to higher animals (Sanders, 1985; Chapters 4 and 6). The copepod *Eurytemora affinis*, barnacle

Table 12.1 Condensed review of literature on food chains with the papers classified according to different metals, showing uptake routes of metals from food, water, sediment, transfer to higher trophic levels and in some cases excretion to water in urine and transfer to sediment in faecal pellets

Food chain	Reference
Arsenic	
Dunaliella marina >> *Artemia salina* >> *Lysmata seticaudata*	Wrench *et al.*, 1979
Fucus spiralis >> *Littorina litoralis* >> *Nucella lapillus*	Klumpp, 1980
Water >> kelp *Ecklonia radiata* >> whiting *Sillago bassensis* Water >> whiting *Sillago maculata*	Edmonds and Francesconi, 1981
Water > *Eurytemora affinis*/*Balanus improvisus*/*Crassostrea virginica* Water >> phytoplankton >> *E. affinis*/*B. improvisus*/*C. virginica*	Sanders *et al.*, 1989
Sediment >> bivalve *Macoma baltica* > water Sediment >> ragworm *Nereis succinea* >> water Sediment >> nemertime *Micrura leidyi* >> water	Riedel *et al.*, 1990
Water > shrimps *Artemia* sp./*Palaemonetes pugio* Water >> phytoplankton > *Artemia* sp.> *P. pugio*	Lindsay and Sanders, 1990
Cadmium	
Water/diet >> euphausiid *Meganyctiphanes norvegica* >> faecal pellets	Benayoun *et al.*, 1974
Water > fish *Pleuronectes platessa*/*Raja clavata* Worm *Nereis diversicolor* > *P. platessa*/*R. clavata*	Pentreath, 1977a
Water >> *Pseudodiaptomus coronatus* Phytoplankton sp. > *P. coronatus*	Sick and Baptist, 1979
Water >> *Carcinus maenas* – *Artemia salina* > C. maenas	Jennings and Rainbow, 1979
Water >> sea skater *Halobates sericeus* >> birds	Cheng *et al.*, 1984
Water >> *Mytilus edulis* – *Phaeodactylum tricornutum* > *M. edulis*	Riisgaard *et al.*, 1987
Water >> snail *Littorina littorea* *Fucus*/*Ulva*/*Porphyra* spp. > *L. littorea*/*Gibbula umbilicalis*	Amiard-Triquet *et al.*, 1987
Caesium	
Water/food ~ fish *Pleuronectes platessa* (diet provides 50%) Water/food ~ fish *Raia clavata* (diet provides 80%)	Jeffries and Hewett, 1971
Chromium	
Water >> *Crassostrea virginica* – *Chlamydomonas* sp. > *C. virginica*	Preston, 1971
Water >> barnacle *Balanus* sp.	Weerelt *et al.*, 1984
Water >> crab *Xantho hydrophilus* – Diet > *Xantho hydrophilus*	Peternac and Legovic, 1986
Water >> *Mytilus edulis*	Chassard-Bouchard *et al.*, 1989

Table 12.1 *Continued*

Food chain	Reference
Water > sea urchin larvae *Pseudechinus novaezealandiae* Water >> alga *Dunaliella primolecta* >> *P. novaezealandiae*	Bremer *et al.*, 1990
Microbial Cr >> bivalves *Potamocorbula amurensis*/*M. balthica*	Decho and Luoma, 1991

Cobalt

Food chain	Reference
Starch pellets >> fish *Pleuronectes platessa* Water > *P. platessa* – *Nereis diversicolor* > *P. platessa*	Pentreath, 1973b
Diatom *Navicula* sp > bivalve *Scrobicularia plana* > *C. maenas* > rat *Rattus rattus* (successive trophic levels show less efficient transfer)	Amiard-Triquet and Amiard, 1975
Water > shrimp *Crangeon crangon* – *M. edulis* >> *C. crangon*	Weers, 1975
Water > *S. plana*/*C. maenas* Diatom *Navicula* sp. >> *S. plana* Worm *Arenicola* sp. >> *C. maenas*	Amiard-Triquet and Amiard, 1976
Duraliella bioculata >> *M. edulis* >> *C. maenas*	Kirchmann *et al.*, 1977
Diatom *Nitzschia closterium* >> clam *Mercenaria mercenaria*	Nakahara and Cross, 1978

Copper

Food chain	Reference
Water >> isopod *Ligia oceanica*/amphipod *Orchestia gammarella*	Weiser, 1967
Polychaete *Nereis diversicolor* >> *Carcinus maenas* Alga sp. > *L. oceanica*/*O. gammarella*	Bryan, 1976a
Water >> *Crassostrea gigas* – Diet > *C. gigas*	Amirad-Triquet *et al.*, 1988, 1992; Amiard *et al.*, 1989
Sedimentation/diatom *Haslea ostrearia* > *C. gigas*	Ettajani *et al.*, 1992

Iron

Food chain	Reference
Water > fish *Pleuronectes platessa* *Nereis diversicolor*/starch pellets >> *P. platessa*	Pentreath, 1973b
Water > *Littorina obtusata*	Young, 1975
Water >> *Fucus serratus* >> *L. obtusata*	
Water > whelk *Nucella lapillus* Barnacle *Balanus balanoides* >> *N. lapillus*	Young, 1977

Lead

Food chain	Reference
Water >> *Mytilus edulis* – *Dunaliella marina* >> *M. edulis*	Schulz-Baldes, 1974

Table 12.1 *Continued*

Food chain	Reference
Fucus/Ulva/Porphyra spp. > *L. littorea/G. umbilicalis*	Amiard-Triquet *et al.*, 1987
Water >> *Crassostrea gigas* – *Diatom* > *C. gigas*	Amiard-Triquet *et al.*, 1988; Amiard *et al.*, 1989

Manganese

Water > lobster *Homarus vulgaris* – Diet >> *H. vulgaris*	Bryan and Ward, 1965
Nereis diversicolor >> *Pleuronectes platessa* – Water > *P. platessa*	Pentreath, 1973a
L. littorea/Chlamys opercularis > *Nassarius reticulatus*	Nott and Nicolaidou, 1990, 1993

Mercury

Large fish/cephalopods >> seals *Halichoerus grypus/Phoca vitulina/Cystophora cristata*	Sergeant and Armstrong, 1973
Water >> alga *Croomonas salina* > copepod *Acartia tonsa* (HgCl$_2$)	Parrish and Carr, 1976
Nereis diversicolor >> *Pleuronectes platessa* (organic Hg) *Nereis diversicolor* > *P. platessa* (inorganic Hg) Starch and gelatin pellets >> *P. platessa* (organic Hg) Starch and gelatin pellets > *P. platessa* (inorganic Hg)	Pentreath, 1976
Water >> *Bacillus licheniformis* >> mosquito larvae *Aedes aegypti* > small fish *Lebistes reticulatus* > large fish *Cichlasoma facetum*	Hamdy and Prabhu, 1979
Food chain >> large fish	Bryan, 1979 (Review)
Water >> *Crangon crangon* (organic Hg) Water > *Crangon crangon* (inorganic Hg) *M. edulis* >> *C. crangon* (organic Hg) *M. edulis* > *C. crangon* (inorganic Hg)	Riisgaard and Famme, 1986
M. edulis >> starfish *Leptasterias polaris*	Pelletier and Larocque, 1987
Alga >> *M. edulis* >> *Platichthys flesus* (organic Hg)	Riisgaard and Hansen, 1990

Plutonium

Water > shrimp – Water >> mussels/worms	Fowler *et al.*, 1975
Water >> seaweed	Hetherington *et al.*, 1975

Table 12.1 *Continued*

Food chain	Reference
Polychaete >> *Carcinus maenas*/*Cancer pagurus*	Fowler and Guary, 1977
Food chain > fish	Guary and Frazier, 1977
Polonium	
Diet >> Euphasiid *Meganyctiphanes norvegica*	Heyraud *et al.*, 1976
Selenium	
Water > *Meganyctiphanes norvegica* – Diet >> *M. norvegica*	Fowler and Benayoun, 1976
Silver	
Alga *Isochrysis* sp. >> *C. virginica* (gut) >> faeces	Abbe and Sanders, 1988
Water >> phytoplankton	Abbe *et al.*, 1988; Abbe and Sanders, 1990
Water >> *C. virginica* – Phtoplankton > *C. virginica*	Sanders *et al.*, 1990
Suspended sediments > *C. virginica*	Abbe and Sanders, 1990
Water >> *C. gigas*/*Mytilus galloprovincialis*/*Chlamys varia* Alga > *C. gigas*/*M. galloprovincialis*/*C. varia*	Metayer *et al.*, 1990
Water >> grass shrimp *Palaemonetes pugio* Water >> diatom *Thalassiosira weissflogii* Plankton *Artemia* sp. > *P. pugio* Detritus *Victorella* sp. > *P. pugio*	Connell *et al.*, 1991
Water >> *Crassostrea gigas* Sediment/diatom *Haslea ostrearia* > *C. gigas*	Ettajani *et al.*, 1992; Amiard-Triquet *et al.*, 1992
Vanadium	
Water >> tunicate *Ciona intestinalis*	Kustin *et al.*, 1975
Water >> *Mytilus galloprovincialis* (shell/byssus) *Dunaliella marina* > *M. galloprovincialis*	Miramand *et al.*, 1980
Water >> *Nereis diversicolor* > *Carcinus maenas* Water >> *C. maenas*	Uensal, 1982
Zinc	
Water/diet >> *Homarus vulgaris*	Bryan, 1964

Table 12.1 *Continued*

Food chain	Reference
Water > flounder *Paralichthys* sp. *A. salina* >> *Paralichthys* sp.	Hoss, 1964
Water/diet >> *C. maenas*/*H. vulcaris*	Bryan, 1967
Water/diet >> mussels	Laumond *et al.*, 1973
Water > *Pleuronectes platessa* *Nereis diversicolor* >> *P. platessa*	Pentreath, 1973a
Water > *Nucella lapillus* *Balanus balanoides* >> *N. Lapillus*	Ireland, 1973; Young, 1977
Water >> *Lysmata seticaudata*/*C. maenas*/*Gobius* sp. *A. salina*/*M. edulis* > *Lysmata seticaudata*/*C. maenas*/*Gobius* sp.	Renfro *et al.*, 1975
Water > winkle *Littorina obtusata* Alga *Fucus serratus* >> *L. obtusata*	Young, 1975
Detritus >> periwinkle *Littorina irrorata* (parenteral uptake)	Ireland, 1983
Water >> alga *Chlamydomas* >> *Artemia* sp. >> fish Water > fish *G. affinis*/*L. xanthurus*	Willis and Sunda, 1984
L. littorea/*Chlamys opercularis* > *Nassarius reticulatus* Barnacle *Balanus balanoides* >whelk *Nucella lapillus*	Nott and Nicolaidou, 1990, 1993
Multiple metals	
Water (Cr Cu Pb Hg Zn) >> phytoplankton	Laumond *et al.*, 1973
Water (Zn Mn Co Fe) > fish *Raja clavata*	Pentreath, 1973c
Metals in prey are in inert compartments and are not available to predators (transfer via diet is inefficient)	Hodson, 1980
Annelids/crustaceans (Cd Pb Cu) > (Zn) >> fish (5 species)	Metayer *et al.*, 1980
Literature review	Baudo, 1981, 1985
Literature review	Fowler, 1982
Microbial exopolymers (Cd Zn Ag) >> *Macoma balthica*	Harvey and Luoma, 1985
Intertidal invertebrates > birds (metals recycled – not retained)	Evans *et al.*, 1987
Literature review	Dallinger *et al.*, 1987
Diet (Cu Pb) > mussels/oysters	Amiard, 1988
Cerithium vulgatum (Mn Fe Co Ni Zn) > *Murex trunculus*	Nott and Nicolaidou, 1989b

Table 12.1 *Continued*

Food chain	Reference
Microbial exopolymers >> microbial consumers	Decho, 1990
Diet >> sharks	Marcovecchio *et al.*, 1991
Gastropods (Mn Ni Cu Zn Ag) > (Cr Cd) >> hermit crab	Nott and Nicolaidou, 1994

From subjective assessments of data and statements in the quoted literature, different transfer stages are marked as either relatively efficient (>>) or inefficient (>) for metal transport. Organisms separated by a slash (/) are on the same trophic level.

Balanus improvisus and oyster *Crassostrea virginica* do not take As from water but they do take it from phytoplankton (Sanders *et al.*, 1989). Phytoplankton is more efficient at taking As from seawater than invertebrates are from the diet (Table 12.1).

Further along the food chain from the primary producers, the larger zooplankton organisms, macroinvertebrates and higher trophic levels have tissues that are differentiated. Metal is taken up by permeable epithelia of the gut and gills and, internally, it is transported, metabolized, stored and excreted by other specialized tissues (Bryan, 1976a; Fowler, 1982; Baudo, 1985; Viarengo and Nott, 1993; Chapters 8–10). Within these tissues, metals are compartmentalized in particular cells and organelles. Highest concentrations can occur in storage tissues that are 'glandular' and they range from the digestive gland/hepatopancreas in crustaceans and molluscs (reviewed in Viarengo and Nott, 1993) to the pyloric caeca in starfish (Pelletier and Larocque, 1987) and the liver in fish (Maage *et al.*, 1991). These compartments can account for widely differing proportions of the total body weight (Figure 12.1).

The weight of the glandular tissue expressed as a percentage of the total soft body weight is termed the hepatosomatic index. The index can be higher than 10% in invertebrates and much lower in higher animals, especially bony fish, where it can be less than 1%. In wild Atlantic salmon the liver contains much higher levels of Cu and Se than any other tissue and high levels of Fe are confined to the liver, spleen and kidney (Maage *et al.*, 1991). This compartmentalization has implications for food chains because, for example, predators that consume salmon flesh without the offal will avoid a dietary intake of metals.

Whole-body analyses of fish and other animals by atomic absorption spectroscopy (AAS) do not reflect high concentrations of metals in the liver and pancreas, which can disrupt normal biochemical processes in these tissues. Also, once saturation of a storage system occurs, overspill into other compartments can disrupt enzyme systems and produce toxic effects, without pro-

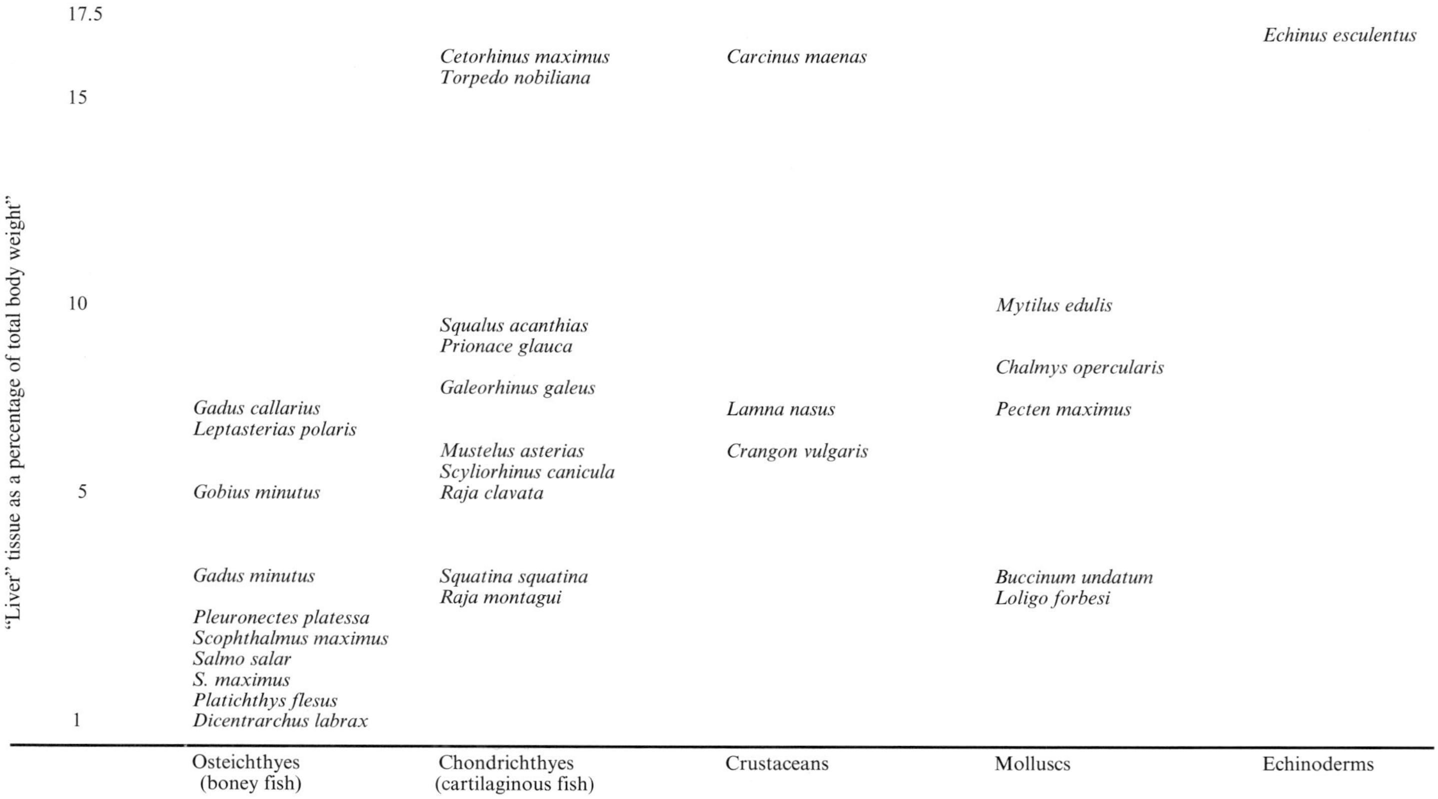

Figure 12.1 Marine animals arranged according to phylum and hepatosomatic index. The index is the weight of hepatic tissue as a percentage of total body weight. Tissues include liver (fish), hepatopancreas/digestive gland (crustaceans and molluscs), pyloric caecum (*Leptasterias*) and intestine (*Echinus*). In crustaceans and molluscs only soft tissues are included in the calculation.

ducing any significant increase in whole-body analyses (Engel and Fowler, 1979; Jenkins and Brown, 1984; Roesijadi, 1992; Chapters 8 and 10).

12.4 BIOMAGNIFICATION

In a summary of food chain work, Bryan (1979) concluded that absorption from food is often the most important route for metal bioaccumulation and transfer along food chains, but there is little evidence that predators at high trophic levels will contain the highest concentrations. Concentration factors for Pu in the field, for example, fall from values of the order of 1000 in seaweeds to 1–100 in fish (Hetherington *et al.*, 1975; Guary and Frazier, 1977). Caesium in fish is an exception in that a high degree of assimilation from prey results in magnification along the food chain (Jefferies and Hewett, 1971; Pentreath, 1977b). Of 18 metals considered by Bryan (1976b) in various organisms, Hg is one of the few where mean levels in fish exceed those in phytoplankton or seaweed as measured on a dry weight basis.

Mercury is rarely amplified between invertebrates and small fish (Knauer and Martin, 1972; Leatherland *et al.*, 1973) but it is sometimes amplified in large fish, where there are effects linked to both trophic level (Ratkowsky *et al.*, 1975) and the age of the animals (Cross *et al.*, 1973; Chapter 5). In marine mammals both factors are important, together with differences in diet (Sergeant and Armstrong, 1973). More Hg occurred in the grey seal *Halichoerus grypus*, which eats large fish and cephalopods, than in the harp seal *Pagophilus groenlandicus*, which eats small fish and crustaceans. Most of the Hg in fish is methylated and it continues to be so in seals except in the liver and kidney. It is suggested that these tissues have the capacity to demethylate the metal (Freeman and Horne, 1973). Mercury, Cu, Zn, Pb and Cr in two different food chain experiments (seawater–plankton–fish, and seawater–phytoplankton–mussel) all had reduced concentration factors at higher trophic levels (Laumond *et al.*, 1973).

In a food chain consisting of the alga *Dunaliella bioculata*, the bivalve *Mytilus edulis* and the decapod *Carcinus maenas*, ^{60}Co is mostly concentrated in the primary producer (Kirchmann *et al.*, 1977). Similarly, in another food chain involving Ag in solution, on suspended particles and in phytoplankton, the major pathway to oysters was directly from the water (Abbe and Sanders, 1990). In a longer chain with bacteria, mosquito larvae, small fish and large fish, Hg was magnified within the first two stages but not the last two (Hamdy and Prabhu, 1979; Table 12.1).

12.5 BIOAVAILABILITY

Amiard (1988) suggested that phytoplankton fixes metals and makes them unavailable to oysters; Amiard-Triquet *et al.* (1992) suggested that speciation of Cu and Ag may determine uptake from the diet (other examples are discussed in Chapter 8). The food pathway can be significant for Cu uptake by oysters but only as a consequence of long-term exposure (Ettajani *et al.*, 1992). The importance of accumulation directly from water is due to the fact that Ag and Cu can be absorbed through the mantle, gill and gut whereas food is only absorbed via the gut (Amiard *et al.*, 1989). This probably also applies to Cd and Pb (Amiard-Triquet *et al.*, 1987) where there is no biomagnification between algae and grazing gastropods. Connell *et al.* (1991) found that the grass shrimp *Palaemonetes pugio* rapidly incorporates Ag dissolved in brackish water but not from dietary sources in planktonic and detrital organisms. They also found that Ag is taken up by the diatom *Thalassiosira weissflogii*, where it remains bound to membranes even after the cells are disrupted by sonication, leached at low pH and treated with digestive enzymes. They conclude that acts of feeding by invertebrates are not likely to dislodge silver from biotic particles.

In the Mediterranean, three species of marine snail living in the same seawater adjacent to a nickel smelting plant accumulate markedly different levels of metals (Nott and Nicolaidou, 1989a). The highest levels occur in the sediment feeder *Cerithium vulgatum* and the lowest levels in the predator *Murex trunculus*, which preys on *C. vulgatum*. It has been established that metals in *C. vulgatum* occur in the digestive gland, where they are accumulated within intracellular phosphate granules and residual lysosomes. The metals are unavailable to the animal in the sense that they are insoluble and within membrane-limited compartments at high concentration. It is proposed that when the digestive glands are consumed by the carnivore the metals remain insoluble and unavailable in the gut and that the detoxification system operating in *C. vulgatum* also protects the carnivore *M. trunculus* (Nott and Nicolaidou, 1989b).

12.6 BIOAVAILABILITY AND ELECTROSTATIC BINDING

The bioavailability hypothesis arising from observations on gastropods in the Mediterranean was tested experimentally on different species at Plymouth, Devon (Nott and Nicolaidou, 1990). Three animals were selected which produce intracellular phosphate granules with sequestered heavy metals (Figure 12.2): the bivalve scallop *Chlamys opercularis* has metal/phosphate granules in the kidney, the gastropod periwinkle *Littorina littorea* has granules in the digestive gland and the barnacle *Balanus balanoides* has granules in the gut parenchymous tissue. Kidney from the scallop and digestive gland from the periwinkle were fed to the carnivorous whelk *Nassarius reticulatus*, and barnacles were grazed on naturally by another carnivorous whelk, *Nucella lapillus*. Granule-containing tissues from the prey and faecal pellets from the predators were prepared for electron microscopy; individual phosphate granules were probed with the beam and metals were detected by X-ray microanalysis (XRMA).

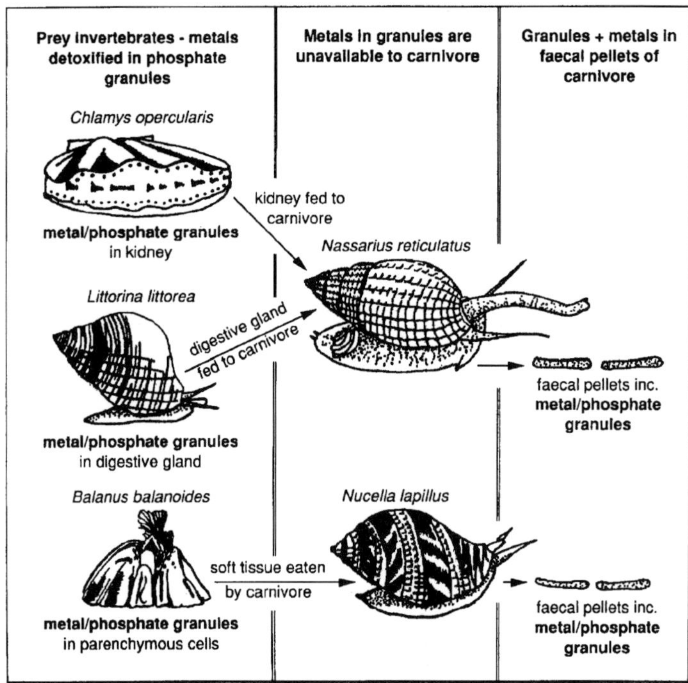

Figure 12.2 Predator/prey food chains. Metals are incorporated in intracellular phosphate granules in the prey (left column). Soft tissues containing the granules are ingested by the carnivore (centre column). The tissues are digested by the carnivore and the residues, including granules, are egested in the faecal pellets (right column). (Reproduced with permission from Nott and Nicolaidou, 1990.)

It was shown (Nott and Nicolaidou, 1990) that metal-containing granules produced in the tissues of the prey still contained the same metals after passing through the gut of a carnivore. This indicates that the detoxification system of the prey also protects the predator by rendering the metals unavailable to its digestive processes. However, comparison of Figures 12.3a and 12.3b shows that the digestive processes in the gut of the carnivorous whelk *N. reticulatus* removed most of the Cl and K from the granules produced by the winkle *L. littorea*. Also, the increase in size of the Ca peak indicates that some Mg and P have been lost. The granules produced by the scallop *C. opercularis* contained significant amounts of Mn and Zn, and those produced by the barnacle *B. balanoides* contained Zn. Both metals remained in the granules after passing through the gut of the predator.

Some *L. littorea* were kept for 16 days in aerated seawater containing 1 mg Zn l^{-1} and others in seawater containing 1 mg Mn l^{-1}. After the treatments the digestive glands were fed to *N. reticulatus*. Granules contained Zn (Figures 12.4a,b) and Mn both before and after passage through the gut of the whelk.

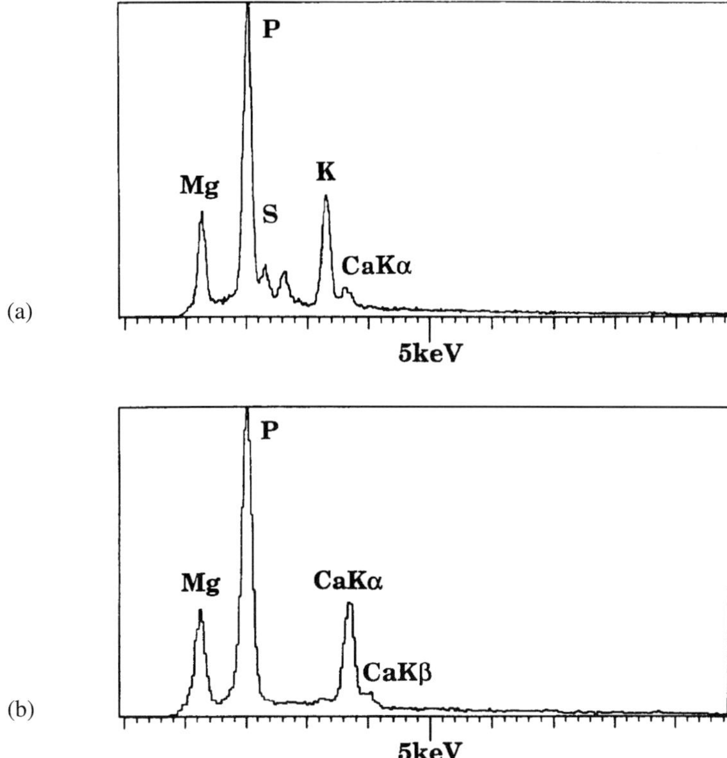

Figure 12.3 X-ray microanalytical spectra of (a) a phosphate granule in the digestive gland of the prey *Littorina littorea*, where it was formed, and (b) a similar granule in the faecal pellet of the carnivore *Nassarius reticulatus* after passing through the gut. Full vertical scale, 2000 X-ray counts; horizontal scale, X-ray energy. (Reproduced with permission from Nott and Nicolaidou, 1990.)

However, when tissues of a marine snail containing Cr were fed to a hermit crab this metal was transferred to the digestive gland of the crab, but Mn, Ni, Cu, Zn and Ag were not transferred (Nott and Nicolaidou, 1994). Also, sea urchin larvae absorbed more Cr from an algal diet than from seawater (Bremer *et al.*, 1990) although the metal was in trivalent form and may not have been as available as the hexavalent form. Uptake of Cr by the crab *Xantho hydrophilus* was more rapid from water than from food and in both cases it was the hexa- valent form that was absorbed internally (Peternac and Legovic, 1986): once in the tissues it was reduced to the trivalent form and complexed with organic molecules. Barnacles (*Balanus* spp.) could accumulate hexavalent Cr from sea- water but not trivalent Cr (Weerelt *et al.*, 1984). Uptake of trivalent Cr by *Mytilus edulis* from seawater was via the gills, with storage as insoluble forms in lysosomes of the kidney; the digestive gland was not involved in storage (Chassard-Bouchaud *et al.*, 1989, 1991). Chassard-Bouchaud *et al.* (1989) have

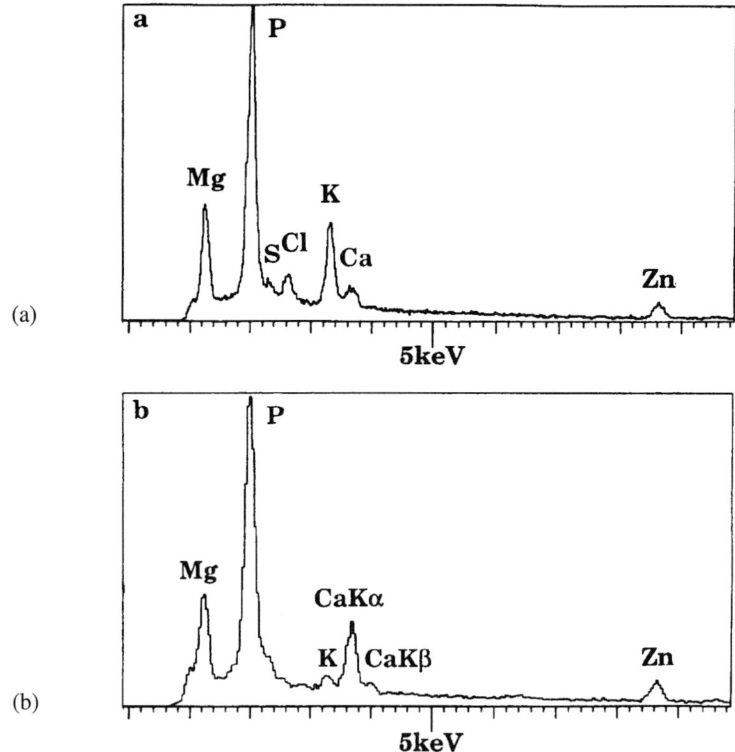

(a)

(b)

Figure 12.4 X-ray microanalytical spectra of (a) a phosphate granule in the digestive gland of *Littorina littorea* after dosing with zinc and (b) a similar *L. littorina* granule in the faecal pellet of *Nassarius reticulatus* after passing through the gut. (Reproduced with permission from Nott and Nicolaidou, 1990.)

carried out extensive XRMA work on metals in aquatic animals and conclude that Cr is metabolized and transported differently from most toxic metals. The oyster *Crassostrea virginica* takes up Cr by direct absorption but ingested material provides the primary source (Preston, 1971).

12.7 BIOAVAILABILITY AND COVALENT BINDING OF METALS

The reactivity of metals, which determines their biochemical reactions within marine organisms, is determined by their electronic configuration (Chapters 1 and 5). The configurations of soft acid, post-transition metals determine that they can bind covalently by sharing electrons. This has important biological implications for detoxification and food chain transfer. Cadmium binds covalently and shows markedly different biochemical processing from that of the transition metals when they are bound electrostatically as phosphates (Viarengo and Nott, 1993).

Cadmium accumulates continuously in marine gastropods at a rate that reflects the concentration in seawater. Furthermore, when animals with accumulated Cd are returned to clean seawater, the metal is not easily excreted. Indeed, *Littorina littorea* that contain Cd lose only insignificant amounts after six months in clean seawater. The question is: how does the high reactivity of the metal relate to its retention in *L. littorea* (Langston and Zhou, 1987; Langston *et al.*, 1989; Chapter 8) and crustaceans (Rainbow, 1988; Rainbow and White, 1989; Chapter 9)? An explanation of retention by gastropods is described in Nott *et al.* (1993) and by decapods in Nott and Nicolaidou (1994). After transfer between tissues in *L. littorea* the metal is finally accumulated in the digestive gland (Langston and Zhou, 1987; Bebianno *et al.*, 1992), where it is released periodically into the lumen when the digestive cells disintegrate (Figure 12.5). Cadmium occurs as a soluble and labile element in the cytosol and is reabsorbed by the digestive epithelium; other metals, which are bound as insoluble compounds and enclosed in membrane-bound vesicles, are excreted. In decapods there is a similar situation where cells of the digestive epithelium disintegrate into the lumen of the hepatopancreas at the end of a cycle of digestion (Hopkin and Nott, 1980; Al-Mohanna and Nott, 1987). Metals in phosphate granules and lysosomes are lost in the faeces but Cd is reabsorbed (Nott and Nicolaidou, 1994) and remains in the food chain.

12.8 ELECTROSTATIC BINDING VERSUS COVALENT BINDING

Effects of the type of chemical binding of metals on food chain transfer have been investigated by XRMA (Nott and Nicolaidou, 1994). In marine invertebrates, accumulations of hard acid and transition metals plus zinc can be bound electrostatically as insoluble phosphate whilst the soft acid, post-transition metals can be bound covalently to high-sulphur ligands, particularly metalloproteins. Metal phosphates occur in the digestive gland of the sediment-feeding snail *Cerithium vulgatum* and post-transition, metal/sulphur compounds in the digestive gland of the carnivorous snail *Murex trunculus*. A third, herbivorous snail *Monodonta mutabilis* does not accumulate transition and post-transition metals (Figure 12.6). All three species accumulate Mg, K and Ca within phosphate granules. Concentrations of metals in the digestive glands were measured by AAS and localizations and associations with phosphate and sulphur were determined by XRMA.

In *Cerithium vulgatum*, Mn, Fe, Co, Ni and Zn were associated with phosphate granules in the digestive epithelium, and in *Murex trunculus* Cu was associated with sulphur as aggregates within pore cells of the connective tissue in the digestive gland (Bouquegneau and Martoja, 1982; Bouquegneau *et al.*, 1984; Mason *et al.*, 1984). Cd, Ag and Cr did not occur in these accumulations. Glands were dissected from the snails and fed to the hermit crab *Clibanarius erythropus*. Digestive glands from the crabs were analysed by AAS before and after the experiment for levels of Cr, Mn, Ni, Cu, Zn, Ag and Cd.

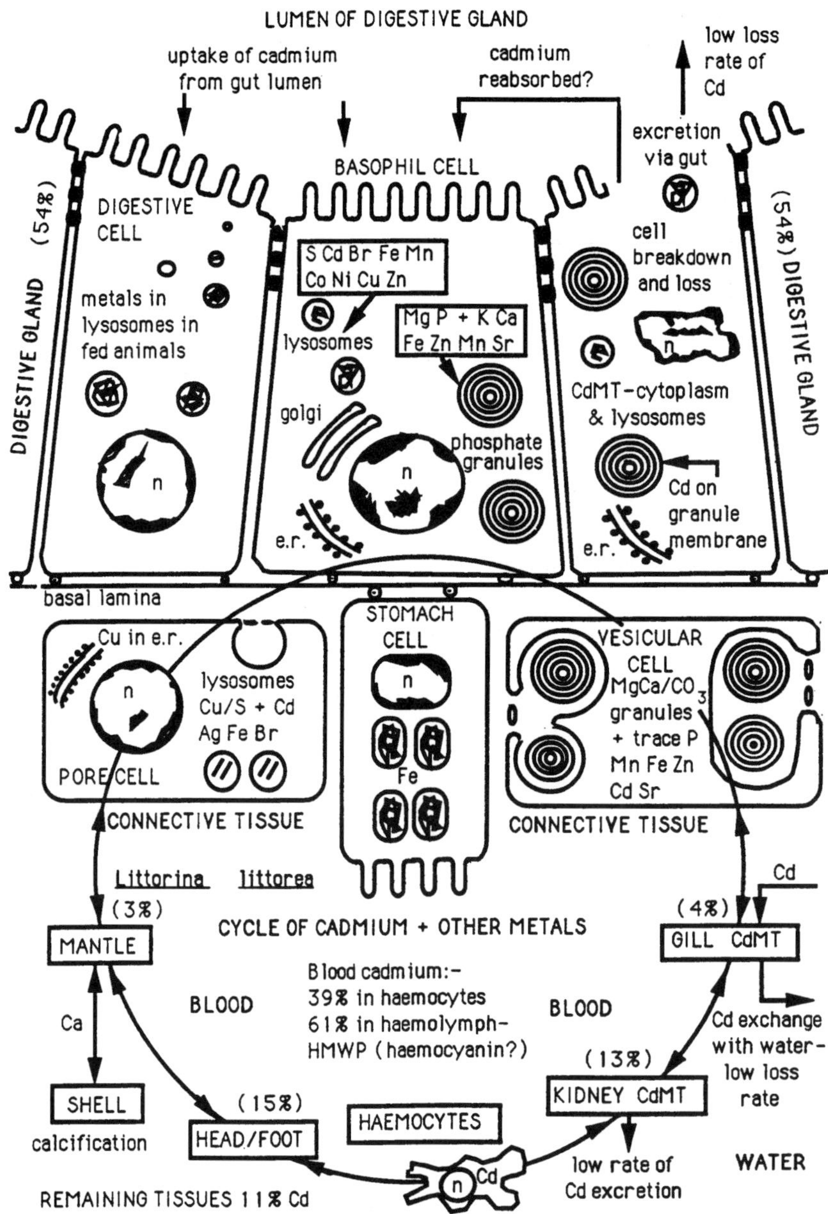

Figure 12.5 Schematic diagram of a gastropod showing sites of uptake, translocation, storage and excretion of cadmium. The metal is mobile between tissues but excretion rates are minimal. Percentages show proportions of total cadmium in tissues. CdMT, cadmium-metallothionein complex; er, endoplasmic reticulum; HMWP, high molecular weight protein; n, nucleus. (Reproduced with permission from Nott *et al.*, 1993.)

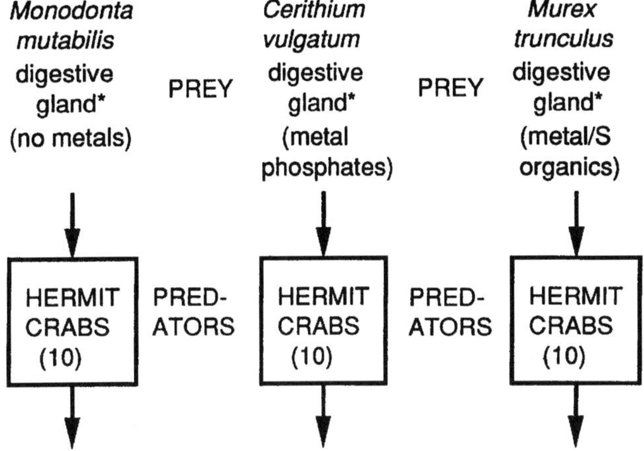

Figure 12.6 Food chains involving metals bound to phosphate and to sulphur. (Reproduced with permission from Nott and Nicolaidou, 1994.)

Phosphate granules extracted experimentally from the digestive gland of *Cerithium vulgatum* and the same granules after passage through the gut of the hermit crab, produced similar XRMA spectra except for loss of Cl and K (Figure 12.7a,b). Likewise, Cu–S granules from pore cells in *Murex trunculus* and the same granules eaten and then voided by the crab both produced similar XRMA spectra except for loss of Mg, Cl and K and addition of Si and Fe (Figure 12.8a,b).

Thus, both forms of binding prevent transfer along a food chain when the metal is within either a mineral granule or a membrane-bound accumulation. The dispersed metals, Cd and Cr, were transferred to the crab digestive gland whereas Ag was not. This could reflect different solubilities based on covalent binding for Ag and Cd and electrostatic binding for Cr. The situation confirms the work of others who found that Ag is mainly taken up from seawater rather than the diet, wherein it is tightly bound to biotic particles and membranes; that Cd can be taken up from seawater and the diet, albeit less efficiently; and that Cr is taken up from the diet, and from seawater according to its valency (for references, see Table 12.1).

12.9 BIPHASIC DIGESTION

In crustaceans (Hopkin and Nott, 1980; Icely and Nott, 1992) and bivalve molluscs (Bayne and Newell, 1983; Chapter 8) ingested food is sorted into

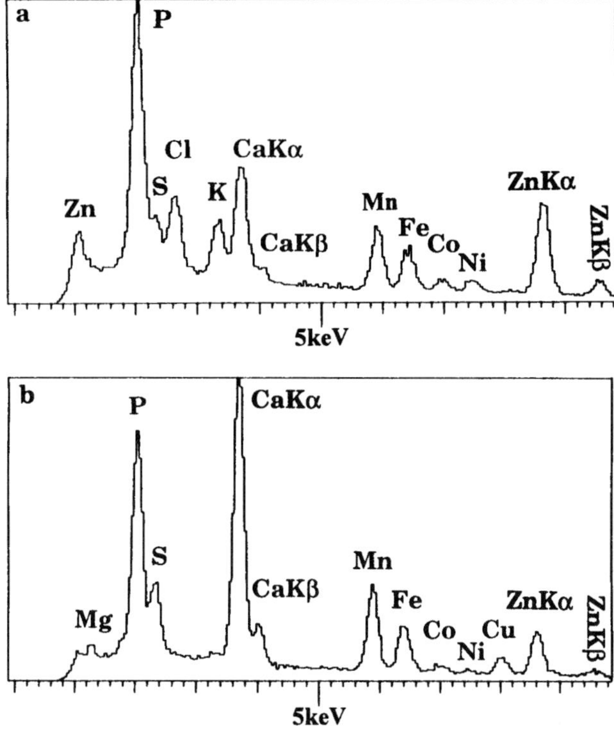

Figure 12.7 X-ray microanalytical spectra of (a) a phosphate granule containing a range of metals in the digestive gland of the gastropod *Cerithium vulgatum* and (b) a similar *C. vulgatum* granule in the faecal pellet of a hermit crab after passing through the gut. (Reproduced with permission from Nott and Nicolaidou, 1994.)

digestible and indigestible components in the stomach. The digestible fraction is transferred into a major diverticulum called the hepatopancreas or digestive gland, where digestion and absorption take place; the indigestible fraction continues along the gut via the intestine and hindgut for elimination in faecal pellets. This biphasic process also takes place in gastropod molluscs and is obvious in the sediment-feeder *Cerithium vulgatum*; particles of silica, clay and calcium carbonate pack the fore-, mid- and hindgut and faecal pellets but the digestive gland contains only organic material that has been separated from the sediment (Nott and Nicolaidou, 1989a)

C. vulgatum in polluted situations accumulates levels of zinc approaching 3500 μg g^{-1} dry weight in the viscera (assumed to be similar to digestive gland) compared with levels that are two orders of magnitude less in the muscle tissue (Nicolaidou and Nott, 1989, 1990). The level of zinc in the digestive gland can reach more than 15 500 μg g^{-1} (Nott and Nicolaidou, 1989a). If the weight of the digestive gland is *c.* 5–10% of the total body weight

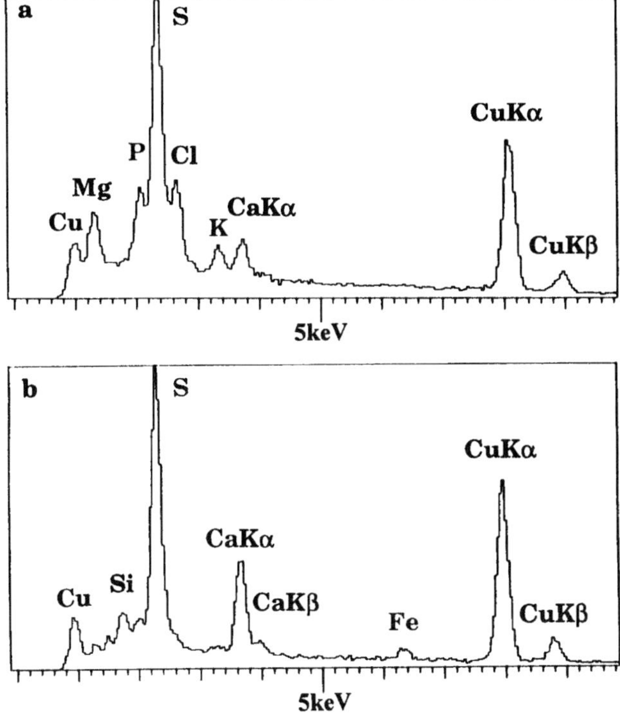

Figure 12.8 X-ray microanalytical spectra of (a) a copper/sulphur in the digestive gland of the gastropod *Murex trunculus* and (b) a similar *M. trunculus* granule in the faecal pellet of a hermit crab after passing through the gut. (Reproduced with permission from Nott and Nicolaidou, 1994.)

(Figure 12.1) the metals in the gland could make a significant contribution to the metal transfer in the food chain.

When Cr bound to bacteria was fed to the bivalves *Potamocorbula anurensi* and *Macoma balthica*, minimal metal was absorbed from the dietary material which was passed straight through the fore-, mid- and hindgut (Decho and Luoma, 1991). However, nearly all the chromium that was passed into the digestive gland for intracellular digestion was absorbed and retained in this tissue. *P. amurensis* absorbed 67% of the Cr in the diet, as opposed to 32% by *M. balthica,* because in the former 75–90% of the ingested metal was subjected to digestion in the gland whereas in the latter the corresponding proportion was 40% (Chapter 8).

This indicates that in animals with the biphasic digestion process, metals associated with the mineral component of the diet will probably pass straight through the gut without absorption. Metals associated with the organic component which is diverted to the digestive gland have an increased chance of absorption. This could account for the increased bioavailability of Ag, Cd and Zn from particles with adsorbed bacterial extracellular polymers by comparison with unamended particles (Harvey and Luoma, 1985). Trivalent Cr is accumulated from sediment by aerobic heterotrophic bacteria (Aislabie and Loutit, 1986); the metal occurs in the extracellular polymers of the bacteria, where it is associated with the polysaccharide fraction of the material. This could be a mechanism of tolerance to Cr and could facilitate entry to the food chain.

12.10 METAL METABOLISM

Metals are available to food chains from seawater and the diet and it is obvious that concentrations of metals in dietary organic material, both living and detrital, are not necessarily similar to concentrations in seawater. Indeed, processes of metabolism establish that both bound and ionic concentrations of metals in tissues can deviate substantially above those in seawater by several orders of magnitude. Also, metabolic processes tend to change the relative proportions of elements compared with environmental levels. In seawater the concentration of Mg, measured as millimoles per kilogram, is 5.2 times higher than Ca, and in *Littorina littorea* the relative concentrations in the blood are similar (Rumsey, 1973). In crustacea the concentrations of Mg and Ca in the blood vary in different species (Burton, 1967).

Metabolic processes rather than inorganic processes control the Mg : Ca ratio in low-magnesium calcite in the shell of *Mytilus edulis* (Lorens and Bender, 1977), which has 120 times less Mg than is expected by precipitation. Under normal conditions *M. edulis* physiologically excludes Mg from its shell-forming fluid but when the concentration of Mg is artificially increased there is a substantial increase in the amount of co-precipitated Mg. In intracellular phosphate granules of molluscs and crustaceans the Mg : Ca ratio is variable and it can be affected by food chains. When the granules are consumed by predators magnesium is extracted by digestive processes in the gut (Figure 12.3). Also, determination of the type of respiratory biochemistry, whether it is based on haemocyanin or haemoglobin, is linked to a type of organism, not to the availability of Cu or Fe, respectively, in seawater. This difference in prey animals controls the relative availabilities of Cu and Fe to food chains. Other metals are required as coenzyme factors and these, like Zn in carbonic anhydrase, are produced despite fluctuating environmental levels.

Thus, inorganic biochemical processes within cells can continue to operate despite elemental deficiencies and excesses in the environment. In addition, metals with similar binding characteristics – for instance, Hg, Cu, Cd and Zn (Viarengo and Nott, 1993) – interact with each other (Chapter 11). Competition for ligands within a tissue can affect the relative concentrations of different metals and affect the balance of metals in a food chain. Organisms that rely on seawater as a source of essential trace metals including V, Cr, Mn, Co, Ni, Cu and Zn must operate efficient methods for the uptake, accumulation and storage. When they are stored they could act as a convenient supply to predators, provided they are available.

In molluscs, cells that accumulate excess metals occur in the digestive glands of gastropods and the kidneys of bivalves and these are apparently not available to either the molluscs themselves or to predators. These cells have a short life cycle in that they form part of a system of cell division, growth, differentiation, maturity (functional state) and, finally, atrophy and disintegration. The inorganic biochemistry of metals is superimposed upon these continual cycles of cell replacement (Nott *et al.*, 1993; Nott and Nicolaidou, 1994) because they accumulate in functional cells and are lost during the disintegration phase. These metals are cleared from the molluscan gut in the faeces and return to the sediments.

In sediments, metals can be retained for some time within faecal pellets. Some species of gastropod, like *Cerithium vulgatum*, produce particularly durable pellets and these can absorb additional metals whilst in sediments (Nott and Nicolaidou, 1996). The longevity of pellets is affected by the type of sediment. Phosphate granules taken from gastropod digestive systems were incubated in sediments; magnesium phosphate granules without transition metals, from *Monodonta mutabilis*, quickly dissolved but calcium phosphate granules with additional metals, from *Cerithium vulgatum*, remained. The latter differentially retained or lost Mn, Fe, Co, Ni and Zn according to the type of sediment.

12.11 SUMMARY

Work with toxic organic compounds like DDT has demonstrated that biomagnification can occur along food chains. Work with metals shows the opposite effect (with the exception of Hg); concentration factors tend to be highest in the primary producers and other organisms which constitute the lower trophic levels. The reason for this is reduced bioavailability. Organisms counter the reactivity of metals and therefore potential toxicity by ligand binding and compartmentalization. The bound forms include insoluble phosphates and sulphur compounds and they are formed and accumulated within membrane-limited vesicles in specialized tissues like the liver and kidney. These metals are also unavailable to predators that consume the tissues because they are not absorbed by the digestive system. Therefore, they are not transferred along the food chain. Mercury is biomagnified, principally as methylmercury – presumably because there are no comparable biochemical processes for adequately reducing availability of lypophilic organomercury (Chapter 5).

ACKNOWLEDGEMENTS

The author is grateful for recent collaboration and assistance from Maria Bebianno, Gary Burt, Bill Langston, Linda Mavin, Artemis Nicolaidou and Keith Ryan.

REFERENCES

Abbe, G.R. and Sanders, J.G. (1988) Accumulation of silver from the alga *Isochrysis* sp. by the American oyster, *Crassostrea virginica*. *Journal of Shellfish Research* **7**, 192.

Abbe, G.R. and Sanders, J.G. (1990) Pathways of silver uptake and accumulation by the American oyster (*Crassostrea virginica*) in Chesapeake Bay. *Estuarine and Coastal Shelf Science* **31** 113–123.

Abbe, G.R., Sanders, J.G. and Bianchi, J.M. (1988) Pathways of silver accumulation by the American oyster (*Crassostrea virginica* Gmelin) in Chesapeake Bay. *Journal of Shellfish Research* **7**, 107.

Al-Mohanna, S.Y. and Nott, J.A. (1987) R-cells and the digestive cycle in *Penaeus semisulcatus* (Crustacea: Decapoda). *Marine Biology* **95**, 129–137.

Aislabie J. and Loutit, M.W. (1986) Accumulation of Cr(III) by bacteria isolated from polluted sediment. *Marine Environmental Research* **20**, 221–232.

Amiard, J.C. (1988) Les mécanismes de transfert des éléments métalliques dans les chaînes alimentaires aboutissant a l'huître et a la moule, mollusques filtreurs; formes chimiques de stockage, conséquences écotoxicologiques. *Océanis* **14**, 283–287.

Amiard, J.C., Amiard-Triquet, C., Ballan-Dufrançais *et al.* (1989) Study of the bioaccumulation at the molecular, cellular and organism levels of lead and copper transferred to the oyster *Crassostrea gigas* Thunberg directly from water or via food, in *Proceedings of the 21st European Marine Biological Symposium*, Gdansk, 1986 (eds R.Z. Klekowski, E. Styczynska-Jurewicz and L. Falkowski), Polish Academy of Sciences, Wroclaw, pp. 521–529.

Amiard-Triquet, C. and Amiard, J.-C. (1975) Etude expérimentale du transfert du cobalt 60 dans une chaîne trophique marine benthique. *Helgoländer wissenschaftliche Meeresuntersuchungen*, **27** 283–297.

Amiard-Triquet, C. and Amiard, J.-C. (1976) L'organotropisme du ^{60}Co chez *Scrobicularia plana* et *Carcinus maenas* en fonction du vecteur de contamination. *Oikos* **27**, 122–126.

Amiard-Triquet, C., Metayer, C., Amiard, J.C. and Berthet, B. (1987) Études in situ et expérimentales de l'écotoxicologie de quatre métaux (Cd, Pb, Cu, Zn) chez des algues et des mollusques gastéropodes brouteurs. *Water, Air and Soil Pollution* **34**, 11–30.

Amiard-Triquet, C., Amiard, J.C., Berthet, B. and Metayer, C. (1988) Field and experimental study of the bioaccumulation of some trace metals in a coastal food chain: seston, oyster (*Crassostrea gigas*), drill (*Oceanebra erinacea*). *Water Science and Technology* **20**, 13–21.

Amiard-Triquet, C., Martoja, R. and Marcaillou, C. (1992) Alternative methodologies for predicting metal transfer in marine food webs including filter-feeders. *Water Science and Techology* **25**, 197–204.

Baudo, R. (1981) Is analytically-defined chemical speciation the answer we need to understand trace element transfer along a trophic chain? in *Trace Element Speciation in Surface Waters and its Ecological Implications*, Proceedings of the NATO-AIOL Workshop, Genova-Nervi, Plenum Publishing Corporation, New York, pp. 275-290.

Baudo, R. (1985) Transfer of trace elements along the aquatic food chain. *Memorie dell'Istituto Italiano di Idrobiologia* **43**, 281–309.

Bayne, B.L. and Newell, R.C. (1983) Physiological energetics of marine molluscs, in *The Mollusca, Vol. 4, Physiology, Part 1*, (eds A.S.M. Saleuddin and K.M. Wilbur), Academic Press, New York, pp. 407–515.

Bebianno, M.J., Langston, W.J. and Simkiss, K. (1992) Metallothionein induction in *Littorina littorea* (Mollusca: Prosobranchia) on exposure to cadmium. *Journal of the Marine Biological Association of the UK* **72**, 329–342.

Benayoun, G., Fowler, S.W. and Oregioni, B. (1974) Flux of cadmium through euphausiids. *Marine Biology* **27**, 205–212.

Bouquegneau, J.M. and Martoja, M. (1982) La teneur en cuivre et son degré de complexation chez quatre gastéropodes marins. Données sur le cadmium et le zinc. *Oceanologica Acta* **5**, 219–228.

Bouquegneau, J.M., Martoja, M. and Truchet, M. (1984) Heavy metal storage in marine animals under various environmental conditions, in *Toxins, Drugs and Pollutants in Marine Animals*, (eds L. Bolis, J. Zadunaisky and R. Gilles), Springer-Verlag, Berlin, pp. 147–160.

Bremer, P.J., Barker, M.F. and Loutit, M.W. (1990) A comparison of the roles of direct absorption and phytoplankton ingestion in accumulation of chromium by sea urchin larvae. *Marine Environmental Research* **30**, 233–241.

Brown, B.E. (1982) The form and function of metal-containing 'granules' in invertebrate tissues. *Biological Reviews* **57**, 621–667.

Bryan, G.W. (1964) Zinc regulation in the lobster *Homarus vulgaris*. 1. Tissue zinc and copper concentrations. *Journal of the Marine Biological Association of the UK* **44**, 549–563.

Bryan, G.W. (1967) The metabolism of Zn and ^{65}Zn in crabs, lobsters and fresh-water crayfish, in *Radioecological Concentration Processes*, (eds B. Åberg and F.P. Hungate), Pergamon Press, Oxford, pp. 1005–1016.

Bryan, G.W. (1976a) Some aspects of heavy metal tolerance in aquatic organisms, in *Effects of Pollutants on Aquatic Organisms*. (ed. A.P.M. Lockwood), Cambridge University Press, pp. 7–34.

Bryan, G.W. (1976b) Heavy metal contamination in the sea, in *Marine Pollution*, (ed. R. Johnston), Academic Press, London and New York, pp. 185–302.

Bryan, G.W. (1979) Bioaccumulation of marine pollutants. *Philosophical Transactions of the Royal Society of London* **B286**, 483–505.

Bryan, G.W. and Ward, E. (1965) The absorption and loss of radioactive and non-radioactive manganese by the lobster, *Homarus vulgaris*. *Journal of the Marine Biological Association of the UK* **45**, 65–95.

Burton, R.F. (1967) Ionic balance in the crustacea. *Nature* **213**, 812–813.

Chassard-Bouchaud, C., Boutin, J.F., Hallegot, P. and Galle, P. (1989) Chromium uptake, distribution and loss in the mussel *Mytilus edulis*: a structural, ultrastructural and microanalytical study. *Diseases of Aquatic Organisms* **7**, 117–136.

Chassard-Bouchaud, C., Kleinbauer, F., Escaig, F. and Boumati, P. (1991) Fate and availability of trace metal pollutants in the mussel *Mytilus* sp. uptake and distribution at the cellular and subcellular levels, in *Mediterranean Action Plan Technical Report, Series number 59*, United Nations Environmental Programme, pp. 163–178.

Cheng, L., Schulz-Baldes, M. and Harrison, C.S. (1984) Cadmium in ocean-skaters, *Halobates sericeus* (Insecta), and in their seabird predators. *Marine Biology* **79**, 321–324.

Connell, D.B., Sanders, J.G., Riedel, G.F. and Abbe, G.R. (1991) Pathways of silver uptake and trophic transfer in estuarine organisms. *Environmental Science and Technology* **25**, 921–924.

Cross, F.A., Hardy, L.H., Jones, N.Y. and Barber, R.T. (1973) Relation between total body weight and concentrations of manganese, iron, copper, zinc and mercury in white muscle of bluefish (*Pomatomus saltatrix*) and a bathyl-demersal fish

Antimora rostrata. Journal of the Fisheries Research Board of Canada **30**, 1287–1291.

Dallinger, R., Prosi, F., Segner, H. and Back, H. (1987) Contaminated food and uptake of heavy metals by fish: a review and a proposal for further research. *Oceologia* **73**, 91–98.

Decho, A.W. (1990) Microbial exopolymer secretions in ocean environments: their role(s) in food webs and marine processes. *Oceanography and Marine Biology. An Annual Review* **28**, 73–153.

Decho, A.W. and Luoma, S.N. (1991) Time-courses in the retention of food material in the bivalves *Potamocorbula amurensis* and *Macoma balthica*: significance to the absorption of carbon and chromium. *Marine Ecological Progress Series* **78**, 303–314.

Edmonds, J.S. and Francesconi, K.A. (1981) Origin and chemical form of arsenic in the school whiting. *Marine Pollution Bulletin* **12**, 92–96.

Engel, D.W. and Fowler, B.A. (1979) Factors influencing cadmium accumulation and its toxicity to marine organisms. *Environmental Health Perspectives* **28**, 81–88.

Ettajani, H., Amiard-Triquet, C. and Amiard, J.-C. (1992) Etude expérimentale du transfert de deux éléments traces (Ag, Cu) dans une chaîne trophique marine: eau-particules (sédiment natural, microalgue) mollusques filtreurs (*Crassostrea gigas* Thunberg). *Water, Air and Soil Pollution* **65**, 215–236.

Evans, P.R., Uttley, J.D., Davidson, N.C. and Ward, P. (1987) Shorebirds (S.Os Charadrii and Scolopaci) as agents of transfer of heavy metals within and between estuarine ecosystems, in *Pollutant Transport and Fate in Ecosystems* (eds P.J. Coughtrey, M.H. Martin and M.S. Unsworth), Special Publication of the British Ecological Society, 6, pp. 337–352.

Fowler, S.W. (1982) Biological transfer and transport processes, in *Pollutant Transfer and Transport in the Sea*, Vol. 2, (ed. G. Kullenberg), CRC Press, Baton Rouge, pp 1–65.

Fowler, S.W. and Benayoun, G. (1976) Selenium kinetics in marine zooplankton. *Marine Science Communications* **2**, 43–67.

Fowler, S.W. and Guary, J.-C. (1977) High absorption efficiency for ingested plutonium in crabs. *Nature* **266**, 827–828.

Fowler, S.W., Heyraud, M. and Beasley, T.M. (1975) Experimental studies on plutonium kinetics in marine biota, in *Impacts of Nuclear Releases into the Aquatic Environment*, International Atomic Energy Agency, Vienna, pp. 157–177.

Freeman, H.C. and Horne, D.A. (1973) Mercury in Canadian seals. *Bulletin of Environmental Contamination and Toxicology* **10**, 172–180.

George, S.G. (1982) Subcellular accumulation and detoxification of metals in aquatic animals, in *Physiological Mechanisms of Marine Pollutant Toxicity*, (eds W.B. Vernberg, A. Calabrese, F.P. Thurberg and F.J. Vernberg), Academic Press, New York, pp. 3–52.

Guary, J.C. and Frazier, A. (1977) Influence of trophic level and calcification on the uptake of plutonium observed, in situ, in marine organisms. *Health Physics* **32**, 21–28.

Hamdy, M.K. and Prabhu, N.V. (1979) Behaviour of mercury in biosystems. III. Biotransference of mercury through food chains. *Bulletin of Environmental Contamination and Toxicology* **21**, 170–178.

Harvey, R.W. and Luoma, S.N. (1985) Effect of adherent bacteria and bacterial extracellular polymers upon assimilation by *Macoma balthica* of sediment-bound Cd, Zn and Ag. *Marine Ecology Progress Series* **22**, 281–289.

Hetherington, J.A., Jefferies, D.F. and Lovett, M.B. (1975) Some investigations into the behaviour of plutonium in the marine environment, in *Impacts of Nuclear Releases into the Aquatic Environment*, International Atomic Energy Agency, Vienna, pp, 193–212.

Heyraud, M., Fowler, S.W., Beasley, T.M. and Cherry, R.D. (1976) Polonium-210 in euphausiids: a detailed study. *Marine Biology* **34**, 127–136.

Hodson, P.V. (1980) Why inorganic metals do not increase in concentration up the food chain. *Thalassia Jugoslavica* **16**, 327.

Hopkin, S.P. and Nott, J.A. (1980) Studies on the digestive cycle of the shore crab *Carcinus maenas* (L.) with special reference to the B cells in the hepatopancreas. *Journal of the Marine Biological Association of the UK* **60**, 891–907.

Hoss, D.E. (1964) Accumulation of zinc-65 by flounder of the genus *Paralichthys*. *Transactions of the American Fisheries Society* **93**, 364–368.

Icely, J.D. and Nott, J.A. (1992) Digestion and absorption: digestive system and associated organs, in *Microscopic Anatomy of Invertebrates, Vol. 10, Decapod Crustacea*, (eds F.W. Harrison and A.G. Humes), Wiley-Liss, Inc., New York, pp. 147–201.

Ireland, M.P. (1973) Result of fluvial zinc pollution on the zinc content of littorial and sub-littorial organisms in Cardigan Bay, Wales. *Environmental Pollution* **4**, 27–35.

Ireland, M.P. (1983) Radioactive zinc uptake in *Littorina irrorata*. *Journal of Molluscan Studies* **49**, 79–80.

Jefferies, D.F. and Hewett, C.J. (1971) The accumulation and excretion of radioactive caesium by the plaice (*Pleuronectes platessa*) and the thornback ray (*Raia clavata*). *Journal of the Marine Biological Association of the UK* **51**, 411–422.

Jenkins, K.D. and Brown, D.A. (1984) Determining the biological significance of contaminant bioaccumulation, in *Concepts in Marine Pollution Measurements*, (ed. H.H. White), Maryland Sea Grant Publication, University of Maryland, pp. 355–363.

Jennings, J.R. and Rainbow, P.S. (1979) Studies on the uptake of cadmium by the crab *Carcinus maenas* in the Laboratory. I. Accumulation from seawater and a food source. *Marine Biology* **50**, 131–139.

Kirchmann, R., Bonotto, S., Bossus, A. *et al.* (1977) Utilisation d'une chaîne trophique expérimentale pour l'étude du transfert du [60]Co. *Revue Internationale d'Océanographie Médicale* **48**, 117–123.

Klumpp, D.W. (1980) Accumulation of arsenic from water and food by *Littorina littoralis* and *Nucella lapillus*. *Marine Biology* **58**, 265–274.

Knauer, G.A. and Martin, J.H. (1972) Mercury in a marine pelagic food chain. *Limnology and Oceanography* **17**, 868–876.

Kustin, K., Ladd, K.V. and McLeod, G.C. (1975) Site and rate of vanadium assimilation in the tunicate *Ciona intestinalis*. *Journal of General Physiology* **65**, 315–328.

Langston, W.J. and Zhou, M. (1987) Cadmium accumulation, distribution and metabolism in the gastropod *Littorina littorea*: the role of metal-binding proteins. *Journal of the Marine Biological Association of the UK* **67**, 585–601.

Langston, W.J., Bebianno M.J. and Zhou, M. (1989) A comparison of metal-binding proteins and cadmium metabolism in the marine molluscs *Littorina littorea* (Gastropoda), *Mytilus edulis* and *Macoma balthica* (Bivalvia). *Marine Environmental Research* **28**, 195–200.

Laumond, F., Neuburger, M., Donnier, B. *et al.* (1973) Experimental investigations, at laboratory, on the transfer of mercury in marine trophic chains. *Revue Internationale d'Océanographie Médicale* **31–32**, 47–53.

Leatherland, T.M., Burton, J.D., Culkin, F. *et al.* (1973) Concentrations of some trace metals in pelagic organisms and of mercury in northeast Atlantic Ocean water. *Deep-sea Research* **20**, 679–685.

Lindsay, D.M. and Sanders, J.G. (1990) Arsenic uptake and transfer in a simplified estuarine food chain. *Environmental Toxicology and Chemistry* **9**, 391–395.

Lorens, R.B. and Bender, M.L. (1977) Physiological exclusion of magnesium from *Mytilus edulis* calcite. *Nature* **269**, 793–794.

Maage, A., Julshamn, K. and Ulgenes, Y. (1991) A comparison of tissue levels of four essential trace elements in wild and farmed Atlantic salmon (*Salma salar*). *Fiskeridirektoratets Skrifter, Serie Ern'ring* **4**, 111–116.

Marcovecchio, J.E., Moreno, V.J. and Perez, A. (1991) Metal accumulation in tissues of sharks from the Bahia Blanca Estuary, Argentina. *Marine Environmental Research* **31**, 263–274.

Mason, A.Z. and Nott, J.A. (1981) The role of intracellular biomineralized granules in the regulation and detoxification of metals in gastropods with special reference to the marine prosobranch *Littorina littorea*. *Aquatic Toxicology* **1**, 239–256.

Mason, A.Z., Simkiss, K. and Ryan, K.P. (1984) The ultrastructural localization of metals in specimens of *Littorina littorea* collected from clean and polluted sites. *Journal of the Marine Biological Association of the UK* **64**, 699–720.

Metayer, C., Amiard, J.-C., Amiard-Triquet, C. and Marchand, J. (1980) Etude du transfert de quelques oligo-éléments dans les chaînes trophiques néritiques et estuariennes: accumulation biologique chez les poissons omnivores et super-carnivores. *Helgoländer Meeresuntersuchungen* **34**, 179–191.

Metayer, C., Amiard-Triquet, C. and Baud, J.P. (1990) Species-related variations of silver bioaccumulation and toxicity to three marine bivalves. *Water Research* **24**, 995–1001.

Miramand, P., Guary, J.C. and Fowler, S.W. (1980) Vanadium transfer in the mussel *Mytilus galloprovincialis*. *Marine Biology* **56**, 281–293.

Nakahara, M. and Cross, F.A. (1978) Transfer of cobalt-60 from phytoplankton to the clam (*Mercenaria mercenaria*). *Bulletin of the Japanese Society of Scientific Fisheries* **44**, 419–425.

Nicolaidou, A. and Nott, J.A. (1989) Heavy metal pollution induced by a ferro-nickel smelting plant in Greece. *The Science of the Total Environment* **84**, 113–117.

Nicolaidou, A. and Nott, J.A. (1990) Mediterranean pollution from a ferro-nickel smelter: differential uptake of metals by some gastropods. *Marine Pollution Bulletin* **21**, 137–143.

Nott, J.A. and Nicolaidou, A. (1989a) The cytology of heavy metal accumulations in the digestive glands of three marine gastropods. *Proceedings of the Royal Society of London* **B237**, 347–362.

Nott, J.A. and Nicolaidou, A. (1989b) Metals in gastropods – metabolism and bioreduction. *Marine Environmental Research* **28**, 201–205.

Nott, J.A. and Nicolaidou, A. (1990) Transfer of metal detoxification along marine food chains. *Journal of the Marine Biological Association of the UK* **70**, 905–912.

Nott, J.A. and Nicolaidou, A. (1993) Bioreduction of zinc and manganese along a molluscan food chain. *Comparative Biochemistry and Physiology* **104A**, 235–238.

Nott, J.A. and Nicolaidou, A. (1994) Variable transfer of detoxified metals from snails to hermit crabs in marine food chains. *Marine Biology* **120**, 369–377.

Nott, J.A. and Nicolaidou (1996) Kinetics of metals in molluscan faecal pellets and mineralized granules, incubated in marine sediments. *Journal of Experimental Marine Biology and Ecology* (in press).

Nott, J.A., Bebianno, M.J., Langston, W.J. and Ryan, K.P. (1993) Cadmium in the gastropod *Littorina littorea*. *Journal of the Marine Biological Association of the UK* **73**, 655–665.

Parrish, K.M and Carr, R.A. (1976) Transport of mercury through a laboratory two-level marine food chain. *Marine Pollution Bulletin* **7**, 90–91.

Pelletier, E. and Larocque, R. (1987) Bioaccumulation of mercury in starfish from contaminated mussels. *Marine Pollution Bulletin* **18**, 482–485.

Pentreath, R.J. (1973a) The accumulation and retention of ^{65}Zn and ^{54}Mn by the plaice, *Pleuronectes platessa* L. *Journal of Experimental Marine Biology and Ecology* **12**, 1–18.

Pentreath, R.J. (1973b) The accumulation and retention of ^{59}Fe and ^{58}Co by the Plaice, *Pleuronectes platessa* L. *Journal of Experimental Marine Biology and Ecology* **12**, 315–326.

Pentreath, R.J. (1973c) The accumulation from sea water of ^{65}Zn, ^{54}Mn, ^{58}Co and ^{59}Fe by the Thornback Ray, *Raja clavata* L. *Journal of Experimental Marine Biology and Ecology* **12**, 327–334.

Pentreath, R.J. (1976) The accumulation of mercury from food by the Plaice, *Pleuronectes platessa* L. *Journal of Experimental Marine Biology and Ecology* **25**, 51–65.

Pentreath, R.J. (1977a) The accumulation of cadmium by the plaice, *Pleuronectes platessa* L. and the thornback ray, *Raja clavata* L. *Journal of Experimental Marine Biology and Ecology* **30**, 223–232.

Pentreath, R.J. (1977b) Radionuclides in marine fish. *Oceanography and Marine Biology Annual Review* **15**, 365–460.

Peternac, B. and Legovic, T. (1986) Uptake, distribution and loss of Cr in the crab *Xantho hydrophilus*. *Marine Biology* **91**, 467–471.

Preston, A., Jefferies, D.F. and Pentreath, R.J. (1972) The possible contributions of radioecology to marine productivity studies. *Symposium of the Zoological Society of London* **29**, 271–284.

Preston, E.M. (1971) The importance of ingestion in chromium-51 accumulation by *Crassostrea virginica* (Gmelin). *Journal of Experimental Marine Biology and Ecology* **6**, 47–54.

Rainbow, P.S. (1988) The significance of trace metal concentrations in decapods. *Symposium of the Zoological Society of London* **59**, 291–313.

Rainbow, P.S. and White, S.L. (1989) Comparative strategies of heavy metal accumulation by crustaceans: zinc, copper and cadmium in a decapod, an amphipod and a barnacle. *Hydrobiologia* **174**, 245–262.

Ratkowsky, D.A., Dix, T.G. and Wilson, K.C. (1975) Mercury in fish in the Derwent Estuary, Tasmania, and its relation to the position of the fish in the food chain. *Australian Journal of Marine and Freshwater Research* **26**, 223–231.

Renfro, W.C., Fowler, S.W., Heyraud, M. and La Rosa, J. (1975) Relative importance of food and water in long-term zinc^{-65} accumulation by marine biota. *Journal of the Fisheries Research Board of Canada* **32**, 1339–1345.

Riedel, G.F., Sanders, J.G. and Osman, R.W. (1990) The role of three species of benthic invertebrates in the transport of arsenic from contaminated estuarine sediment. *Journal of Experimental Marine Biology and Ecology* **134**, 143–155.

Riisgaard, H.U. and Famme, P. (1986) Accumulation of inorganic and organic mercury in shrimp, *Crangon crangon*. *Marine Pollution Bulletin* **17**, 255–257.

Riisgaard, H.U. and Hansen, S. (1990) Biomagnification of mercury in a marine grazing food-chain: algal cells *Phaeodactylum tricornutum*, mussels *Mytilus edulis* and flounders *Platichthys flesus* studied by means of a stepwise-reduction-CVAA method. *Marine Ecology Progress Series* **62**, 259–270.

Riisgaard, H.U., Bjoernestad, E. and Moehlenberg, F. (1987) Accumulation of cadmium in the mussel *Mytilus edulis*; kinetics and importance of uptake via food and sea water. *Marine Biology* **96**, 349–353.

Roesijadi, G. (1992) Metallothioneins in metal regulation and toxicity in aquatic animals. *Aquatic Toxicology* **22**, 81–114.

Rumsey, T.J. (1973) Some aspects of osmotic and ionic regulation in *Littorina littorea* (L.) (Gastropoda, Prosobranchia). *Comparative Biochemistry and Physiology* **45A**, 327–344.

Sanders, J.G. (1985) Arsenic transport, reactivity, and toxicity in the Chesapeake Bay, in *The Fate and Effects of [ollutants: a Symposium*, Technical Report, Maryland University Sea Grant Program, p. 55.

Sanders, J.G., Osman, R.W. and Riedel, G.F. (1989) Pathways of arsenic uptake and incorporation in estuarine phytoplankton and the filter-feeding invertebrates *Eurytemora affinis*, *Balanus improvisus* and *Crassostrea virginica*. *Marine Biology* **103**, 319–325.

Sanders, J.G., Abbe, G.R. and Riedel, G.F. (1990) Silver uptake and subsequent effects on growth and species composition in an estuarine community. *Science of the Total Environment* **97–98**, 761–769.

Schulz-Baldes, M. (1974) Lead uptake from sea water and food, and lead loss in the common mussel *Mytilus edulus*. *Marine Biology* **25**, 177–193.

Sergeant, D.E. and Armstrong, F.A.J. (1973) Mercury in seals from eastern Canada. *Journal of the Fisheries Research Board of Canada* **30**, 843–846.

Sick, L.V. and Baptist, G.J. (1979) Cadmium incorporation by the marine copepod *Pseudodiaptomus coronatus*. *Limnology and Oceanography* **24**, 453–462.

Suedel, B.C., Boraczek, J.A., Peddicord, R.K. *et al.* (1994) Trophic transfer and biomagnification potential of contaminants in aquatic ecosystems. *Reviews of Environmental Contamination and Toxicology* **136**, 21–89.

Taylor, M. and Simkiss, K. (1984) Inorganic deposits in invertebrate tissues. *Environmental Chemistry* **3**, 102–138.

Uensal, M. (1982) Transfer pathways and accumulation of vanadium in the crab *Carcinus maenas*. *Marine Biology* **72**, 279–282.

Viarengo, A. and Nott, J.A. (1993) Mechanisms of heavy metal cation homeostasis in marine invertebrates. *Comparative Biochemistry and Physiology* **104C**, 355–372.

Weerelt, M. van, Pfeiffer, W.C. and Fiszman, M. (1984) Uptake and release of [51]Cr(VI) and [51]Cr(III) by barnacles (*Balanus* sp). *Marine Environmental Research* **11**, 201–211.

Weers, A.W. van (1975) Uptake of cobalt-60 from sea water and from labelled food by the common shrimp *Crangon crangon* (L.), in *Impacts of Nuclear Release into the Aquatic Environment*, International Atomic Energy Agency, Vienna, pp. 349–361.

Wieser, W. (1967) Conquering terra firma: the copper problem from the isopod's point of view. *Helgoländ wissenschaftliche Meeresuntersuchungen* **15**, 282–293.

Willis, J.N. and Sunda, W.G. (1984) Relative contributions of food and water in the accumulation of zinc by two species of marine fish. *Marine Biology* **80**, 273–279.

Wrench, J., Fowler, S.W. and šnlü, M.Y. (1979) Arsenic metabolism in a marine food chain. *Marine Pollution Bulletin* **10**, 18–20.

Young, M.L. (1975) The transfer of [65]Zn and ^{59}Fe along a *Fucus serratus* (L.)

13 *Metal accumulation and detoxification in humans*

HING MAN CHAN

13.1 INTRODUCTION

Metals are probably the oldest known toxins. About 80 of the 105 elements in the periodic table are regarded as metals, but fewer than 30 have been reported to produce toxicity in humans (Goyer, 1996). Many metals are important as micronutrients and play an essential role in tissue metabolism and growth. They include Co, Cu, Cr, Fe, Mn, Ni, Mo, Se, Sn, V and Zn. Insufficient intake of these metals results in diseases or growth retardation. Other metals, such as Pb, Cd and Hg, are non-essential, i.e. have no known biological functions and no perceived effect on deficiency. Arsenic has been proposed as an essential metal in vertebrates and some mammals (Nielsen and Uthus, 1984) but there is no known deficiency in humans. Overexposure to both essential and non-essential metals results in toxicity. Therefore, over the course of evolution, humans have developed highly regulated metabolic pathways to maintain the essential metals at optimal concentration ranges and detoxification mechanisms for many of the non-essential toxic metals.

A brief overview of accumulation and detoxification of four toxic metals (As, Cd, Pb and Hg) will be presented in this chapter. The rationale for choosing these four metals is that they are the pollutants of major concern and are listed among the top twenty most important hazardous environmental contaminants in the United States.

Sources of exposure to metals in humans include inhalation from air, ingestion through the gastrointestinal tract and dermal exposure. The absorption, accumulation and metabolism of metals can be very different from different routes of exposure. We shall only discuss the toxicokinetics of metals entering the body from oral ingestion of drinking water and aquatic animals and plants. Emphasis will be put on data from epidemiological and human

Metal Metabolism in Aquatic Environments. Edited by William J. Langston and Maria João Bebianno. Published in 1998 by Chapman & Hall, London. ISBN 0 412 80370 4

studies. Results from animal studies will only be used when human data is not available.

Chemical speciation affects properties such as toxicity, biological uptake and fate. This phenomenon is best illustrated by the inorganic and organic form of As and Hg (Chapters 4 and 5). The organic forms of As are usually less toxic than the inorganic forms. The As species commonly encountered in dissolved form in the aquatic environment are arsenate [As(V)], arsenite [As(III)], methylarsonic acid (MMA) and dimethylarsinic acid (DMA) (Andreae, 1977; Maher and Butler, 1988; Cullen and Reimer, 1989; Chapter 4). The methylated species of As are at least 1000 times less toxic than the inorganic ones. In marine animals, the most abundant form of As is arseno-betaine (Lawrence *et al.*, 1986; Chapter 4), which is thought to be non-toxic to humans. Arsenic in aquatic plants is associated with lipids and sugar (Culler and Reimer, 1989; Chapter 4). Toxicity of these As compounds is not known. Mercury in ocean waters exists mainly as inorganic Hg (Hg^{2+}) complexed with chloride ions (Chapter 5). Speciation in fresh water is poorly understood but organic Hg (methylmercury) may constitute approximately 1–6% of the total Hg (Canada-Ontario Steering Committee, 1983). Proportion of methylmercury varies in aquatic species, being highest (> 90%) in fish and marine mammal meat and lowest in organs (about 50%) (WHO, 1990). Organic Hg in the environment is thought to be more toxic than inorganic Hg because the absorption rate is much higher and it crosses the blood–brain barrier and the placenta. Accordingly, the accumulation and metabolism of different metal species will be discussed separately.

13.2 ABSORPTION

Toxicity of metals depends on the internal dose, i.e. the amount of ingested metals absorbed in the gastrointestinal tract. The rate of absorption is usually determined from the amount of ingested metals being excreted in the faeces and urine. Unabsorbed metals pass through the gastrointestinal tract and are excreted in the faeces. In contrast, absorbed metals are usually excreted in the urine eventually.

Both arsenates and arsenites can be completely absorbed across the gastrointestinal tract in humans. Less than 5% of oral dose of arsenite was recovered in the faeces (Bettley and O'Shea, 1975). Urinary excretion accounted for 55–80% of daily oral intakes of arsenate or arsenite (Buchet *et al.*, 1981b; Crecelius, 1977; Mappes, 1977; Tam *et al.*, 1979). Similarly, both methylarsonic acid and dimethylarsinic acid are efficiently absorbed (at least 75–85%) across the gastrointestinal tract (Buchet *et al.*, 1981a; Marafante *et al.*, 1987). Arsenobetaine ingested from seafood is excreted mainly in the urine (Cannon *et al.*, 1983; Luten *et al.*, 1982), indicating a high absorption rate. In contrast, the gastrointestinal absorption rates are much lower for less soluble forms of arsenicals such as triselenide (AsSe) and result in no increase in urinary excretion (Mappes, 1977).

Most ingested Cd passes through the gastrointestinal tract without being absorbed (Kjellstrom *et al.*, 1978). Comparisons of the body burden of Cd in non-smokers with estimated daily intakes from the diet provide estimates of Cd absorption from food of 3–5% (Ellis *et al.*, 1979; Morgan and Sherlock, 1984). Results of two experiments performed on healthy adults show that the absorption rates of Cd administered in food or drinking water range from 4.6 to 6% (McLellan *et al.*, 1978; Rahola *et al.*, 1973).

Cadmium absorption is affected by several factors, including nutritional status; subjects with low Fe stores (assessed by serum ferritin levels) had an average absorption of 8.9%, while those with adequate Fe stores had an average absorption of 2.3% (Flanagan *et al.*, 1978). Similar results were reported by Nordberg *et al.* (1985). The chemical complexation of Cd has a slight influence on absorption in humans; whole-body retention rate for crab metallothionein bound Cd was 2.7% (Newton *et al.*, 1984), compared with values of 4.6–6% obtained using dissolved Cd ion (McLellan *et al.*, 1978; Rahola *et al.*, 1973). However, some populations with high dietary Cd exposure from bluff oysters (McKenzie-Parnell *et al.*, 1988) or seal meat (Hansen *et al.*, 1985) did not show elevated blood Cd levels, suggesting that the particular form of Cd in these foods may not be bioavailable. A recent study found a lower rate of Cd absorption from shellfish compared with a mixed diet (Vahter *et al.*, 1996) but their results were confounded by differences in Fe status among participating women.

The rate of Pb absorption depends on age; absorption in children is 50%, compared with 8% (Hammond, 1982) or 15% (Chamberlain *et al.*, 1978) gastrointestinal Pb absorption in adults. The solubility of a particular Pb salt in gastric acid and a number of dietary factors, including chemical interactions, will affect the extent and rate of gastrointestinal absorption of Pb. Fasting also has a pronounced effect on the absorption of Pb, which can be as high as 45% in adults under fasting conditions (Chamberlain *et al.*, 1978). Specific data regarding the gastrointestinal absorption of alkyl Pb compounds in humans are not available.

Inorganic Hg is absorbed on a limited basis. Oral absorption of metallic Hg has been estimated to be approximately 0.1% (Friberg and Nordberg, 1973). Approximately 15% of a trace dose of mercuric nitrate in an aqueous solution or bound to calf liver protein is absorbed from the gastrointestinal tract (Weiss *et al.*, 1973).

In contrast, organic Hg compounds are readily absorbed in humans. Based on retention and excretion studies in humans, approximately 95% of an oral tracer dose of aqueous methylmercuric nitrate is absorbed (Aberg *et al.*, 1969). Similar absorption rates of Hg were reported in volunteers who received doses of methylmercury bound to protein (Miettinen, 1973), and in victims who ate bread contaminated with a fungicide that contained methylmercury (Al-Shahristani *et al.*, 1976).

13.3 BODY DISTRIBUTION

The body burden of a particular chemical is the total amount of that chemical found in the body. It represents the difference between cumulative lifetime absorption from all sources and total excretion. The distribution of metals in the body depends initially on the rate of delivery by the bloodstream to various organs and tissues. A subsequent redistribution may then occur, based on the relative affinity of tissues for the element (EPA, 1986). Understanding of the body distribution of metals will allow the identification of target organs (e.g. Cd in kidneys) and the use of particular tissues for dosimetric studies (e.g. Hg in hair). The ratio of metal concentrations in maternal and cord blood and levels in the placenta can be used as indicators of potential fetotoxicity of the metal.

Analysis of tissues taken at autopsy from people who had been exposed to background levels of As in food and water revealed that As is present in all tissues of the body (Liebscher and Smith, 1968). Most tissues had about the same concentration (0.05–0.15 μg g^{-1}), while levels in hair (0.65μg g^{-1}) and nails (0.36 μg g^{-1}) were somewhat higher, indicating that there is little tendency for As to accumulate preferentially in any internal organs. No studies were located on the distribution of organic arsenicals in people after oral exposure. However, in hamsters, MMA and DMA formed *in vivo* by methylation of inorganic As appear to be distributed to all tissues (Yamauchi and Yamamura, 1985; Takahashi *et al.*, 1988). Studies of tissue levels of As in fetuses and newborn babies in Japan show that the total amount of As in a fetus tends to increase during gestation, indicating placental transfer. The placental transfer of As is further supported by the results of a study of pregnant women in the United States which showed similar As levels in cord blood and maternal blood (Kagey *et al.*, 1977).

Cadmium is present in virtually all tissues in adults, with the greatest concentrations found in liver and kidney (Chung *et al.*, 1986; Sumino *et al.*, 1975; ATSDR, 1993). About a third of the total body burden in a non-smoking male is in the kidney and about a quarter in the liver and muscles (WHO, 1992). Body burden of Cd concentrations increases with age. Cadmium concentrations in kidney are near zero at birth, and rise roughly linearly with age to a peak (typically around 40–50 μg g^{-1} wet weight) between ages 50 and 60, after which kidney concentrations plateau or decline (Hammer *et al.*, 1973; Lauwerys *et al.*, 1984; Chung *et al.*, 1986). Liver Cd concentrations also begin near zero at birth; they increase to 1–2 μg $^{-1}$g wet weight by age 20–25 and then increase only slightly thereafter (Hammer *et al.*, 1973; Sumino *et al.*, 1975; Lauwerys *et al.*, 1984; Chung *et al.*, 1986).

The placenta may act as a partial barrier to fetal exposure to Cd. Cadmium concentration has been found to be approximately half as high in cord blood as in maternal blood in several studies which included both smoking and non-smoking women (Lauwerys *et al.*, 1978; Kuhnert *et al.*, 1982; Truska *et al.*, 1989). Accumulation of Cd in the placenta at levels about 10 times higher than maternal blood Cd concentration has been found in women in Belgium (Roels *et al.*, 1978), the United States (Kuhnert *et al.*, 1982) and Poland (Baranowska, 1995). Cadmium in blood occurs mainly in red blood cells, and plasma concentrations are very low. Cadmium levels in human milk are 5–10% of levels in blood, possibly due to inhibited transfer from blood (Radisch *et al.*, 1987).

In human adults, 95% of the total body burden of Pb is found in the bones. However, bone Pb levels depend on age. For example, bone Pb accounts for only 73% of the body burden in children (Barry, 1975, 1981). Drasch *et al.* (1987) analysed the bone Pb content at autopsy in a total of 240 adults with no known occupational exposure to Pb and found that the Pb content of the temporal bone increased steadily with age, whereas the Pb content of the mid-femur and that of the pelvic bone reached a plateau in middle age followed by a decline with advancing age. The levels of Pb in teeth, especially in the dentine, are also known to increase with age. This age-related increase may occur as a function of exposure (Steenhout and Pourtois, 1981), which supports the use of dentine Pb measurements as an indicator of exposure (Needleman and Shapiro, 1974).

The large pool of Pb in adults can serve to maintain blood Pb levels long after periods of exposure have ended (O'Flaherty *et al.*, 1982). Lead in bones appears to be stored in two physiological compartments. In one compartment, bone Pb is essentially inert, having a half-life of several decades. Another more labile compartment allows for a more dynamic equilibrium of Pb between bone and soft tissue or blood (Rabinowitz *et al.*, 1976, 1977). Increased mobilization of Pb from human bone can occur during the physiological stresses of pregnancy and lactation (ATSDR, 1988). Zaric *et al.* (1987) found that women living in a smelter region had higher blood levels of Pb during pregnancy, and Manton (1985) reported increased blood levels of Pb in women during lactation.

The age effect is not observed for Pb in most soft tissues in humans over 20 years old (Barry, 1975, 1981). The exception to this trend is the renal cortex, which may retain Pb due to the formation of Pb nuclear inclusion bodies (Indraprasit *et al.*, 1974), and the aorta, in which entrapped Pb may be present in atherosclerotic plaque deposits (Barry, 1975, 1981). Brain tissue does not show an age-related increase in Pb accumulation, but increased brain Pb levels have been measured in subjects with known or suspected occupational exposure to Pb (Barry, 1975). Several authors have shown that Pb is selectively accumulated in the hippocampus in both children and adults (EPA, 1986).

Since blood Pb levels are commonly used as bioindicators for Pb exposure, many studies have been conducted on the partitioning of Pb in blood. Under steady-state conditions, over 99% of blood Pb is associated with the erythrocytes (Everson and Patterson, 1980). Over 50% of this erythrocyte Pb pool is bound to haemoglobin, with lesser amounts bound to other proteins (Bruenger *et al.*, 1973). Fetal haemoglobin appears to have a greater affinity for Pb than does adult haemoglobin (Ong and Lee, 1980). At blood Pb levels less than 40 pg dl^{-1}, blood Pb and serum Pb levels have a positive linear relationship; whereas at higher blood Pb levels they assume a curvilinear relationship. The ratio of Pb in serum to that in whole blood increases dramatically at blood Pb levels higher than 40 pg dl^{-1} (Manton and Cook, 1984). The departure from linearity of this relationship at blood Pb levels higher than 40 pg dl^{-1} may be caused by altered cell morphology at these higher concentrations, resulting in reduced availability or stability of Pb binding sites in the erythrocytes (EPA, 1986).

The half-life of Pb in adult human blood has been measured as 36 days by Rabinowitz *et al.* (1976) and 28 days by Griffin *et al.* (1975). The biological half-life of Pb in the blood of 2-year-old children was reported to be 10 months (Succop *et al.*, 1987).

Transplacental transfer of Pb in humans has been demonstrated in a number of studies. An extensive survey (sample size of larger than 11 000) conducted by Bellinger *et al.* (1987), found the mean ± SD of Pb concentration in umbilical cord blood was 6.6 ± 3.2 pg dl^{-1}. Barltrop (1969) and Horiuchi *et al.* (1959) reported that fetal uptake of Pb occurs by the 12th week of development and increases throughout development. These authors measured the highest Pb levels in fetal bone, kidney and liver tissue, and lesser amounts in the brain and heart. There is no metabolic barrier to the uptake of Pb by the fetus (ATSDR, 1988); therefore, exposure of women to Pb during pregnancy results in uptake by the fetus. Moreover, because of the distribution of Pb in mineralizing tissue of humans, the physiological stress of pregnancy may result in mobilization of Pb from maternal bone, further increasing the uptake of Pb by the fetus. Thus, fetal uptake of Pb can occur from a mother who was exposed to Pb before pregnancy, even if no Pb exposure occurred during pregnancy.

Metallic Hg in solution in the body is highly diffusible and lipophilic. It is distributed to all tissues and reaches peak levels within 24 hours except in the brain, where peak levels are achieved within 2–3 days. Highest levels are found in the kidney, with lower concentrations found in the chest area and the lung (Hursh *et al.*, 1976).

Distribution of inorganic Hg compounds resembles that of metallic Hg but, in humans, distribution is preferentially to the kidney, liver and intestine, and levels in the brain are substantially lower as these compounds have a lower lipophilicity. The concentrations in the kidney are orders of magnitude higher than in other tissues (Rothstein and Hayes, 1964). Brain tissue retains Hg for the longest time. Japanese workers who died 10 years after their last Hg exposure still had high residual levels of Hg in the brain (Takahata *et al.*, 1970). The transport of mercuric ions is limited at the placental barrier by the presence of high-affinity binding sites (Dencker *et al.*, 1983).

In blood, the mercuric ion exists as a diffusible or a non-diffusible form. The non-diffusible form is the mercuric ion that is bound to protein or is part of a high molecular weight complex. The non-diffusible form exists in equilibrium with the diffusible form. In plasma, the mercuric ion is predominantly non-diffusible (Berlin and Gibson, 1963). This form binds to albumin (Clarkson *et al.*, 1961) and globulins (Cember *et al.*, 1968). After mercuric salt administration, levels of mercuric ions in the plasma are similar to levels of mercuric ions in the red blood cells.

Methylmercury is distributed readily to all tissues in the body after absorption from the gastrointestinal tract. The distribution is very uniform because of its ability to cross diffusion barriers and penetrate all membranes without difficulty (Aberg *et al.*, 1969; Miettinen, 1973). Because of methylmercury's mobility, tissue concentrations relative to blood levels tend to remain constant. Therefore, blood levels are good indicators of tissue concentrations, independent of dose (Nordberg, 1976). Although distribution is generally uniform, highest levels are still found in the kidney. Methylmercury is transformed into an inorganic form in tissues other than blood. The liver (Magos *et al.*, 1976) and the kidney (Norseth and Clarkson, 1970) have been suggested as potential sites of biotransformation. The fraction present as inorganic Hg depends on the duration of exposure to methylmercury and the time after cessation of exposure.

Methylmercury can readily traverse the placental barrier. Concentrations of methylmercury in fetal blood are higher than in maternal blood (Kuhnert *et al.*, 1982). Methylmercury is also secreted in mother's milk (Bakir *et al.*, 1973), but much of it in its inorganic form.

Hair Hg level is commonly used as a bioindicator for methylmercury exposure. Mercury accumulates in the hair at the time of its formation in the follicle. Once the follicle is formed, the concentration of Hg in the hair is proportional to the concentration of Hg in the blood. Correlations can be drawn to determine blood concentrations of Hg from its concentration in the hair (Phelps *et al.*, 1980).

13.4 EXCRETION

Toxicity of the metals depends on how readily they are excreted. Direct measurements of As excretion in humans who have ingested known amounts of arsenite or arsenate indicate that very little is excreted in the faeces (Bettley and O'Shea, 1975), and that 45–85% is excreted in urine within 1–3 days (Crecelius, 1977; Mappes, 1977; Tam *et al.*, 1979; Buchet *et al.*, 1981a). Therefore, whole-body clearance is very rapid, with a half-life of 40–60 hours in humans (Mappes, 1977; Buchet *et al.*, 1981b). Studies in humans indicate that ingested MMA and DMA are excreted mainly in the urine (75–85%), and this occurs mostly within 1 day (Buchet *et al.*, 1981a; Marafante *et al.*, 1987). Arsenobetaine ingested from seafood is not biotransformed appreciably in the human body, and is excreted mainly in urine within hours (Luten *et al.*, 1982; Cannon *et al.*, 1983; Yamauchi and Yamamura, 1984).

The major proportion of orally ingested Cd is excreted in the faeces (Kjellstrom *et al.*, 1978). However, almost all excreted Cd represents material that was not absorbed from the gastrointestinal tract. Most absorbed Cd is excreted very slowly, with urinary and faecal excretion being approximately equal (Kjellstrom and Nordberg, 1978). Daily faecal and urinary excretion are estimated to be 0.007% and 0.009% of the body burden, respectively (Kjellstrom and Nordberg, 1978, 1985). Half-life of Cd has been estimated to be 20–50 years (WHO, 1992).

Dietary Pb not absorbed by the gastrointestinal tract is eliminated in the faeces. Rosen (1985) reported that 50–60% of the absorbed fraction of Pb in adults in a steady-state condition with regard to Pb intake/output was excreted in the short term. Chamberlain *et al.* (1978) found the half-life of this short-term fraction to be 19 days. The long-term fraction stored in bones has a half-life of several decades (Rabinowitz *et al.*, 1976, 1977). Infants have a lower total excretion rate for Pb. Young children (infants from birth to 2 years of age) retain 34% of the total amount of Pb absorbed, based on a study by Ziegler *et al.* (1978), whereas data by Rabinowitz *et al.* (1977) demonstrate only a 1% retention of an absorbed dose of Pb in adults.

The urine and faeces are the main excretory pathways of inorganic Hg in humans but exhalation and excretion in saliva, bile and sweat also contribute (Joselow *et al.*, 1968; Lovejoy *et al.*, 1974). In humans, after a brief exposure to Hg, urinary excretion accounts for 13% of the total body burden. After long-term exposure, urinary excretion increases to 58%. Levels in the urine do not parallel those seen in plasma. Thus, urinary Hg comes from a body pool of Hg as opposed to the glomerular filtrate of the plasma (Cherian *et al.*, 1978). Half-life of inorganic Hg is 60 days (Hursh *et al.*, 1976).

The faecal pathway is the predominant excretory route for organic Hg compounds (Norseth and Clarkson, 1970). Less than one-third of total Hg excretion is through the urine. In humans, all Hg in the faeces is in its inorganic form after organic Hg administration. The conversion to the inorganic form is believed to be conducted by the intestinal flora (Nakamura *et al.*, 1977; Rowland *et al.*, 1980). Elimination of organic Hg compounds generally follows first-order kinetics. In the case of methylmercury, whole-body clearance and clearance from the blood take much longer than the inorganic compounds. There is also some evidence of sex differences in the elimination of organic Hg. Females tend to excrete organic Hg compounds faster than males (Aberg *et al.*, 1969; Miettinen, 1973). Half-life for methylmercury has been estimated as 71–79 days (Miettinen, 1973).

Table 13.1 summarizes the biological half-life of the different species of the four metals mentioned above. Approximately five times the half-life period will be required to eliminate the absorbed metals completely. Therefore, the toxic symptoms are sometimes expressed even after the cessation of exposure, e.g. with Cd and Pb.

Table 13.1 Biological half-life of toxic metals in humans

Metal	Species/Organ	Half-life	Reference
Arsenic	As(III) and As(V)	40–60 hours	Buchet *et al.*, 1981b
	Methylated As	< 1 day	
	Arsenobentaine	6 hours	Yamauchi and Yamamura, 1984
Cadmium		20–50 years	WHO, 1992
Lead	Skeleton	> 20 years	Rabinowitz *et al.*, 1976, 1977
	Soft tissue	19 days	Chamberlain *et al.*, 1978
Mercury	Inorganic	60 days	Hursh *et al.*, 1976
	Organic	71–79 days	Miettinen, 1973

13.5 TOXICOLOGY

Oral exposure to metals may produce injury in a number of different body tissues or systems (termed as systemic effects) or they may affect a particular organ where they tend to accumulate (target organ). A brief summary of the toxicity of As, Cd, Pb and Hg is presented in this section.

Ingestion of large doses of As (70–180 mg) may be fatal (ATSDR, 1992). While As has no specific target organ, chronic exposure to inorganic As compounds may lead to neurotoxicity of both the peripheral and central nervous systems, liver injury and peripheral vascular disease. The EPA (1987) and IARC (1987) classify As as a human carcinogen that can increase the risk of liver, bladder, kidney, skin and lung cancer.

The primary toxic action of inorganic As(III) is via reactions with sulfhydryl groups of proteins and subsequent enzyme inhibition (Eisler, 1994). Inorganic As(III) interrupts oxidative metabolic pathways, and sometimes causes morphological changes in liver mitochondria. Arsenite *in vitro* reacts with protein-SH groups to inactivate enzymes such as dihydrolipoyl dehydrogenase and thiolase, inhibiting oxidation of pyruvate, and beta oxidation of fatty acids (Belton *et al.*, 1985). In comparison, As(V) inhibits adenosine triphosphate (ATP) synthesis by uncoupling oxidative phosphorylation, which results in the inhibition of energy metabolism (Andreae, 1986). Arsenate may form arsenate esters by replacing phosphate in monosaccharides, such as glucose-6-phosphate leading to the formation of glucose-6-arsenate. Arsenobetaine has no substantial acute toxicity; its LD_{50} (oral administration) in mice has been estimated to be in excess of 10 g kg^{-1} (Kaise *et al.*, 1985).

Ingestion of high doses of Cd causes severe irritation to the stomach, leading to vomiting and diarrhaea, but such high exposures are extremely rare. Of greater concern is the chronic Cd toxicity usually caused by long-term exposure to low levels of environmental Cd (WHO, 1992). The major target organ of chronic Cd toxicity is the kidney. Renal tubular damage is the best identified health effect of chronic Cd exposure (WHO, 1977). Bone disorder is another manifestation of chronic Cd toxicity. Osteomalacia was diagnosed together with renal tubular dysfunction among people in Japan exposed to high levels of

Cd (WHO, 1992). Bone disorder was believed to be secondary to renal failure rather than directly an effect of Cd on bone, because levels of Cd required for bone disorder were higher than those for renal damage (Cherian and Goyer, 1989). However, there is evidence showing that Cd accumulates directly in bone even when the kidneys function normally, indicating that the toxic effect may be primary to the bone rather than solely secondary to renal failure (Krishnan *et al.*, 1990; Whelton *et al.*, 1994). Cadmium has been classified as a human carcinogen (IARC, 1994); however, the classification is based primarily on evidence of pulmonary tumours resulting from inhalation of Cd.

The multiple toxic effects of Pb poisoning have been known for a long time (ATSDR, 1990). Children are most susceptible to Pb toxicity: over-exposure to Pb can cause overt encephalopathy, hearing deficit, IQ deficit and peripheral neuropathy (Goyer, 1996). The mechanism of Pb neurotoxicity is reviewed by Silbergeld (1992). Lead affects the development of the nervous system by impairing the timed programming of cell-to-cell connections, resulting in the modification of neuronal circuitry. It can interfere with synaptic mechanisms of transmitter release. Lead can also cause multiple haematological effects, mainly by inhibiting the enzyme pyrimidine-5-nucleosidase, and can induce anaemia (Paglia *et al.*, 1975). The kidney is also a target organ for Pb. Other toxic effects of Pb include increases in blood pressure and reproductive effects associated with sterility and neonatal deaths in humans. Lead causes kidney cancer in animals (EPA, 1989) but there is no conclusive evidence of its carcinogenicity in humans. Thus, it is classified as a possible human carcinogen (IARC, 1987).

Long-term exposure to either inorganic or organic Hg can permanently damage the brain, kidneys and developing fetuses (ASTDR, 1989). The form of Hg determines which of these health effects will be most severe. For example, inorganic Hg salts eaten in contaminated food or drunk in water may cause greater harm to the kidneys, whereas organic Hg eaten in contaminated fish may cause greater harm to the brain and developing fetuses. The mechanism of neurotoxicity caused by methylmercury has been one of the major topics of investigation in metal toxicology but it remains unclear which molecular species (the mercurials, or any radicals formed by these mercurials) is the proximate toxic species actually causing neurological injury and which cellular molecule is the primary target of Hg toxicity (Suzuki *et al.*, 1991). Possible mechanisms include interaction of DNA and RNA, disruption of protein synthesis and free radical generation, etc. The main sites of accumulation of inorganic Hg are in the cortex and outer medulla of the kidney; it accumulates along the entire length of the proximal tubule and causes necrosis (Zalups and Lash, 1994).

13.6 METABOLISM AND DETOXIFICATION

Toxicity of metals depends on the duration of exposure and availability of the metals to react with the target tissues or cellular component. Detoxification can be achieved if the metal is excreted from the body rapidly or stored in the body in an inert form. Excretion can sometimes be enhanced if the metal is metabolized and biotransformed to another form which may be more or less toxic than the mother compound.

The metabolism of inorganic As has been extensively studied in humans. Two processes are involved: reduction/oxidation reactions that interconvert arsenate and arsenite, and methylation reactions that convert arsenite to methylarsonic acid and dimethylarsinic acid. A large portion of As(III) or As(V) is methylated to methylarsonic acid and dimethylarsinic acid after entering the human body. Methylation of As greatly reduces toxicity and is regarded as a detoxification process (Eisler, 1994). Methylarsonic acid and dimethylarsinic acid are nearly 1000 times less toxic to animals than is potassium arsenite or calcium arsenate. Moreover, methylation tends to result in lower tissue retention of inorganic As (Marafante and Vahter, 1984, 1986; Marafante *et al.*, 1985; Vahter and Marafante, 1987). Before methylation (which occurs largely in the liver) As(V) is reduced to As(III), with the kidney being an important site for this transformation (Belton *et al.*, 1985). Both the arsenate reduction and the methylation processes are fast. Methylated arsenicals are cleared rapidly from all tissues except the thyroid (Marafante and Vahter, 1984; Marafante *et al.*, 1985). As(III), As(V) and DMA have been detected in urine (Braman and Foreback, 1973; Lakso and Peoples, 1975; Crecelius, 1977; Charbonneau *et al.*, 1978; Tam *et al.*, 1979; Yamauchi and Yamamura, 1984). DMA accounted for 50% of the ingested As, MMA 14 to 21% and the inorganic As species were less than 30%. Complete urinary clearance required from 3 to 5 days (Crecelius, 1977; Mappes, 1977).

Because methylation is an enzymic process, the dose of As that saturates the methylation capacity of an organism is an important issue. Limited data from studies in humans suggest that methylation may begin to be limiting at doses of about 0.2–1 mg per day (0.003–0.015 mg kg^{-1} per day) (Buchet *et al.*, 1981b; Marcus and Rispin, 1988).

Methylmercury is converted to inorganic Hg, assumed to be Hg^{2+}, in mammals (WHO, 1976). An average of 80% of the Hg in the occipital lobe cortex of autopsy cases in Sweden was found to be inorganic Hg (3–22 ng g^{-1} wet weight) (Friberg *et al.*, 1986, Nylander *et al.*, 1987). Studies by Suda and Takahashi (1986) indicate that macrophage cells, such as those present in the spleen, are capable of converting methylmercury to inorganic Hg. The reaction may involve the production of oxygen free-radicals. The conversion of methylmercury to Hg^{2+} may be a key step in the processes of excretion. The faecal pathway accounts for about 90% of the total elimination of Hg in humans and other mammals after exposure to methylmercury (WHO, 1976). Virtually all the Hg in human faeces is in the inorganic form (Turner *et al.*, 1975). The process of faecal elimination begins with the biliary secretion of inorganic and organic Hg, complexed mainly, if not entirely, with glutathione (GSH) (Refsvik and Norseth, 1975) or other sulfhydryl peptides (Norseth and

Clarkson, 1971; Ohsawa and Magos, 1974). Most of the inorganic Hg (approximately 90%) secreted in bile passes directly into the faeces. Methylmercury secreted into the intestinal contents is in large part reabsorbed into the bloodstream and may subsequently contribute to biliary secretion, thereby forming a secretion–reabsorption cycle (Norseth and Clarkson, 1971). This cycle (also called enterohepatic circulation) increases the amount of methylmercury passing through the intestinal contents and thus provides a continuous supply of methylmercury to serve as a substrate for the intestinal microflora. These microorganisms are capable of converting methylmercury to inorganic Hg, which then becomes the major contributor to total faecal elimination in the rat (Rowland *et al.*, 1980). Presumably about 10% of the inorganic Hg produced by the intestinal microflora is absorbed into the bloodstream and contributes to the inorganic Hg concentrations in tissues, plasma, bile, breast milk and urine. To what extent this model of enterohepatic circulation and intestinal conversion to inorganic Hg applies to humans is not yet known (WHO, 1990). Considerable species differences exist in rates of biliary excretion (Naganuma *et al.*, 1991).

Cadmium, Pb and inorganic Hg are not known to undergo any direct metabolic conversion such as oxidation, reduction or alkylation. Detoxification of Cd and inorganic Hg involves the protein metallothionein.

Metallothionein is a low molecular weight protein, very rich in cysteine, which is capable of binding divalent ions such as Cd and Hg and decreasing their toxicity (Chapters 8–11). Metallothionein is inducible in most tissues by exposure to Cd, Zn, inorganic Hg (but not organic Hg) and other metals, as well as organic compounds and a variety of other physiological stresses (irradiation, food deprivation, exercise, hypothermia and inflammation) (Cherian and Chan, 1993). Cadmium-induced metallothionein shows a paradoxical role in Cd toxicity (Nordberg *et al.*, 1975). Intracellularly induced metallothionein decreases Cd toxicity by sequestering Cd ions and forming Cd-bound metallothionein (Cd-MT) (Cherian, 1980; Goering and Klaassen, 1984). Exogenous Cd-MT, however, is more toxic than inorganic Cd (Nordberg *et al.*, 1975; Sendelbach *et al.*, 1988; Chan *et al.*, 1992). Dudley *et al.* (1985) suggested that Cd nephrotoxicity may be caused by Cd-MT released from the liver. This theory is supported by Chan *et al.* (1993) with a study of liver transplantation. It was found that Cd can be released from the liver, exist in the form of Cd-MT in the blood stream, and accumulate in the kidney. When Cd-MT is transported to the kidney, it is readily diffusible and filterable at the glomerulus and may be effectively reabsorbed from the glomerular filtrate by the proximal tubule cells (Foulkes, 1978). Exogenous metallothionein is degraded in lysosomes; this process may release Cd, which may induce fresh metallothionein synthesis in the proximal tubule (Squibb *et al.*, 1984). Cadmium-induced renal toxicity is probably associated with Cd not bound to metallothionein (Goyer *et al.*, 1989; Nomiyama and Nomiyama, 1986), though brush-border membranes of the renal tubule may be damaged by Cd that is bound to metallothionein (Suzuki and Cherian, 1987). Nevertheless, renal damage is believed to occur if the localization of Cd or an excessive concentration of Cd prevent it from becoming bound to metallothionein. The protective effects of metallothionien for Cd are further supported by a recent study which showed that transgenic mice without metallothionein were much more susceptible to Cd toxicity (Zheng *et al.*, 1996).

Unlike Cd-MT, Hg bound metallothionein (Hg-MT) does not have a long biological half-life in the renal cortex. It has been suggested that this is due to the high but unspecific affinity of Hg for thiol compounds; Hg-MT can be more easily dissociated and released Hg bound to other -SH compounds like glutathione (Foulkes, 1993). Pre-induction of MT by Zn injection was found to have a protective effect on the nephrotoxicity caused by subsequent doses of inorganic Hg (Zalups and Cherian, 1992). The protective mechanism is unknown but a shift in the intrarenal accumulation of inorganic Hg from the pars recta to the S1 and S2 segments of the proximal tubule has been suggested (Zalups and Cherian, 1992).

Arsenic also induces metallothionein synthesis (Kreppel *et al.*, 1993) but there is little evidence suggesting that MT plays a major role in As detoxification.

Lead is not bound by metallothionein. However, excess Pb is believed to be stored in the kidney as a Pb-protein complex which appears in renal tubular cells as inclusion bodies (Goyer *et al.*, 1970). The protein is acidic and contains large amounts of aspartic and glutamic acids and little cystine. It is suggested that Pb binds loosely to the carboxyl groups of the acidic amino acids (Goyer, 1996). Treatment of Pb-exposed animals with chelating agents such as EDTA is accompanied by a sudden increase of urinary Pb, which is at a maximum 12 to 24 hours after treatment (Goyer and Wilson, 1975). Also, no inclusion bodies are found in people after treatment with EDTA. The bodies may be found intact in the urinary sediment of workmen with heavy exposure to Pb. The inclusion bodies account for the major fraction of intracellular Pb and may provide a major pathway for the cellular excretion of Pb (Goyer, 1971). The origin of the protein forming the inclusion bodies is uncertain. Egle and Shelton (1986) have shown that a nuclear matrix protein termed p32/6.3 is the most abundant protein component of the inclusion bodies. Lead forms inclusion bodies in the cytoplasm of kidney cells grown in culture and tends to migrate secondarily into nuclei (McLaughlin *et al.*, 1980).

The production of oxygen free-radicals has been suggested as a possible mechanism for the toxicity of the four metals discussed (Goyer, 1996). Therefore, the cellular levels of antioxidants such as catalase, peroxidase, α-tocopherol, ascorbic acid and particularly glutathione can affect the cellular response to metal toxicty. The role of glutathione on metal detoxifica-

Table 13.2 Exposure levels for public health consideration (μg/day)

Metal	Typical oral intake level (non-exposed general population)	Tolerable intake level[1]	Lowest observable adverse effect level (LOAEL)[4]	Lethal dose	Reference
Arsenic	25–50	140[2]	1000–1500	50 000	ATSDR, 1985
Cadmium	10–59	70	140–160	N/A	WHO, 1992
Lead	< 20	250	1400	N/A	ATSDR, 1990
Mercury	4–10	50[3]	210–490	N/A	WHO, 1990

[1]Data from Joint Food and Agriculture Organization/World Health Organization Expert Committee on Food. All data converted for adult of 70 kg body weight.
Note: WHO guidelines for drinking water are: As 10 μg l^{-1}; Cd 3 μg l^{-1}; Pb 10 μg l^{-1}; Hg (total) 1 μg l^{-1}.
[2]Inorganic arsenic only
[3]Total mercury (inorganic + organic)
[4]All data are converted for an adult of 70 kg body weight

tion has been a major research topic and was recently reviewed by Maines (1994).

13.7 PUBLIC HEALTH CONSIDERATIONS

A major problem in the area of public health is that the health implications of chronic exposure to low doses (environmental levels) of toxic metals are usually not known. This is partly due to the large individual variability in response to metal exposure. It is known that toxic metals interact with other essential minerals such as Zn, Ca, Fe and Se, and so the uptake, metabolism, excretion and toxicity of the metals can be enhanced or decreased by the levels of the essential minerals in the diet as well as the nutritional status of the person. Maines (1994) has reviewed the multiple modulating factors that determine interindividual differences in response to metals. They include sex, age, inherited disorders, pregnancy, occupation, drugs, season, environmental chemicals, exercise, duration of exposure, diet, stress, disease state, gastrointestinal function, renal function, temperature, plasma and cellular binding proteins, and genetic constitution. The levels in public health regulation guidelines are usually set to protect the most susceptible group of people. Therefore, it is a common practice to set the regulation guidelines – sometimes known as tolerable daily intake (TDI) – at a level below the lowest observable effect levels (LOAEL) obtained from either animal or human studies by a certain factor in order to allow for the error of uncertainty. The usual ranges of metal exposure, the TDI, the LOAEL and the lethal doses for As, Cd, Pb and Hg are summarized in Table 13.2. Little is known about the health risk when exposure levels exceed the TDI but are below the LOAEL. This problem is exemplified in the many populations with elevated As exposure from drinking water containing higher levels of As, and in the fish-eating populations around the world with elevated Hg exposure. More basic and epidemiological studies are clearly required.

13.8 FURTHER READING

This chapter can only provide a brief outline of the toxicokinetics and some examples of detoxification mechanisms for four toxic metals. Those interested in the mechanisms of the toxicity of As, Cd, Pb and Hg or other metals may refer to the excellent review written by Goyer (1996). The International Program on Chemical Safety (IPCS) of the World Health Organization has published a series of monographs which summarize current knowledge, at the time of publication, on the health effects of many environmental contaminants including the four metals discussed in this chapter (WHO, 1981, 1989, 1990, 1992). The Agency for Toxic Substances and Disease Registry (ATSDR) of the Department of Health and Human Services in the United States has also published extensive toxicological profiles for the four metals (ATSDR, 1988, 1989, 1990, 1992) and much information is available at their WEB page at http://atsdr1.atsdr.cdc.gov:8080/.

REFERENCES

Aberg, B., Ekman, R., Falk, U. *et al.* (1969) Metabolism of methylmercury (203Hg) compounds in man: excretion and distribution. *Arch. Environ. Health* **19**, 478–484.

Al-Shahristani, J., Shihab, K.M. and Al-Haddad, J.K. (1976) Mercury in hair as an indicator of total body burden. *Bull. WHO* (Suppl.) **53**, 105–112.

Andreae, M.O. (1977) Determination of arsenic species in natural waters. *Anal. Chem.* **49**, 820–823.

Andreae, M.O. (1986) Organoarsenic compounds in the environment, in *Organometallic Compounds in the Environment*, (ed. P.T. Craigh), Longman Group, pp. 199–228.

ATSDR (1988) *The Nature and Extent of Lead Poisoning in Children in the United States: a Report to Congress*, Agency for Toxic Substances and Disease Registry, Public Health Service, Department of Health and Human Services, Atlanta, Ga.

ATSDR (1989) *Toxicological Profile for Mercury*, Agency for Toxic Substances and Disease Registry, Public Health Service, Department of Health and Human Services, Atlanta, Ga.

ATSDR (1990) *Toxicological Profile for Lead*, Agency for Toxic Substances and Disease Registry, Public Health Service, Department of Health and Human Services, Atlanta, Ga.

ATSDR (1992) *Toxicological Profile for Arsenic*, Agency for Toxic Substances and Disease Registry, Public Health Service, Department of Health and Human Services, Atlanta, Ga.

ATSDR (1993) *Toxicological Profile for Cadmium*, Agency for Toxic Substances and Disease Registry, Public Health Service, Department of Health and Human Services, Atlanta, Ga.

Bakir, F., Damluji, S.F., Amin-Zaki, L. *et al.* (1973) Methylmercury poisoning in Iraq. *Science* **181**, 230–241.

Baranowska, I. (1995) Lead and cadmium in human placentas and maternal and neonatal blood (in a heavily polluted area) measured by graphite furnace atomic absorption spectrometry. *Occup. Environ. Med.* **52**, 229–232.

Barltrop, D. (1969) Transfer to lead to the human fetus, in *Mineral Metabolism in Pediatrics*, (eds D. Barltrop and W.L. Burland), Davis Co, Philadelphia, Pa, pp. 135–151.

Barry, P.S.I. (1975) A comparison of concentration of lead in human tissue. *Br. J. Ind. Med.* **32**, 119–139.

Barry, P.S.I. (1981) Concentrations of lead in the tissue of children. *Br. J. Ind. Med.* **38**, 61–71.

Bellinger, D.C., Levitron, A., Waternaux, C. *et al.* (1987) Longitudinal analyses of prenatal and postnatal lead exposure and early cognitive development. *New Engl. J. Med.* **316**, 1037–1043.

Belton, J.C., Benson, N.C., Hanna, M.L. and Taylor, R.T. (1985) Growth inhibitory and cytotoxic effects of three arsenic compounds on cultured Chinese hamster ovary cells. *J. Environ. Sci. Health* **20A**, 37–72.

Berlin, M. and Gibson, S. (1963) Renal uptake, excretion and retention of mercury: Part I. A study in the rabbit during infusion of mercuric chloride. *Arch. Environ. Health* **6**, 56–63.

Bettley, F.R. and O'Shea, J.A. (1975) The absorption of arsenic and its relation to carcinoma. *Br. J. Dermatol.* **92**, 563–568.

Braman, R.S. and Foreback, C.C. (1973) Methylated forms of arsenic in the environment. *Science* **182**, 1247–1249.

Bruenger, F.W., Stevens, W. and Stover, B.J. (1973) The association of ^{210}Pb with constituents of erythrocytes. *Health Phys.* **25**, 37–42.

Buchet, J.P., Lauwerys, R. and Roels, H. (1981a) Comparison of the urinary excretion of arsenic metabolites after a single oral dose of sodium arsenite, monomethyl arsonate or dimethyl arsinate in man. *Int. Arch. Occup. Environ. Health* **48**, 71–79.

Buchet, J.P., Lauwerys, R. and Roels, H. (1981b) Urinary excretion of inorganic arsenic and its metabolites after repeated ingestion of sodium meta arsenite by volunteers. *Int. Arch. Occup. Environ. Health* **48**, 111–118.

Canada-Ontario Steering Committee (1983) *Mercury Pollution in the Wabifon–English River System of Northwestern Ontario, and Possible Remedial Measures*, summary of a technical report, Canada-Ontario Steering Committee, Provincial Ministry of the Environment, Toronot.

Cannon, J.R., Saunders, J.B. and Toia, R.F. (1983) Isolation and preliminary toxicological evaluation of arsenobetaine, the water-soluble arsenical constituent from the hepatopancreas of the western rock lobster. *Sci. Total Environ.* **31**, 181–185.

Cember, H., Gallagher, P. and Faulkner, A. (1968) Distribution of mercury among blood fractions and serum proteins. *Am. Ind. Hyg. Assoc. J.* **29**, 233–237.

Chamberlain, A., Heard, C., Little, M.J. *et al.* (1978) *Investigations into Lead from Motor Vehicles*, Report No. AERE-9198, United Kingdom Atomic Energy Authority, Harwell (cited in EPA 1986a).

Chan, H.M., Satoh, M., Zalups, R.K. and Cherian, M.G. (1992) Exogenous metallothionein and renal toxicity of cadmium and mercury in rats. *Toxicology* **76**, 15–26.

Chan, H.M., Zhu, L.F., Zhong, R. *et al.* (1993) Nephrotoxicity in rats following liver transplantation from cadmium-exposed rats, *Toxicol. Appl. Pharmacol.* **123**, 89–96.

Charbonneau, S.M., Spencer, K., Bryce, F. *et al.* (1978) Arsenic excretion by monkeys dosed with arsenic-containing fish or with inorganic arsenic. *Bull. Environ. Contam. Toxicol.* **20**, 470–477.

Cherian, M.G. (1980) The synthesis of metallothionein and cellular adaptation to metal toxicity in primary rat kidney epithelial cell cultures. *Toxicology* **17**, 225–231.

Cherian, M.G. and Chan, H.M. (1993) Biological functions of metallothionein – a review, in *Metallothionein III: Biological Roles and Medical Implications*, (eds K.T. Suzuki, N. Imura and M. Kimura).

Cherian, M.G. and Goyer, R.A. (1989) Cadmium toxicity. *Comments Toxicol.* **3**, 191–206.

Cherian, M.G., Hursh, J.G., Clarkson, T.W. *et al.* (1978) Radioactive mercury distribution in biological fluids and excretion in human subjects after inhalation of mercury vapor. *Arch. Environ. Health* **33**, 190–114.

Chung, J., Nartey, N.O. and Cherian, M.G. (1986) Metallothionein levels in liver and kidney of Canadians – a potential indicator of environmental exposure to cadmium. *Arch. Environ. Health* **41**, 319–323.

Clarkson, T.W., Gatzy, J. and Dalton, C. (1961) *Studies on the Equilibration of Mercury Vapor with Blood*, University of Rochester Atomic Energy Project, Division of Radiation Chemistry and Toxicology, Rochester, New York.

Crecelius, E.A. (1977) Changes in the chemical speciation of arsenic following ingestion by man. *Environ. Health Perspect.* **19**, 147–150.

Cullen, W.R. and Reimer, K.J. (1989) Arsenic speciation in the environment. *Chem. Rev.* **89**, 713–764.

Dencker, L., Danielsson, B., Khayat, A. *et al.* (1983) Deposition of metals in the embryo and fetus, in *Reproductive and Developmental Toxicity of Metals*, (eds T.W. Clarkson, G.G. Nordberg and P.R. Sager), Plenum Press, New York, pp. 607–631.

Drasch, G.A., Bohm, J. and Baur, C. (1987) Lead in human bones. Investigation of an occupationally non-exposed population in southern Bavaria (F.R.G.). 1. Adults. *Sci. Total Environ.* **647**, 303–315.

Dudley, R.E., Gammal, L.M. and Klaassen, C.D. (1985) Cadmium-induced hepatic and renal injury in chronically exposed rats: likely role of hepatic cadmium-metallothionein in nephrotoxicity. *Toxicol. Appl. Pharmacol.* **77**, 414–426.

Egle, P.M. and Shelton, K.R. (1986) Chronic lead intoxication causes a brain-specific nuclear protein to accumulate in the nuclei of cells lining kidney tubule. *J. Biol. Chem.* **261**, 2294–2298.

Eisler, R. (1994) A review of arsenic hazards to plants and animals with emphasis on fishery and wildlife resources, in *Arsenic in the Environment, Part II: Human Health and Ecosystem Effects*, (ed. J.O. Nriagaru), John Wiley and Sons Inc., pp. 185–259.

Ellis, K.J., Vartsky, D., Zanzi, I. *et al.* (1979) Cadmium: *in vivo* measurement in smokers and nonsmokers. *Science* **205**, 323–325.

EPA (1986) *Air Quality Criteria for Lead*. June 1986 and Addendum, September 1986, EPA 600/8-83-018F, Office of Research and Development, Office of Health and Environmental Assessment, Environmental Criteria and Assessment Office, Environmental Protection Agency, Research Triangle Park, N.C.

EPA (1987) *Special Report on Ingested Inorganic Arsenic: Skin Cancer and Nutritional Essentiality. Risk Assessment Form*, Environmental Protection Agency, Washington, DC.

EPA (1989) *Evaluation of the Potential Carcinogenicity of Lead and Lead Compounds*, EPA-600/8-89/045A, Environmental Protection Agency.

Everson, J. and Patterson, C.C. (1980) 'Ultra-clean' isotope dilution/mass spectrometric analyses for lead in human blood plasma indicate that most reported values are artificially high. *Clin. Chem.* **26**, 1603–1607.

Flanagan, P.R., McLellan, J., Haist, J. *et al.* (1978) Increased dietary cadmium absorption in mice and human subjects with iron deficiency. *Gastroenterology* **74**, 841–846.

Foulkes, E.C. (1978) Renal tubular transport of cadmium-metallothionein. *Toxicol. Appl. Pharmacol.* **45**, 505–512.

Foulkes, E.C. (1993) Metallothionein and glutathione as determinants of cellular retention and extrusion of cadmium and mercury. *Life Sciences* **52(20)**, 1617–1620.

Friberg, L., Kullman, L., Lind, B. and Nylander, M. (1986) Mercury in the central nervous system in relation to amalgam fillings. *Lakartidningen* **83**, 519–522.

Friberg, L. and Nordberg, F. (1973) Inorganic mercury – a toxicological and epidemiological appraisal, in *Mercury, Mercurials and Mercaptans*, (eds M.W. Miller and T.W. Clarkson), Charles C. Thomas, Springfield, Illinois, pp. 5–22.

Goering, P.K. and Klaassen, C.D. (1984) Zinc-induced tolerance to cadmium hepatotoxicity. *Toxicol. Appl. Pharmacol.* **74**, 299–307.

Goyer, R.A. (1971) Lead toxicity: a problem in environmental pathology. *Am. J. Pathol.* **64**, 167–182.

Goyer, R.A. (1996) Toxic effects of metals, in *Casarett and Doull's Toxicology*, 5th edn, (ed. C.D. Klaassen), McGraw-Hill, New York, pp. 691–736.

Goyer, R.A. and Wilson, M.H. (1975) Lead-induced inclusion bodies: results of EDTA treatment. *Lab. Inest.* **32**, 149–156.

Goyer, R.A., Leonard, D.L., Moore, J.F. *et al.* (1970) Lead dosage and the role of the intranuclear inclusion body. An experimental Study. *Arch. Environ. Health* **20**, 705–711.

Goyer, R.A., Miller, C.R., Zhu, S.Y. *et al.* (1989) Non-metallothionein-bound cadmium in the pathogenesis of cadmium nephrotoxicity in the rat. *Toxicol. Appl. Pharmacol.* **101**, 232–244.

Griffin, T.B., Coulston, F. and Wills, H. (1975) Biological and clinical effects of continuous exposure to airborne particulate lead. *Arh. Hig. Toksikol. (Yugoslavia)* **26**, 191–208.

Hammer, D.I., Calocci, A.V., Hasselblad, V. *et al.* (1973) Cadmium and lead in autopsy tissues. *J. Occup. Med.* **15**, 956–964.

Hammond, P.B. (1982) Metabolism of lead, in *Lead Absorption in Children: Management, Clinical and Environmental Aspects*, (eds J.J. Chisolm and D.M. O'Hara), Urban and Schwarzenberg, Baltimore, Md, pp. 11–20.

Hansen, J.C., Wulf, H.C., Kromann, N. *et al.* (1985) Cadmium concentrations in blood samples from an East Greenlandic population. *Dan. Med. Bull.* **32**, 277–279.

Horiuchi, K., Horiguchi, S. and Suekane, M. (1959) Studies on industrial lead poisoning. 1: Absorption, transportation, deposition and excretion of lead. 6: The lead contents in organ-tissues of the normal Japanese. *Osaka City Med. J.* **5**, 41–70.

Hursh, J.B., Clarkson, T.W., Cherian, M.G. *et al.* (1976) Clearance of mercury (Hg-197, Hg-203) vapor inhaled by human subjects. *Arch. Environ. Health* **31**, 302–309.

IARC (1987) *Monograph on the Evaluation of Carcinogenicity: an Update of IARC Monographs*, Vol. 1-42, Suppl. 7,

IARC (1994) *Monograph on the Evaluation of Risks to Humans. Cadmium, Mercury, Beryllium and the Glass Industry*, Vol. 58, International Agency for Research on Cancer, World Health Organization, Lyons.

Indraprasit, S., Alexander, G.V. and Gonick, H.C. (1974) Tissue composition of major and trace elements in uremia and hypertension. *J. Chronic Dis.* **27**, 135–161.

Joselow, M.M., Ruiz, R. and Goldwater, L. (1968) Absorption and excretion of mercury in man. XIV. Salivary excretion of mercury and its relationship to blood and urine. *Arch. Environ. Health.* **17**, 35–38.

Kagey, B.T., Bumgarner, J.E. and Creason, J.P. (1977) Arsenic levels in maternal-fetal tissue sets, in *Trace Substances in Environmental Health XI*, (ed. O.D. Hemphill), University of Missouri Press, Columbia, pp. 252–256.

Kaise, T., Watanabe, S. and Itoh, K. (1985) The acute toxicity of arsenobetaine. *Chemosphere* **14**, 1327–1332.

Kjellstrom, T. and Nordberg, G.F. (1978) A kinetic model of cadmium metabolism in the human being. *Environ. Res.* **16**, 248–269.

Kjellstrom, T. and Nordberg, G.F. (1985) Kinetic model of cadmium metabolism, in *Cadmium and Health: a Toxicological and Epidemiological Appraisal. Vol.I. Exposure, Dose, and Metabolism*, (eds L. Friberg, C.G. Elinder, T. Kjellstrom and G.F. Nordberg), CRC Press, Boca Raton, FL, pp. 179–197.

Kjellstrom, T., Borg, K. and Lind, B. (1978) Cadmium in feces as an estimator of daily cadmium intake in Sweden. *Environ. Res.* **15**, 242–251.

Kreppel, H., Bauman, J.W., Liu, J. *et al.* (1993) Induction of metallothionein by arsenicals in mice. *Fund. Appl. Toxicol.* **20**, 184–189.

Krishnan, S.S., Lui, S.M.W., Jervis, R.E. and Harrison, J.E. (1990) Studies of cadmium uptake in bone and its environmental distribution. *Biol. Ttrace Elem. Res.* **26–27**, 257–261.

Kuhnert, P.M., Kuhnert, B.R., Bottoms, S.F. *et al.* (1982) Cadmium levels in maternal blood, fetal cord blood, and placental tissues of pregnant women who smoke. *Am. J. Obstet. Gynecol.* **142**, 1021–1025.

Lakso, J.U. and Peoples, S.A. (1975) Methylation of inorganic arsenic by mammals. *J. Agric. Food Chem.* **23**, 674–676.

Lauwerys, R., Buchet, J.P., Roels, H. *et al.* (1978) Placental transfer of lead, mercury, cadmium, and carbon monoxide in women. I. Comparison of the frequency distributions of the biological indices in maternal and umbilical cord blood. *Environ. Res.* **15**, 278–289.

Lauwerys, R., Hardey, R., Job, M. *et al.* (1984) Environmental pollution by cadmium and cadmium body burden: an autopsy study. *Toxicol. Lett.* **23**, 287–289.

Lawrence, J.J., Michalik, P., Tam, G. and Conacher, H.B.C. (1986) Identification of arsenobetaine and arsenocholine in Canadian fish and shellfish by high-performance liquid chromatography with atomic absorption detection and confirmation by fast atom bombardment mass spectrometry. *J. Agric. Food Chem.* **34**, 315–319.

Liebscher, K. and Smith, H. (1968) Essential and nonessential trace elements: a method of determining whether an element is essential or nonessential in human tissue. *Arch. Environ. Health* **17**, 881–890.

Lovejoy, H.B., Bell, Z.G. and Vizena, T.R. (1974) Mercury exposure evaluations and their correlation with urine mercury excretion. *J. Occup. Med.* **15**, 590.

Luten, J.B., Riekwel-Booy, G. and Rauchbaar, A. (1982) Occurrence of arsenic in plaice (*Pleuronectes platessa*), nature of organoarsenic compound present and its excretion by man. *Environ. Health Perspect.* **45**, 165–170.

Magos, L., Bakir, F., Clarkson, T.W. *et al.* (1976) Tissue levels of mercury in autopsy specimens of liver and kidney. *Bull. WHO* **53**, 93–96.

Maher, W.A. and Butler, E. (1988) Arsenic in the marine environment. *Appl. Organomet. Chem.* **2**, 191–214.

Maines, M.D. (1994) Modulating factors that determine interindividual differences in response to metals, in *Risk Assessment of Essential Elements*, (eds W. Mertz *et al.*), ILSI Press, Washington, DC.

Manton, W.I. (1985) Total contribution of airborne lead to blood lead. *Br. J. Ind. Med.* **42**, 168–172.

Manton, W.I. and Cook, J.D. (1984) High-accuracy (stable isotope dilution) measurements of lead in serum and cerebrospinal fluid. *Br. J. Ind. Med.* **41**, 313–319.

Mappes, R. (1977) [Experiments on excretion of arsenic in urine.] *Int. Arch. Occup. Environ. Health* **40**, 267–272 (in German).

Marafante, E. and Vahter, M. (1984) The effect of methyltransferase inhibition on the metabolism of [^{74}As] arsenite in mice and rabbits. *Chem. Biol. Interact.* **50**, 49–57.

Marafante, E. and Vahter, M. (1986) The effect of dietary and chemically induced methylation deficiency on the metabolism of arsenate in the rabbit. *Acta Pharmacol. Toxicol.* **59** (Suppl. 7), 35–38.

Marafante, E., Vahter, M. and Envall, J. (1985) The role of the methylation in the detoxication of arsenate in the rabbit. *Chem. Biol. Interact.* **56**, 225–238.

Marafante, E., Vahter, M., Norin, H. *et al.* (1987) Biotransformation of dimethylarsinic acid in mouse, hamster and man. *J. Appl. Toxicol.* **7**, 111–117.

Marcus, W.L. and Rispin, A.S. (1988) Threshold carcinogenicity using arsenic as an example, in *Risk Assessment and Risk Management of Industrial and Environmental Chemicals*, Vol XV, (eds C.R. Cothern, M.A. Mehlman and W.L. Marcus), Princeton Scientific Publishing Co., Princeton, NJ, pp. 133–158.

McKenzie-Parnell, J.M., Kjellstrom, T.E., Sharma, R.P. *et al.* (1988) Unusually high intake and fecal output of cadmium and fecal output of other trace elements in New Zealand adults consuming dredge oysters. *Environ. Res.* **46**, 1–14.

McLaughlin, J.R., Goyer, R.A. and Cherian, M.G. (1980) Formation of lead-induced inclusion bodies in primary rat kidney epithelial cell cultures: effect of actinomycin D and cycloheximide. *Toxicol. Appl. Pharmacol.* **56**, 418–431.

McLellan, J.S., Flanagan, P.R., Chamberlain, M.J. *et al.* (1978) Measurement of dietary cadmium absorption in humans. *J. Toxicol. Environ. Health* **4**, 131–138.

Miettinen, J.K. (1973) Absorption and elimination of dietary (Hg++) and methylmercury in man, in *Mercury, Mercurials, and Mercaptans*, (eds M.W. Miller and T.W. Clarkson), Charles C. Thomas, Springfield, IL, pp. 233–243.

Morgan, H. and Sherlock, J.C. (1984) Cadmium intake and cadmium in the human kidney. *Food Addit. Contam.* **1**, 45–51.

Naganuma, A., Tanaka, T., Urano, and Imura, N. (1991) Role of glutathione in mercury disposition, in *Advances in Mercury Toxicology*, (eds T. Suzuki, N. Imura and T.W. Clarkson), Plenum Press, New York, p.111–120.

Nakamura, I., Hosokawa, K., Tamra, H. *et al.* (1977) Reduced mercury excretion with feces in germfree mice after oral administration of methylmercury chloride. *Bull. Environ. Contam. Toxicol* **17**, 5.

Needleman, H.L. and Shapiro, I.M. (1974) Dentine lead levels in asymptomatic Philadelphia school children: subclinical exposure in high and low risk groups. *Environ. Health Perspect.* **7**, 27–31.

Newton, D., Johnson, P., Lally, A.E. *et al.* (1984) The uptake by man of cadmium ingested in crab meat. *Hum. Toxicol.* **3**, 23–28.

Nielsen, F.H. and Uthus, E.O. (1984) Arsenic, in *Biochemistry of the Essential Ultratrace Elements*, (ed. E. Frieden), Plenum Press, New York, pp. 319–40.

Nomiyama, K. and Nomiyama, H. (1986) Critical concentrations of 'unbound' cadmium in the rabbit renal cortex. *Exerientia* **42**, 149.

Nordberg, G.F., Goyer, R. and Nordberg, M. (1975) Comparative toxicity of cadmium-metallothionein and cadmium chloride on mose kidney. *Arch. Pathol.* **99**, 192–197.

Nordberg, G. (1976) *Effects and Dose–response of Toxic Metals*, Elsevier/North Holland Biomedical Press, New York.

Nordberg, G., Kjellstrom, T. and Nordberg, M. (1985) Kinetics and metabolism, in *Cadmium and health: a Toxicological and Epidemiological Appraisal. Vol. I. Exposure, Dose, and Metabolism*, (eds L. Friberg, C.G. Elinder, T. Kjellstrom *et al.*), CRC Press, Boca Raton, FL, pp. 103–178.

Norseth, T. and Clarkson, T.W. (1970) Studies on the biotransformation of Hg-203-labelled methylmercury chloride. *Arch. Environ. Health* **21**, 717–727.

Norseth, T. and Clarkson, T.W. (1971) Intestinal transport of Hg-203-labeled methyl mercury chloride. Role of biotransformations in rats. *Arch. Environ. Health* **22**, 668–577.

Nylander, N., Friberg, L. and Lind, B. (1987) Mercury concentrations in the human brain and kidneys in relation to exposure from dental amalgams. *Swed. Dent. J.* **11**, 179–187.

O'Flaherty, E.J., Hammond, P.B. and Lerner, S.I. (1982) Dependence of apparent blood lead half-life on the length of previous lead exposure in humans. *Fund. Appl. Toxicol.* **2**, 49–54.

Ohsawa, M. and Magos, L. (1974) The chemical form of methylmercury complex in rat bile. *Biochem. Pharmacol.* **23**, 1903–1906.

Ong, C.N. and Lee, W.R. (1980) High affinity of lead for fetal hemoglobin. *Br. J. Ind. Med.* **37**, 292–298.

Paglia, D.E., Valentine, W.N. and Dahlgner, J.G. (1975) Effects of low level lead exposure on pyrimidine-5'-nucleotidase and other erythrocyte enzymes. *J. Clin. Invest.* **56**, 1164–1169.

Phelps, R.W., Clarkson, T.W., Kershaw, T.G. *et al.* (1980) Interrelationships of blood and hair mercury concentrations in a North American population exposed to methylmercury. *Arch. Environ. Health* **35**, 161–168.

Rabinowitz, M.B., Wetherill, G.W. and Kopple, J.D. (1976) Kinetic analysis of lead metabolism in healthy humans. *J. Clin. Invest.* **58**, 260–270.

Rabinowitz, M.B., Wetherill, G.W. and Kopple, J.D. (1977) Magnitude of lead intake from respiration by normal man. *J. Lab. Clin. Med.* **90**, 238–248.

Radisch, B., Luck, W. and Nau, H. (1987) Cadmium concentrations in milk and blood of smoking mothers. *Toxicol. Lett.* **36**, 147–152.

Rahola, T., Aaran, R-K. and Miettenen, J.K. (1973) Retention and elimination of [115m]Cd in man, in *Health Physics Problems of Internal Contaminations*, Akademia, Budapest, pp. 213–218.

Refsvik, T. and Norseth, T. (1975) Methylmercuric compounds in rat bile. *Acta Pharmacol. Toxicol.* **36**, 67–78.

Roels, H.A., Hubermont, G., Buchet, J.P. *et al.* (1978) Placental transfer of lead, mercury, cadmium, and carbon monoxide in women. III. Factors influencing the accumulation of heavy metals in the placenta, and the relationship between maternal concentration in the placenta and in maternal and cord blood. *Environ. Res.* **16**, 236–247.

Rosen, J.F. (1985) Metabolic and cellular effects of lead: a guide to low-level lead toxicity in children, in *Dietary and Environmental Lead: Human Health Effects*, (ed. K.R. Mahaffey), Elsevier Science Publishers, pp. 157–185.

Rothstein, A. and Hayes, A.L. (1964) The turnover of mercury in rats exposed repeatedly to inhalation of vapor. *Health Phys.* **10**, 1099–1113.

Rowland, I., Davies, M. and Evans, J. (1980) Tissue content of mercury in rats given methylmercury chloride orally: influence of intestinal flora. *Arch. Environ. Health.* **35**, 155.

Sendelbach, L.E. and Klaassen, C.D. (1988) Kidney synthesizes less metallothionein than liver in response to cadmium chloride and cadmium-metallothionein. *Toxicol. Appl. Pharmacol.* **92**, 95–102.

Silbergeld, E.K. (1992) Mechanisms of lead neurotoxicity, or looking beyond the lamppost. *FASEB J.* **6**, 3201–3206.

Squibb, K.S., Pritchard, J.B. and Fowler, B.A. (1984) Cadmium-metallothionein nephropathy: relationships between ultrastrutural/biochemical alterations and intracellular cadmium binding. *J. Pharamcol. Exp. Therap.* **229**, 311–321.

Steenhout, A. and Pourtois, M. (1981) Lead accumulation in teeth as a function of age with different exposures. *Br. J. Ind. Med.* **38**, 297–303.

Succop, P.A., O'Flaherty, E.J., Bornschein, R.L. *et al.* (1987) A kinetic model for estimating changes in the concentration of lead in the of young children, in *International Conference: Heavy Metals in the Environment*, Vol. 2, September, (eds S.E. Lindberg and T.C. Hutchinson), CEP Consultants, New Orleans, LA.

Suda, I. and Takahashi, H. (1986) Enhanced and inhibited bio-transformation of methylmercury in the rat spleen. *Toxicol. Appl. Pharmacol.* **82**, 45–52.

Sumino, K., Hayakawa, K. and Shibata, T. *et al.* (1975) Heavy metals in normal Japanese tissues. *Arch. Environ. Health* **30**, 487–494.

Suzuki, C.A.M. and Cherian, M.G. (1987) Renal toxicity of cadmium-metallothionein and enzymuria in rats. *J. Pharmacol. Exp. Ther.* **240**, 314–319.

Suzuki, T., Imura, N. and Clarkson, T.W. (1991) Overview, in *Advances in Mercury Toxicology*, (eds T. Suzuki, N. Imura and T.W. Clarkson), Plenum Press, New York, pp. 1–32.

Takahashi, K., Yamauchi, H., Yamato, N. *et al.* (1988) Methylation of arsenic trioxide in hamsters with liver damage induced by long-term administration of carbon tetrachloride. *Appl. Organomet. Chem.* **2**, 309–314.

Takahata, N., Hayashi, H., Watanabe, B. *et al.* (1970) Accumulation of mercury in the brains of two autopsy cases with chronic inorganic mercury poisoning. *Folia Psychiatr. Neurol. Jpn* **24**, 59–69.

Tam, G.K., Charbonneau, S.M., Bruce, F. *et al.* (1979) Metabolism of inorganic arsenic (^{74}As) in humans following oral ingestion. *Toxicol. Appl. Pharmacol.* **50**, 319–322.

Truska, P., Rosival, L., Balazova, G. *et al.* (1989) Blood and placental concentrations of cadmium, lead, and mercury in mothers and their newborns. *J. Hyg. Epidemiol. Microbiol. Immunol.* **33**, 141–147.

Turner, M.D., Kilpper, R.W., Smith, J.C. *et al.* (1975) Studies on volunteers consuming methylmercury in tuna fish. *Clin. Res.* **23**, 2.

Vahter, M. and Marafante, E. (1987) Effects of low dietary intake of methionine, choline or proteins on the biotransformation of arsenite in the rabbit. *Toxicol. Lett.* **37**, 41–46.

Vahter, M., Berglund, M., Nermell, B. and Akesson, A. (1996) Bioavailability of cadmium from shellfish and mixed diet in women. *Toxicol. Appl. Pharmacol.* **136(2)**, 332–341.

Weiss, S.H., Wands, J.R. and Yardley, J.H. (1973) Demonstration by electron defraction of mercuric sulfide (b-HgS) in a case of 'melanosis coli and black kidneys' caused by chronic inorganic mercury poisoning. *Lab. Invest*, 401–402 (abstract).

Whelton, B.D., Bhattacharyya, M.H., Peterson, D.P. *et al.* (1994) Skeletal changes in multiparous and uniparous mice fed a nutrient-deficient diet containing cadmium. *Toxicology* **91**, 235–251.

WHO (1976) *Environmental Health Criteria 1: Mercury*, World Health Organization, Geneva, 132 pp.

WHO (1977) *Environmental Health Aspects of Cadmium*, WHO Task Group, Geneva.

WHO (1981) *Environmental Health Criteria 18: Arsenic*, World Health Organization, Geneva.

WHO (1989) *Environmental Health Criteria 85: Lead – Environmental Aspects*, World Health Organization, Geneva.

WHO (1990) *Environmental Health Criteria 101: Methylmercury*, World Health Organization, Geneva.

WHO (1992) *Environmental Health Criteria 134: Cadmium*, World Health Organization, Geneva.

Yamauchi, H. and Yamamura, Y. (1984) Metabolism and excretion of orally ingested trimethylarsenic in man. *Bull. Environ. Contam. Toxicol.* **32**, 682–687.

Yamauchi, H. and Yamamura, Y. (1985) Metabolism and excretion of orally administered arsenic trioxide in the hamster. *Toxicology* **34**, 113–121.

Yamauchi, H., Takahashi, K. and Yamamura, Y. (1986) Metabolism and excretion of orally and intraperitoneally administered gallium arsenide in the hamster. *Toxicology* **40**, 237–246.

Zalups, R.K. and Cherian, M.G. (1992) Renal metallothionein metabolism after a reduction of renal mass. II. Effect of zinc pretreatment on the renal toxicity and intrarenal accumulation of inorganic mercury. *Toxicology* **71**, 103–117.

Zalups, R.K. and Lash, L.H. (1994) Advances in understanding the renal transport and toxicity of mercury. *J. Toxicol. Environ. Health* **42**, 1–44.

Zaric, M., Prpic-Majic, D., Kostial, K. and Piasek, M. (1987) Exposure to lead and reproduction, in *Summary Proceedings of a Workshop: Selected Aspects of Exposure to Heavy Metals in the Environment. Monitors, Indicators, and High Risk Groups, April, 1985*, National Academy of Sciences, Washington, DC; Council of Academies of Sciences and Arts, Yugoslavia; pp. 119–126 (cited in ATSDR, 1988).

Zheng, H., Liu, J., Choo, K.H. *et al.* (1996) Metallothionein-I and -II knock-out mice are sensitive to cadmium-induced liver mRNA expression of c-jun and p53. *Toxicol. Appl. Pharmacol.* **136**, 229–235.

Ziegler, E.E., Edwards, B.B., Jensen, R.L. *et al.* (1978) Absorption and retention of lead by infants. *Pediatr. Res.* **12**, 29–34.